Remote Sensing in Hydrology and Water Management

Springer
*Berlin
Heidelberg
New York
Barcelona
Hong Kong
London
Milano
Paris
Singapore
Tokyo*

Gert A. Schultz · Edwin T. Engman (Eds.)

Remote Sensing in Hydrology and Water Management

With 184 figures and 22 tables

Springer

Prof. Gert A. Schultz
Ruhr-Universität Bochum
Lehrstuhl für Hydrologie, Wasserwirtschaft und Umwelttechnik
44780 Bochum
Germany

Dr. Edwin T. Engman
NASA/Goddard Space Flight Center
Greenbelt, MD
USA

ISBN 3-540-64075-4 Springer Verlag Berlin Heidelberg New York

Cataloging-in-Publication Data applied for

Schultz, Gert A.:
Remote sensing in hydrology and water management / G. A. Schultz ; E. T. Engman. – Berlin ; Heidelberg ;
New York ; Barcelona ; Hong Kong ; London ; Milano ; Paris ; Sinagpore ; Tokyo : Springer, 2000

ISBN 3-540-64075-4

This work is subject to copyright. All rights are reserved, whether the whole or part of the material is concerned, specifically the rights of translation, reprinting, reuse of illustrations, recitation, broadcasting, reproduction on microfilm or in other ways, and storage in data banks. Duplication of this publication or parts thereof is permitted only under the provisions of the German Copyright Law of September 9, 1965, in its current version, and permission for use must always be obtained from Springer-Verlag. Violations are liable for prosecution under German Copyright Law.

Springer-Verlag is a company in the BertelsmannSpringer publishing group

© Springer-Verlag Berlin Heidelberg 2000
Printed in Germany

The use of general descriptive names, registered names, trademarks, etc. in this publication does not imply, even in the absence of a specific statement, that such names are exempt from the relevant protective laws and regulations and therefore free for general use.

Coverdesign: design & production, Heidelberg
Typesetting: Camera-ready copy by author
SPIN: 10508018   61/3020 · Printed on acid-free paper – 5 4 3 2 1 0 –

Preface

The authors of this book have been approached by consulting engineers: "You know about remote sensing and thus you can obtain hydrological information where no data exist, even without ground measurements, so that I can design the required capacity of a drinking water supply reservoir in a developing country." On the other hand, the authors have been told: "Remote sensing is not of much use since it is still impossible to estimate ground water resources or surface water flows from remote sensing data." The major aim of this book is to correct such unjustified illusions as well as exaggerated criticism by providing the reader with sound information on the potential – and the limitations – of remote sensing in the field of hydrology and water management.

The book is meant to be a reference and text; it is not a collection of papers from some meeting. The book is intended to provide methods to help the readers solve their own problems in hydrology and water management. Therefore, scientific issues are presented only as far as they are necessary for the application of remote sensing. The reader will see that in some fields, (e.g. evapotranspiration, soil moisture, hydrological modeling) the scientific development is still on the way, while in others operational techniques are already available (e.g. snow melt runoff-model, land use classification and detection of land use changes, flood forecasting and control). Furthermore, the reader will realize, that in many cases remote sensing data are only auxiliary information to other data, e.g. digital maps, digital elevation models, modern hydrometric data, etc. In almost all cases remote sensing will not replace, but will augment traditional data.

The application of remote sensing in hydrology and water management requires not only remote sensing data, but also access to the required hardware and software. Hopefully, the reader will realize that using and manipulating remote sensing data has become greatly simplified and cheaper in recent years. Tasks that had to be done on expensive specialized computers a few years ago can now be accomplished on a desk top PC and high performance computer systems and the powerful software are available at reasonable costs. It can also be seen that although some of the algorithms for transforming remote sensing data into hydrological information are becoming more and more complex (to handle data from multiple satellites or merging satellite and traditional data), they are being distributed in a very user-friendly format. Furthermore, a combination of remote sensing data, ground truth, (e.g. from modern hydrometric equipment) and powerful software domains such as GIS give the hydrologist or water resource manager extremely powerful tools to help solve their problems.

The readership of this book is expected to consist of hydrologists and water managers, i.e. mainly civil engineers, environmental engineers, geo-scientists and agricultural engineers. The information provided emphasizes possible applications of remote sensing data to hydrological monitoring and modeling as well as for water resources management decisions. New techniques are provided, which – due to the unique characteristics of remote sensing information – have a structure differing from conventional hydrological and water management data and models. However, this structure provides new and unique information not generally available from conventional measurements and models.

Additionally, we observe a growing number of university departments, (e.g. civil engineering, agricultural and geosciences) offering courses in remote sensing theory and applications. This book should also be of value to students in these fields because it provides an up-to-date compendium of the state of the science of remote sensing applications in hydrology and water resources management.

In a time of impressive growth of knowledge in all fields of natural and technical sciences and of rapid changes in relevant technologies, the editors realized that much of this information is scattered throughout many discipline oriented journals and publications. Based on their own scientific work and their long-term international activities the editors realized that it would be beneficial to the hydrology and water resources communities if this information could be pulled together under one cover. They carefully selected the 25 authors from many institutions in nine different countries. Although the book covers almost 500 pages, it is by no means comprehensive, a reason, why it is important to provide an extensive list of references, altogether more than 700.

The editors wish to thank not only their authors, with whom they co-operated for more than four years, but also Joachim Geyer, who assisted patently with the compilation of the chapters and Springer-Verlag which organized the production and publication of the book.

Gert A. Schultz
Ruhr-University Bochum, Germany

Edwin T. Engman
NASA/Goddard Space Flight Center
Greenbelt, MD, USA

About the Editors

Gert A. Schultz is a professor of hydrology, water management and environmental techniques in the department of civil engineering at the Ruhr-University, Bochum, Germany. He is Past-President of the "International Commission of Remote Sensing" in the "International Association of Hydrological Sciences" (IAHS). He was Vice-President of the "International Water Resources Association" (IWRA) and former rapporteur on remote sensing in hydrology to UNESCO. He has published about 150 international and national publications and was/is associate editor of several international and national scientific journals. He was chairman of the IAHS/WMO Working Group for GEWEX and convener/co-convener of a number of international symposia and workshops.

Edwin T. Engman is head of the Hydrological Science Branch, Laboratory for Hydrospheric Processes, NASA/Goddard Space Flight Center in Greenbelt, Maryland, USA. He is one of the co-authors of the book "Remote Sensing in Hydrology" (Chapman and Hall, 1991) and author or co-author of more than 150 technical papers. For the past twenty years or so, he has been actively involved in promoting and conducting research in remote sensing applications to hydrology, first with the U.S. Department of Agriculture, Agricultural Research Service, Hydrology Lab, and recently with the Hydrological Sciences Branch at NASA's Goddard Space Flight Center.

Authors

Chris G. Collier
Telford Institute of Environmental Systems, Department of Civil and Environmental Engineering, University of Salford, United Kingdom
Author Chapter 6

Kevin P. Czajkowski
Laboratory for Global Remote Sensing Studies and Department of Geography, University of Maryland, Maryland, USA
Co-Author Chapter 5

Ralph O. Dubayah
Laboratory for Global Remote Sensing Studies and Department of Geography, University of Maryland, Maryland, USA
Co-Author Chapter 5

Edwin T. Engman
Head, Hydrological Sciences Branch, Laboratory for Hydrospheric Processes, NASA/Goddard Space Flight Center, Greenbelt, Maryland, USA
Editor. Co-Author Chapter 1, 4, 5, 9 and 20

Barry Goodison
Climate Research Branch, Atmospheric Environmental Service, Downsview, Ontario, Canada
Co-Author Chapter 11

Ben Gorte
Geoinformatics and Spatial Data Acquisition, International Institute for Aerospace Survey and Earth Sciences (ITC), Enschede, The Netherlands
Author Chapter 7

John Jensen
Department of Geography, University of South Carolina, Columbia, South Carolina, USA
Author Chapter 3

Geoff Kite
International Water Management Institute, THAEM, Menemen, Izmir, Turkey
Co-Author Chapter 10

Chris M.M. Mannaerts
Department of Earth Resources Surveys, International Institute for Aerospace Survey and Earth Sciences (ITC), Enschede, The Netherlands
Co-Author Chapter 15

Nandish Mattikalli
Cambridge Research Associates, Virginia, USA
Co-Author Chapter 4

Allard M.J. Meijerink
Department of Earth Resources Surveys, International Institute for Aerospace Survey and Earth Sciences (ITC), Enschede, The Netherlands
Author Chapter 14 and Co-Author Chapter 15

Massimo Menenti
The Winand Staring Centre for Integrated Land, Soil and Water Research, Wageningen, The Netherlands
Author Chapters 8 and 17

Ioannis Papadakis
Consulting Engineer, Hattingen, Germany
Co-Author Chapter 18

Alain Pietroniro
> National Water Research Institute, NHRC, Environment Canada, Saskatoon, Saskatchewan, Canada
> *Co-Author Chapter 10*

Albert Rango
> Hydrology Laboratory, USDA Agricultural Research Service, Beltsville, Maryland, USA
> *Co-Author Chapter 11*

Joshua Rhoads
> Laboratory for Global Remote Sensing Studies and Department of Geography, University of Maryland, Maryland, USA
> *Co-Author Chapter 5*

Jerry Ritchie
> USDA-ARS Hydrology Laboratory, Agricultural Research Service, BARC-West, Beltsville, Maryland, USA
> *Author Chapter 12 and Co-Author Chapter 13*

Helmut Rott
> Institute of Meteorology and Geophysics, University of Innsbruck, Austria
> *Author Chapter 2*

Frank R. Schiebe
> SST Development Group Inc., Stillwater, Oklahoma, USA
> *Co-Author Chapter 13*

Gert A. Schultz
> Institute of Hydrology, Water Resources and Environmental Techniques, Department of Civil Engineering, Ruhr-University Bochum, Germany
> *Editor. Co-Author Chapters 1, 16, 18, 19 and 20*

Andreas Schumann

Institute of Hydrology, Water Resources and Environmental Techniques, Department of Civil Engineering, Ruhr-University Bochum, Germany
Co-Author Chapter 20

Anne E. Walker

Climate Research Branch, Atmospheric Environmental Service, Downsview, Ontario, Canada
Co-Author Chapter 11

Eric F. Wood

Water Resources Program, Department of Civil Engineering, Princeton University, Princeton, New Jersey, USA
Co-Author Chapter 5

Mark Zion

NASA Goddard Space Flight Center, Greenbelt, Maryland, USA
Co-Author Chapter 5

Contents

Preface V

About the Editors VII

Authors IX

Section I: **Overview and Basic Principles** 1

Chapter 1   **Introduction** 3
- 1.1   Introduction 3
- 1.2   Remote Sensing Defined 3
- 1.3   The Nature of Remote Sensing Data 4
- 1.4   Satellite Systems 6
- 1.4.1   Remote Sensing Platforms 6
- 1.4.2   Remote Sensing Sensors 9
- 1.4.3   Spatial Resolution 10
- 1.4.4   Temporal Resolution 12
- 1.5   Remote Sensing and Hydrology 12
- 1.6   Structure of the Book 13

Chapter 2   **Physical Principles and Technical Aspects of Remote Sensing** 15
- 2.1   Introduction 15
- 2.2   The Electromagnetic Spectrum and Radiation Laws 15
- 2.3   Atmospheric Propagation 21
- 2.4   Reflection and Emission Characteristics of Natural Media ... 26
- 2.5   Sensor Principles 30
- 2.6   Summary of Current and Future Earth Observation Missions 37

Chapter 3   **Processing Remotely Sensed Data: Hardware and Software Considerations** 41
- 3.1   Image Processing System Characteristics 41
- 3.1.1   The Central Processing Unit (CPU): Personal Computers, Workstations and Mainframes 41
- 3.1.2   Number of Analysts on a System and Mode of Operation ... 44
- 3.1.3   Serial versus Parallel Image Processing, Arithmetic Coprocessor, and Random Access Memory (RAM) 44
- 3.1.4   Operating System and Software Compilers 46
- 3.1.5   Mass Storage 47
- 3.1.6   Screen Display Resolution 48

3.1.7	Screen Color Resolution	49
3.1.8	Image Scanning (Digitization) Considerations	49
3.2	Image Processing and GIS Software Requirement	50
3.2.1	Preprocessing	52
3.2.2	Display and Enhancement	52
3.2.3	Remote Sensing Information Extraction	53
3.2.4	Photogrammetric Information Extraction	54
3.2.5	Metadata and Image/Map Lineage Documentation	54
3.2.6	Image and Map Cartographic Composition	57
3.2.7	Geographic Information Systems (GIS)	57
3.2.8	Utilities	57
3.3	Commercial and Publicly Available Digital Image Processing Systems	58
3.4	Summary	58

Chapter 4 **Integration of Remotely Sensed Data into Geographical Information Systems** ... 65

4.1	Introduction	65
4.2	General Approach	67
4.2.1	Raster and Vector Data Structures	67
4.2.2	Current Approaches to the Integration	70
4.2.3	Errors Associated with Geographical Processing	71
4.3	Current Applications	72
4.3.1	Watershed Database Development	72
4.3.2	Integrated Use of Elevation Data	73
4.3.3	Land-use/Land-cover Change Detection	74
4.3.4	Modeling Watershed Runoff	75
4.3.5	Monitoring and Modeling of Water Quality	76
4.3.6	Soil Erosion Monitoring	77
4.4	Future Perspectives	78

Section II: **Remote Sensing Application to Hydrologic Monitoring and Modeling** ... 83

Chapter 5 **Remote Sensing in Hydrological Modeling** ... 85

5.1	Introduction	85
5.2	Remote Sensing in Operational Hydrologic Modeling	87
5.3	Remote Sensing in Coupled Water-Energy Balance Modeling	90
5.4	Remote Sensing Approach	92
5.4.1	Solar radiation	92
5.4.2	Downwelling longwave	93
5.4.3	Precipitation	94
5.4.4	Air Temperature	94
5.4.5	Surface Air Humidity	95

5.5	Modeling Example: The Red River Arkansas Basin	96
5.6	Future Directions	97

Colour Plates of Chaps. 2-5 ... 103

Chapter 6 **Precipitation** ... 111
- 6.1 Introduction ... 111
- 6.2 General Approach ... 112
- 6.2.1 Ground-based radar ... 112
- 6.2.2 Use of visible and infrared satellite data ... 114
- 6.2.3 Use of passive microwave satellite data ... 114
- 6.2.4 Space-borne radar ... 115
- 6.3 Current Techniques ... 115
- 6.3.1 Single polarisation radar measurements of rainfall ... 115
- 6.3.2 Measurement of snowfall and hail ... 118
- 6.3.3 Multi-parameter radar ... 120
- 6.3.4 Satellite cloud indexing and life history methods of rainfall estimation ... 121
- 6.3.5 Bispectral techniques ... 123
- 6.3.6 Passive microwave estimates of rainfall from space ... 124
- 6.3.7 Sampling errors ... 126
- 6.4 The potential for improvement ... 127
- 6.4.1 Current performance levels ... 127
- 6.4.2 The future ... 128

Chapter 7 **Land-use and Catchment Characteristics** ... 133
- 7.1 Introduction ... 133
- 7.2 Land cover Mapping with Remote Sensing ... 134
- 7.3 Vegetation Indices ... 135
- 7.3.1 Simple Vegetation Indices ... 136
- 7.3.2 Normalized Difference Vegetation Index (NDVI) ... 138
- 7.3.3 Refined estimates ... 139
- 7.3.4 Multi-temporal Vegetation Index ... 140
- 7.4 Thematic Classification ... 140
- 7.4.1 Image Classification Methods ... 142
- 7.4.2 Maximum Likelihood Classification ... 145
- 7.4.3 Discussion ... 147
- 7.4.4 Probability estimation refinements ... 147
- 7.4.5 Segmentation ... 149
- 7.4.6 Case study in the Pantanal Area, Brazil ... 150
- 7.5 Radar ... 152

Chapter 8 **Evaporation** ... 157
- 8.1. Introduction ... 157
- 8.1.1 General ... 157
- 8.1.2 Remote sensing of land evaporation ... 158

8.2	Evaporation and radiometric variables	160
8.2.1	Potential Evaporation	160
8.2.2	Actual Evaporation	162
8.3	Remote Sensing of Land Evaporation: Applications and Modelling Approaches	165
8.3.1	General	165
8.3.2	Linear relationships between evaporation and land surface temperature [1]	166
8.3.3	Improved linear relationships [2]	167
8.3.4	Relationships between evaporation, surface, temperature and spectral indices [3]	168
8.3.5	Soil Vegetation Atmosphere Transfer (SVAT) models [4]	169
8.3.6	Integrated SVAT and Planetary Boundary Layer (PBL) models [5]	170
8.4	Current trends: improved observations and improved parameterizations	171
8.4.1	Local maximum evaporation and land surface temperature [6]	171
8.4.2	Improved observation of land surface variables [7]	174
8.5	Spatial variability	177
8.6	Accuracy	178
8.7	Applications	179
8.8	Current and Future Observations	180
8.9	Summary and Conclusions	181

Colour Plates of Chaps. 6-8 189

Chapter 9 **Soil Moisture** 197
    9.1   Introduction 197
    9.2   General Approach 198
    9.3   Sensor-Target Interactions 202
    9.4   Hydrologic Examples 209
    9.5   Future Microwave Remote Sensing of Soil Moisture 212

Chapter 10 **Remote Sensing of Surface Water** 217
    10.1   Introduction 217
    10.2   Surface Water Detection 218
    10.3   Lake and Reservoir Area Estimates 220
    10.4   Wetlands 223
    10.5   Lake Levels 224
    10.6   River Levels and Flows 226
    10.7   Flood Extent 230
    10.8   Conclusion 233

Chapter 11	Snow and Ice	239
11.1	Role of Snow and Ice	239
11.2	General Approach	240
11.2.1	Gamma Radiation	240
11.2.2	Visible Imagery	242
11.2.3	Thermal Infrared	244
11.2.4	Passive and Active Microwave	244
11.2.5	Related Applications	248
11.3	Current Applications	249
11.3.1	NOHRSC– Snow Cover and Snow Water Equivalent Products	249
11.3.2	Canadian Prairie Snow Water Equivalent Mapping	250
11.3.3	Snowmelt Runoff Forecast Operations	252
11.4	Future Directions	255
11.4.1	Improved Resolution in the Passive Microwave	255
11.4.2	Improved Algorithms in the Passive Microwave	256
11.4.3	Outlook for Radar Applications	256
11.4.4	Integration of Various Data Types	257

Colour Plates of Chaps. 9-11 263

Chapter 12	Soil Erosion	271
12.1	Introduction	271
12.2	Basis for using Remote Sensing	273
12.3	Applications	274
12.4	Case Studies	276
12.4.1	Photointerpretation/Photogrammetry	277
12.4.2	Model/GIS Inputs	279
12.4.3	Spectral Properties	280
12.4.4	Topographic Measurements	281
12.5	Future Directions	282

Chapter 13	Water Quality	287
13.1	Introduction	287
13.2	Basis for using Remote Sensing	288
13.3	Application	290
13.4	Case Studies	291
13.4.1	Suspended Sediments	291
13.4.2	Chlorophyll	294
13.4.3	Temperature	297
13.4.4	Oils	298
13.5	Future Directions	299

Chapter 14	Groundwater	305
14.1	Introduction	305
14.2	Conceptualization of the hydrogeology	306

	14.2.1	The three dimensional hydrogeologic situation 306
	14.2.2	Groundwater surface.................................. 309
	14.2.3	Flow systems 310
	14.3	Aspects of water budgets 312
	14.3.1	Groundwater irrigation drafts 312
	14.3.2	Recharge.. 313
	14.4	Hard rock terrain and lineaments 319
	14.5	Groundwater management and conclusions 321
	14.6	Conclusions and future perspectives 322

Section III: Water Management with the Aid of Remote Sensing Data 327

Chapter 15   Introduction to and General Aspects of Water Management with the aid of Remote Sensing 329

	15.1	Introduction 329
	15.2	Potential of remote sensing in water management.......... 329
	15.2.1	Surveying and mapping 330
	15.2.2	Spatial analysis and regionalization..................... 332
	15.2.3	Monitoring and forecasting 332
	15.3	River basin planning with the aid of remote sensing........ 334
	15.3.1	Introduction 334
	15.3.2	Hydrologic monitoring & forecasting 334
	15.3.3	Upstream-downstream interrelationships in river basins 335
	15.4	Watershed management with the aid of remote sensing..... 338
	15.4.1	Introduction 338
	15.4.2	Hydrologic photo-interpretation for watershed management 338
	15.5	Small-scale water resource development and remote sensing 340
	15.5.1	Introduction 340
	15.5.2	Runoff water harvesting with the aid of remote sensing..... 340
	15.5.3	Flood spreading and groundwater recharge 341
	15.6	Irrigation water management and remote sensing 341
	15.7	Decision support systems for water management 342
	15.7.1	Introduction 342
	15.7.2	Expert and decision support systems 342

Colour Plates of Chaps. 12-15 349

Chapter 16   Flood Forecasting and Control 357

	16.1	Introduction 357
	16.2	General Approach.................................... 358
	16.2.1	Modeling Philosophy 358

16.2.2	Remote Sensing Data, Types and Acquisition	360
16.2.3	Determination of Hydro-meteorological Information from Remote Sensing Data	360
16.2.4	Transformation of Area Precipitation into a Real-time Forecast of a Runoff Hydrograph	362
16.3	Real-time Flood Control with the Aid of Flood Forecasts Based on Remote Sensing Data – an Example	365
16.3.1	Basic Principle	365
16.3.2	Radar Rainfall Measurements in the Günz River Catchment	367
16.3.3	Quantitative Precipitation Forecast (QPF)	368
16.3.4	Rainfall-Runoff-Model Application for Flood Forecasting	368
16.3.5	Optimum Reservoir Operation Based on Forecast Flood Hydrographs	370
16.4	Flood Forecasting and Control in an Urban Environment	372
16.5	Future Perspectives	375

Chapter 17 Irrigation and Drainage 377

17.1	Introduction	377
17.1.1	Current non-remote sensing approaches and limitations	378
17.1.2	Reviews of remote sensing applications in irrigation and drainage	379
17.2	General Approach	380
17.2.1	Applications versus Observables and Algorithms	380
17.2.2	Theory and conceptual approach	380
17.2.3	Examples of applications	386
17.3	Current Applications	387
17.3.1	General	387
17.3.2	High resolution mapping of irrigated lands	389
17.3.3	Crop water requirements – Visible and Near Infrared	390
17.3.4	Crop water stress – Thermal Infrared	391
17.3.5	Catchment hydrology	392
17.3.6	Detection of saline areas	392
17.3.7	Irrigation management	393
17.4	Current and future observations	394
17.5	Future Directions and Potential	395

Chapter 18 Computation of Hydrological Data for Design of Water Projects in Ungauged River Basins 401

18.1	Introduction	401
18.2	General Approach	403
18.2.1	MODUL I: Satellite system, data processing	403
18.2.2	MODUL II: Assessment of the monthly area precipitation on the basis of multi-temporal satellite imagery	406

	18.2.3	MODUL III: Estimation of runoff values 409
	18.3	Application.. 410
	18.3.1	Study area and data used 410
	18.3.2	Assessment of the monthly area precipitation with the aid of multi-temporal B2-Meteosat satellite imagery 411
	18.3.3	Rainfall – Runoff Model 413
	18.4	Further Applications................................... 414
	18.5	Summary and Discussion............................... 416

Chapter 19		**Detection of Land Cover Change Tendencies and their Effect on Water Management** 419
	19.1	General Remarks....................................... 419
	19.2	Hydrological Modelling and Land Cover Change.......... 422
	19.3	A Case Study: Land Use Change Detection by Remote Sensing in the Sauer River Basin, Western Europe......... 424
	19.4	Summary 432

Colour Plates of Chaps. 16-19... 435

Section IV: Future Perspectives 443

Chapter 20		**Future Perspectives** 445
	20.1	Introduction 445
	20.2	Status of Hydrologic Research and Modeling 446
	20.3	Water Management..................................... 448
	20.4	Data Issues in Hydrology and Water Resources Management 449
	20.5	Intensive Field Campaigns 452
	20.6	Existing Sensors and Platforms.......................... 453
	20.7	Planned and Proposed Sensors and Platforms 454
	20.8	Remote Sensing and Future Needs in Hydrology 456

Appendix 20.1 Existing and Future Remote Sensing Satellites and Sensors Relevant to Hydrological Applications 458

Appendix 20.2 Specification for Sensors Listed in Appendix 20.1 461

List of Acronyms 471

Index 475

Section I

Overview and Basic Principles

1 Introduction

Gert A. Schultz[1] and Edwin T. Engman[2]

[1]Ruhr University Bochum, 44780 Bochum, Germany
[2]Hydrological Science Branch, Code 974, Laboratory for Hydrospheric Processes-NASA/Goddard Space Flight Center, Greenbelt, MD 29771, USA

1.1 Introduction

The Nilometer in Cairo may be considered as an ancient device of remote sensing, since it measures - for millennia - the water level of the river Nile not in the river itself, but rather in a historic tower building of very fine architecture near the river Nile. The type of remote sensing discussed in this book is, however, of a rather different type: It deals with techniques and methodologies the electronic age can offer to hydrologists and water managers. At present a rather small international community of hydrological scientists have developed - and are still developing - methods for application of remote sensing information to the solution of hydrological and water management problems. Although many of these techniques are already far advanced - several of them are operational - unfortunately many practitioners responsible for hydrological networks and water resources development are still reluctant to use these methods. This situation is due to several reasons, e.g. unavailability of relevant hardware and software, lack of knowledge in the application of remote sensing techniques, reluctance to change conventional and well established methods etc. It is the aim of this book to overcome these barriers and show to practitioners the potential of remote sensing and, hopefully convince them of the advantages of these techniques for their future work. They will soon see, that several of their problems which could not be treated at all so far will become tractable and other problems dealt with at present with difficulty may be solved in a much more elegant way.

1.2 Remote Sensing Defined

The literature provides many definitions of remote sensing and it may suffice here to cite only one (Ritchie and Rango, 1996), "Remote sensing has been defined as the science and art of obtaining information about an object, area, or phenomenon through the analyses of data acquired by a sensor that is not in direct contact with the target of investigation". It must be recognized at the outset that remote sensing data are different from traditional hydrologic data. Compared to conventional hydrological measurements remote sensing has certain significant advantages, but also some disadvantages. One of the disadvantages of remote sensing is the fact, that hydrological parameters are almost never measured directly. RS means always the acquisition of data from the electromagnetic spectrum. This implies the necessity, that, in order to use RS data for hydrology or water management, the RS data have to be transformed into hydrologically relevant information. This requires the development and application of certain methodology and algorithms suitable for the purpose. In many cases we are limited to inferring the hydrologic information.

Remote sensing uses measurements of the electromagnetic spectrum to characterize the landscape, or infer properties of it, or in some cases, actually measure hydrologic state variables. Aerial photography in the visible wavelengths is the remote sensing technique that most hydrologists are familiar with; however, modern remote sensing is centered around satellite systems and most of the discussions will emphasize satellite data. Over the years remote sensing techniques have expanded to the point that they now include most of the electromagnetic spectrum. Different sensors can provide unique information about properties of the surface or shallow layers of the Earth. For example, measurements of the reflected solar radiation give information on albedo, thermal sensors measure surface temperature, and microwave sensors measure the dielectric properties and hence, the moisture content, of surface soil or of snow. Remote sensing and its continued development has added new techniques that hydrologist can use in a large number of applications.

Because remote sensing data are different from traditional hydrologic data, the hydrologist must recognize what these differences are and take advantage of their strengths and not be discouraged by their weeknesses. For example, we usually deal with a finite resolution element known as a PIXEL whose basic dimensions may vary from a few meters to kilometers. Obviously one looses some detail compared to point samples. This can perhaps best be explained by considering a very typical remote sensing application: measuring various classes of land use (i.e., pasture, forests, urban, etc.). In many cases one will have a pixel (say 100 m square) that will not be pure forest or pure pasture if it straddles the boundary. One has to classify it as either forest or pasture, in either case it will be technically incorrect. In remote sensing applications, one seldom duplicates detailed land use statistics exactly. For example, a study by the Corps of Engineers (Rango et al., 1983) estimated that an individual pixel may be incorrectly classified about one-third of the time. However, by aggregating land use over a significant area, the misclassification of land use can be reduced to about two percent which is too small to affect a hydrologic application such as computing the runoff coefficient and the resulting flood statistics.

1.3 The Nature of Remote Sensing Data

When considering how remote sensing data may be used in hydrology and water management, it is necessary to consider the characteristics of remote sensing data and how these may be used to improve our understanding and operational techniques. There are four characteristics of remote sensing data that make it a potentially very powerful tool for advancing hydrologic sciences. Each of these characteristics are discussed below:

Measuring System States. Thermal infrared and microwave remote sensing, because of their unique responses to surface properties important to hydrology, such as surface temperature, soil moisture and snow water content, have the capability to measure these system states directly. However, using system-state data will require new models to incorporate the new data types. Such models would structurally resemble contemporary simulation models but would be more capable of accounting for spatial variability and changes. Also, the subprocess algorithms (infiltration, evapotranspiration, etc.) would be designed to use remote sensing data as well as the more traditional inputs.

Area versus Point Data. The use of data representing an area in which the spatial variability of specific parameters of the area have been integrated may help provide one of the keys to understanding scaling and scale interdependence in hydrologic systems. The capability to aggregate up in scale or disaggregate down in scale by electronic means may provide a perspective of scaling that may instill new insight to answering the scale questions that dominate scientific hydrology.

Temporal Data. Remote sensing data from a satellite platform can provide unique time series data for hydrologic use. The actual frequency of observation can vary from continuous to once every two weeks or so, depending upon the sensors and type of orbit. This approach is appealing because it may be a very cost-effective method to monitor various hydrologic states over very large areas as well as monitor the dynamic properties in hydrology. Temporal data may provide a means for imparting a hydrologic interpretation to certain observations. For example, observing the time changes in soil moisture may provide information on soil types and even hydraulic properties such as hydraulic conductivity. In fact the interpretation of soil properties as a remote sensing signature could be extremely useful for hydrology because it would represent an areal value rather than a point value determined in a laboratory or with a field measurement.

New Data Forms. Entirely new data types may be formed by merging several data sets of different wavelengths, polarizations, look angles, etc. to provide entirely new hydrologic parameters that are developed from the unique characteristics of remote sensing. New data forms could also be considered to be combinations of remote sensing data combined with other spatial data (such as soil maps) and even point data through a data assimilation scheme or sophisticated GIS (Geographical Information System).

Weather radar is a good example of a new remote sensing data form that combines an areal signature and a temporal signature. The weather radars produce a nearly continuous picture of the space-time changes of rainfall rates over the radar's operational area.

These and other ideas need to be explored through research that combines remote sensing and hydrologic modeling. Each presents a unique opportunity for hydrologists and water managers to apply remote sensing in ways other than simple extensions of photogrammetry. Remote sensing can produce an integrated measurement that is simultaneously observing several factors. It is also giving us a view that is uncommon to our past thinking in that it looks at a relatively large area and somehow integrates information from the entire scene. A great deal of research is needed to learn how to properly interpret the complex response obtained from the various remote sensing instruments. To use these data effectively, we also must develop new concepts and change our historical way of conceptualizing hydrologic processes.

1.4 Satellite Systems

This book addresses the subject mostly from the perspective of satellite sensors because these observations are almost universally available worldwide. Although the choice of satellite orbit and design are beyond the scope of this book, a practicing hydrologist should understand the basics of sensors and orbits of the major satellite systems because they can influence the choice of data. The existing satellite systems provide very good coverage of the Earth and give the hydrologist a number of options for satisfying data needs. The choice of which satellite system to use depends upon the requirements for the data, which translate into the need for specific spectral bands, spatial requirements, temporal coverage, and the possible need for stereo coverage, all of which are related to the satellite platform. Each of these is discussed more thoroughly below:

1.4.1 Remote Sensing Platforms

The choice of the remote sensing platform is also important to the hydrologist. Platforms include ground based (usually truck or tower mounted), aircraft and the space shuttle, in addition to the satellite systems.

Generally truck mounted and ground based systems are used for sensor development, investigating sensor-target interactions, and algorithm development. These systems enable one to control very precisely what the sensor is "seeing". Figure 1.1 is an example of truck mounted instruments being used for soil moisture experiments over a controlled target.

The aircraft and space shuttle provide an intermediate step before going to a satellite for further instrument and algorithm validation. Aircraft, however, also provide a very useful platform for coverage of relatively small areas and nonrepetitive missions such as aerial photography, multispectral and thermal imaging missions and side-looking airborne radar surveys. Figure 1.2 is the NASA C-130 that has been used for many aircraft campaigns. This aircraft is essentially a flying laboratory and is designed to collect data from multiple instruments at any time.

The Space Shuttle (Fig. 1.3) is frequently used as a space borne platform for proof of concept and testing of new instruments. The Shuttle Imaging Radar (SIR-C) is a good example of this. SIR-C was a 1994 experiment with a two frequency Synthetic Aperture Radar (SAR) for measuring a number of Earth science characteristics, including snow (see Chap. 11) and soil moisture (see Chap. 9).

Fig. 1.1. A cherry picker boom truck being used to make controlled measurements over a known target. There are two passive microwave radiometers (L and C band) mounted on the instrument package at the end of the boom. (Courtesy of Peggy O'Neill, NASA-GSFC)

Fig. 1.2. The NASA C-130 is a typical aircraft platform that has been used in many hydrologic remote sensing campaigns

Fig. 1.3. The Space Shuttle Discovery during a launch of a multi-day science mission

Satellites are an ideal platform (Fig. 1.4) for remote sensing because they can provide essentially global coverage if they are polar-orbiting or continuous coverage if they are geostationary (Fig. 1.5). Polar-orbiting satellites generally fly in a low Earth orbit (hundreds of km) and provide relatively high resolution measurements with repeat times of days to tens of days. Typical polar orbiting satellites are the NOAA-AVHRR (Advanced Very High Resolution Radiometer), the French SPOT (Systeme probatoire d'Observation de la Terre), and the U.S Landsat and TM (Thematic Mapper) series.

Geostationary satellites orbit the earth with the earth's rotation so that they observe the same point on the Earth continuously, but from a much higher altitude approaching 36,000 km. Geostationary satellites are the primary meteorological observation platforms and provide continuous but somewhat coarser spatial data. The European community's Meteosat, the U.S. GOES series, and the Japanese GMS are typical of the geostationary satellites.

There are also satellites that are neither polar orbiting nor geosynchronous. In these cases the orbit path has been chosen to meet a specific science requiremement. The Tropical Rainfall Mapping Mission (TRMM) is a good example of this. TRMM's orbital path takes it to + and − 35 degrees above and below the equator to improve the sampling frequency of tropical rainfall.

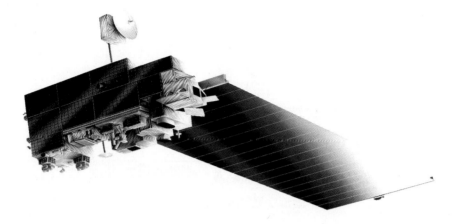

Fig. 1.4. An artists rendition of TERRA, the EOS-AM satellite scheduled to be launched in the second half of 1999

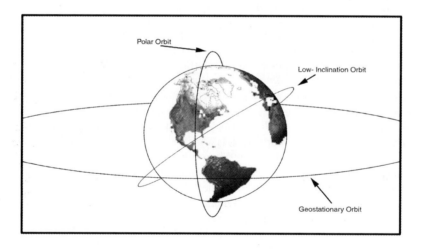

Fig. 1.5. A schematic illustrating common orbits used for Earth remote sensing

1.4.2 Remote Sensing Sensors

Just about all regions of the electromagnetic spectrum (Fig. 1.6) are used in remote sensing. RS data are acquired in predetermined spectral bands (wave lengths). Visible and near infrared spectral bands (which can be displayed as colors) are chosen to amplify or separate specific earth features such as vegetation and water. Figure 1.7 illustrates the relative response (reflectance) of common

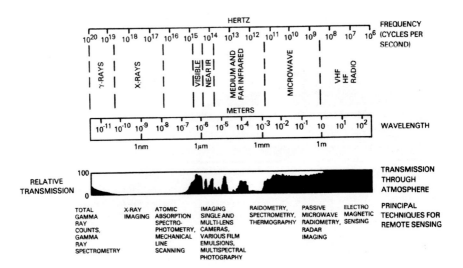

Fig. 1.6. A schematic diagram of the electromagnetic spectrum illustrating the various regions of the spectrum (i.e., visible, near IR, etc.), the wavelength and frequency. The figure also illustrates the relative transmission of regions of the spectrum through the atmosphere

Earth targets for the visible and near infrared portion of the spectrum. Notice how one can separate a chosen land feature from other land features by choice of the wavelength. For example, water has a very low reflectance at a wavelength of 1.1 μm (less than 10%) compared to that of vegetation (about 50%).

In many other cases data from the thermal infrared bands are of high interest, particularly since the thermal infrared data is a measure of the surface temperature and can also be obtained during the night. Microwave data (active and passive) are of particular relevance for certain hydrological variables such as soil moisture and precipitation, and because they can be obtained during the night and are not restricted to cloud free conditions. Table 1.1 lists some representative applications for different spectral bands available from existing satellites.

The application that the hydrologist or water resource manager is addressing will dictate the region of the spectrum that will provide the proper information and thus guide the selection of the sensor(s).

1.4.3 Spatial Resolution

A hydrologist also has to define the spatial resolution needed. This choice depends to a great deal on the nature of his problem and the details needed in his model. In some cases for very large basins, one would not need or want high resolution data. The spatial resolution varies very much from sensor to sensor. In general, the satellites in higher orbits are not able to provide high spatial resolution data, if all other aspects are equal. However, we will see that this is very much dependent upon what parts of the spectrum are used. There are satellites providing data with

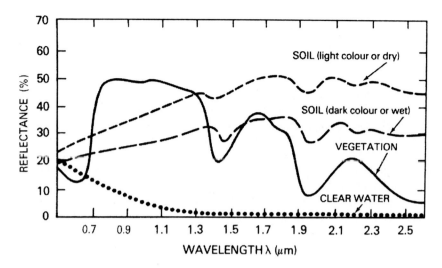

Fig. 1.7. Schematic diagram illustrating the relative reflectances of some common Earth targets

a high resolution of e.g. 10 m (SPOT satellite) and others having a rather coarse resolution of 5 km (e.g. geostationary satellites like Meteosat, GOES, GMS) or even up to 25 km (e.g. satellite sensors in the passive microwave channel).

Table 1.1. Remote sensing Applications for Different spectral bands

Spectral band m	Applications
Blue (0.45 - 0.50)	Water penetration, land use, vegetation characteristics, sediment
Green (0.50 – 0.60)	Green reflectance of healthy vegetation
Red (0.60 – 0.70)	Vegetation discrimination because of red chlorophyll absorption
Panchromatic (0.50 –0.75)	Mapping, land use, stereo pairs
Reflective Infrared (0.75 –0.90)	Biomass, crop identification, soil-crop, land-water boundaries
Mid-infrared (1.5 -.1.75)	Plant turgidity, droughts, clouds, snow-ice discrimination
Mid-infrared (2.0 – 2.35)	Geology, rock formations
Thermal infrared (10 – 12.5)	Relative temperature, thermal discharges, vegetation classification, moisture studies, thermal inertia
Microwave – Short wave (0.1 – 5 cm)	Snow cover, depth, vegetation water content
Microwave-Long wave (5 – 24 cm)	Melting snow, soil moisture, water-land boundaries, penetrate vegetation

1.4.4 Temporal Resolution

RS data are acquired with a given resolution in time. Also here the resolution varies very much from sensor to sensor and satellite system. Ground-based weather radar data can be acquired every 5 minutes, geostationary satellites provide data every half hour and some polar orbiting satellites provide data as seldom as every 16 days (e.g. Landsat) or longer with some of the narrow swath SARs. The hydrologist considering use of remote sensing data has to choose data that match the needs of his analysis. In some cases of dynamic processes and small basins, the data may be needed daily of more often. In other cases for less dynamic processes and large basins, data on two week or longer may be satisfactory. An example of this might be snow melt runoff prediction in a large drainage basin. Then again, there are some needs that have little or no temporal criteria. Examples of this would be delineating stream channels where maps do not exist or land use which changes very slowly.

Appendix 20.1 lists the currently available and future satellite systems that are of interest to hydrologists and water managers. Appendix 20.2 is a similar table that lists the specific sensor and orbit characteristics that are of great interest to hydrologists and water managers.

1.5 Remote Sensing and Hydrology

It should be kept in mind, that RS data are not only used for monitoring of hydrological state variables, but also as the basis for parameter estimation of hydrological models. Remote sensing, particularly from various satellites in various spectral bands, can provide information on catchment characteristics (e.g. landcover, landuse, slope, vegetation), from which the parameters of hydrological models can be gathered. Particularly in combination with other spatial information, such as digital elevation models, digital terrain models, digital soil maps. RS will allow the spatial estimation of hydrological model parameters, e.g. the maximum soil water storage capacity in a river basin.

Another important facet of remote sensing is the fact, that such data can be acquired in remote areas, where no measurements are feasible or can be carried out only under very difficult circumstances which cause high costs. For these measurements particularly airplanes and satellites are suitable. Furthermore satellite remote sensing allows coverage of the whole globe, which is highly relevant in the development of global coupled atmospheric-hydrological models for weather and flood forecasting as well as for long-term analysis and forecast of climate conditions. This property makes RS data particularly valuable for all activities within the framework of the world climate research program (e.g. GEWEX, CLIVAR, ACSYS etc.).

Looking at the historical development of hydrological modeling one comes easily to the conclusion, that all hydrological models are data limited. Models are generally not built in the way which would be scientifically most sound, but rather according to data availability. A significant example of this deficiency of existing models is the fact, that rainfall-runoff models as well as water balance models are all calibrated with the aid of observed data, e.g. rainfall, runoff, evaporation etc. The hydrological variable having the most significant influence on hydrological processes is, however, the state of soil moisture in space and time. Soil moisture determines, how much water goes up from an area element (evaporation, transpiration, interception), how much goes down (infiltration, percolation) and how much water moves laterally (surface runoff, interflow, groundwater flow). Due to the limited data situation in hydrology, this most important parameter soil moisture is almost never measured in the field. Therefore it occurs in hydrological models in the form of a residual. This fact is certainly one of the reasons, why the performance of so many hydrological models is unsatisfactory. It can be hoped, that in the near future remote sensing will allow to measure soil moisture with an acceptable resolution in time and space and with an appropriate accuracy. If this becomes operational, the time has arrived for the development of completely new - and hopefully better - hydrological models, in which soil moisture becomes the central parameter instead of a residual and the information required for model calibration and validation will come from remote sensing sources. It can also be expected, that in many other fields of hydrological modeling the existing or expected availability of remote sensing information will lead to the development of much more efficient hydrological models.

1.6 Structure of the Book

It is surprising that there are very few books or compendiums that cover the complete subject of remote sensing in hydrology and water management, where as there are many books on hydrology and water resources management. The number dealing with hydrology can be easily counted on one hand, Wiesnet et al, 1979, Schultz and Barrett, 1989, Engman and Gurney, 1991, Rango, 1994, and a soon to be published compendium by Rango and Shalaby. All except the Engman and Gurney book have been sponsored by International organizations such as the UN and WMO. Only the Schultz and Barrett publication addressed water resources management. There have been many advances both in instrumentation and applications since the last of these have been published and there is an urgent need to provide an updated source to the hydrology and water management communities.

Since this book is not meant to be a textbook with problem and examples, but rather a reference book in which information on any aspect of interest can easily be found, the book is sub-divided into four general sections, each of which is again sub-divided into various chapters. The sections are as follows:

Section I: Overview and Basis Principles
Section II: RS Application to Hydrological Monitoring and Modeling
Section III: Water Management with the Aid of RS Data
Section IV: Future Perspective.

The first section provides general principles and techniques which should be understood in order to recognize what remote sensing can - and cannot - deliver. Section II is devoted to hydrology, while Section III deals with water management problems. Section IV gives some information on potential future developments in the field of remote sensing and the applications of those new data in the fields of hydrology and water management.

Section II on hydrology deals with the various fields of hydrology, in which remote sensing can be applied. These fields are discussed in the chapters on precipitation, landuse and catchment characteristics, evapotranspiration, soil moisture, surface water, snow and ice, soil erosion, water quality, groundwater. In Section III on water management the following major problems are discussed in different chapters: flood forecasting and control, irrigation and drainage, hydrological data in ungauged river basins and detection of landuse changes and their effect on water management.

References

Engman, E.T., and R.J. Gurney, 1991, Remote Sensing in Hydrology, Chapman and Hall, London, 225pp.

Rango, A., A. Feldman, T.S. George, III, and R.M. Ragan, 1983. Effective use of Landsat data in hydrologic models. Water Resour. Bull. 19, 165-174

Rango, A. 1994. Applications of remote sensing by satellite, radar, and other methods to hydrology. World Meteorological Organization, WMO-No. 804, Operational Hydrology Report No. 39, Geneva Switzerland, 33pp.

Ritchie, J.C. and Rango, A., 1996 Remote sensing applications to hydrology: Introduction. Hydrological Sciences Journal 41(4):429-431

Schultz, G.A., and E.C. Barrett, 1989. Advances in remote sensing for hydrology and water resources management. UNESCO, International Hydrological Programme, Paris, 102pp.

Wiesnet, D.R., V.G. Konovalov and S.I. Solomon, 1979. Applications of Remote Sensing to Hydrology. World Meteorological Organization, WMO-No. 513, Operational Hydrology Report No. 12, 52pp.

2 Physical Principles and Technical Aspects of Remote Sensing

Helmut Rott
Institut für Meteorologie und Geophysik
Universität Innsbruck. A-6020 Innsbruck, Austria

2.1 Introduction

Remote sensing refers to the extraction of information about an object without physical contact. The discussion in this chapter is restricted to remote sensing by means of electromagnetic radiation, which is the main carrier of information for earth observation from space. *Passive sensing* utilizes natural radiation sources: the radiation emitted from the earth's surface and atmosphere or reflected solar radiation. For *active sensing* the object of interest (the target) is illuminated by an artificial source of electromagnetic radiation, the reflected signal received by the sensor is recorded and analyzed.

Remote sensing involves the interpretation of measurements of electromagnetic radiation which is received by a detector after interacting with a target. In general this represents an *inverse problem* with no unique solution because the observed medium is characterized by a number of physical parameters, various states of which may lead to the observed radiation signature. This means that some kind of a priori knowledge is crucial for solving the inverse problem. For this purpose it is necessary to understand the interaction mechanisms of electromagnetic radiation with the objects of interest.

2.2 The Electromagnetic Spectrum and Radiation Laws

The main spectral regions for earth observation from space are the visible (VIS), the infrared (IR), and the microwave (MW) region. The corresponding wavelengths and frequencies are shown in Fig. 2.1. Various, more or less arbitrary, subdivisions of the IR region are used, including solar (or shortwave) IR (0.75-4μm) and terrestrial (or thermal) IR ($>4\mu$m), or near, middle, and far IR. In the microwave region various schemes of letter designation for bands exist. The most common set of microwave band designations is shown in Fig. 2.1. Spaceborne imaging radars operate in the L-, S-, C-, and X-bands, whereas microwave radiometry for earth surface observations utilizes mainly the K-and W-bands.

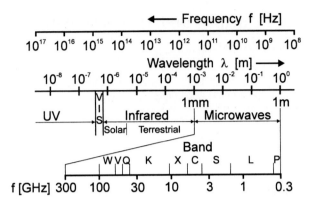

Fig. 2.1. Frequency, wavelength, and nomenclature of the electromagnetic spectrum from the ultraviolet (UV) to the microwave region

Wavelength, λ, and frequency, ν, are related by the equation $\lambda = c/\nu$ where c is the velocity of the wave. The electromagnetic waves are characterized by λ (or ν), the direction of propagation, the amplitude, A, and the polarization (Fig. 2.2). The electric field vector is used to define the direction of wave polarization. In remote sensing, the surface upon which the wave is incident is usually taken as the reference plane. A wave is *horizontally polarized* if the electric field vibrates in a plane which is parallel to the surface, and *vertically polarized* if it vibrates perpendicular to that plane. Solar radiation and thermal emission are usually specified in terms of intensity, which within an isotropic medium is proportional to the square of the amplitude of the electrical field.

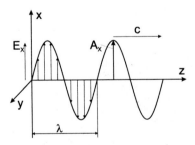

Fig. 2.2. The electric field component E of a plane electromagnetic wave, linearly polarized in x-direction, propagating in z-direction

Radiation Laws. Any material emits radiation in dependence of its temperature and physical properties. As a convenient concept the ideal thermal emitter, called *blackbody*, describes the maximum rate of emitted energy at a given temperature and wavelength. According to *Planck's law* the blackbody spectral exitance $M_{B,\lambda}$ (Wm$^{-2}\mu$m^{-1}) is a function of thermodynamic temperature (T) in degrees Kelvin and can be written in dependence of wavelength:

$$M_{B,\lambda}(T) = \frac{2\pi hc^2/\lambda^5}{\exp(hc/\lambda kT) - 1} \qquad (2.1)$$

where h is Planck's constant, c is the speed of light in vacuum, and k is Boltzmann's constant. $M_{B,\lambda}$ represents the spectral emission from a unit area into the hemisphere. Figure 2.3 shows the spectral radiant exitance for blackbodies at 330 K, 270 K, and 210 K, temperatures observable on the earth's surface. The wavelength, λ_{max}, for which the blackbody spectral exitance reaches its maximum, depends on the temperature, and is described by the *Wien displacement law*: $\lambda_{max}T = const$. The exitance of a blackbody integrated over all wavelengths is described by the *Stefan-Boltzmann law*:

$$M_B = \sigma T^4 \qquad (2.2)$$

where $\sigma = 5.6698 \cdot 10^{-8} Wm^{-2}K^{-4}$. In remote sensing the *radiance*, L, is frequently used, specifying the flux of radiant energy per unit time across a unit area into a cone defined by the unit solid angle (steradian, sr). For an ideal diffuse radiator (a *Lambertian radiator*) the relation between spectral radiance, L_λ (W m^{-2} sr$^{-1}\mu$m^{-1}), and spectral exitance is given by

$$L_\lambda = M_\lambda/\pi \qquad (2.3)$$

In the microwave region the spectral exitance emitted by the earth's surface is several orders of magnitude smaller than at the spectral maximum.

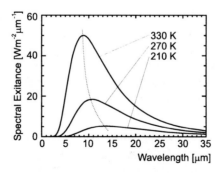

Fig. 2.3. Spectral radiant exitance of blackbodies at 3 different temperatures

In this case $h\nu/kT \ll 1$ is valid, and the spectral radiance can be approximated by the *Rayleigh-Jeans law* which is linear in the temperature radiance relationship:

$$L_\nu = 2kT(\nu/c)^2 = 2kT(1/\lambda)^2 \qquad (2.4)$$

L_ν is the spectral radiance in $(\mathrm{Wm^{-2}sr^{-1}Hz^{-1}})$.

Natural materials emit less energy at a given temperature than a blackbody. The *spectral emissivity*, $\epsilon(\lambda)$, describes the capability to emit radiation as ratio of the spectral radiant exitance of a material to that of a blackbody:

$$\epsilon(\lambda) = M_\lambda(material)/M_{B,\lambda} \qquad (2.5)$$

As long as a material does not undergo phase changes, $\epsilon(\lambda)$ shows little dependence on temperature, but the observation geometry plays a role. Remote sensing applies radiometric methods for measuring surface temperature. The *brightness temperature*, T_B, is the temperature calculated from the measured exitance in a narrow spectral band by inverting the Planck radiation formula under the assumption $\epsilon(\lambda) = 1$, thus

$$T_B(\lambda) = \epsilon(\lambda)T \qquad (2.6)$$

Radiative Transfer. The radiative transfer theory can be applied to describe the propagation of radiation through a bounded medium such as the atmosphere (Liou, 1980), water, or for microwave radiation a layer of snow or soil. The spectral radiance, L_λ, passing trough a layer of thickness, dz, changes by dL_λ (Fig. 2.4). This is expressed by the *radiative transfer equation*

$$\frac{dL_\lambda}{dz} = -k_{e,\lambda}L_\lambda + J_\lambda \qquad (2.7)$$

where the first term on the right-hand side refers to the losses and the second term to the increase of radiance. $k_{e,\lambda}$ $(\mathrm{m^{-1}})$ is the *spectral extinction coefficient* which accounts for absorption losses, described by the spectral absorption coefficient, $k_{a,\lambda}$, and for scattering losses, described by the spectral scattering coefficient, $k_{s,\lambda}$:

$$k_{e,\lambda} = k_{a,\lambda} + k_{s,\lambda} \qquad (2.8)$$

The ratio $k_{s,\lambda}/k_{e,\lambda}$ is called the *single scattering albedo*, $\omega_0(\lambda)$. The source function, J_λ $(Wm^{-3}sr^{-1}\mu m^{-1})$, accounts for the increase of radiation due to thermal emission, given by the source function for emission, $J_{a,\lambda}$, and for the increase due to scattering of radiation in direction of the observation, given by the source function for scattering, $J_{s,\lambda}$; thus $J_\lambda = J_{a,\lambda} + J_{s,\lambda}$.

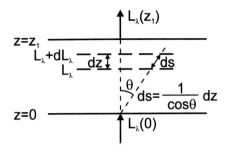

Fig. 2.4. Propagation of radiation in a plane-parallel medium and the relation between vertical and slanting path

If the increase of radiance due to scattering and emission can be neglected ($J_\lambda \approx 0$), we can integrate the transfer equation for propagation through a plane-parallel medium with the thickness z_1 (Fig. 2.4). With the radiance $L_\lambda(0)$ incident at $z = 0$, we obtain at $z = z_1$:

$$L_\lambda(z_1) = L_\lambda(0) \exp[-\tau_\lambda(0, z_1)] \tag{2.9}$$

where $\tau_\lambda(0, z_1)$ is the spectral *optical thickness* of the medium between $z = 0$ and $z = z_1$:

$$\tau_\lambda(0, z_1) = \int_0^{z_1} k_{e,\lambda} dz \tag{2.10}$$

For a homogeneous medium, $k_{e,\lambda}$ is independent of distance, and the optical thickness becomes $\tau(0, z_1) = k_e z_1$. For non-vertical incidence Eq. 2.10 is integrated along path S with $ds = (1/\cos\theta)dz$ (Fig. 2.4). The spectral *transmissivity*, t_λ, between $z = 0$ and $z = z_1$ is defined as

$$t_\lambda(0, z_1) = L_\lambda(z_1)/L_\lambda(0) = \exp[-\tau_\lambda(0, z_1)] \tag{2.11}$$

The Radar Equation. Radar (radio detection and ranging) sensors generate microwave radiation and send it out towards a target. The reflected signal is received and recorded. For explanation of the radar principle, the geometry of a bistatic radar is shown, for which the transmitting and the receiving antenna are separated (Fig. 2.5).

The *radar equation*, which is a form of the radiative transfer equation under specific assumptions, describes the fundamental relation between the properties of the radar, the target, and the received signal. For a single surface-reflecting target the radar equation can be written as

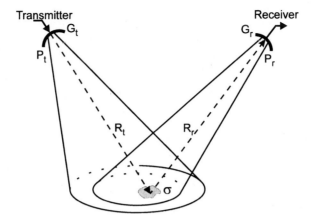

Fig. 2.5. Geometry of a bistatic radar illuminating the surface of an object with radar cross section σ

$$\frac{P_r}{P_t} = \frac{\lambda^2 G_t G_r}{(4\pi)^3 R_t^2 R_r^2} \sigma \qquad (2.12)$$

where P_t is the transmitted power, P_r is the received power, λ is the wavelength of the radar beam, G is the antenna gain, and R is the distance (range) between target and antenna, with the indices t and r referring to the transmitter and the receiver. The power received from a scattering object decreases with the fourth power of range. σ (m^2), the *radar scattering cross section* of an object (also called scattering coefficient) in the observed direction, is the ratio of the total power scattered by an equivalent isotropic scatterer to the power density of a plane wave incident on the object. σ is a function of the dielectric properties and shape of the scattering object and depends also on the observation geometry (Ulaby et al., 1982).

For remote sensing radars the transmitter and receiver are usually co-located. These are *monostatic* radars, for which $G_t = G_r$ and $R_t = R_r$. If the system parameters P_t, G, and λ, and the distance are known, σ is directly proportional the received power P_r.

In environmental remote sensing most of the targets are distributed objects. A soil or water surface can be considered as an object composed of many randomly distributed scatterers with cross sections σ_i, covering differential areas dA_i. If a distributed target is homogeneous within a sensor resolution element, A_0 (m^2), we can replace σ in Eq. 2.12 by $\sigma^\circ A_0$. The dimension-less *radar cross section per unit area*, σ° (m^{2}m^{-2}), is defined as:

$$\sigma^\circ = \left\langle \frac{d\sigma_i}{dA_i} \right\rangle_{A_0} \qquad (2.13)$$

The symbol $\langle\rangle$ denotes the statistical average.

The *scattering cross section per unit volume*, σ_v (m^{2}m^{-3}), is used to describe scattering from a random medium with many individual scatterers with the cross section σ_i, for example a rain cell:

$$\sigma_v = \left\langle \frac{d\sigma_i}{dV_i} \right\rangle_{V_0} \tag{2.14}$$

where V_0 is the resolution volume. In the radiative transfer approach σ_v corresponds to k_s (Eq. 2.8). If the attenuation between the antenna and the scattering volume can be neglected, σ in Eq. 2.12 can be substituted by $\sigma_v V_0$. If the diameter D of a particle is small ($D \lesssim 0.05\lambda$), σ_i can be described by the Rayleigh scattering approximation:

$$\sigma_i = \frac{2\pi^5}{3\lambda^4} |K|^2 D^6 \tag{2.15}$$

where $K = (\epsilon_s - 1)/(\epsilon_s + 2)$, and ϵ_s is the relative dielectric constant of the scattering sphere. In order to derive σ_v, the size distribution of the particles has to be taken into account (Ishimaru, 1978). With $N(D)$ particles per unit volume having a diameter between D and $D + dD$, we obtain in case of Rayleigh scattering

$$\sigma_v = \int_0^\infty \sigma_i(D) N(D) dD = \frac{2\pi^5}{3\lambda^4} |K|^2 \int_0^\infty N(D) D^6 dD = \frac{2\pi^5}{3\lambda^4} |K|^2 Z \tag{2.16}$$

In radar meteorology the *radar reflectivity factor*, Z, is commonly used to describe backscattering from clouds or precipitation (Sauvageot, 1992). It represents the reflectivity for a specific radar system and specific dielectric target properties (e.g. water or ice). Though scattering from a single sphere increases with D^6, this is usually not the case for a scattering volume, because the number density decreases significantly for large particles. For a snow volume σ_v typically increases with about D^3, because the number of ice particles in the unit volume decreases with about the third power of the mean particle size (Mätzler, 1987). Methods for measurement of precipitation by means of weather radar are described in Chap. 6.

2.3 Atmospheric Propagation

Electromagnetic radiation propagating through the atmosphere is attenuated due to absorption and scattering along its path. At the same time, thermal emission and scattering from other directions contribute to the observed radiance. In order to obtain the emission or reflection on the earth's surface from measurements at satellite altitude, the atmospheric effects have to be eliminated. On the other hand, the interaction processes within the propagation path offer the opportunity to retrieve atmospheric properties.

Propagation in the Visible and Solar Infrared. The main factor for atmospheric attenuation of visible light is scattering by air molecules and by aerosols. Molecular scattering can be modelled accurately using the *Rayleigh scattering* approximation (Eq. 2.15). The atmospheric optical thickness, τ_λ, due to Rayleigh scattering is proportional λ^{-4}. This means that molecular scattering in the atmosphere is very important at short wavelengths (UV and visible blue) and of little relevance in the infrared.

The magnitude and angular distribution of *scattering by aerosols* are highly variable, depending on total aerosol content and on the size distribution, the dielectric properties, and the shape of the particles. Due to this variability, effects of aerosol scattering are difficult to correct. Because the aerosol particles are much larger than the air molecules, aerosol scattering is less dependent on λ. In an atmosphere with average turbidity, aerosol scattering dominates over molecular scattering for $\lambda \gtrsim 0.5$ μm.

In the infrared region the absorption due to various gases (primarily H_2O and CO_2) dominates over the scattering losses. The spectral curve of irradiance at the earth's surface in Fig. 2.6 was calculated with the computer code LOWTRAN-7 (Kneizys et al., 1988), assuming a standard atmosphere and clean air (tropospheric) aerosol. In the infrared the observation of the surface is limited to the *atmospheric window regions* between the absorption bands. The main window regions in the solar IR are at $\lambda < 1.1\mu m$, $1.2 \leq \lambda \leq 1.3\mu m$, $1.5 \leq \lambda \leq 1.7\mu m$, and $2.0 \leq \lambda \leq 2.3\mu m$.

Figure 2.7 illustrates the different contributions to the reflected spectral radiance, $L_{\infty,\lambda}$, observed by a satellite sensor:

$$L_{\infty,\lambda} = L_{p,\lambda} + L_{r,\lambda} + L'_{r,\lambda} \qquad (2.17)$$

where $L_{p,\lambda}$ is the radiance scattered within the atmosphere in direction of the sensor (the path radiance), $L_{r,\lambda}$ is the radiance coming from the target at the earth's surface with reflectance $r_{s,\lambda}$, and $L'_{r,\lambda}$ is the reflected radiance from the areas adjacent to the target. For quantitative analysis the spectral surface reflectance, $r_{s,\lambda}$, is required which represents the ratio of the reflected radiance at the surface, $L_{s,\lambda}$, over the global irradiance (the sum of the direct, $E_{dir,\lambda}$, and diffuse irradiance, $E_{dif,\lambda}$):

$$r_{s,\lambda} = \pi L_{s,\lambda}/(E_{dir,\lambda} + E_{dif,\lambda}) \qquad (2.18)$$

$L_{s,\lambda}$ is attenuated in the atmosphere with the optical thickness $\tau_\lambda(0,\infty)$; at the sensor arrives

$$L_{r,\lambda} = L_{s,\lambda} \exp[-\tau_\lambda(0,\infty)/\cos\theta_r] \qquad (2.19)$$

From Eq. 2.17 to Eq. 2.19 it is obvious that information on atmospheric scattering and absorption properties is required for calculating the surface reflectivity from satellite measurements. The radiative transfer code 6S enables the calculation of the surface reflectance from satellite measurements for

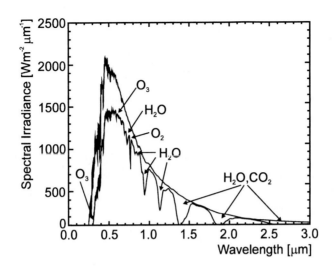

Fig. 2.6. Solar irradiance outside the atmosphere (upper curve) and after passing vertically through a standard atmosphere (lower curve)

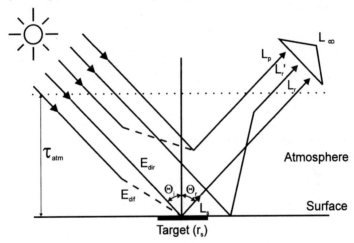

Fig. 2.7. Contributions to satellite-observed radiance of reflected solar radiation. The radiation fluxes are explained in the text

various types of reflectance functions (Vermote et al., 1997). If no information on atmospheric properties is available, empirical atmospheric corrections can be applied, relying on targets with known and temporally stable reflectivity (Schott, 1997).

Propagation in the Thermal Infrared. The transmissivity of the atmosphere in the infrared region between 3 μm and 20 μm is shown in Fig. 2.8, calculated with LOWTRAN-7 for a vertical path assuming standard atmospheric conditions. CO_2, H_2O, N_2O and O_3 are the main absorbing gases (Liou, 1980). Measurements in the spectral regions around 4.3 μm (CO_2), 4.5 μm (N_2O) and 13-15 μm (CO_2) are used for sounding atmospheric temperature profiles, the 6-7μm region for water vapor soundings (Stephens, 1994). The earth's surface can be observed in the narrow window centered at 3.8 μm and in the broad window between 8.5 and 12.5 μm. But also in the window regions the surface emitted radiance, $L_{0,\lambda}$, is affected by atmospheric absorption and emission, primarily by the gases H_2O and CO_2. For the *upwelling radiance* from the local vertical, $L_{\infty,\lambda}$, here conveniently assuming a non-reflecting surface (emissivity $\epsilon_s = 1$), we can write:

$$L_{\infty,\lambda} = L_{0,\lambda} t_\lambda(0, \infty) + \int_{z=0}^{z=\infty} L_{B,\lambda}\left[T(z)\right] \frac{\partial t_\lambda(z, \infty)}{\partial z} dz \qquad (2.20)$$

where t_λ is the spectral transmissivity. The first term on the right-hand side accounts for the surface contribution, the second term for the atmospheric emission, where $L_{B,\lambda}$ is the spectral blackbody radiance. If $\epsilon_s \neq 1$, the reflected downwelling radiance has also to be taken into account. In order to derive the radiance and temperature at the surface, the atmospheric effects have to be eliminated. A common method applies numerical radiative transfer calculations, for example with the public domain software LOWTRAN (Kneizys et al., 1988) or MODTRAN (AFGC, 1989). Input data for these calculations are vertical profiles of temperature T(z) and water vapor.

Another possibility to correct atmospheric effects for deriving the surface temperature and radiance is the split-window technique which is based on measurements in adjacent spectral intervals of the atmospheric window with different transmissivities. The Advanced Very High Resolution Radiometer

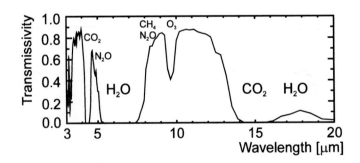

Fig. 2.8. Atmospheric transmissivity in the infrared region through a standard atmosphere at vertical incidence

(AVHRR) of the NOAA satellites has this capability (Cracknell, 1997; Francois & Ottlé, 1996).

Propagation in the Microwave Region. The microwave region offers the advantage to observe the surface through clouds and precipitation. The atmospheric transmissivity between 3 GHz and 300 GHz (10 cm $\geq \lambda \geq$ 1 mm) is shown in Fig. 2.9 for a clear standard atmosphere and for two different types of clouds, based on radiative transfer calculations for a non-scattering atmosphere. Imaging radars are operating in the spectral region below 10 GHz, which is almost unaffected by clouds. Strong precipitation may cause disturbances for radar propagation in this region, but does not inhibit surface imaging. The interaction of radar signals with hydrometeors is the basis for precipitation monitoring by means of weather radar.

Microwave radiometry is applied for surface observations at frequencies up to 40 GHz and in the atmospheric window between 80 and 100 GHz (Ulaby et al., 1981). Above 15 GHz the atmospheric contribution becomes increasingly important, in particular in case of clouds and rain (Liebe, 1985). For a non-scattering atmosphere the brightness temperature $T_{B,\infty}$ at a given frequency measured at satellite altitude includes the following contributions

$$T_{B,\infty} = t(0,\infty) \left[\epsilon_s T_s + (1-\epsilon_s)T_B^\downarrow + (1-\epsilon_s)t(\infty,0)T_c \right] + T_B^\uparrow \qquad (2.21)$$

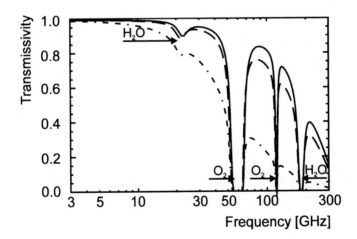

Fig. 2.9. Atmospheric transmissivity in the microwave region at vertical incidence, for clear standard atmosphere (full line), with stratus of 0.5 km thickness (broken line), and with stratocumulus of 2 km thickness (dotted line)

where ϵ_s is the surface emissivity, T_s is the surface temperature, and T_c is the brightness temperature of the cosmic background (2.7 K). Whereas the infrared emissivity of most natural materials is $0.9 \leq \epsilon_{IR} < 1.0$, ϵ_s in the microwave region is much more variable, as explained in Sect. 2.4. Because of the linear temperature-radiance relationship (Eq. 2.4), the radiative transfer calculations can be carried out in terms of T_B. The upward $T_B^\uparrow$ and downward $T_B^\downarrow$ radiating brightness temperatures of the atmosphere can be calculated from the atmospheric temperature profile and the vertical derivative of the transmissivity in analogy to Eq. 2.20. Thus

$$T_B^\uparrow = \int_{z=0}^{z=\infty} T(z) \frac{\partial t(z,\infty)}{\partial z} dz \qquad (2.22)$$

Atmospheric emission and scattering at frequencies between 18 and 40 GHz and in the 80-100 GHz window are the basis for estimating precipitation rates from satellite radiometric measurements.

2.4 Reflection and Emission Characteristics of Natural Media

When an electromagnetic wave hits the interface between two media with different dielectric properties, it is partly reflected back into medium 1 and partly refracted and proceeding into medium 2. The penetrating component may be partly or completely absorbed and scattered. Part of the scattered radiation passes back through the interface into medium 1. The reflection coefficients depend on the polarization of the incident wave, which means that the polarization state of a reflected wave is changed.

The percentages of reflection and penetration at a given incidence angle can be calculated from the dielectric properties of the two media using the Fresnel equations (Schanda, 1986). For a homogeneous lossy material the dielectric properties can be described by the complex *dielectric constant*

$$\varepsilon_r = \varepsilon_r' - i\varepsilon_r'' \qquad (2.23)$$

The real part, ε_r', corresponds to the dielectric constant for a lossless medium, the imaginary part, ε_r'', accounts for the losses. Here the dielectric constant, ε_r, refers to the permittivity relative to vacuum. In the optical region the *refractive index, n*, is commonly used, the two parameters are related by $n = \sqrt{\varepsilon_r}$.

Reflectivity in the Visible and Infrared. In the visible and infrared the penetration into solid materials is in most cases limited to a very thin layer at the surface. Among the exceptions are snow with a typical penetration of several centimeters and ice with penetration of several meters in the visible.

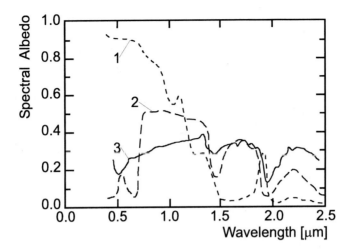

Fig. 2.10. Characteristic spectral reflectivity in the visible and shortwave infrared for (1) fresh snow, (2) a soybean leave, and (3) wet clay

Visible transmissivity of water ranges from centimeters to tens of meters, depending on turbidity. Information on phytoplankton and particulate matter can be derived from the spectral properties of the radiation scattered back from the liquid volume. The absorption in water increases strongly in the near infrared.

Main land cover classes can be discriminated due to characteristic spectral reflectivities. As examples, Fig. 2.10 shows the reflectivity of clean fresh snow (Grenfell et al., 1981), a green soybean leave (Jacquemoud and Baret, 1990), and wet clay (Jacquemoud et al., 1992). The strong increase of reflectivity at 0.7 μm is characteristic for vegetation. The near IR reflectivity enables monitoring of vegetation health, the reflectivity is reduced in case of plant diseases and water stress. Accurate spectral measurements offer excellent capabilities for a wide range of environmental applications such as monitoring of water quality, of vegetation type and state, of soil and rock types, and of snow and ice properties. This was the reason for developing imaging spectrometers.

Microwave Emissivity. Microwaves are able to penetrate into solid material. The *penetration depth*, d_p, represents the distance into the medium so that the intensity is attenuated to the 0.37 - fold of the value at the surface. If $\varepsilon_r'' \ll \varepsilon_r'$, as observed for many natural media, the penetration depth can be approximated by

Table 2.1. Examples for penetration depth, d_p, of soil, ice, and snow with different liquid water content, V_w (in parts by volume)

	d_p (1GHz)	d_p (10GHz)
Soil, V_w=0.05	25 cm	2.5 cm
Soil, V_w=0.25	8 cm	0.5 cm
Ice, V_w=0.00	30 m	3 m
Snow, V_w=0.01	3 m	10 cm
Snow, V_w=0.05	70 cm	2 cm

$$d_p = \frac{\lambda_o}{2\pi} \frac{\sqrt{\varepsilon'_r}}{\varepsilon''_r} \qquad (2.24)$$

where λ_o is the wavelength in free space. Because of high dielectric losses, the presence of water strongly affects the penetration. Examples for penetration into soil and snow with different liquid water content, calculated according to Eq. 2.24, are given in Table 2.1. It is obvious that long wavelengths are required to measure sub-surface soil properties.

The microwave emissivities of natural media depend on the dielectric properties, the surface roughness, and the internal structure. For many media the emissivity shows pronounced variations with frequency and polarization. For this reason multi-channel microwave radiometry is a useful tool for classification and for retrieving target properties.

Examples for characteristic emissivities in the frequency range from 4 to 100 GHz are shown in Fig. 2.11. The data for soil, grass, and snow represent mean values of many measurements carried out in Switzerland by Mätzler (1994). The emissivity of vegetated areas is in general high and varies little with frequency and polarization, as evident from the signature of grass. The bare soil data represent the mean of 9 situations with water contents V_w from 0.12 to 0.40 parts by volume. For wet soil the emissivity increases with frequency and shows a pronounced polarization difference ($\epsilon_v - \epsilon_h$) at frequencies $\leq$ 20 GHz. At low microwave frequencies the emissivity is quite sensitive to the liquid water content (Wang & Choudhury, 1995). The snow cover emissivities represent the mean of 12 dry snow situations with water equivalents between 10 and 25 cm. Because of scattering in the snow volume, the emissivity decreases with increasing frequency. The polarization difference is comparatively high due to reflections at boundaries between snow layers. For comparison, the emissivity for a smooth water surface has been calculated with the Fresnel equation. As opposed to snow, the emissivity of water increases strongly with frequency.

Radar Backscattering Signatures. Active microwave signals are sensitive to the dielectric properties of a medium and to its geometrical structure at the surface and in the volume. Single channel radars, operating at a given

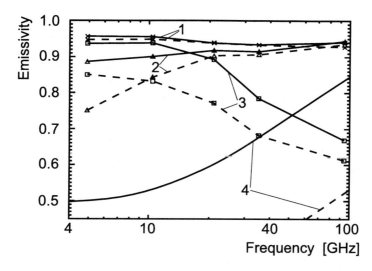

Fig. 2.11. Spectral emissivities at an incidence angle of 50° at horizontal polarization (broken lines) and vertical polarization (full lines) for (1) grass, (2) bare soil, (3) dry alpine snow, and (4) a smooth water surface

frequency and polarization, as well as polarimetric radars are in use. *Polarimetric radars* transmit and receive signals at various polarizations and monitor also the phase relationships between these channels. For specifying the polarization state of a backscatter signal, the first index refers to the receive, the second to the transmit polarization (e.g. HV: horizontally polarized received, vertically polarized transmitted).

Figure 2.12 illustrates the main *backscattering mechanisms* for a radar beam under oblique incidence: 1) Backscatter from a rough surface; a significant part of the energy is reflected incoherently in all directions, therefore the backscatter signal is comparatively high. 2) Backscatter from a smooth surface; the coherent (specular) contribution dominates, only a very small part is scattered back to the illuminating antenna. 3) Double bounce reflection; the beam is reflected at the soil surface and at vertical structures (stalks, trunks), resulting in pronounced phase shifts between HH and VV polarized beams. 4) Direct backscatter from a volume scattering medium (e.g. a vegetation canopy or a snowpack). 5) Indirect scattering contribution due to forward scattering at the soil surface and diffuse scattering in the vegetation, or the reverse process.

Surface roughness is an important parameter for backscattering. A criterion to classify the roughness of a surface has to consider the relation between the standard deviation of surface height, h_s, and the radar wavelength, λ. The *Fraunhofer criterion* states that a surface is smooth if $h_s < \lambda/(32\cos\theta_i)$,

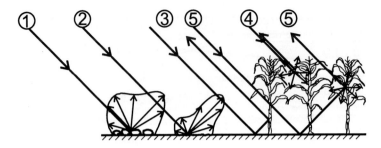

Fig. 2.12. Geometry of the principle radar scattering mechanisms for bare surface and vegetated areas. The individual backscatter contributions are described in the text

where θ_i is the incidence angle of the radar beam. Figure 2.13 shows examples of angular backscatter functions. The bare soil data (curves 1 and 2) clearly demonstrate the decrease of the angular dependence of the backscattering coefficient[1], σ^o, with increasing roughness, in both cases for wet soil. For dense vegetation (curve 3) the signal contribution from the soil below the canopy is small at C-band and higher frequencies, the diffuse backscattering of the vegetation canopy shows comparatively small angular variations. This is different for the wet snowpack (curve 4) with smooth surface and high dielectric losses, for which σ^o is high only at and near vertical incidence due to specular reflection.

In addition to roughness, radar measurements are sensitive to soil moisture, because the dielectric constant, and consequently the reflectivity, depend on the soil water content. If the soil is covered by vegetation, the capability to derive soil moisture depends on the transmissivity of the canopy and decreases towards higher frequencies. L-band shows the best capability for sensing soil moisture below a vegetation canopy. However, also in this frequency band the structure and water content of the vegetation influence the scattering properties and consequently affect the analysis of soil moisture (Dubois et al., 1995).

2.5 Sensor Principles

Categories of spaceborne instruments for earth observation are listed in Table 2.2. For surface observations mainly imaging sensors are used, but also non-imaging sensors such as altimeters, measuring along the nadir-track. For

---

[1] σ^o is commonly expressed in decibels (dB), which is 10 times the logarithm to the base 10 of the ratio of two quantities (in case of σ^o the ratio of reflected to incident power).

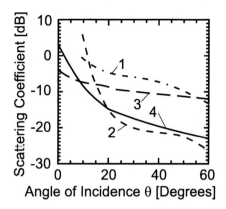

Fig. 2.13. Angular dependence of backscattering coefficients σ^o for bare soil with standard deviation of surface height $h_s = 3.0$ cm (1) and $h_s = 0.4$ cm (2) at 1.5 GHz, HH polarizations; (3) soybean canopy at 4.25 GHz, σ^o_{VV}; (4) wet snow at 10.3 GHz, $(\sigma^o_{HH} + \sigma^o_{VV})/2$. (1) to (3) from Ulaby et al. 1986; (4) from Mätzler, 1987

atmospheric observations imaging as well as sounding sensors are applied. Sounders measure the emitted radiation in narrow spectral intervals originating from different atmospheric layers. Satellite sensors are comprehensively described in the special literature (e.g. Kramer, 1996), a short introduction follows.

Visible and Near-Infrared Sensors. A wide range of sensors is available for observing the earth's surface in the visible and solar infrared (Schott, 1997 and Sect. 2.6). *Photographic cameras* are still used on manned platforms and also on unmanned satellites, providing high resolution images for mapping. For repeat applications *electro-optical scanners*, either applying mechanical or electronical scanning mechanisms, are preferable, because they offer better spectral and radiometric capabilities and allow for real-time data transmission. With mechanical scanners, such as the Thematic Mapper (TM) on board of Landsat, the sideways scanning is achieved by a rotating mirror with the axis of rotation in direction of satellite motion. At any instant the sensor views a given area on the earth's surface (a pixel) through a telescope, the intensity of the radiation from the pixel is measured in each spectral band by one or several detectors.

More recent imaging sensors apply push-broom systems with CCD (charge coupled detector) arrays. A single scan line is imaged at a given time by the CCD array. As for mechanical scanners, the advance from one scan line to

Table 2.2. Categories of earth observation sensors on board of satellites. Acronyms: VIS - visible, SIR - solar infrared, TIR - terrestrial infrared, MW - microwaves

Instrument Type	Spectral Region	Horizontal Resolution	Main Application
Photographic camera	VIS	3-10 m	Land surface mapping
Multispectral scanner	VIS, SIR	6-80 m	Earth surface
Multidirectional scanner	VIS	1-20 m	Topographic mapping
Imaging spectrometer	VIS, SIR	0.25-1 km	Vegetat.,geology,water
Medium resol. scanner	VIS,SIR,TIR	0.5-5 km	Earth surface, clouds
Atmospheric lidar	Active IR	0.1-1 km	Atmosph. properties
Radiation budget radiom.	VIS,SIR,TIR	20-200 km	Radiation balance
Atmospheric sounder	TIR, MW	10-100 km	Atmospheric profiles
Limb sounder	TIR, MW	>300 km	Atmosph. trace gases
Scanning MW radiom.	MW	10-100 km	Land, ocean, atmosph.
Imaging radar	Active MW	10-30 m	Land, oceans, ice
Scatterometer	Active MW	25-50 km	Wind over oceans
Cloud/rain radar	Active MW	5 km	Clouds, precipitation
Altimeters	Active MW	1-5 km	Ocean&ice topography

the next is achieved by motion of the satellite. Significant advancements have also been achieved in the spectral capabilities. In *imaging spectrometers* the incoming radiation is spectrally dispersed onto two-dimensional CCD arrays. Spectral resolutions are typically of the order of a few nm, the number of spectral bands ranges from about 20 to more than 200, depending on the sensor type and data storage capability.

Medium resolution scanners have been designed primarily for meteorological applications, because they provide synoptic coverage of the earth within one day or even shorter intervals. Due the short repeat cycle these sensors are also of interest for observing dynamic features on the earth's surface. Lidars, operating on aircrafts and on the Space Shuttle, have been applied for atmospheric research to measure aerosols, cloud properties, and the wind field. Satellite missions are in preparation.

Thermal Infrared Sensors. Imaging sensors in the thermal infrared region apply mechanical scanning techniques. The infrared radiation emitted by a surface resolution element is received by a photoelectric detector which has to be cooled, controlled radiant temperature sources are viewed at the end of each scan line for calibration (Schott, 1997). Whereas spectral bands of spaceborne thermal imagers are comparatively broad, infrared sounders operate in narrow spectral intervals but with low spatial resolution. Vertical sounders, operating on meteorological satellites, are used for measuring atmospheric temperature and water vapor profiles. Special sounding instruments are applied for monitoring atmospheric trace gases (Harris, 1994).

Passive Microwave Sensors. *Microwave radiometers* measure the emitted radiation. The receiver consists in principle of a high gain antenna, a switching device, one or several noise sources of known temperature used as calibration reference, a bandpass filter, amplifiers, and a detector. The signal received at the antenna is compared with the signals of the reference sources by switching between the antenna input and the reference loads. According to Eq. 2.4 the microwave radiometric measurements can conveniently be expressed in terms of temperature. The radiometric resolution, ΔT, characterizes the performance of a microwave radiometer which may be expressed in the general form (Ulaby et al., 1981):

$$\Delta T = M/\sqrt{Bt} \qquad (2.25)$$

where $B = \Delta \nu$ is the bandwidth in Hz, t is the integration time in sec, and M is the radiometric figure of merit, which is a constant for a given receiver configuration, depending on its technical design. It is obvious that the radiometric resolution has to be traded off versus spectral resolution (given by B) and integration time. Because of the low intensity comparatively broad bandwidths are used for scanning microwave radiometers to obtain radiometric resolutions of the order of several tenths of a degree. Improved spectral resolution is achieved for atmospheric sounders by means of heterodyne techniques, by which the observed signal and the signals of local oscillators are mixed and downconverted to lower frequencies before passing spectral filters.

Another limiting factor of microwave radiometry is the spatial resolution which can be defined by the *half-power beamwidth*, $\beta_{1/2}$. For an antenna with circular aperture the ideal half-power beamwidth in radian is given by

$$\beta_{1/2} = \lambda/d \qquad (2.26)$$

where d is the aperture diameter. Due to practical limits in the physical size of an antenna, the resolution of spaceborne scanning microwave radiometers is ≥ 10 km and varies with the wavelength λ. Imaging microwave radiometers apply either mechanical scanning mechanisms with a rotating antenna (Fig. 2.14) or electronic beam steering. Conical scanning mechanisms enable constant incidence angle on the surface. As an example, the *Special Microwave Sensor/Imager (SSM/I)* on board the DMSP satellite covers a swath of 1400 km width, the elliptical IFOV varies from 69km x 43km at 19 GHz to 15km x 13km at 85 GHz.

Active Microwave Sensors. Active microwave sensors transmit electromagnetic waves towards the target; the reflected wave, incident on the receiver, is recorded and analyzed in order to derive information on physical structure and dielectric properties of the target (Elachi, 1987; Ulaby et al., 1982). For spaceborne radars, as well as for ground-based weather radar, the transmitting and receiving antennas are co-located. The signal is transmitted

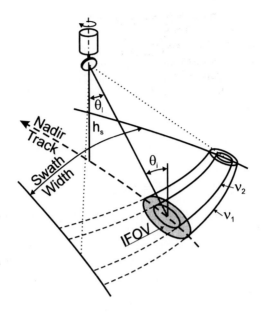

Fig. 2.14. Imaging geometry of a scanning microwave radiometer. IFOV - instantaneous field of view, h_s - satellite height, θ_l - antenna look angle, θ_i - incidence angle, ν_1, ν_2 - measurement frequencies ($\nu_1 < \nu_2$)

in form of short pulses with duration Δt, the distance between the antenna and the target is calculated from the time difference between transmittance and reception of the pulse. The resolution in direction of the radar beam (the range resolution, r_r) is equal to

$$r_r = c\Delta t/2 \qquad (2.27)$$

where c is the propagation velocity. c is slightly affected by atmospheric properties, in particular water vapor content. For accurate range measurements (radar altimetry) these effects have to be corrected.

Spaceborne radar systems can be separated in the categories: imaging radars, altimeters, scatterometers, rain radars. *Altimeters* are used for accurate surface height measurements along the satellite nadir track. The main application is topographic mapping of ocean, lake, and ice surfaces. *Scatterometers* measure accurately the surface backscatter across a swath of several hundred km width, but the spatial resolution is low. The main application is monitoring of wind velocity over the oceans which is derived from the backscatter measurements. Recent investigations demonstrate the usefulness of scatterometry for large scale observations of soil moisture, vegetation, and cryospheric properties. *Rain radars*, operating in the K_u-band, are scatterometers, optimized for measuring backscatter from water drops and ice particles in precipitating clouds with vertical resolution of several hundreds of meters and horizontal resolution of a few kilometers. The first spaceborne rain radar is flown on the Tropical Rainfall Measuring Mission (TRMM).

Spaceborne *imaging radars* (Elachi, 1987) provide surface images with high spatial resolution (between about 10 m and 100 m) over a swath of about 100 km to a few hundreds km width. The across track resolution is determined by the pulse length (Eq. 2.27), high resolution is achieved by means of pulse modulation. The *ground resolution across track*, r_g, varies with the incidence angle: $r_g = r_r/sin\theta_i$ (Fig. 2.15). Synthetic aperture techniques are applied to obtain high resolution along track. The Doppler shift of the return signal, which varies along track within the illuminated footprint, is utilized to synthesize an antenna of 1 km length or more. Radar echos from many hundred pulses are added coherently to calculate the return for one pixel. Theoretically, the *along track resolution*, r_a, depends only on the antenna length ($r_a = L/2$) and thus is independent of the distance between antenna and surface. However, there are practical limits resulting from the power required to detect a return signal. Typical antenna lengths of spaceborne SARs are about 10 m. Swath widths with conventional antenna design are $\leq$ 100 km. Distributed array technology (e.g. on Radarsat) applies electronic beam steering to cover a wider swath with reduced resolution or to select a narrow sub-swath with full resolution.

The radar echo from an illuminated surface area or volume element is the coherent addition of the returns from the individual scatterers. Therefore the resulting return signal of a distributed target that consists of many point scatterers fluctuates randomly in successive observations as the radar moves along. The statistics of the backscattering power, P, which is proportional to σ^o, is described by an *exponential probability density function* (Ulaby et al., 1982). The standard deviation of the power is

$$\sigma_P = \overline{P}/\sqrt{N} \qquad (2.28)$$

where N is the number of independent looks and $\overline{P}$ is the mean power. $N = 1$ is valid for pixels in full spatial resolution. In order to measure σ^o of a distributed target accurately, averaging or low pass filtering over many looks is necessary.

Whereas single channel SARs are operating on the European Remote Sensing Satellite (ERS) and on Radarsat, the experimental three-frequency polarimetric Spaceborne Imaging Radar-C (SIR-C)/X-SAR has been operating during two Shuttle missions in 1994. Polarimetric SARs are also in operation on aircrafts. Radar polarimetry offers improvements for quantitative inversion of target properties and for classification. For example, polarimetric measurements enable the separation of surface roughness and moisture effects and therefore are of interest for soil moisture monitoring (Dubois et al., 1995).

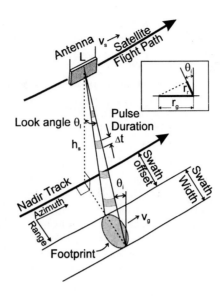

Fig. 2.15. Geometry of a side-looking imaging radar. r_r - slant range resolution, r_g - ground range resolution

SAR interferometry (InSAR) is a new technique of interest for a wide range of applications in geoscience and environmental monitoring (Massonet and Feigl, 1998). With spaceborne SAR an interferogram is calculated from two images of repeat orbits which are slightly displaced in the across-track direction. A condition for interferogram generation is the preservation of the signal phase (coherence) between the two images. Coherence is affected by temporal changes of backscattering (e.g. due to snowmelt or rain). The phase shifts in a repeat pass interferogram result from topography and from differential motion. If there is no motion, a digital elevation model can be derived from a single interferogram. In case of motion (vertical and/or horizontal) at least two interferograms (calculated from three or four images) are needed to separate the topographic and motion-related phase shifts. This technique, called *differential interferometry*, is very sensitive. Displacements of the order of fractions of a wavelength between the acquisition times of two images can be detected. Applications include measurements of glacier motion, monitoring of subsidence due to ground-water withdrawal or mining activities, detection of volcanic inflation, and mapping of seismic deformation.

Colour plates 2.A and 2.B show an example of interferometric analysis based on SIR-C data of the L-band channel (λ = 24.3 cm), acquired on 9 and 10 October 1994 over Moreno Glacier, Southern Patagonia, Argentina (Rott et al., 1998). The coherence image shows a high degree of coherence over the snow- and ice-free land surfaces. Over water the signal decorrelates completely, over glaciers and snow-covered areas the degree of coherence was reduced due to surface melt, but still sufficient for calculating the interferogram. One colour cycle (one fringe) in the interferogram represents a phase shift of 2π between the two images. This corresponds to a differential motion of 21.7 cm on a horizontal surface. As derived from the distance of 29 m between the antenna positions of the two images and the wavelength, one fringe corresponds to an altitude difference of 550 m for a non-moving target. Because only one interferogram was available, the topographic phase was calculated from digital elevation data in order to generate the motion map of the glacier terminus. Only the motion component in direction of the radar beam (range) can be obtained by means of interferometry. Therefore the direction of motion on the glacier was derived from flowlines in optical and SAR images.

2.6 Summary of Current and Future Earth Observation Missions

Many earth observation instruments are presently flown on satellites or are due to be launched in the next few years. A selection of un-manned earth observation missions is listed in Appendix 20.1 and sensor characteristics are specified in Appendix 20.2. Comprehensive information on satellites and sensors is provided by Kramer (1996). The majority of the satellites has near-polar orbits, with the lowest altitudes (around 250 km) for photographic camera missions, and typical orbital altitudes between 700 km and 900 km for the other missions. Geostationary satellite series, such as GOES and Meteosat, are imaging the earth disk in close time intervals (typically 1/2 hour) from an altitude of 35800 km, but are not covering the high latitudes. Important experimental earth observation missions have been flown on manned spacecrafts, including the NASA Space Shuttle and the Russian space station MIR. Of particular interest for the development of SAR applications were the Spaceborne Radar Laboratory Missions SRL-1 and SRL-2 in 1994 with the multi-frequency (L-, C-, X-band) polarimetric SIR-C/X-SAR.

Sensors with medium spatial resolution (1-5 km) on polar orbiting and geostationary satellites, such as AVHRR on POES (NOAA), VISSR on GOES, and MVIRI on Meteosat, enable global imaging with high temporal resolution. Data from various of these sensors can be received in real time with reasonable technical efforts, making them very useful tools for operational meteorology and hydrology. There is also a broad choice of high resolution optical sensors, ranging from photographic cameras to various types of multispectral scanners. Data reception for these sensors is usually restricted to special ground stations. Some of the agencies (e.g. SPOT Image) offer possibilities for fast delivery. Cameras are flown on a number Russian satellites in low orbits; the spatial resolution and areal coverage varies with the orbit altitude. SARs became important tools for earth observation with the launch of ERS-1 in 1991, JERS-1 in 1992, ERS-2 and Radarsat in 1995. Fast delivery SAR products are disseminated through digital networks. Radar altimeter data are available from several satellites since 1978 (Seasat, Geosat, ERS-1, ERS-2, TOPEX/POSEIDON). Though the altimeters have been designed for ocean surface measurements, they are also applied for measuring the level of inland waters and large-scale surface topography over Antarctica and Greenland.

New imaging sensors on scheduled missions will include imaging spectrometers with high spectral resolution and pushbroom scanners with very high spatial resolution (< 5m). The imaging spectrometers MODIS on EOS AM and MERIS on ENVISAT will have moderate spatial resolution (250-300 m), and spectral resolution between 2.5 and 10 nm. MODIS can acquire 36 spectral channels at the same time, and MERIS 15, the position of the channels is programmable. In addition, the launch of hyperspectral imagers with high spectral and spatial resolution, but narrow swath coverage, is planned, e.g. on the NMP/EO-1 mission of NASA with 30 m resolution. The Advanced SAR (ASAR), to be launched on ENVISAT, will be also a single frequency systems, but with flexible pointing capabilities and with various polarization modes. Several satellites with very high resolution imagers are due to be launched in the near future, including 1m resolution sensors on Ikonos and on QuickBird.

References

AFGC (1989) MODTRAN 3 User Manual. AFGL-TR-89-0122, Air Force Geophysics Laboratory, Hanscom AFB, MA., USA

Cracknell A.P. (1997) The Advanced Very High Resolution Radiometer. Taylor and Francis, London

Dubois P.C., van Zyl J. & Engman T. (1995) Measuring soil moisture with imaging radars. IEEE Trans. Geosc. Rem Sens. 33(4), 915 - 926

Elachi C. (1987) Spaceborne Radar Remote Sensing: Applications and Techniques. IEEE Press, New York

Francois C. & Ottlé C. (1996) Atmospheric corrections in the thermal infrared: Global and water vapor dependent split-window algorithms - applications to ATSR and AVHRR data. IEEE Trans. Geosc. Rem Sens. 34(3), 457 - 470

Grenfell T.C., Perovich D.K. & Ogren J.A. (1981) Spectral albedos of an alpine snowpack. Cold Regions Science and Technology 4, 121-127

Harris J.E. (1994) Earthwatch, the Climate from Space. John Wiley and Sons, Chichester, New York

Ishimaru A. (1978) Wave Propagation and Scattering in Random Media. Academic Press, New York, San Francisco, London

Jacquemoud S. & Baret F. (1990) PROSPECT: A model of leaf optical properties spectra. Remote Sens. Environ. 34, 75-91

Jacquemoud S., Baret F. & Hanocq J.F. (1992) Modeling spectral and bidirectional soil reflectance. Remote Sens. Environ. 41, 123-132

Kramer H.J. (1996) Observation of the Earth and it's Environment, Survey of Missions and Sensors. Springer-Verlag, Berlin, 3. edition

Kneizys F.X. (1988) Users Guide to LOWTRAN. Technical report AFGL-TR-88-0177, Air Force Geophysics Laboratory, Hansom AFB, MA, USA

Liebe H.J. (1985) An updated model for millimeter wave propagation in moist air. Radio Science 20(5), 1069-1089

Liou K.-N. (1980) An Introduction to Atmospheric Radiation. Academic Press, San Diego, New York

Massonet D. K.L. Feigl (1998) Radar interferometry and its application to changes in the Earth's surface. Rev. Geophys. 36, 441-500, 1998

Mätzler C. (1987) Applications of the interaction of microwaves with the natural snow cover. Remote Sensing Reviews 2, 259-387

Mätzler C. (1994) Passive microwave signatures of landscapes in winter. Meteorology and Atmospheric Physics 54, 241-260

Rott H., Stuefer M., Siegel A., Skvarca P. & Eckstaller A. (1998) Mass fluxes and dynamics of Moreno Glacier, Southern Patagonia Icefield. Geophys. Res. Letter 25(9), 1407-1410

Sauvageot H. (1992) Radar Meteorology. Artech House, Boston - London

Schanda E. (1986) Physical Fundamentals of Remote Sensing. Springer-Verlag, Berlin, Heidelberg, New York, Tokio

Schott J.R. (1997) Remote Sensing, The Image Chain Approach. Oxford Univ. Press, New York, Oxford

Stephens G. L. (1994) Remote Sensing of the Lower Atmosphere. Oxford Univ. Press, New York, Oxford

Ulaby F.T., Moore R.K. & Fung A.K. (1981) Microwave Remote Sensing, Vol.1: Fundamentals and Radiometry. Addison-Wesley, Reading, MA

Ulaby F.T., Moore R.K. & Fung A.K. (1982) Microwave Remote Sensing, Vol. 2: Radar Remote Sensing and Surface Scattering and Emission Theory. Addison-Wesley, Reading, MA

Ulaby F.T., Moore R.K. & Fung A.K. (1986) Microwave Remote Sensing, Vol. 3: Active and Passive, from Theory to Applications. Artech House, Dedham, MA

Vermote E.F., Tanré D., Deuzé J.L., Herman M. & Morcette J.-J. (1997) Second simulation of the satellite signal in the solar spectrum, 6S: an overview. IEEE Trans. Geosc. Rem Sens. 35(3), 675-686

Wang J.R. & Choudhury B.J. (1995) Passive microwave radiation from soil: Examples of emission models and observations. In: Passive Microwave Remote Sensing of Land-Atmosphere Interactions, ed. B.J. Choudhury, Y.H. Kerr, E.G. Njoku, P. Pampaloni, VSP Int. Science Publishers, Zeist, 423-460

3 Processing Remotely Sensed Data: Hardware and Software Considerations

J. R. Jensen
Department of Geography, University of South Carolina, Columbia, SC 29208

Analog and digital remotely sensed data are used operationally in many hydrologic and water management applications. Analog remotely sensed data (e.g. aerial photographs) are routinely analyzed visually using optical instruments such as stereoscopes. Digital remote sensor data are analyzed using a digital image processing system that consists of both hardware and software. This chapter introduces fundamental:

- image processing system characteristics,
- image processing (and some GIS) software requirements, and
- commercial and public providers of digital image processing systems.

3.1 Image Processing System Characteristics

Hydrologists and other scientists must make decisions about the characteristics of the digital image processing system used to process digital remote sensor data. Important factors include: the type of computer, number of analysts and mode of operation, serial versus parallel processing, arithmetic coprocessor, random-access memory (RAM), compiler(s), amount and type of mass storage, spatial and color display resolution desired, and image processing applications software. It is instructive to review these factors.

3.1.1 The Central Processing Unit (CPU): Personal Computers, Workstations and Mainframes

Hydrologic studies often require the analysis of large watersheds or other geographically extensive phenomena. Scientists also are taking advantage of higher spatial and spectral resolution remote sensor data (even hyperspectral involving hundreds of bands). Processing these massive remote sensor datasets requires a large number of computations to be performed. The type of computer selected dictates how fast (efficient) these computations or operations can be performed. The central processing unit (CPU) is the computing part of the computer and consists of a control unit and an arithmetic logic unit (ALU) (Freedman, 1995). The CPU performs a) numerical calculations, and b) input–output to mass storage devices, color monitors, plotters, etc. Its efficiency is often measured in terms of how many millions-of-instructions-per-second (MIPS) it can process, e.g. 450 MIPS.

Personal computers (16- to 32-bit CPUs) are routinely used to perform digital image processing and GIS analysis. These complex-instruction-set-computers (CISC) have CPUs with 16- to 32-bit registers (word size) that compute integer arithmetic expressions at greater clock speeds and process significantly more MIPS than their 8-bit predecessors. 16-bit CPUs can process two 8-bit bytes at a time while 32-bit CPUs can process four bytes at a time. Table 3.1 documents the historical development of the Intel family of CPUs used in IBM compatible personal computers. Scientists often populate laboratories with PC-based digital image processing systems because the hardware and software are relatively inexpensive per unit and hardware maintenance is low. Typical personal computers cost <$3,000 with 24-bits per pixel of image processor memory (with a 16.7 million color lookup table) and a high resolution color monitor.

Table 3.1. Historical development of the Intel family of CPUs used in numerous IBM compatible personal computers (Freedman, 1995; Spooner, 1999)

CPU (word size)	Clock Speed (Mhz)	MIPS
8088 (16)	5	0.33
8086 (16)	5 – 10	0.33 - 0.66
286 (16)	6 – 12	1.2 - 2.4
386DX (32)	16 – 40	6 - 15
486DX (32)	25 – 100	20 - 80
Pentium I	60 – 200	100 - 250
Pentium II	300 – 400	300 - 400
Pentium III Celeron	>466	> 450

Computer Workstations (≥32-bit CPUs) usually consist of a 32 to 64-bit reduced-instruction-set-computer (RISC) that can address substantially more random access memory than personal computers. The RISC chip is faster than the traditional CISC and is designed and built more economically (Freedman, 1995). Workstations also usually have a bank of 8- or 24-bit image processing memory and a very high resolution color monitor (Denning, 1993). Figure 3.1 summarizes the components found in a typical digital image processing workstation laboratory. RISC workstations can function independently or be networked to a file-server as shown. Some RISC workstations have multiple CPUs that allow remotely sensed data to be processed in parallel and at great speed. Such configurations make the distinction between mainframes and workstations fuzzy (Berry, 1993). RISC workstations application software and hardware maintenance costs are usually higher than personal computer based image processing systems.

Mainframe Computers (≥32-bit CPU) are generally more efficient than workstations, especially parallel mainframe computers such as a CRAY (Earnshaw and Wiseman, 1992). Mainframes are ideal for intensive CPU-dependent tasks such as spatial/frequency filtering, image rectification, mosaicking of numerous scenes,

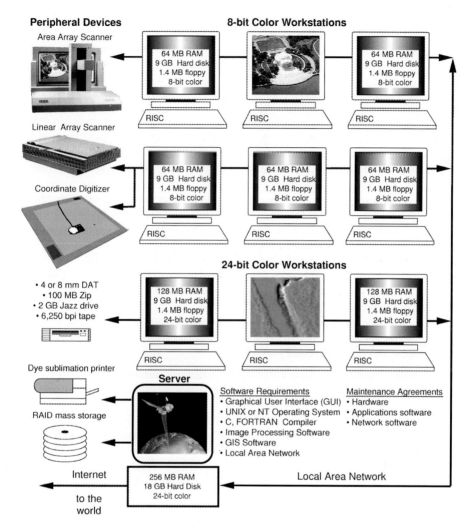

Fig. 3.1. A digital image processing lab consisting of reduced-instruction-set-computers (RISC) and peripheral devices. There are six 8-bit displays and four 24-bit color displays. The workstations communicate locally via a local area network (LAN) and with the world via the Internet. Each workstation has 64 to 128 MB random access memory (RAM), 9 GB hard disk space, compact disk, and a floppy disk. UNIX or NT is the operating system of choice. Image processing software and remote sensor data may reside on each workstation (increasing the speed of execution), but may also reside on the server

classification, or complex spatial GIS modeling. The output from the intensive mainframe processing can be passed to a workstation or personal computer for subsequent less expensive processing if desired (Davis, 1993). Mainframe computer systems are expensive to purchase and maintain.

3.1.2 Number of Analysts on a System and Mode of Operation

The ideal digital image processing environment is when a single user sits in front of single workstation. Unfortunately, this is not always possible due to cost constraints. The sophisticated workstation lab shown in Fig. 3.1 might be ideal for research, but ineffective for education or short course instruction where many analysts (e.g. > 20) must be served.

It is well known that the best scientific visualization environment takes place when the digital image processing system uses an interactive graphical-user-interface (GUI) (Mazlish, 1993; Miller and DeCampo, 1994). Two effective graphical user interfaces include ERDAS Imagine's intuitive point and click icons (Fig. 3.2a, b) and ENVI's hyperspectral data analysis interface (Fig. 3.3). Late night non-interactive batch processing is of value for time consuming processes (e.g. resampling during image rectification) and helps to free-up lab workstations during peak demand.

3.1.3 Serial versus Parallel Image Processing, Arithmetic Coprocessor, and Random Access Memory (RAM)

Some computers have multiple CPUs that operate concurrently. Parallel processing software allocates the CPUs to perform efficient digital image processing (Faust et al., 1991). For example, consider a 512 node (CPU) parallel computer. If a remote sensing dataset consisted of 512 pixels (columns) in a line, each of the 512 CPUs could be programmed to process an individual pixel, speeding up the processing of a single line of data by 512 times. If 512 bands of hyperspectral data were available, each processor could be allocated to a single band to perform independent processing. Many vendors are developing digital image processing code that takes advantage of parallel architecture.

An *arithmetic coprocessor* is a special mathematical circuit that performs high-speed floating point operations while working in harmony with the CPU. Most sophisticated image processing software often will not function without a math coprocessor. If substantial resources are available, then an array processor is ideal. It consists of a bank of memory dedicated to performing simultaneous computations on elements of an array (matrix) of data in n dimensions (Freedman, 1995). Remotely sensed data are collected and stored as arrays of numbers so array processors are especially well suited to image enhancement and analysis operations.

Random access memory (RAM) is the computer's primary temporary workspace. RAM chips require power to maintain their content. Therefore, all information stored in RAM must be saved to a hard disk (or other media) before turning off the computer. The computer should contain sufficient RAM for the operating system, image processing applications software, and any remote sensor data that must be held in memory while calculations are performed. Computers with 64-bit

3 Processing Remotely Sensed Data: Hardware and Software Considerations   45

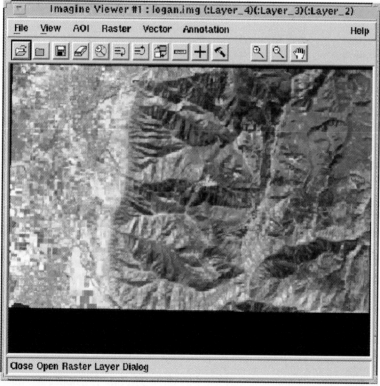

Fig. 3.2. The ERDAS Imagine graphical user interface consists of point-and-click icons (**a**) that are used to select various types of image processing and GIS analysis in the Imagine viewer interface (**b**)

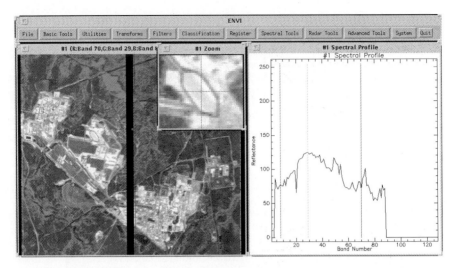

Fig 3.3. ENVI's graphical user interface allows complex operations to be applied to hyperspectral data. Here, the cursor is located on a clay capped waste unit on the Savannah River Site. The three bands used to make the color composite (shown here in black and white) are depicted as bars in the spectral profile plot. The spectral reflectance curve at the cursor location is displayed in the spectral profile plot. This dataset consisted of 88 bands of 10 x 10 m TRWIS hyperspectral data

CPUs can address more RAM than 32-bit machines, etc. Figure 3.1 depicts individual workstations with 64 to 128 MB of RAM and a server with 256 MB of RAM. One can never have too much RAM for image processing applications. Fortunately, RAM prices continue to decline.

3.1.4 Operating System and Software Compilers

The operating system is the master control program that runs the computer. It is the first program loaded into memory (RAM) when the computer is turned on, and its main part, called the kernel, resides in memory at all times (Freedman, 1995). The operating system sets the standards for the image processing application programs that are executed by it. All programs must 'talk to' the operating system. The difference between an operating system and a network operating system is its multi-user capability. DOS, Windows 98, and Macintosh OS are single-user operating systems designed for one person at a desktop computer. Windows NT and UNIX are network operating systems because they are designed to manage multiple user requests at the same time and handle the related security. The operating system provides the user interface and controls multi-tasking. It handles the input and output to the disk and all peripheral devices such as printers, plotters, and color displays.

A *compiler* is the software that translates a high-level programming language such as C or FORTRAN into machine language. A compiler usually generates assembly language first and then translates the assembly language into machine language. The programming languages most often used in the development of digital image processing software are Assembler, C, JAVA, and FORTRAN. Many digital image processing systems provide a toolkit that programmers can use to compile their own digital image processing algorithms. The toolkit consists of primitive subroutines such as reading a line of image data into RAM or writing a vector to the screen. Also many software packages provide high-level macro languages, or libraries of image processing and GIS software components. These tools allow the development of complex applications for specific purposes.

3.1.5 Mass Storage

Digital remote sensing research often requires substantial mass storage resources. For example, during the early part of a project it is not uncommon to load an entire 7-band Landsat TM scene onto a hard disk at one time (3,000 x 3,000 pixels x 7 bands = 63 MB) to obtain an appreciation of the geographic extent of the image and its apparent quality. Therefore, the storage media should have rapid access times, have longevity (i.e. last for a long time), and be inexpensive (Rothenberg, 1995). Digital remote sensor data (and other ancillary raster GIS data) are normally stored in a matrix band sequential (BSQ) format in which each spectral band of imagery (or GIS data) is stored as an individual file. Each picture element of each band is represented in the computer by a single 8-bit byte (with values from 0 to 255). The best way to make the brightness value available to the computer rapidly is to place the data on a hard (or optical) disk where each pixel of the data matrix may be accessed at random and at great speed (within microseconds). The cost of hard disk storage per gigabyte continues to decline rapidly. It is common for digital image processing laboratories to have gigabytes of mass storage associated with each workstation as suggested in Fig. 3.1. In fact, many image processing laboratories now use RAID (redundant arrays of inexpensive disks) technology in which two or more drives working together provide increased performance and various levels of error recovery and fault tolerance (Freedman, 1995).

Figure 3.4 depicts several types of analog and digital remote sensor data mass storage devices and the average time to physical obsolescence, i.e. when the media begin to deteriorate and information is lost. Properly exposed, washed, and fixed black and white aerial negatives have considerable longevity, often more than 100 years. Color negatives with their respective dye layers have longevity, but not as long as the black and white negatives. Similarly, black and white paper prints have greater longevity than color prints (EDC, 1995; Kodak, 1995). Hard and floppy magnetic disks have relatively short longevity, often less than 20 years. Magnetic tape media (e.g. 3/4" tape, 8-mm tape, and 1/2" tape shown in Fig. 3.4) can become unreadable within 10 to 15 years if not rewound and properly stored in a cool, dry environment (EDC, 1995).

Optical disks can now be written to, read, and written over again at very high speeds (Normile, 1996). The technology used in re-writeable optical systems is magneto-optics (MO), where data is recorded magnetically like disks and tapes, but the bits are much smaller because a laser is used to etch the bit. The laser heats

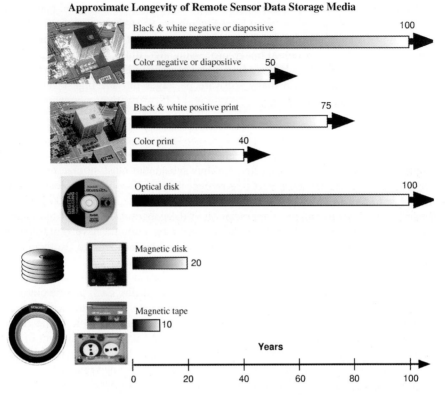

Fig. 3.4. Analog photography is an excellent storage media with black and white negatives or diapositives having the longest usable life. Optical disk media is superior to magnetic media for storing digital remote sensor data (Lulla, 1995; Jensen et al., 1996)

the bit to 150 C at which temperature the bit is realigned when subjected to a magnetic field. In order to record new data, existing bits must first be set to zero. Only the optical disk provides relatively long-term storage potential (>100 years). In addition, optical disks store large volumes of data on relatively small media. Advances in optical compact disc (CD) technology promise to increase the storage capacity from the current 780 MB per disc to more than 17 GB using new rewriteable digital video disc (DVD) technology (Normile, 1996). It is important to remember when archiving remote sensor data, however, that sometimes it may be the loss of a) the read-write software, and/or b) the read-write hardware (drive mechanism and heads) that are the problem and not the digital media itself (Rothenberg, 1995; Jensen et al., 1996).

3.1.6 Screen Display Resolution

Display resolution is the number of rows and columns of pixels that can be displayed on a CRT screen at one time. Most earth scientists prefer a regional perspective when performing terrain analysis using remote sensor data. Therefore, the image processing system should be able to display at least 512 rows x 512 columns of pixels and preferably more (e.g., 1024 x 1024) on the CRT at one time. This allows larger geographic areas to be examined at one time and places the terrain of interest in its regional context.

3.1.7 Screen Color Resolution

CRT screen color resolution is the number of gray-scale tones or colors (e.g., 256) that a pixel may be displayed in on a screen out of a palette of available colors. Most sophisticated image processing systems provide a tremendous number of displayable colors from a large color palette (e.g., 16.7 million). The primary reason for these significant color requirements is that image analysts must often display a color composite of several individual images at one time on the screen. For example, to display the false-color composite of Hurricane Andrew in the Gulf of Mexico (Colour Plate 3.A) it was necessary to place three separate 8-bit images in three distinct planes of *image processor memory* [AVHRR band 5 (11.5 - 12.5 µm) thermal infrared data were placed in the blue image processor memory plane; band 2 (0.725 - 1.10 µm) near-infrared data were placed in the green memory plane; and visible band 1 (0.58 - 0.68 µm) data were placed in the red image plane]. Thus, each pixel could take upon itself any of 2^{24} possible color combinations (16,777,216). Such true color systems are expensive because every pixel location is *bit mapped*, i.e. a specific location in memory keeps track of the blue, green, and red brightness values.

The network configured in Fig. 3.1 has six 8-bit color workstations and four 24-bit color workstations. Everyone does not require access to a 24-bit color display at once because many image processing functions such as black and white image display and the creation of final color thematic maps can be performed quite well in 8 bits (Busbey et al., 1992).

3.1.8 Image Scanning (Digitization) Considerations

Many hydrologic projects require the analysis of analog (hard-copy) aerial photography such as color-infrared U.S. National Aerial Photography Program (NAPP) imagery. Also, many studies make use of multiple dates of historical panchromatic black and white aerial photography (e.g. 1:20,000 U. S. Agricultural Stabilization and Conservation Service photography obtained in the 1930s and 1940s). Such data are often digitized, rectified, and then analyzed in a digital image processing system. To be of value, careful decisions must be made about the spatial and radiometric resolution of the digitized image data. Tables that summarize the relationship between input image scale, digitizer detector instantaneous-field-of-view (IFOV) measured in dots-per-inch (dpi) and micrometers (µm), and output image spatial resolution in meters are found in Jensen (1995b; 1996).

Images to be used for simple illustrative purposes may be scanned at low resolutions such as 100 dpi (254 µm). For vertical aerial photography obtained at 1:20,000 scale, this yields images with a spatial resolution of 5.08 x 5.08 m that retain sufficient detail for orientation purposes. For a 9 x 9" photograph, this produces a 0.81 megabyte file (9 x 9 x 100 x 100 dpi = 810,000 bytes) for black and white photographs and a 2.43 MB file for color aerial photography. This assumes that the black and white images are scanned at 8-bit radiometric resolution and that color images are scanned at 24-bit resolution. Images that are to be used for scientific purposes are usually scanned at much higher spatial resolution such as 500 dpi (12.7 μm) to retain the subtle reflectance and/or emittance information found in the original imagery. For 1:20,000 scale photography, this yields a spatial resolution of 1.02 x 1.02 m per pixel and results in a 20.25 MB file for a single black and white photograph and a 60.75 MB file for a single color photograph (Jensen, 1995b; 1996).

Most aerial photography is collected using a 9 x 9" metric camera. The ideal scanning system digitizes the entire image and any ancillary 'titling' information on the periphery of the image at one time. Therefore, the digitizer of choice should have a field of view of at least 9 x 9". While it is possible to use inexpensive 8.5 x 11" desktop scanners, this requires that the imagery be broken up into two parts that must then be mosaicked together. This introduces radiometric and geometric error and should be avoided. Advances in the 'desktop publishing' industry have spurred the development of flatbed, 11 x 14" desktop linear array digitizers that can be used to digitize negatives, diapositives, and paper prints at 100 to 3,000 dots-per-inch (Foley et al., 1994). Some scanners digitize color photographs with one color filter then repeat the process with the other color filters. This can result in color misregistration and loss of image quality. Ideally, color aerial photography is scanned in a single pass and converted into three registered red, green, and blue (RGB) files.

Area array charge-coupled-device (CCD) digital camera technology has been adapted for hard-copy image digitization (Fig. 3.1). Typical area array CCD systems digitize from 160 dpi to 3,000 dpi (approximately 160 μm to 8.5 μm) over a 10 x 20" image area (254 μm x 508 μm). They scan the original negative or positive transparency as a series of rectangular image tiles. The scanner then illuminates and digitizes a reseau grid which is an array of precisely located crosshatches etched into the glass of the film carrier. The reseau grid coordinate data are used to locate the exact orientation of the CCD camera during scanning and to geometrically correct each digitized 'tile' of the image relative to all others. Radiometric calibration algorithms are then used to compensate for uneven illumination encountered in any of the tile regions. Area array digitizing technology has obtained geometric accuracy of < 5 μm over 23 x 23 cm images when scanned at 25 μm per pixel and repeatability of < 3 μm (Jensen, 1996).

3.2 Image Processing and GIS Software Requirements

A variety of digital image processing and geographic information system (GIS) functions are required to analyze remotely sensed data for hydrology and water resource management applications. Some of the most important functions are summarized in Table 3.2. It is useful to briefly identify characteristics of the most important functions. The reader is encouraged to review textbooks on digital image processing and GIS to obtain specific information about the algorithms (Jahne, 1991; Lillesand and Kiefer, 1995; Jensen, 1996).

Table 3.2. Image processing functions required to analyze remote sensor data for hydrology and water management applications.

Preprocessing
1. *Radiometric correction* of error introduced by the sensor system electronics and/or environmental effects (includes relative image-to-image normalization and absolute radiometric correction of atmospheric attenuation)
2. *Geometric correction* (image-to-map rectification or image-to-image registration)

Display and Enhancement
3. Black & white (8-bit)
4. Color-composite display (24-bit)
5. Black & white or color density slice
6. Magnification, reduction, roam
7. Contrast manipulation (linear, non-linear)
8. Color space transformations (e.g. RGB to IHS)
9. Image algebra (band ratioing, image differencing, etc.)
10. Linear combinations (e.g., Kauth transform)
11. Spatial filtering (e.g. high, low, band-pass)
12. Edge enhancement (e.g. Sobel, Robert's, Kirsch)
13. Principal components (e.g. standardized, unstandardized)
14. Texture transforms (e.g. min-max, texture spectrum, fractal dimension)
15. Frequency transformations (e.g. Fourier, Walsh)
16. Digital elevation models (e.g. analytical hill shading)
17. Animation (e.g. movies of change detection)

Remote Sensing Information Extraction
18. Pixel brightness value
19. Transects
20. Univariate and multivariate statistical analysis (e.g. mean, covariance)
21. Feature (band) selection (graphical and statistical)
22. Supervised classification (e.g. minimum distance, maximum likelihood)
23. Unsupervised classification (e.g. ISODATA)
24. Contextual classification
25. Incorporation of ancillary data during classification
26. Expert system image analysis
27. Neural network image analysis
28. Fuzzy logic classification
29. Hyperspectral data analysis
30. Radar image processing
31. Accuracy assessment (descriptive and analytical)

Photogrammetric Information Extraction
32. Soft-copy extraction of digital elevation models
33. Soft-copy production of orthoimages

Metadata and Image/Map Lineage Documentation
34. Metadata
35. Complete image and GIS file processing history

Image/Map Cartography
36. Scaled Postscript Level II output of images and maps

Geographic Information Systems (GIS)
37. Raster (image) based GIS
38. Vector (polygon) based GIS (must allow polygon overlay)

Integrated Image Processing and GIS
39. Complete image processing systems (Functions 1 through 36 plus utilities)
40. Complete image processing systems and GIS (Functions 1 through 43)

Utilities
41. Network (Internet, local talk, etc.)
42. Image compression (single image, video)
43. Import and export of various file formats

3.2.1 Preprocessing

Remote sensor data must be carefully preprocessed before information can be extracted from it. The scientist must radiometrically correct the remote sensor data to remove a) system introduced error (e.g. systematic noise or stripping) and/or b) environmentally introduced image degradation (e.g. due to atmospheric haze). Many remote sensing projects do not require detailed radiometric correction. However, projects dealing with water quality, differentially illuminated mountainous terrain, and subtle differences in aquatic vegetation health and vigor do require careful radiometric correction (e.g. Jensen et al., 1995; Bishop et al., 1998). Therefore, it is important that the application software provide a robust suite of radiometric correction alternatives.

Hydrologists and water management scientists require that most spatial information derived from remote sensor data be reprojected into a standard map projection suitable for input to a GIS. This involves rectification of the remote sensor data to a Universal Transverse Mercator (UTM) or other map projection using nearest-neighbor, bilinear interpolation, or cubic convolution resampling logic. The image processing software should allow ground control points (GCPs) to be easily and interactively identified on the base map and in the unrectified imagery. The GCP coordinates are used to compute the coefficients necessary to warp the unrectified image to a planimetric map projection. The accuracy of the image-to-map rectification or image-to-image registration is specified in root-mean-square-error (RMSE) units, e.g. the pixels in the image are within ±10 m of their true planimetric location. The user must also be able to specify the geoid and datum (e.g. NAD83 refers to the North American Datum 1983 to which all U.S. digital orthophoto quarterquads must be referenced).

Change detection projects are especially dependent upon accurate geometric rectification (or registration) of multiple date images. Therefore, it is imperative that the image processing system software perform accurate geometric rectification.

3.2.2 Display and Enhancement

Digital image processing systems must be able to display individual black and white images (usually 8-bit) and color composites of three bands at one time (24-

bit). The analyst can then density slice, magnify, roam, or manipulate the contrast of the individual bands or color composites. Through the years, algebraic and linear combinations of bands of remote sensor data have proven useful for hydrologic research. Therefore, the system must be able to build simple algebraic statements (e.g. ratio bands 4/5) or linear combinations of bands (e.g. the Kauth transform) to produce more sophisticated vegetation and hydrologic transformations of the remote sensor data.

Spatial and frequency filtering algorithms can be used to enhance and display subtle high and low frequency features and edges of these features in the remote sensor data. Texture algorithms enhance areas of uniform texture (e.g. coarse, smooth, rough). Some bands of remote sensor data are highly correlated with other bands, therefore, there is redundant information. Principal components analysis is often applied to reduce the dimensionality (number of redundant bands) used in the analysis while still maintaining the critical essence of the data.

Digital elevation models (DEMs) are critical to successful modeling and understanding of many landscape processes. The analyst must be able to display a DEM in a planimetric (vertical) view using analytical hill shading as well as in a pseudo 3-dimensional perspective view. Ideally, it is possible to drape thematic information such as a hydrologic network on top of the hill-shaded DEM. The DEMs and orthophotos are produced using photogrammetric principles as discussed in the soft-copy photogrammetry information extraction section.

Finally, it is important to be able to monitor change in the landscape by displaying multiple dates of imagery in an animated fashion. Change information can be used to gather information about the processes at work.

3.2.3 Remote Sensing Information Extraction

The analyst must be able to read the brightness values (z) at any x,y location in the image and along user-specified transects. The user must also be able to draw polygons around objects of interest on the screen and extract fundamental area, perimeter, and even volume information. Ideally, the polygon and its attribute information may be saved in a standard format for subsequent processing, e.g. in ERDAS or ArcInfo coverage formats. *Heads-up on-screen* image interpretation and digitization is becoming more important as very high spatial resolution satellite remote sensor data becomes available and fundamental photo-interpretation techniques are merged with automated feature extraction (Jensen, 1995a; Firestone et al., 1996).

Reputable digital image processing systems allow the user to classify multispectral remote sensor data using supervised or unsupervised classification techniques. In a supervised classification the analyst supervises the training of the algorithm. In an unsupervised classification the analyst relies on the computer to identify pixels in the terrain that have approximately the same multispectral characteristics. These 'clusters' are then labeled by the analyst to produce a thematic map. Most image processing systems still do not easily allow the incorporation of contextual or other types of ancillary data in the classification. Furthermore, only a few systems allow the user to apply expert system, neural network, or fuzzy classification logic to extract thematic information from remotely sense data (Jensen and Qiu, 1998). However, this is changing. The terrain usually grades from one land cover into another without 'hard' partitions. In fact, the 'fuzzy transition interface' between homogeneous terrain elements is often where the greatest species diversity of plants and animals exists. Therefore, 'fuzzy' classification algorithms will likely see dramatic utility in the future.

In the past decade there has been a tremendous increase in the amount of active microwave (radar) imagery available for hydrologic investigations. The launch of the Canadian RADARSAT in 1995 stimulated research even more. Only a few image processing systems provide the software necessary to remove or adjust the speckle associated with the raw radar data and to geometrically warp it into a ground-range map projection (as opposed to the original slant-range geometry).

3.2.4 Photogrammetric Information Extraction

This is one of the most exciting areas of digital image processing. For 50 years it has been necessary to use expensive and time consuming analytical stereoplotters to a) extract digital elevation models from stereoscopic vertical aerial photography, and b) produce orthophotographs. With advances in desktop computers and soft-copy photogrammetry software, it is now possible to extract DEMs from both aerial photography (Fig. 3.5) and satellite digital data and then produce accurate orthoimages (e.g., Greve et al., 1992; Jensen, 1995b; 1997). Some of the new software can produce true orthophotos where the building footprint is in its proper planimetric location over the foundation and all relief displacement has been removed (Fig. 3.6).

Hydrologists, natural resource managers, and water management groups can now produce accurate DEMs and orthophotos on demand for their local modeling purposes rather than being dependent on the dreadfully slow cycle of government DEM and orthophoto production. This should be of significant value for urban hydrologic studies of pervious versus impervious areas and their elevation, slope, and aspect. Appendix 3.1 identifies several vendors that provide photogrammetric image processing software.

3.2.5 Metadata and Image/Map Lineage Documentation

Metadata is data about data. Many countries have adopted rigorous national standards about the content, accuracy, and transmission associated with map and image spatial data. In the United States, the Federal Geographic Data Committee (FGDC) has developed stringent metadata standards for all data produced for government use. All digital image processing systems and GIS in the future will eventually provide detailed metadata and image lineage (genealogy) information about the processing applied to each image or map (Lunetta et al., 1991; Lanter, and Veregin, 1992; Jensen and Narumalani, 1992; FGDC, 1998). The image lineage information is indispensable when the products derived from the analysis of remotely sensed data are subjected to intense public scrutiny or litigation.

Phtogrammetrically Derived Digital Elevation Model and Orthoimage

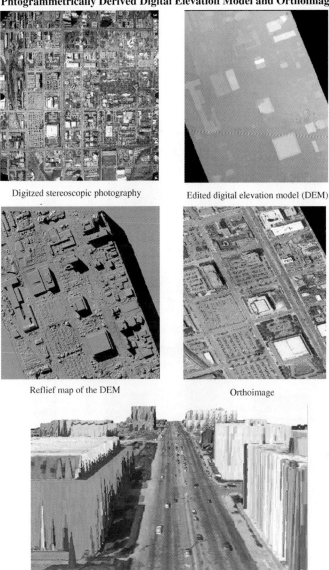

Digitzed stereoscopic photography

Edited digital elevation model (DEM)

Reflief map of the DEM

Orthoimage

Orthophoto draped over DEM in 3-d perspective view

Fig. 3.5. Soft-copy photogrammetric digital image processing software can extract digital elevation models from stereoscopic aerial photography or satellite high resolution remote sensor data. The DEMs can then be used to create scaled orthoimages. The orthoimages can then be draped over the DEM and viewed in perspective. Animated movies of such views allow the analyst to literally walk through a virtual reality city or landscape. This example of Columbia, SC is based on 1:6,000 photography

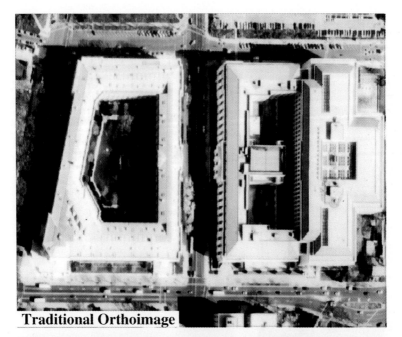

Fig. 3.6. Comparison of a traditional orthophoto versus a true orthophoto produced using softcopy photogrammetric digital image processing. Note that the sides of the buildings in the true orthoimage are in their proper planimetric position

3.2.6 Image and Map Cartographic Composition

Popular image processing software (e.g. Photoshop) and graphics programs (e.g. Freehand) can be used to produce useful unrectified images and diagrams. However, if scaled images and/or thematic maps in a map projection are required, then a full-function digital image processing system or GIS must be utilized that produce quality cartographic products that can be printed to Postscript level II output devices.

3.2.7 Geographic Information Systems (GIS)

Information derived from remote sensor data often fulfills its promise best when used in conjunction with other ancillary data (e.g., soils, elevation, slope, aspect, depth-to-ground-water) stored in a GIS (Lunetta et al., 1991). Therefore, the ideal integrated system performs both digital image processing and GIS spatial modeling and considers map data as image data (and vice-versa) (Cowen et al., 1995). The GIS analytical capabilities are hopefully based on 'map algebra' logic that can easily perform linear combinations of GIS operations to model the desired process. The GIS must also be able to perform raster-to-vector and vector-to-raster conversion accurately and efficiently.

3.2.8 Utilities

The digital image processing system should have the ability to network not only with colleagues and computer databases in the building but with those throughout the world. Therefore, efficient transmission lines and communication software (protocol) must be available. Much information is now routinely served on the world-wide-web (WWW) (Ubois, 1993). Some have suggested that the Internet is the sales channel of the future for imagery (Thorpe, 1996).

The type of *data compression* algorithm used to store the image data can have a serious impact on the amount of mass storage required. The basic idea of image compression is to remove redundancy from the image data, hopefully, without sacrificing valuable information. This is usually done by mapping the image to a set of coefficients. The resulting set is then quantized to a number of possible values that are recorded by an appropriate coding method. Most commonly used image compression methods are based on the discrete cosine transform such as the JPEG algorithm (Joint Photographic Experts Group), on vector quantization, on differential pulse code modulation, and on the use of image pyramids (Lammi and Sarjakoski, 1995).

One must decide on whether to use a *loss-less* or *lossy* data compression algorithm (Sayood, 1996). When an image is compressed using lossy logic and then uncompressed, it may appear similar to the original image but it does not contain all of the subtle multispectral brightness value differences present in the original (Bryan, 1995; Nelson, 1996). Imagery that has been compressed using a lossy algorithm may be suitable as an illustrative image where cursory visual photo-interpretation is all that is required. If lossy compression is absolutely necessary, the algorithm of choice at the present time appears to be JPEG which has a compression ratio of about 1:10 for color photos without considerable degradation in the visual or geometric quality of the image for photogrammetric applications (Lammi and Sarjakoski, 1995). Unfortunately, such lossy data may be unsuitable for scientists performing quantitative analysis of the data. Therefore, it is good practice to store the archive image using a loss-less data compression algorithm based on a) JPEG differential pulse code modulation - DPCM or b) run-length encoding (e.g. the simple UNIX compress command) so that both novice users and scientists have access to the best reproduction of the original data. The optimum lossy and/or loss-less image compression algorithm(s) are still being debated, e.g., fractal, wavelet, quadtree, run-length encoding (Russ, 1992; Pennebaker and Mitchell, 1993; Jensen, 1996; Nelson, 1996; Sayood, 1996). Robust multiple frame video data compression algorithms now exist that are of benefit for remote sensing image animation projects (e.g. MPEG).

The image processing system must have the capability to import and export remote sensing and GIS data files stored in a variety of standard formats. The system must be able to read at least the following file formats: encapsulated postscript file - EPSF, tagged interchange file format - TIFF, Macintosh PICT, ERDAS, ESRI coverages, GeoTIFF, and CompuServe GIF.

3.3 Commercial and Publicly Available Digital Image Processing Systems

The development and marketing of digital image processing systems is a multi-million dollar industry. The image processing software/hardware may be used for non-destructive evaluation of items on an assembly line, medical image diagnosis, and/or analysis of remote sensor data. Some vendors provide only the software while others provide both proprietary hardware and software. Several of the most widely used systems that are used to analyze remotely sensed data are summarized in Appendix 3.1. Their capabilities are cross-referenced to the general image processing functions summarized in Table 3.2.

Universities and public government agencies have developed digital image processing software. Several of the most widely used and publicly available digital image processing systems are summarized in Appendix 3.1. COSMIC at the University of Georgia is a clearinghouse for obtaining NASA sponsored digital image processing software for the cost of media duplication.

3.4 Summary

Analysis of remotely sensed data and GIS information for hydrologic applications requires access to sophisticated digital image processing system and geographic information system software. This chapter identified typical a) computer hardware/software characteristics, b) image processing functions required to analyze remote sensor data for hydrologic and water management applications, and c) selected commercial and public digital image processing systems and their generic functions available for earth resource mapping. The Earth Observing System

(EOS) and several commercial enterprises are about to provide dramatically improved spatial and spectral resolution remote sensor data (e.g. NASA's Moderate Resolution Imaging Spectrometer - MODIS, France's improved Spot Image, and EOSAT Space Imaging's IKONOS (Asar et al., 1996; Jensen et al., 1998). Continually improving digital image processing software will hopefully be developed to extract the required biophysical information from these data for hydrologic applications.

References

Asrar, G. and R. Greenstone: Mission to planet Earth EOS reference handbook, Washington, DC: NASA, 1996
Berry, F. C.: Inventing the future: how science and technology transform our world. Washington, DC: Brassey's 1993
Bishop, M. P: Scale-dependent analysis of satellite imagery for characterization of glaciers in the Karakoram Himalaya. Geomorphology, 21:217-232 (1998).
Bryan, J.: Compression scorecard. Byte 20, 107-111 (1995)
Busbey, A. B. et al.: Image processing approaches using the Macintosh. Photogrammetric Engineering & Remote Sensing 58, 1665–1668 (1992)
Cowen, D. J. et al.: The design and implementation of an integrated geographic information system for environmental applications. Photogrammetric Engineering & Remote Sensing 61, 1393-1404 (1995)
Davis, B.: Hydrology and topography mapping from digital elevation data, CRAY Channels 15, 4–7 (1993)
Denning, P. J.: RISC architecture. American Scientist 81,7–10 (1993)
Earnshaw, R. A. et al.: An introductory guide to scientific visualization. New York: Springer-Verlag 1992
EDC: Correspondence with EROS data center personnel, Sioux Falls, S. D. (1995)
Faust, N. L. et al.: Geographic information systems and remote sensing future computing environment. Photogrammetric Engineering and Remote Sensing 57, 655–668 (1991)
Firestone, L. et al.: Automated feature extraction: the key to future productivity. Photogrammetric Engineering & Remote Sensing 62, 671-674 (1996)
FGDC: The Value of METADATA. Reston, DC: Federal Geographic Data Committee 1998
Foley, J. D. et al.: Introduction to computer graphics. N. Y.: Addison-Wesley 1994
Freedman, A.: The computer glossary. N. Y.: American Mgt. Assoc., 7th ed. 1995
Greve, C. W. et al.: Image processing on open systems. Photogrammetric Engineering & Remote Sensing 58, 85–89 (1992)
Jahne, B.: Digital image processing. Heidelberg: Springer-Verlag 1991
Jensen, J. R.: President's inaugural address. Photogrammetric Engineering & Remote Sensing 10, 835-840 (1995a)
Jensen, J. R.: Issues involving the creation of digital elevation models and terrain corrected orthoimagery using soft-copy photogrammetry. Geocarto International - A Multidisciplinary Journal of Remote Sensing and GIS 10, 5-21 (1995b)
Jensen, J. R.: Introductory digital image processing: a remote sensing perspective. Saddle River, NJ: Prentice-Hall, 2nd ed. (1996)
Jensen, J. R.: Issues involving the creation of digital elevation models and terrain corrected orthoimagery using soft-copy photogrammetry. In: Greve, C. (Ed): Manual of Photogrammetry Addendum. Bethesda, MD: American Society for Photogrammetry & Remote Sensing (1997)
Jensen, J. R. and F. Qiu: A neural network based system for visual landscape interpretation using high resolution remotely sensed imagery. Proceedings, Annual Meeting of the American Society for Photogrammetry & Remote Sensing, Tampa, FL: CD:15 (1998)

Jensen, J. R. et al.: Remote sensing image browse and archival systems. Geocarto Intl.- A Multidisciplinary Journal of Remote Sensing and GIS 11, 33-42 (1996)

Jensen, J. R. et al.: Improved remote sensing and GIS reliability diagrams, image genealogy diagrams, and thematic map legends to enhance communication. International Archives of Photogrammetry and Remote Sensing 6, 125-132 (1992)

Kodak: Correspondence with Kodak aerial systems division, Rochester, NY (1995)

Lanter, D. P. et al.: A research paradigm for propagating error in layer-based GIS. Photogrammetric Engineering & Remote Sensing 58, 825–833 (1992)

Lammi, J. et al.: Image compression by the JPEG algorithm. Photogrammetric Engineering & Remote Sensing 61, 1261-1266 (1995)

Lulla, K.: Availability of NASA's earth observation images via electronic media. Geocarto Intl. - A Multidisciplinary Journal of Remote Sensing & GIS 10, 65-66 (1995)

Lunetta, R. S. et al.: Remote sensing and geographic information system data integration: error sources and research issues. Photogrammetric Engineering & Remote Sensing 57, 677–687 (1991)

Mazlish, B.: The fourth discontinuity: the co-evolution of humans and machines. New Haven, CN: Yale University Press 1993

Miller, R. L. et al.: C coast: a PC-based program for the analysis of coastal processes using NOAA CoastWatch data. Photogrammetric Engineering & Remote Sensing 60, 155–159 (1994)

Nelson, L. J.: Wavelet-based image compression: commercializing the capabilities. Advanced Imaging 11, 16-18 (1996)

Normile, D.: Get set for the super disc. Popular Science 96, 55-58 (1996)

Pennebaker, W. B. et al.: JPEG: still image data compression standard. New York: Van Nostrand Reinhold 1993

Rothenberg, J.: Ensuring the longevity of digital documents. Scientific American 272, 42-47 (1995)

Russ, J. C.: The image processing handbook. Boca Raton, FL: CRC Press 1992

Sayood, K.: Introduction to data compression. San Francisco: Morgan Kaufmann Publishers 1996

Spooner, J. G., Next Celeron to merge graphics, audio, PC Week, April 26, 1999

Thorpe, J.: Aerial photography and satellite imagery: competing or complementary? Earth Observation Magazine June, 35-39 (1996)

Ubois, J.: The internet today. SunWorld April, 90–95 (1996)

Appendix 3.1 Selected commercial and public image processing systems sued for earth resource mapping and their functions
(Operating systems: W = Microsoft Windows 98; U = Unix; NT = Microsoft NT, Mac = Macintosh; Functions: ● = significant capability; ○ = moderate capability; no symbol = little or no capability) [updated from Jensen, 1996]

System	Operating System	Pre Processing	Display & Enhancement	Info Extraction	Soft-Copy Photo	Metadata Lineage	Image/ Map Cartography	GIS	IP/GIS
Commercial									
Applied Analysis	W U NT			●					
CORE HardCore	W U	○	○	○					
CORE ImageNet	W U		●	○					
Dragon	W	●	●	●			●		
EarthView	W	●	●	●			●		
EIDETIC	W	○	●	○					
ESRI Image Analyst	Mac W U NT	●	●	○			●	●	●
ENVI	Mac W U NT	●	●	●			●	○	
ERDAS Imagine	W U NT	●	●	●	●	●	●	●	●
ERIM	U	●	●	●	●		●	●	●
ER Mapper	U NT	●	●	●	●		●	●	●
GAIA	Mac	○	○	○					
GENASYS	W U	●	●	●			●	●	●
GenIsis	W	○	○	○	○				
Global Lab Image	W		●	○					
GRASS	U	●	●	●		●	●	●	●
Helava Associates	W U	●	●	●	●		●	●	●
IDRISI	W NT	●	●	●			●	●	●
Intergraph	W NT U	●	●	●	●	●	●	●	●
Leica	W U	●	●	●	●		●	●	●
PCI	Mac W NT U	●	●	●	●		●	●	●
Photoshop	Mac W NT U	○	●	○					
MacSadie	Mac	●	●	●					

System	Operating System	Pre Pro-cessing	Display & Enhan-cement	Info Extrac-tion	Soft-Copy Photo	Meta-data Lineage	Image/Map Carto-graphy	GIS	IP/GIS
Jandel MOCHA	W	●	●	●					
OrthoView	U	●	●		●			○	
R-WEL	W	●	●	●	●		●	●	●
GDE Socet Set	NT U	●	●	●	●		●	●	●
Vision Softplotter	W U	●	●	●	●		●	○	○
Microi-mage TNTmips	W NT U	●	●	●	●	●	●	●	●
Trifid TruVue	W U	●	●	●	●		●	●	●
Vexel IDAS	U	●	●	●	●		●	●	●
VISILOG	U	●	●	●	●				
Public Systems									
C-Coast	W		●	●					
Cosmic VICAR	U	●	●	●			●	●	●
EPPL7	W	○	●	○			○	●	○
MultiSpec	Mac	●	●	●					
NIH-Image	U		○						
NOAA	U	○	○						
XV	U		●						

Sources of information for Appendix 3.1:

Applied Analysis Inc., 46 Manning Road, Suite 201, Billerica, MA 01821; subpixel processing.
ArcView/ArcGrid/ArcInfo/Image Analyst - ESRI, 380 New York St., Redlands, CA 92373; http://www.esri.com/
C-Coast, JA20 Building 1000, Stennis Space Center, MS 29519
CORE, Box 50845, Pasadena, CA 91115
Cosmic, University of Georgia, Athens, GA 30602
Dragon, Goldin-Rudahl Systems, Six University Dr. Suite 213, Amherst, MA 01002
EarthView, Atlantis Scientific Systems Group, 1827 Woodward Dr. Ottawa, Canada K2C 0P9; www.atlsci.com
ENVI, Research Systems, Inc., 2995 Wilderness Place, Boulder CO 80301; www.rsinc.com
EPPL7, Land Management Information Center, 300 Centennial Building, 638 Cedar St., St. Paul, MN 55155
ERIM, Box 134001, Ann Arbor, MI 48113-4001
ERDAS Imagine, 2801 Buford Hwy., NE, Suite 300, Atlanta, GA 30329
ER Mapper, 4370 La Jolla Village Dr., San Diego, CA 92122
GAIA, 235 W. 56th St., 20N, New York, NY, 10019
Global Lab, Data Translation, 100 Locke Dr., Marlboro, MA 01752-1192
GRASS - http://www.cecer.army.mil/announcements/grass.html

Helava Associates, Inc., 10965 Via Frontera, #100, San Diego, CA 92127-1703.
IDAS - Vexcel Imaging, Inc., 3131 Indian Road, Boulder, CO 80301.
IDRISI, Graduate School of Geography, Clarke Univ. 950 Main, Worcester, MA 01610
Intergraph, Huntsville, AL, 35894
Leica, Inc., 2 Inverness Drive East, #108, Englewood, CO 80112.
MOCHA Jandel Scientific, 2591 Kerner Blvd., San Rafael, CA 94901
MultiSpec, Dr. David Landgrebe, Purdue Research Foundation, W. Lafayette, IN 47907
NIH-Image, National Institutes of Health, Washington, D.C.
OrthoView, Hammon-Jensen-Wallen, 8407 Edgewater Dr., Oakland, CA 94621
PCI, 50 W. Wilmot, Richmond Hill, Ontario Canada L4B 1M5; www.pci.on.ca
PHOTOSHOP, Adobe Systems Inc., 1585 Charleston Road, Mountain View, CA 94039
R-WEL Inc., Box 6206, Athens, GA 30604
SOCET SET, GDE Systems, Inc., Sand Diego, CA and Helava Associates, Inc., 10965 Via Frontera, #100, San Diego, CA 92127-1703.
SPHIGS (Simple Programmer's Hierarchial Interactive Graphics System) and SRGP (Simple Raster Graphics Package);
Terra-Mar Resource Information Services, Inc., 1937 Landings Dr., Mountain View, CA 94043.
TNTmips, MicroImages, 201 N. 8th St., Lincoln, NB 68508; info@microimages.com
TruVue, Trifid Corp., 680 Craig Rd., Suite 308, St. Louis, MO 63141.
VISILOG, NOESIS Vision, Inc., 6800 Cote de Liesse, Suite 200, St. Laurent, Quebec, H4T 2A7
Vision International Inc., Division of Autometric, Inc., 81 Park Street, Bangor, ME 04401.
VI2STA, International Imaging Systems, Inc., 1500 Buckeye Drive, Milpitas, CA 95035.
XV, image viewer program written by John Bradley; http://phoenix.csc.calpoly.edu/CSL/cobra/xv.html

4 Integration of Remotely Sensed Data into Geographical Information Systems

Nandish M. Mattikalli[1], Edwin T. Engman[2]

[1]Cambridge Research Associates, 1430 Spring Hill Road, Suite 200 McLean, Virginia 22102, USA
[2]NASA-Goddard Space Flight Center, Greenbelt, MD 20771, USA

4.1 Introduction

Remotely sensed data and information derived from them have a wide range of applications in hydrology and water resources management (Schultz, 1988; Engman and Gurney, 1991). Remote sensing and its associated image processing technology provide access to spatial and temporal information on watershed, regional, continental and global scales. Further, new sensors and imaging technology are increasing the capability of remote sensing to acquire information at a variety of spatial and temporal scales. Management and efficient utilization of such information is going to be one of the major challenges of the coming decade. With the advent of space programs such as the Earth Observing System (EOS), this problem is going to become even more complex especially because a variety of new sensors are employed to cover the full range of the electromagnetic spectrum. Effective utilization of this large spatial data volume is dependent upon existence of an efficient, geographic handling and processing system that will transform these data into usable information. A major tool for handling spatial data is the Geographical Information System (GIS).

GIS provides appropriate methods for efficient storage, retrieval, manipulation, analysis and display of large volumes of spatially referenced data. Accordingly, GIS consists of four basic components: data input and editing, storage of geographic databases, data analysis and spatial modeling, and data visualization and presentation (Fig. 4.1). The data may be collected from fieldwork, extraction of map data, air photo interpretation, and interpretation and classification of remotely sensed images. Data input may be carried out by manual digitization or computer assisted semi-automatic methods. Collected data are then organized into a series of spatially geo-registered layers, with each layer relating to a particular theme (e.g., vegetation, soils, geology, topography etc.) or a set of layers relating to temporal variation of a theme (e.g., changes in land-use or variation of soil moisture etc.). Data input and editing (i.e., to correct digitizing errors, establishing topological relationships etc.) are the most time-consuming and labor intensive tasks. Data analysis and spatial modeling capability are the most important characteristics of a GIS. Conventional analysis and manipulation operations include retrieval, reclassification procedures (e.g., reclassifying soils map into a permeability map), map overlay (e.g., merging of various data layers to calculate soil erosion), proximity

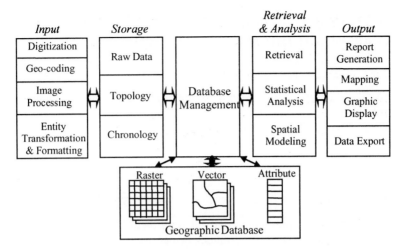

Fig. 4.1. Sub-modules of a GIS for input, storage, retrieval, analysis, modeling and presentation of spatial and non-spatial data

analysis (e.g., determining the distance from stream network), optimum corridor and other modeling techniques. Output from a GIS include maps, graphs, tabular statistics, and reports, which may be the end products or may be employed as input to further analysis.

Remotely sensed data can be best utilized if they are incorporated in a GIS that is designed to accept large volumes of spatial data. Figure 4.2 shows a procedure of deriving both spatial and non-spatial data from remotely sensed data for input into a GIS. When combined with up-to-date data from remote sensing a GIS can assist in automation of several operations (e.g., interpretation, change detection, map revisions etc.). A major feature of a GIS is its ability to overlay layers of spatially geo-referenced data. This enables the user to determine both, graphically and analytically, how spatial structures and objects (such as stream network, river discharge, and land use pattern) interact with each other. A basin hydrological system is a dynamic entity, and information stored in a GIS is only a *static* representation of the real world and therefore data has to be updated for temporal coverage on a regular basis. Remotely sensed satellite data offer excellent inputs in this context to provide repetitive, synoptic, and accurate information of the changes in a watershed, and offer the potential to monitor these dynamic changes. Further, successful applications of remote sensing in hydrology have influenced hydrologists to modify existing hydrological models or develop new types of models to incorporate widely available spatial data. In many cases, remotely sensed data alone are not sufficient for hydrological purposes and such data have to be merged with ancillary information such as soils, geology and elevation etc. GIS offers an appropriate technology for merging various spatial data layers. Integration of remotely sensed data with a GIS will greatly enhance modeling and analyzing capability of the GIS.

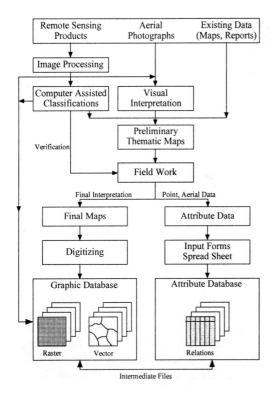

Fig. 4.2. Derivation of remotely sensed products for input into a GIS; Spatial and non-spatial data generated by remote sensing are imported into a GIS for data analysis and spatial modeling

4.2 General Approach

4.2.1 Raster and Vector Data Structures

Remote sensing systems use raster (or grid-based) format for collection and acquisition of data. In this format, information (ordinal or nominal) is stored as a collection of picture elements (or pixels), each holding only one value for the information at specific spatial coordinates. Raster structure, besides being more compatible with modern input/output hardware, has an advantage that the order of elements, as stored in digital form, is dictated by their geographical positions (see Chap. 3). This preserves many intrinsic spatial interrelationships among the data elements (as contrasted to points, lines and polygon entities of vector formats), and makes data retrieval on the basis of these spatial relationships a straightforward task.

Commonly employed GISs (e.g., Arc/Info) mainly use vector format to store data. In this format, data are collected as points, lines, and polygons, where each structure holds information for a specific region (Fig. 4.3). The standard approach is to represent polygon boundaries as lines (normally a sequence of lines that make up the polygon boundary). Lines are represented as a sequence of very short, straight line segments which are, in turn, represented by an ordered sequence of points representing the end points of short line segments. Thus, digital representation of line and polygon information generally consists of an ordered sequence of [x, y] coordinate pairs. Point, line and polygon features are associated with unique identifiers that are assigned with descriptive attributes. Both, the vector and raster structures have advantages and difficulties which are well described in the literature (e.g., Peuquet, 1984; Burrough, 1990), yet their fundamental differences, as described below, can make data conversion between them a complicated task (Piwowar et al., 1990).

In the recent past, many commercial GISs have been adapted to offer raster image display and handling capabilities (e.g., Arc/Info Version 6.0 or later), and several others offer both raster and vector capabilities (e.g., GRASS) (see Chap. 3). Integration of remotely sensed data with GIS data occurs naturally in a raster GIS because data structures are approximately the same for both sources. In a vector system, the integration requires more effort, and several technical problems need to be overcome for the true integration. Important problems in the integration are the raster/ vector dichotomy, generalization and accuracy of digital information (Piwowar et al., 1990; Lunetta et al., 1991). In addition, there are two difficulties to be addressed. First, there may be a necessity for reduction or reclassification of categories based on a range of attribute values, such as slope, cover type, cover density etc. For example, a digital map layer in a GIS can have thousands of classes (e.g., a forest cover map can have more than 1800 different classes, Goodenough, 1988). But, many remote sensing products have less than

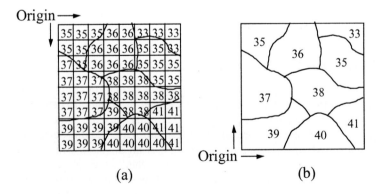

Fig. 4.3. Representation of spatial data in a GIS: **(a)** raster formatted data consists of a sequence of orderly placed pixels (or picture elements); **(b)** vector formatted data consists of polygon entities to represent features

256 classes because classification algorithms operate with a cost proportional to n^2 where n are the number of classes. Second, GIS class labels may not correspond to detectable remote sensing classes, which would need remotely sensed data of finer resolution and incorporation of contextual knowledge into a classification algorithm. Conversely, some classes easily identified by remote sensing may not correspond to classes desired by a hydrologist, which may have to be grouped into similar and fewer classes.

Ehlers et al. (1989) discuss strategies for efficient integration of Remotely Sensed Data Processing System (RSDPS) with a GIS, and classify these strategies into three approaches. First-level integration, *separate but equal,* mainly consists of development of a common data exchange format giving the user a transparent transformation of data from one environment to other (Fig. 4.4a). Second-level integration, *seamless,* provides a full integration of processing components of both RSDPS and GIS (Fig. 4.4b). Such systems will have capabilities to handle both, raster and vector data simultaneously for further processing without the need for time-consuming and costly examination and transformation of data formats. Third-level integration, *unity,* views RSDPS and GIS as one system. Ideally, such a system could provide an operational capability of performing analytical procedures on any type of data structure. Design and development of such a system would involve representation of raster *and* vector data in a hierarchical structure and construction of an integrated spatial data model capable of storing information derived from both, remote sensing and GIS (Fig. 4.4c).

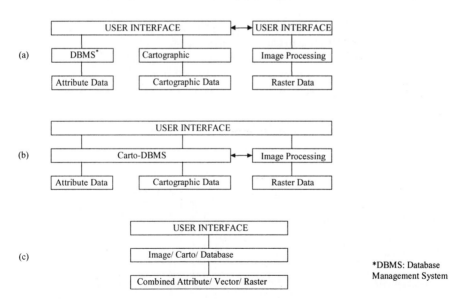

Fig. 4.4. Three levels of integration of remotely sensed data processing system with a GIS.
(a) Level I: *separate but equal* - two software modules with data exchange formats linking them; **(b)** Level II: *seamless* - two software modules with a common user interface and simultaneous display; and **(c)** Level III: *unity* - one software unit with combined processing

4.2.2 Current Approaches to the Integration

Most previous studies have handled the integration through data transfer between separate RSDPS and GIS. However, recent advances offer integrated solutions and/or standard interfaces between different systems to facilitate data integration (Piwowar et al., 1990). Examples of such systems that have some capabilities include GRASS, Arc/Info Version 6.0 onwards, ERDAS IMAGINE, PCI etc (Chap. 3 presents an exhaustive list of image processing systems and their GIS capability). Although the raster/vector dichotomy is a major impediment for a true integration, a significant advancement has been made to resolve the issue (e.g., Conese et al., 1992; van der Laan, 1992). To achieve a true integration these studies have employed a variety of approaches including quadtrees, object-oriented methods, knowledge-based systems, expert systems, artificial intelligence etc. (Goodenough et al., 1987; McKeown, 1987; Molenaar and Janssen, 1992).

Integration of raster and vector data types requires an efficient raster-to-vector (and vice versa) conversion routine. Several routines, such as the line following and polygon-capturing, are available for raster-to-vector conversion of cartographic data (Fulford, 1981). However, such routines require sophisticated computer hardware and software systems to perform the task, and require considerable manual labor to structure resulting data. Further, currently available routines suffer from certain shortcomings when processing boundary pixels because of interpolation methods involved in the data conversion procedure. This raises the question of accuracy of the results.

Mattikalli et al. (1995) developed a methodology for the *separate but equal* type of integration, in which the key process is a raster-to-vector (and vice versa) conversion (Fig. 4.5). This methodology does not require sophisticated systems and is independent of the problems encountered with other routines. The procedure makes use of some built-in routines commonly available in most vector-GISs, and some intermediate data formats viz., lattice (or grid) and SVF (Single Variable File). First, a raster image is transformed into lattice data structure. Lattice data

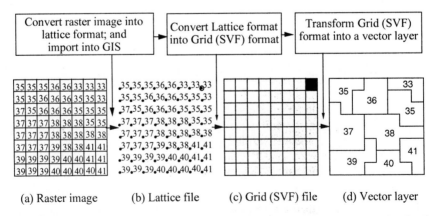

Fig. 4.5. Steps involved in a simple raster-to-vector (and vice versa) conversion routine for the integration of remote sensing and GIS

can be easily read and imported by most GIS. The second step is to transform imported lattice data into a polygon vector layer. This is achieved using the SVF format. Built-in routines are available in a vector GIS to convert lattice format into SVF format (e.g., LATTICEGRID routine of Arc/Info), and then SVF format into a polygon vector layer (e.g., GRIDPOLY) (ESRI, 1991). Polygons in the resulting vector layer will be built from groups of contiguous pixels having the same pixel value. Normally, these routines create lines along pixel borders thus forming closed polygons. Mattikalli (1995) employed this approach to integrate satellite data derived from both, fine and coarse resolution sensors with data digitized from maps.

This methodology has both, advantages and minor limitations. Main advantages are that it does not require sophisticated computer hardware and software systems, and that the result does not introduce new errors because no data interpolation is involved in the conversion. One limitation is that the process is relatively data expensive (in terms of storage) since polygon boundaries follow pixel borders, which could be a major difficulty for use with raw satellite images. However, this is not a severe problem for watershed applications that normally use classified image products.

Errors contained in the original data are carried-on and propagate through geographical processing performed in a hydrological analysis. Although it is not possible to eliminate errors in an integrated application, they can at least be managed and kept to an acceptable minimum. The following section briefly discusses the common and most obvious sources of errors and their propagation.

4.2.3 Errors Associated with Geographical Processing

Spatial data accuracy is governed by user requirements, inherent characteristics of data source, and instruments used to create digital data. Errors can arise at every stage of using a GIS, from collection of the original data to the output and use of resulting information (Walsh et al., 1987; Lunetta et al., 1991). Burrough (1990) discusses various sources of errors in a GIS. Needless to say, positional accuracy in a GIS cannot exceed those of the original data source whether it is a large-scale map sheet that has been digitized or data collected from an orbital remote sensor. Original maps and their cartographic features inevitably contain errors, no matter how small, and precision of the input data depends on the scale of the original map. GIS layers derived from maps of varying scales are subject to such errors.

Remotely sensed data integrated into a GIS are no exception. Remotely sensed data are invariably classified to derive thematic products. Although, classifications often result in a fair degree of accuracy, such products will have some errors associated with them. In addition, remotely sensed data will be associated with errors due to geometric correction and rectification. Positional accuracy of spatial entities recorded in a GIS are of critical importance because of the problems that can be generated during overlay operations. Within the limits of accuracy of a given input layer, a variety of accuracy levels may be utilized depending upon the application needs. Information derived (e.g., area calculations rather than precise determination of boundary locations) from original layers can frequently be acceptable with a significantly lower level of accuracy.

Errors compound and propagate through spatial processing of digital layers. Rasterizing a vector data layer (such as soils or watershed boundary) is one obvious source of error that invariably leads to generalization and loss of accuracy. This may be critical depending on the application. Vector-to-raster conversion can be considered as a form of point sampling, in which a pixel is assigned a value of that attribute which occurs at the center of pixel (see Fig. 4.3). Resolution, size and volume of the resulting map depend on pixel dimensions. Pixel dimension also has a strong influence on positional errors, boundary formation between map units, and errors in calculated areas of individual map units. Frolov and Maling (1969) and Switzer (1975) present a basis for a mathematical treatment of errors resulting from vector-to-raster conversion. In addition to pixel dimension, map complexity (i. e., number of boundary pixels) affects the magnitude of error. Crapper (1980) discusses such a problem in estimating areas of land cover from Landsat images. If a conversion procedure and its processing options can be judiciously selected to suit a particular problem, it is possible to convert data between two formats to minimize any errors (Piwowar et al., 1990).

Errors compound during commonly employed overlay operations (Newcomer and Szajgin, 1984). MacDougall (1975) discusses the result of overlaying six maps each of which on its own was considered to be of acceptable accuracy, and find that the resulting map was not significantly different from a random map. Although this may be an extreme case study, the implications of error propagation in map overlay can be serious (Bailey, 1988). The magnitude of error in the end product is directly related to the number of data layers employed in overlay operation (Walsh et al., 1987). Therefore, such errors can be minimized in some applications by employing a smaller number of data layers in overlay operations at any one time (Mattikalli, 1995).

4.3 Current Applications

The potential of integrating remotely sensed data with a GIS has been demonstrated in many areas of hydrology and water management. Although it is not possible to cover all areas of application, a few promising examples of current applications are presented in the following paragraphs. An exhaustive discussion of applications is not provided here, but the cited references provide fuller details.

4.3.1 Watershed Database Development

Development of an accurate and up-to-date watershed database is the first and an important stage of a hydrological study. Information on hydro-meteorological variables and watershed characteristics are stored as thematic geo-registered layers. Typically, these data layers include raw data such as digital elevation data, multi-spectral satellite imagery, soils map, watershed boundary etc. The database also includes derived data layers such as terrain slope and aspect, upslope area, land cover classification, soil erodibility, evapotranspiration etc.

Multi-spectral satellite image data are processed to generate thematic maps (see Chap. 3). SPOT stereoscopic data are suitable for digital topographic, thematic map production at scales of 1:50,000 and better (Gugan and Dowman, 1988; Swann et al. 1988). Data derived from various sensors can be integrated to derive improved map products with higher detail of cartographic information (about 15-20%) than from an individual image (Welch and Ehlers, 1988).

Two or more data layers can be employed in an overlay operation to merge spatial information. Figure 4.6 illustrates the overlay concept for both raster and vector data layers. Schultz (1993) applied this overlay technique to derive maximum soil water storage information by merging plant root depth data (derived from land-use classification of Landsat image) and soil porosity data (derived from digitized soil maps) for the Ennepe reservoir basin (see also Color Plate 5.A). During the overlay operation arithmetic and/ or Boolean operations can be performed on input layers (Mattikalli, 1995). Overlay operation is invariably utilized in most of the GIS applications including the examples presented below.

4.3.2 Integrated Use of Elevation Data

Digital Elevation Model (DEM) data are useful to derive a variety of information including slope, aspect, curvature etc. Fett et al. (1990) utilized 50 m x 50 m DEM data resampled to correspond to 30 m resolution of Landsat and generated land slope (relevant to flow velocity) and exposure to the Sun (relevant to evapotranspiration) in the Volme River basin, Germany. Slope angle (β) and upslope drainage area (α) derived from DEM are employed to calculate soil moisture deficit using the TOPMODEL topographic index, $\ln(\alpha / T \tan \beta)$, where T is the local saturated transmissivity (Romanowicz et al., 1993).

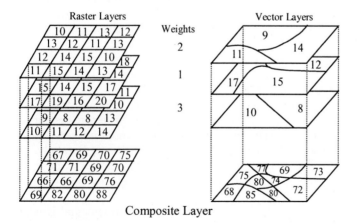

Fig. 4.6. Illustration of an overlay operation on data layers in a GIS. Mathematical operations can be performed on input thematic layers to derive new information as well as to conduct complex spatial analysis

A DEM can also be employed in conjunction with satellite reflectance data to estimate and model hydrological processes. Dubayah (1992) employed a DEM, Landsat TM data and a radiative transfer algorithm to model spatial variability of net solar radiation at a fine spatial resolution. Most of the examples presented in this chapter integrate DEM derived data with remotely sensed information for various applications.

Remotely sensed data (viz. land-use information), DEM (slope data) and digitized soil map can be merged to determine Hydrological Similar Units (HSU), which sub-divide a drainage basin into areas that have hydrologically similar behavior (Schultz, 1994). Su et al. (1992) utilized overlay operation on input layers to derive HSU for the Nims River basin in Germany (Color Plate 4.A).

Realistic 3D perspective views can be generated using a DEM to visualize terrain variation. Colour Plate 4.B shows a 3D view of elevation variation across the Little Washita watershed, Oklahoma. Remotely sensed images can be integrated with such 3D views for improved understanding of spatial and temporal variability of certain hydrological parameters. Colour Plate 4.C illustrates daily sequence of spatial distribution of remotely sensed microwave brightness temperature for the Little Washita watershed during June 10-18, 1992. These data are useful to derive near-surface soil moisture and relate the spatial and temporal variations to soil properties (Mattikalli et al., 1998).

4.3.3 Land-use/ Land-cover Change Detection

Remote sensing offers multi-temporal repetitive data for identification and quantification of land surface changes, and therefore, greatly enhances capability of a GIS in updating map information on a regular basis (Eckhardt et al., 1990). Michalak (1993) reviews current examples of integrated approach to land-use change analysis. Mattikalli (1995) reports a Boolean-logic technique applied to vector formatted layers for automatic analysis of historical land-use dynamics. This technique performs overlay operation to merge land-use data acquired on two dates, and then carries out Boolean operations to generate change map and associated statistics. Such a methodology is useful to analyze large amounts of data derived from remote sensing in conjunction with information derived from maps and aerial photographs archived in a GIS.

An expert system's approach to change detection has been implemented (Wang and Newkirk, 1987). Automation of integrated GIS using an expert system involves three separate tasks, viz. classification of remotely sensed imagery, detection of change, and expert rules for updating the GIS database. Currently, no commercially available GIS offers such a level of integration. However, there are several experimental applications incorporating some of the principles of automated land-use change analysis.

Examples of land-use change detection with the aid of multi-temporal Landsat imagery are presented in Chap. 19.

4.3.4 Modeling Watershed Runoff

Historically, runoff modeling at the river basin scale has lumped rainfall, infiltration and other hydraulic parameters to apply everywhere in the basin. With the advent of distributed modeling, a basin is subdivided into computational elements at a smaller scale. A distributed simulation model allows a user to simulate spatially variable parameters without lumping. However, setting up such a model with spatially distributed data and parameters is a time consuming and laborious task. If a GIS is integrated with the model these chores become much easier and often transparent to the user. An additional advantage of integrating distributed numerical models with a GIS includes calculation and display of runoff flow depths across watershed sub-basins.

The runoff curve number (CN) approach (USDA, 1972) to rainfall-runoff modeling is appealing for an integrated remote sensing and GIS environment. This approach estimates volume of direct runoff (Q) in terms of volume of rainfall (P) and potential maximum storage (S). S is derived from the CN, a coefficient that is directly related to watershed land-use, land management and soil properties. Since land-use can be routinely monitored using remote sensing, it is possible to analyze the effects of land-use changes (e.g., urbanization) on watershed runoff. Figure 4.7 shows various stages of computation of this approach implemented within a GIS. Mattikalli et al. (1996) employed Arc/Info to store various input parameters as thematic layers and generated flood hydrographs in a predominantly rural watershed. This approach has also been used to generate single event flood hydrographs and synthetic flood frequency curves (Muzik and Chang, 1993).

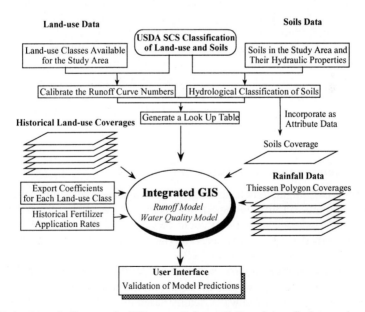

Fig. 4.7. A schematic diagram of a GIS approach for prediction of river discharge using the SCS Curve Numbers, and water quality using the export coefficient model (Mattikalli et al., 1996)

In urban watersheds, the spatial analysis capabilities of a GIS can be used for hydrological analysis. Watershed attributes such as soils information (infiltration rates, hydraulic conductivity, and storage capacities), surface characteristics (pervious, impervious, slope, roughness), geometry and dimensions of flow planes, routing lengths (overland, gutter, and sewer) and geometry and characteristics of routing segments can be efficiently stored and utilized for urban runoff calculations. Most of earlier studies have used GISs to derive parameters of lumped models. For example, Johnson (1989) used a GIS for generation of input data for digital map-based modeling system that supported lumped parameter models such as unit hydrograph, time-area, and cascade of reservoirs. Moeller (1991) determined input parameters for the HEC-1 model while Sircar et al. 1991) derived time-area curves. Djokic and Maidment (1991) used Arc/Info with a rational method to determine inlet and pipe capacity of an urban storm sewer system. Greene and Cruise (1996) employed Arc/Info to derive urban watershed feature attributes (location coordinates, parameters of runoff generating polygons, gutters and storm drains) for input into a hydrologic modeling procedure to estimate runoff.

Vieux (1991) developed a procedure for modeling direct surface runoff using a combination of a finite element method and Arc/Info Triangular Irregular Network (TIN) module. An internal integration of this model (viz. *r.water.fea*) allows *seamless* simulation of storm water runoff using flow networks (derived from DEM) and spatially distributed parameters. Schultz (1994) presents three different examples on hydrological modeling using remote sensing within the framework of ILWIS and Arc/Info. These examples demonstrate merging of Landsat TM and Meteosat geostationary image products and ancillary data (viz. DEM and its derived products) for rainfall/ runoff modeling and water balance parameter computation at 30 m, 5 km and HSU scales. Ott et al. (1991) employed the HSU concept in a rainfall-runoff model for flood forecasting and for impact assessment of changing land-use in the Mosel basin. Fett et al. (1990) simulated rainfall-runoff process in a vegetated hillslope area in the humid climate of the Volme river basin. This study utilized input parameters derived from remote sensing and elevation data, and employed a distributed hydrologic model on a pixel by pixel basis to derive 3-day hydrographs. Romanowicz et al. (1993) employed the TOPMODEL within Water Information System and simulated hourly runoff in the Severn basin, UK.

4.3.5 Monitoring and Modeling of Water Quality

Applications of GIS and remote sensing to water quality have mainly concentrated on Non Point Source (NPS) pollution. This is because remotely sensed data products (such as land-use/ land-cover) could be directly utilized in NPS modeling. Several watershed models have been interfaced with GIS including the export coefficient model, AGNPS and DRASTIC.

Agricultural Non Point Source pollution (AGNPS) model estimates nitrogen, phosphorus and chemical oxygen demand concentration in runoff and assesses agricultural impact on surface water quality based on spatially varying controlling parameters (e.g., topography, soils, land-use etc.). Srinivasan and Engel (1994) developed a linkage between AGNPS and a spatially distributed model. Osmond et al. (1997) employed this linkage and added capabilities that allowed users to enter point source, pesticide and channel information in a decision support system, WATERSHEDSS (WATER, Soil, and Hydro-Environmental Decision Support System). Using such a system a user can determine critical areas within a watershed and evaluate effects of alternative land treatment scenarios on water quality. Kim and Ventura (1993) managed and manipulated land-use data for modeling NPS pollution of an urban basin using an empirical urban water quality model.

Another approach uses an export coefficient model to calculate nutrient losses from catchment to surface water mainly in terms of areal extent of different land-use/ land-cover and their associated fertilizer application rates. Mattikalli et al. (1996) derived historical land-cover data from airborne and satellite sensors, and implemented the model using Arc/Info to estimate historical nitrogen and phosphorus loading in the River Glen watershed in the UK (see Fig. 4.7 for the methodological approach).

GIS have also been used in aspects of groundwater management and modeling (e.g., Evans and Myers, 1990; Maidment, 1993). DRASTIC (depth to water, D; net recharge, R; aquifer media, A; soil media, S; slope, T, impact of the vadose zone, I, and hydraulic conductivity, C) is an empirical model used to evaluate regional groundwater pollution potential. Evans and Myers (1990) implemented this model in ERDAS around the Rehoboth Beach, Delaware, and generated groundwater pollution risk and hazard assessment maps. Methodology consisted of additive overlay process of input layers assigned with certain weights. Most of these examples have implemented spatial models designed to evaluate groundwater vulnerability to contamination. However, these approaches have not employed data derived from remote sensing, probably because of specific nature of the input parameters.

4.3.6 Soil Erosion Monitoring

Soil erosion monitoring and/ or prediction is a popular application of integrated remote sensing and GIS (see also Chap. 12). Soil erosion potential is typically computed using the Universal Soil Loss Equation (USLE):

$$A = S \cdot L \cdot R \cdot P \cdot C \cdot K$$

where A = mean annual soil loss per unit area, S = slope steepness factor, L = slope length factor, R = rainfall factor, P = erosion control practice factor, C = land-use/ land-cover (or cropping management) factor, and K = soil erodibility factor. S- and L- factors can be derived using slope and aspect information generated from a DEM. R- factor can be assigned using a TIN structure created for rainfall gauging stations. P- and C- factors can be estimated using remotely sensed data via land-use/ land-cover classification and associated land management information.

Pelletier (1985) employed a GIS framework to implement the USLE formulation. Figure 4.8 shows various data sources and files managed by a GIS. This study used data from Landsat (MSS and TM) sensors to determine land-cover, and

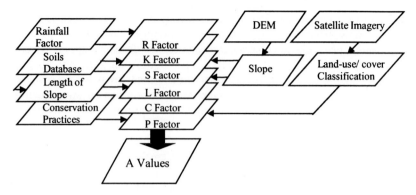

Fig. 4.8. A database schema for the universal soil-loss equation implementation in a GIS

subsequently the *C*- factor. Similarly, Jurgens and Fander (1993) employed Landsat TM data and performed USLE calculations in a GIS to derive soil erosion rates and soil erosion risk maps. They demonstrated usage of GIS data for sensitivity analysis of the USLE parameters. In the revised USLE (Renard et al., 1991), *L*-factor has been modified for influence of profile convexity/ concavity using segmentation of irregular slopes of a complex terrain. Mitasova et al. (1996) integrated regularized spline with tension for computation of *S*- and *L*- factors, and used an approach based on the unit stream power and directional derivatives for modeling spatial distribution of areas with topographic potential for erosion or deposition.

4.4 Future Perspectives

The potential of multi-sensor remote sensing as one of the main sources of data input and monitoring in hydrology and water management has been recognized. Remote sensing systems generate large volumes of spatial data because of their capability of providing (i) fine spatial resolution data; (ii) coarse resolution data on regional, continental and global scales; and (iii) high temporal resolution data. GIS offers an appropriate technology not only for efficient storage and retrieval of large volumes of spatially referenced data, but also to carry out data manipulation and complex spatial analysis required in distributed modeling. With the advent of EOS suite of platforms and sensors, it is expected that volume of data being received would require usage of fully integrated spatial information systems supported by knowledge based techniques in all facets of data handling.

Future development of both, GIS and RSDPS is controlled by the state of technology, and therefore assessing probable developments in the integration of GIS and remote sensing data is a difficult task. At the present time, current level of integration provides an environment largely free of the logistical considerations of moving data between the two systems. Over the next few years, more efforts need to be focussed on the fundamental aspects of integration such as data generalization and accuracy specification.

To fully realize the potential of multi-sensor remote sensing for GIS in hydrology, more advanced integration techniques need to be developed. Several problems of a *true* integration of remote sensing and GIS could probably be solved by recognition that GIS data and remote sensing data process and manage spatial information at different levels of representation. Ultimately, a GIS and remote sensing should be viewed as one entity that will be concerned with handling and analyzing spatial data. Unification of these technologies will lead to a synergistic integration of spatial data handling, and the final system would have more capabilities than just the sum of the two.

Hydrologic processes interact in three-dimensions (3D) between atmosphere, land surface and sub-surface. Although many GIS can generate a 3D plot, no commercial system has *true* 3D geometry and topology. Future developments in GIS should concentrate on design and construction of a *real* 3D system so that disparate databases can be integrated and analyzed in 3D as well as they are in 2D.

Hydrologic processes have a significant dynamic component. Recent applications have recognized the importance of change in hydrologic processes or input model parameters over time. Time can be characterized as the fourth dimension of the physical space-time continuum. However, we still cannot adequately represent temporal dimension in GIS modeling (Langran, 1992), because no system currently handles chronology. We typically illustrate effects of temporal change as slices of time for discrete intervals, but we need to show dynamic change over continuous time. The ultimate solution would be to handle change in space as well as change in time. An ideal GIS that handles time as a fourth dimension (4D GIS) will have chronology treated much like topology; *before and after* taking on the same importance as *left and right* in 2D space or *above and below* in 3D space. Such a 4D GIS would be of immense value for a number of research areas in hydrology including soil moisture modeling, groundwater modeling etc. because of their inherent *four-dimensional* nature.

The 4D GIS technology should enable transparent access to heterogeneous hydrologic datasets and processing resources in intra- and inter-networked environments. A key component of such an interoperating system is the open interface specification upon which the interoperating components are developed. The OpenGIS Consortium is developing a comprehensive suite of OpenGIS specifications that provide a common Open Geodata Model framework and OpenGIS Services Model to solve both technical and institutional non-interoperability problems (Buehler and McKee, 1998). Future developments in the GIS world should employ interoperating specifications to fully integrate and exploit both the data and processing resources.

References

Bailey, R. G. (1988). Problems with using overlay mapping for planning and their implications for geographic information systems. *Environmental Management*, 12(1), 11-17

Buehler, K., and McKee, L. (eds.) (1998). *The OpenGIS Guide - Introduction to Interoperable Geoprocessing and the OpenGIS Specification.* The Open GIS Consortium

Burrough, P. A. (1990). *Principles of geographical information systems for land resources assessment.* (Oxford: Clarendon)

Conese, C., Maracchi, G., Maselli, F., Romani, M., and Bottai, L. (1992). Integration of remotely sensed data into a GIS for the assessment of land suitability. *EARSeL Advances in Remote Sensing*, 1, 173-179

Crapper, P. F. (1980). Errors incurred in estimating an area of uniform land cover using Landsat. *Photogrammetric Engineering and Remote Sensing*, 46(10), 1295-1301

Djokic, D., and Maidment, D. R. (1991). Terrain analysis for stormwater modeling. *Hydrological Processes*, 5(1), 115-124

Dubayah, R. (1992). Estimating net solar radiation using Landsat Thematic Mapper and digital elevation data. *Water Resources Research*, 28(9), 2469-2484

Ehlers, M., Edwards, G., and Bedard, Y. (1989). Integration of remote sensing with geographic information systems: A necessary evolution. *Photogrammetric Engineering and Remote Sensing*, 55(11), 1619-1627

Eckhardt, D. W., Verdin, J. P., and Lyford, G. R. (1990). Automated update of an irrigated lands GIS using SPOT HRV imagery. *Photogrammetric Engineering and Remote Sensing*, 56(11), 1515-1522

Engman, E. T., and Gurney, R. J. (1991). *Remote sensing in hydrology.* (London: Chapman and Hall)

ESRI (1991). ARC/INFO Version 6.0 Data Conversion User's Manual. Environmental Systems Research Institute, Redlands, California

Evans, B. M., and Myers, W. L. (1990). A GIS-based approach to evaluating regional ground-water pollution potential with DRASTIC. *Journal of Soil and Water Conservation*, 45(2), 242-245

Fett, W., Neumann, P., and Schultz, G. A. (1990). Hydrological model based on satellite imagery and GIS. Proce. Int. Symp. on Remote Sensing and Water Resources (Enschede, The Netherlands), 347-357

Frolov, Y. S., and Maling, D. H. (1969). The accuracy of area measurement by point counting techniques. *Cartographic Journal*, 6(1), 21-35

Fulford, M. C. (1981). The fastrak automatic digitizing for line drawings. *International Journal of Pattern Recognition*, 14, 65-72

Goodenough, D. G. (1988). Thematic Mapper and SPOT integration with a geographic information system. *Photogrammetric Engineering and Remote Sensing*, 54(2), 167-176

Goodenough, D. G., Goldberg, M., Plunkett, G., and Zelek, J. (1987). An expert system for remote sensing. *IEEE Transactions on Geoscience and Remote Sensing*, 25, 349-359

Greene, R. G., and Cruise, J. F. (1996) Development of a geographic information system for urban watershed analysis. *Photogrammetric Engineering and Remote Sensing*, 62(7), 863-870

Gugan, D. J., and Dowman, I. J. (1988). Accuracy and completeness of topographic mapping from SPOT imagery. *Photogrammetric Record*, 12(72), 787-796

Jurgens, C., and Fander, M. (1993). Soil erosion assessment and simulation by means of SGEOS and ancillary digital data. *International Journal of Remote Sensing*, 14(15), 2847-2855

Johnson, L. E. (1989). MAPHYD - A digital map based hydrologic modeling system. *Photogrammetric Engineering and Remote Sensing*, 55(6), 911-917

Kim, K., and Ventura, S. (1993). Large-scale modeling of urban nonpoint source pollution using a geographical information system. *Photogrammetric Engineering and Remote Sensing*, 59(10), 1539-1544

Langran, G. (1992). *Time in GIS.* Taylor and Francis, New York

Lunetta, R. S., Congalton, R. G., Fenstermaker, L. K., Jensen, J. R., McGwire, K. C., and Tinney, L. R. (1991). Remote sensing and geographic information system data integration: error sources and research issues. *Photogrammetric Engineering and Remote Sensing*, 57, 677-687

MacDougall, E. B. (1975). The accuracy of map overlays. *Landscape Planning*, 2, 23-30

Maidment, D. R. (1993). GIS and hydrologic modeling. In: Goodchild, M., Parks, B., and Steyaert. L. (eds.) *Environmental Modeling with GIS*, Oxford University Press, New York, 147-167

Mattikalli, N. M., Engman, E. T., Jackson, T. J., and Ahuja, L. R. (1998). Microwave remote sensing of temporal variations of brightness temperature and near-surface soil water content during a watershed-scale field experiment, and its application to the estimation of soil physical properties. *Water Resources Research*, 34(9), 2289-2299

Mattikalli, N. M. (1995). Integration of remotely sensed raster data with vector based geographical information system for land-use change detection. *International Journal of Remote Sensing*, 16(15), 2813-2828

Mattikalli, N. M., Devereux, B. J., and Richards, K. S. (1995). Integration of remotely sensed satellite images with a geographical information system. *Computers and Geosciences*, 21(8), 947-956

Mattikalli, N. M., Devereux, B. J., and Richards, K. S. (1996). Prediction of river discharge and surface water quality using an integrated geographical information system approach. *International Journal of Remote Sensing*, 17(4), 683-701

McKeown, D. (1987). The role of artificial intelligence in the integration of remotely sensed data with geographic information systems. *IEEE Transactions on Geoscience and Remote Sensing*, 25, 330-348

Michalak, W. Z. (1993). GIS in land use change analysis: integration of remotely sensed data into GIS. *Applied Geography*, 13, 28-44

Mitasova, H., Hofierka J., Zlocha, M., and Iverson, L. (1996). Modeling topographic potential for erosion and deposition using GIS. *International Journal of Geographical Information Systems*, 10(5), 629-641

Moeller, R. A. (1991). Application of a geographic information system to hydrologic modeling using HEC-1. In: Stafford, D. B. (ed.) *Civil Engineering applications of remote sensing and GIS*, ASCE, 269-277

Molenaar, M., and Janssen, L. L. F. (1992). Integrated processing of remotely sensed and geographic data for land inventory purposes. *EARSeL Advances in Remote Sensing*, 1, 113-121.

Muzik, I., and Chang, C. (1993). Flood simulation assisted by a GIS. IAHS Publication no. 211, 531-539

Newcomer, J. A., and Szajgin, J. (1984). Accumulation of thematic map errors in digital overlay analysis. *American Cartographer*, 11, 58-62

Osmond, D. L., Gannon, R. W., Gale, J. A., Line, D. E., Knott, C. B., Phillips, K. A., Turner, M. H., Foster, M. A., Lehning, D. E., Coffey, S. W., and Spooner, J. (1997). WATERSHEDSS: A decision support system for watershed-scale nonpoint source water quality problems. *Journal of American Water Resources Association*, 33(2), 327-341

Ott, M., Su, Z., Schumann, A. H., and Schultz, G. A. (1991). Development of a distributed hydrological model for flood forecasting and impact assessment of landuse change in the international Mosel basin. IAHS Publication no. 201

Pelletier, R. E. (1985). Evaluating non-point pollution using remotely sensed data in soil erosion models. *Journal of Soil Water Conservation*, 40, 332-335

Peuquet, D. J. (1984). A conceptual framework and comparison of spatial data models. *Cartographica*, 21(4), 66-113

Piwowar, J. M., LeDrew, E. F., and Dudycha, D. J. (1990). Integration of spatial data in vector and raster formats in a geographic information system environment. *International Journal of Geographical Information Systems*, 4, 429-444

Renard, G. K., Foster, G. R., Weesies, G. A., Porter, J. P. (1991). RUSLE--Revised universal soil loss equation. *Journal of Soil Water Conservation*, 46, 30-33

Romanowicz, R. Beven, K., and Freer, J. (1993). TOPMODEL as an application module within WIS. IAHS Publication no. 211, 211-223

Schultz, G. A. (1988). Remote sensing in hydrology. *Journal of Hydrology*, 100, 239-265.

Schultz, G. A. (1993). Application of GIS and remote sensing in hydrology. IAHS Publication No. 211, 127-140

Schultz, G. A. (1994). Meso-scale modeling of runoff and water balances using remote sensing and other GIS data. *Hydrological Sciences Journal*, 39(2), 121-142

Sircar, J. K., Ragan, R. M., Engman, E. T., and R. A. Fink (1991). A GIS based geomorphic approach for the computation of time-area curves. In: Stafford, D. B. (ed.) *Civil Engineering applications of remote sensing and GIS*, ASCE, 287-296

Srinivasan, R., and Engel, B. A. (1994). A spatial decision support system for assessing agricultural nonpoint source pollution. *Water Resources Bulletin*, 30(3), 441-462

Su, Z., Neumann, P., Fett, W., Schumann, A., and Schultz, G. A. (1992). Application of remote sensing and geographical information system in hydrological modeling. *EARSeL Advances in Remote Sensing*, 1(3), 180-185

Swann, R., Hawkins, D., Westwell-Roper, A., and Johnstone, W. (1988). The potential for automated mapping from geocoded digital image data. *Photogrammetric Engineering and Remote Sensing*, 54(2), 187-193

Switzer, P. (1975). Estimation of the accuracy of qualitative maps. In: Davis and MacCullagh (eds.). *Display and analysis of spatial data* by), Wiley, New York, 1-13

USDA (United States Department of Agriculture) (1972). *National Engineering Handbook*. Soil Conservation Service. (Washington, DC: Government Printing Press)

van der Laan, F. B. (1992). Integration of remote sensing in a raster and vector GIS environment. *EARSeL Advances in Remote Sensing*, 1, 71-80

Vieux, B. E. (1991). Geographic information systems and non-point source water quality and quantity modelling. *Hydrological Processes*, 5, 101-113

Walsh, S. J., Lightfoot, D. R., Butler, D. R. (1987). Recognition and assessment of error in geographic information systems. *Photogrammetric Engineering and Remote Sensing*, 53, 1423-1430

Wang, F., and Newkirk, R. T. (1987). Design and implementation of a knowledge based system for remotely change detection. *Journal of Imaging Technology*, 13, 116-122

Welch, R., and Ehlers, M. (1988). Cartographic feature extraction from integrated SIR-B and Landsat TM images. *International Journal of Remote Sensing*, 9(5), 873-889

Section II

Remote Sensing Application to Hydrologic Monitoring and Modeling

5 Remote Sensing in Hydrological Modeling

Ralph O. Dubayah[1], Eric F. Wood[2], Edwin T. Engman[3], Kevin P. Czajkowski[1], Mark Zion[2] and Joshua Rhoads[1]

[1]Laboratory for Global Remote Sensing Studies and Department of Geography, University of Maryland, College Park, MD 20742, USA
[2]Water Resources Program, Department of Civil Engineering, Princeton University, Princeton, NJ 08544, USA
[3]NASA-Goddard Space Flight Center, Greenbelt, MD 20771, USA

5.1 Introduction

Hydrology is a science built on observations and measurements. Operational hydrology and water resources engineering have utilized these measurements for the design and operation of water resource systems and the forecasting of hydrologic systems. There has been a long recorded history of hydrologic data collection in support of operational hydrology going back to ancient Chinese and Egyptian times. In modern industrialized countries, hydrologic data collection has focused on streamflow, precipitation and basic surface meteorological data which are sufficient for the design and forecasting needs of the water resource engineers: primarily the design of water supply and flood protection works, which requires long-term records for river flows, and the forecasting of floods, which requires (spatially) accurate precipitation measurements.

To fully understand the data needs for operational hydrology, consider the primitive water balance equation:

$$\frac{\Delta S}{dt} = P - E - Q \tag{5.1}$$

where $\frac{\Delta S}{dt}$ is the change in soil moisture over a specified time interval, P is precipitation, E represents evapotranspiration which is the sum of evaporation from bare soil, E_S, and transpiration from vegetation, E_V, and Q is runoff which is the sum of surface, or direct storm runoff and subsurface or base flow. For water supply and/or flood protection design where long-term reliability is critical, the dynamics of Eq. (5.1) are unimportant. Thus, the important measurements are time series of runoff and possibly precipitation, and a climatological estimate of monthly evapotranspiration. Changes in soil moisture over the long-term are assumed zero. Similarly, for flood forecasting, evaporation can be ignored, soil moisture is only relevant to the extent that initial abstractions (or losses) can be estimated, and riverflow to the extent that comparisons can be made between forecasts and observations.

The measurement needs implied by Eq. (5.1) have guided both the station-based observation program run by national hydrometeorological organizations and the use of remote sensing by the operational hydrologic and water resource engineering practitioners. In the first portion of this chapter, the traditional application of remote sensing in operational hydrology will be briefly reviewed. More comprehensive treatments of this area can be found in texts such as Engman and Gurney (1991). More recently, sensitivity studies with Atmospheric General Circulation Models (AGCMs) have shown that land surface processes affect climate at regional to global scales. These studies include the effects of albedo, soil moisture anomalies, surface roughness (which affects evaporation), and land cover change.

Further understanding the role of the terrestrial hydrosphere-biosphere in Earth's climate system, and understanding the nature and effects of possible changes to the terrestrial water balance as a result of changing climate and land surface characteristics, is of central importance to the World Climate Research Program (WCRP) of the World Meteorological Organization. It is expected that these questions will be answered using process-based, terrestrial water and energy balance models. But, it is recognized that questions regarding terrestrial hydrology within the climate system cannot be answered through ground-based observations alone, due to the scarcity of land surface observations and difficulties in representing hydrological processes at large scales.

Remote sensing potentially may provide the required inputs for hydrological modeling at regional to global scales. As a result, remote sensing initiatives have included field experiments (e.g. the First International Satellite Land Surface Climatology Field Experiment, FIFE) that have linked ground measurements with remote sensing algorithm development. In addition, there are now consistent, long-term remote sensing data archives from satellites such as AVHRR (Agbu, 1993), GOES (Young, 1995) and SSM/I (Hollinger et al., 1992).

By the late 1990's new and enhanced meteorological satellites and higher spectral resolution land surface sensors being launched under NASA's Earth Observing System mission, combined with faster computer networking and data handling capabilities, will give operational hydrologists access to new types of land surface and hydrologic data. In the second part of this chapter, we discuss the potential for utilizing these data for hydrological modeling.

To fully understand both the historical and potential use of remote sensing for hydrological modeling, it is important to recognize that Eq. (5.1) is not directly useable as a hydrologic model. Each of the terms is often parameterized in terms of the catchment characteristics or conditions. For example, a change in soil moisture is the result of either evaporation or infiltration depending upon whether a dry or rain period is being considered. These processes (models) require information on land surface/cover, soil texture, initial soil moisture, and perhaps topography. Historically then, operational hydrology has used conceptual models for which there is the possibility that remote sensing can provide some of the model parameters. For modeling land-atmospheric hydrologic interactions, there is the desire that remote sensing provide both model parameters (even those which may change with time) and meteorological data, like surface air temperature, humidity, precipitation and radiation, which would permit model simulations based solely on remote sensing data.

5.2 Remote Sensing in Operational Hydrologic Modeling

Runoff. Runoff cannot be directly measured by remote sensing techniques. However, there are two general areas where remote sensing can be used in hydrologic and runoff modeling: (1) determining watershed geometry, drainage network, and other map-type information for distributed hydrologic models and for empirical flood peak, annual runoff or low flow equations; and (2) providing input data such as snow cover, soil moisture or delineated land use classes that are used to define runoff coefficients.

Watershed Geometry. Remote sensing data can be used to obtain almost any information that is typically obtained from maps or aerial photography. In many regions of the world, remotely-sensed data, and particularly Landsat TM or SPOT data, are the only source of good cartographic information. Drainage basin areas and stream networks are easily obtained from good imagery, even in remote regions. There have also been a number of studies to extract quantitative geomorphic information from Landsat imagery (Haralick, et al., 1985).

Topography is a basic need for any hydrologic analysis and modeling. Remote sensing can provide quantitative topographic information of suitable spatial resolution to be extremely valuable for model inputs. For example, stereo SPOT imagery can be used to develop a DEM with 10 m horizontal resolution and vertical resolution approaching 5m in ideal cases (Case, 1989). A new technology using interferometric Synthetic Aperture Radar (SAR) has been used to demonstrate similar horizontal resolutions with approximately 2m vertical resolution (Zebker et al., 1992).

Empirical relationships. Empirical flood formulae are useful for making estimates of peak flow when there is a lack of historical streamflow data. Generally these equations are restricted in application to the size range of the basin and the climatic/hydrologic region of the world in which they were developed.

Most of the empirical flood formulae relate peak discharge to the drainage area of the basin; see for example United Nations Flood Control Series No. 7 (United Nations, 1955). Landsat data are used to improve empirical regression equations of various runoff characteristics. For example, Allord and Scarpace (1979) have shown how the addition of Landsat-derived land cover data can improve regression equations based on topographic maps alone.

Runoff Models. One of the first applications of remote sensing data in hydrologic models used Landsat data to determine both urban and rural land use for estimating runoff coefficients (Jackson et al., 1976). Land use is an important characteristic of the runoff process that affects infiltration, erosion, and evapotranspiration. Distributed models, in particular, need specific data on land use and its location within the basin. Most of the work on adapting remote sensing to hydrologic modeling has involved the Soil Conservation Service (SCS) runoff curve number model (U.S. Department of Agriculture, 1972) for which remote sensing data are used as a substitute for land cover maps obtained by conventional means (Jackson et al., 1977, Bondelid et al., 1982).

In remote sensing applications, one seldom duplicates detailed land use statistics exactly. For example, a study by the Corps of Engineers (Rango et al., 1983) estimated that an individual pixel may be incorrectly classified about one-third of the time. However, by aggregating land use over a significant area, the misclassification of land use can be reduced to about two percent, which is too small to affect the runoff coefficient or the resulting flood statistics.

Studies have shown (Jackson et al., 1977) that for planning studies the Landsat approach is cost effective. The authors estimated that the cost benefits were on the order of 2.5 to 1 and can be as high as 6 to 1, in favor of the Landsat approach. These benefits increase for larger basins or for multiple basins in the same general hydrological area. Mettel et al., (1994) demonstrated that the recomputation of Probable Maximum Flood (PMF) for the Au Sable River using HEC-1 and updated and detailed land use data from Landsat TM resulted in 90% cost cuts in upgrading dams and spillways in the basin.

Other types of runoff models that are not based only on land use are beginning to be developed. For example Strübing and Schultz (1983) have developed a runoff regression model that is based on Barrett's (1970) indexing technique. The cloud area and temperature are the satellite variables used to develop a temperature weighted cloud cover index. This index is then transformed linearly to mean monthly runoff. Rott (1986) also developed a daily runoff model using Meteosat data for a cloud index. Recently, Papadakis et al., (1993) have used a cloud cover index from satellite imagery to estimate monthly area precipitation. A series of non-linear reservoirs then transforms the precipitation into monthly runoff values. This approach was successfully demonstrated for the large (16000 sq. km.) Tano River Basin in Africa and illustrates the value of remote sensing data when conventional data are not readily available (see also Chap. 18). Ottle et al. (1989) have shown how satellite-derived surface temperatures can be used to estimate ET and soil moisture in a model that has been modified to use these data.

Integration with GIS. The pixel format of digital remote sensing data makes it ideal for merging with Geographical Information Systems (GIS). Remote sensing can be incorporated into the system in a variety of ways: as a measure of land use, impervious surfaces, for providing initial conditions for flood forecasting, and for monitoring flooded areas (Neumann et al., 1990). The GIS allows for the combining of other spatial data forms such as topography, soils maps and hydrologic variables such as rainfall distributions or soil moisture. This approach was demonstrated by Kouwen et al., (1993) where their Grouped Response Unit (GRU) included satellite based land use and lies within a computational element that may be either a sub-basin or an area of uniform meteorological forcing. In HYDROTEL, Fortin and Bernier (1991) propose combining SPOT DEM data with satellite-derived land use and soils mapping data to define Homogeneous Hydrologic Units (HHU). In a study of the impact of land use change on the Mosel River Basin, Ott et al. (1991) and Schultz (1993) have defined Hydrologically Similar Units (HSU) by DEM data, soils maps and satellite-derived land use. They also used satellite data to determine a vegetation index (NDVI) and a leaf water content index (WCI) which are combined to delineate areas where subsurface supply of water is available to vegetation. Mauser (1991) has shown how multi-temporal SPOT and TM data can be used to derive plant parameters for estimating ET in a GIS based model.

Soil Moisture. There continues to be speculation about the potential value for soil moisture data as in input variable in hydrologic models, either to establish the initial conditions for simulating storm runoff, or as a descriptor of hydrologic processes and much progress is beginning to appear as some of the aircraft experimental data become available (see also Chap. 9).

Aircraft data taken during the FIFE campaign were used to map the spatial pattern of soil moisture resulting from drainage and ET in a 37.7 ha watershed (Wang et al., 1989). These patterns were seen to map the results of a simple slab model and identified the region contributing base flow to the channel (Engman, et al., 1989). Attempts to use passive microwave measurements in a small watershed showed good correlation with the ground data and may yield a reliable technique for calibrating the model (Wood et. al., 1993). Also, even the relatively low-resolution passive data can improve the water budget calculations of a small basin (Lin, et. al., 1994). Goodrich et al., (1994) studied the pre-storm soil moisture at various scales of basin runoff. They concluded that initial values were important but that the resolution of the final remote sensing product was not a limitation.

Soil water storage capacity. All hydrological catchment models, rainfall-runoff models as well as water balance models contain a component dealing with the soil water storage process. As long as there are difficulties to measure soil water storage by remote sensing directly a substitute is often used, which uses remote sensing information coupled with other information for the determination of the soil water storage capacity. If this is known, the soil water storage process in time and space can be simulated. Here an example will be briefly given, how this can be done.

It is assumed, that the soil water storage in the upper soil zone is determined by soil type and vegetation type. As can be seen in Color Plate 5A (bottom left) a soil map giving information on soil type is used in order to determine the effective soil porosity as can be seen in Color Plate 5.A (bottom center). Furthermore Landsat imagery is used (Color Plate 5.A, top left) in order to determine the landuse by a suitable landuse classification technique (Color Plate 5.A, top center). The knowledge of the landuse allows to determine the root depth over the catchment area (Color Plate 5.A, top right). Under the assumption that root depth and soil porosity determine the maximum soil storage capacity in a catchment area it is possible to merge the digital information contained in the maps of soil porosity and root depth in Color Plate 5.A in order to determine the soil storage capacity as shown in Color Plate 5.A, bottom right. As can be seen from the soil storage capacity map in Color Plate 5.A the storage capacity varies considerably over the catchment area. This information can be used in hydrological modeling in different ways. It is possible, e.g. to base the vertical and lateral flow modules of the hydrological model on a pixel by pixel simulation of evapotranspiration, infiltration and lateral flows, or it is possible to generate distribution functions of maximum soil storage capacity for the total catchment area or sub-areas in the form of HRU's or GRU's as mentioned above. Figure 5.1 shows such maximum soil water storage capacity

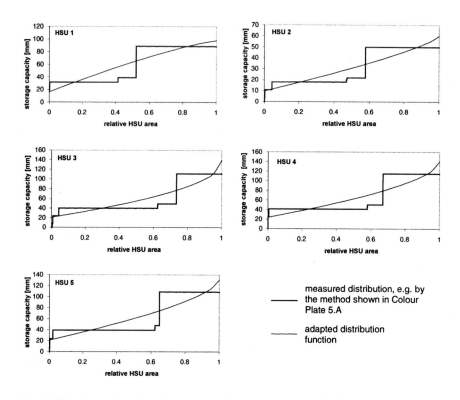

Fig. 5.1. Distribution functions of maximum soil water storage capacity for various HSU's in the Ennepe Dam catchment (HSU's defined by different soil types)

distributions for a catchment area in Germany for various HSU's (Hydrologically Similar Units) which are determined by different soil types.

5.3 Remote Sensing in Coupled Water-Energy Balance Modeling

As stated earlier, understanding the role of the terrestrial hydrosphere-biosphere in Earth's climate system, and the nature and effects of possible changes to the terrestrial water balance as a result of changing climate and land surface characteristics requires analyses using process-based, terrestrial water and energy balance models. These models will be applied at regional to global scales. There have been some successes in using large-scale coupled (water-energy) hydrological models to reproduce the hydrographs of large continental rivers (Liston et. al., 1994; Nijssen et al., 1995), and for estimating the seasonal and inter-annual variability of basin evapotranspiration (Abdulla et. al., 1995).

Although these recent studies are encouraging, performing large scale applications of energy and water balance models is greatly complicated by the scarcity of land surface observations needed to force them. For example, the simulations of Abdulla et al. (1995) for the Red River-Arkansas basin were performed in conjunction with GCIP, which is focused within the Mississippi River basin. It was selected because of the wealth of historical observational data. Nonetheless, there were only 26 stations within the Red River-Arkansas basin with long-term records of wind and relative humidity, and even fewer with solar and longwave radiation observations. Elsewhere globally, the availability of forcing variables over large areas (especially radiation) is even more limited and will increasingly be a factor as the emphasis on continental and global modeling increases.

Remote sensing offers a potentially attractive alternative to the use of ground observations as the forcings for hydrological modeling given:
(1) the recent availability of consistent, long-term remote sensing records, such as the AVHRR (Agbu, 1993), GOES (Young, 1995) and SSM/I (Hollinger et al., 1992) Pathfinder data sets, as well as other compilations of remote data such as part of the ISLSCP (Sellers et al., 1995) initiative;
(2) recent advances in remote sensing algorithms for deriving forcing variables that have been traditionally measured on the ground, such as radiation, humidity, and temperature;
(3) the development of new remote sensing instruments such as rain radar, and the future NASA Earth Observing System suite of sensors, that may be used to directly or indirectly estimate the required forcing fields, hopefully with increased accuracy.

There is a need to develop a predominantly remote sensing approach to macroscale hydrological modeling. To achieve this, land surface hydrologic models must be developed that are capable of utilizing remotely-sensed data; and, to develop and test remote sensing algorithms appropriate for generating data for hydrologic modeling. Equation (5.1) gave the primitive water balance equation. The corresponding energy balance equation is:

$$R_n = \lambda E + H + G \quad (5.2)$$

where R_n is net surface radiation (solar and longwave), λE is the latent heat, H the sensible heat and G the ground heat flux. Equations (5.1) and (5.2) are not directly usable for determining the terrestrial water and energy fluxes and states. To make them usable, the flux and state terms ($\frac{\Delta S}{dt}$, P, E, Q, H and G) must be parameterized in terms of the state variables which are the soil moisture profile, the surface, ground and near-surface air temperatures, and near-surface humidity. This parameterization leads to so-called Soil-Vegetation-Atmosphere-Transfer (SVAT) models.

Figure 5.2 shows the data required for such land surface models. The inputs to the model can be divided into three categories:
(i.) Forcing variables needed to drive the model. These include precipitation (both liquid and solid), incoming solar (shortwave and near-infrared) radiation from the atmosphere, and downwelling longwave (thermal) radiation from the atmosphere.
(ii.) Surface meteorology that is required as part of the parameterization of the evaporation, transpiration and sensible heat variables. The required meteorology includes surface air temperature, surface humidity and surface wind.

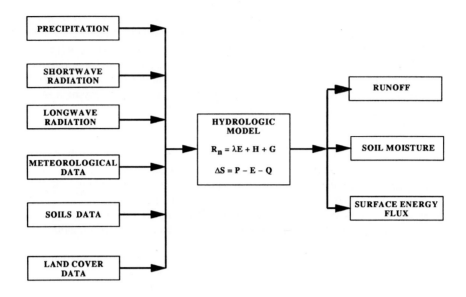

Fig. 5.2. Required forcing and parameterization variables for the water and energy balance formulations used in the VIC-3L model

(iii.) Land surface, soil and vegetation variables whose values are needed for the parameterization of water balance processes. These include interception of precipitation by vegetation and net throughfall, infiltration, bare soil evaporation, surface runoff, and base flow recession. These variables are needed for canopy-scale transpiration and ground heat flux, including the energy processes associated with snow melt and frozen soils. The output from the water and energy balance models includes soil moisture, surface temperature, runoff, and latent, sensible and ground heat fluxes.

The following sections give some recent results in the development of the remote sensing products needed for hydrological modeling.

5.4 Remote Sensing Approach

There has been progress in the estimation of radiation variables, surface air temperature, and surface humidity. In addition, some discussion on the use of remote sensing of precipitation will be provided, however, it is still poorly measured from space.

5.4.1 Solar radiation

Incoming and net solar radiation may be derived over large areas using sensors aboard operational meteorological satellites, such as Geostationary Operational Environmental Satellite (GOES) which combines high temporal resolution (30 minutes) with coarse (> 1 km) spatial resolution. Several physically based algorithms exist for both clear and cloudy sky conditions, including Gautier et al. (1980), Cess and Vulius (1989), Frouin et. al (1989), Bishop and Rossow (1991), Pinker and Laszlo (1992), and Gautier and Landsfeld (1997). These algorithms typically use the at-sensor radiance along with climatological, surficial and atmospheric information to determine incoming and net solar radiation.

There are several issues with GOES data that are worth mentioning. First, the resulting 1 km insolation fields generated by the radiation codes are generally not appropriate for driving macroscale models. A particular terrain element will not just see the portion of the sky corresponding to the field of view of the sensor, but will receive energy from the entire overlaying hemisphere. Secondly, clouds move with time. Hence, the instantaneous flux as determined from the GOES data are not representative of the time-integrated fluxes usually used to drive models (e.g. hourly-integrated fluxes from pyranometers). Lastly, the navigation of GOES data generally is not precise so that the exact location on the ground corresponding to particular pixel is not known. For all these reasons, some form of aggregation of the data is required (Dubayah and Loechel, 1997)

5.4.2 Downwelling longwave

Downwelling longwave radiation $(LW \downarrow)$ from the atmosphere is difficult to estimate from remotely-sensed data. It is dependent on several factors including clouds, air temperature, surface temperature, surface and atmospheric emissivity, near surface humidity, and the water vapor and temperature lapse rates. Several empirical and semi-empirical relationships exist which relate air temperature and humidity at shelter height, and in some cases cloudiness, to downwelling longwave (e.g. Brutsaert 1975; Unsworth and Monteith 1975). Zhong et al. (1990) have shown that differences among some of these techniques are small, on the order of 6 to 10% of the fluxes. A simple formulation is given by Bras (1992):

$$LW \downarrow = cc \varepsilon_a \sigma T^4 \qquad (5.3)$$

where cc is cloud correction factor, ε_a is the emissivity of the atmosphere given as a function of near-surface water vapor pressure, σ is the Stephan-Boltzmann constant, T is the air temperature in degrees Kelvin. Air temperature and humidity can be obtained from ground or satellite observations.

Physically based methods use the vertical profile of water vapor, temperature, and cloud information, as obtained from sounders or radiosondes, to estimate the downwelling flux with radiative transfer models. Radiances obtained from infrared sounders can be correlated with downwelling longwave (Wu and Cheng, 1989). Another approach would be to use total precipitable water vapor (either from GOES or AVHRR) and air temperature from AVHRR as above, along with assumed vertical vapor and temperature lapse rates in a radiative transfer algorithm to estimate clear-sky downwelling longwave. However, according to the AVHRR Atmospheric Pathfinder Working Group Report (1993), it is not thought possible to estimate downwelling longwave with sufficient accuracy (~ 20 W m^{-2}) using AVHRR data alone due to difficulties in remotely estimating near-surface air temperature, near-surface humidity, and cloud base pressure.

5.4.3 Precipitation

Estimation of precipitation by satellite remote sensing remains problematic at the time scales required by hydrologic models. While missions such as TRMM have great promise for improving climatological estimates of precipitation, and perhaps for estimating areal precipitation at seasonal to annual time steps, the current (and planned) platforms suffer from an inability to observe the diurnal cycle directly. In the U.S., the new NOAA WSR-88D weather radars are now producing archived precipitation products at 4 km resolution. The most widely available (level 3) products effectively incorporate some observing station data as well. Arola et al. (1994) have used these products with some success over part of the Arkansas-Red basin. Known problems with these data products, such as a radial bias in the estimated intensities, are currently being addressed (e.g., Smith, 1996). In the central U.S. and other areas without complex topography, it seems likely that such products will eventually become a standard forcing for hydrologic models. In mountainous areas, such as the western U.S., radar-based precipitation products do not appear to be viable for macroscale modeling. More detailed information on the potential of utilizing remote sensing for precipitation estimation is given in Chaps. 6, 16 and 18.

5.4.4 Air Temperature

An innovative approach has been developed that estimates near surface air temperature from AVHRR data. The Temperature/Vegetation Index (TVX) utilizes the relationship between the normalized difference vegetation index (NDVI) and surface skin temperature (Goward et al. 1994, Prince and Goward 1995, Prihodko and Goward 1997, Czajkowski et al. 1997, Prince et al. 1998). The surface temperature sensed by AVHRR is some function of canopy and soil background contributions. A 9 km x 9 km (or other sufficiently sized) window centered on the pixel for which an air temperature is to be estimated is used to generate a regression between NDVI and surface temperature and extended to a maximum NDVI. The window is moved to cover each pixel of the image. Due to the low thermal capacity of leaves which prevents them from reaching temperatures much higher than about 2°C from air temperature (Gates 1980), the temperature of the canopy at maximum NDVI will approximately be air temperature. An underlying assumption is that air temperature does not vary as rapidly in space across the spatial array. For areas where the TVX relationship is contaminated by clouds or the variability of NDVI in the landscape is low, such as during winter, no air temperature may be inferred.

There are several limitations to using AVHRR data to derive air temperature. The daily curve of air temperature cannot be estimated from AVHRR observations directly as these are only available at overpass time during mid-afternoon. One approach is to use the overpass temperature to scale an "average" diurnal curve obtained from climatology (Fig. 5.3). The actual form of the curve will differ from climatology whenever factors affecting air temperature change within a day, e.g. a precipitation event, passage of a cold or warm air mass etc. Similarly, cloudy conditions present another limitation because surface temperature cannot be sensed, and therefore no TVX relationship may be formed. One approach is to interpolate

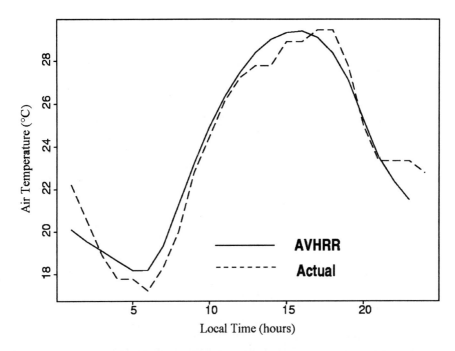

Fig. 5.3. Diurnal air temperature estimation. Air temperature is derived at the AVHRR afternoon overpass time. This mid-afternoon temperature is then used to scale an average diurnal curve for the location in question using climatology. Although useful for days that are typical climatologically, the method will result in errors when actual conditions depart from these, e.g. when cloud conditions change within a day. This plot is for 1 June 87 for a station in Missouri, within the Red-Arkansas basin

between clear areas. However, this will then cause the retrieved temperatures to be biased towards clear conditions.

5.4.5 Surface Air Humidity

The differential absorption of water vapor by channels 4 and 5 of AVHRR used to derive surface temperature may also be exploited to find total column precipitable water. Others have noted the relationship between precipitable water and the channel 4-5 temperature difference (e.g. Eck and Holben 1994), however, this relationship is incomplete and rarely exploited. A confounding factor is that the temperature difference is also a function of surface temperature, and not just precipitable water (Prabakara et al. 1979). As shown by Dubayah (see Fig. 5 in Prince and Goward, 1995), the slope of the line relating the temperature difference as a function of surface temperature may be used to predict precipitable water. Using simulations with an atmospheric radiative transfer code (LOWTRAN7, Kneizys et al. 1988) the following relationship was created relating precipitable water to the temperature difference (Prince and Goward, 1995):

$$U = \left(17.32\left(\frac{\Delta T - 0.6831}{T_s - 291.97}\right)\right) + 0.5456 \tag{5.4}$$

where U is the precipitable water (in cm), T_S is the surface temperature (derived using the split-window technique), and ΔT is the channel 4-5 temperature difference.

Surface humidity may be derived from total precipitable water using an empirical relationship derived by Smith (1966) which relates dew point temperature as a function of precipitable water and latitude. Dew point temperature may be converted to saturation vapor pressure using a standard formulation, which is then combined with air temperature (derived from AVHRR) to predict vapor pressure deficit.

5.5 Modeling Example: The Red River Arkansas Basin

We illustrate the remote sensing approach outlined above over the Red River Arkansas basin. The Arkansas and Red Rivers head on the eastern slope of the Rocky Mountains and flow to the Mississippi River near Little Rock, AR and Shreveport, LA, respectively (Fig. 5.4). The modified Variable Infiltration Capacity model (VIC-3L) (Liang et al. 1994, 1996a,b), was used to estimate the water and energy fluxes for the month of June 1987. Ground data needed to derive the required forcing and parameterization fields were acquired from 26 meteorological stations scattered throughout the basin, interpolated and gridded to 1° resolution. The cor-

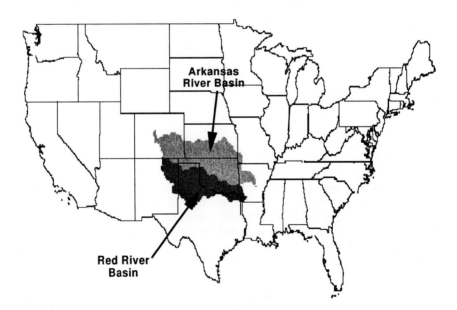

Fig. 5.4. The Red River and Arkansas River basins. These basins were used in this study because they are so similar

responding remotely-sensed fields were also generated. AVHRR-derived variables (air temperature, humidity and downwelling longwave radiation) were time-interpolated to produce diurnal curves, as described earlier. The model was run at a 3-hour time-step, first using ground data alone, and then using remotely-sensed inputs.

A summary of the June, 1987 basin average energy balance results for the hydrologic model runs is presented in Table 5.1. The increased incoming radiation of the remotely-sensed forcings causes generally higher surface energy fluxes. The vast majority of this increased net incoming energy is partitioned to the sensible heat flux. Color Plates 5.B, 5.C and 5.D show the spatial variations in the components of the energy balance over the basin. Part (a) shows the energy fluxes using the ground based forcing data; part (b) shows the fluxes using the remotely-sensed forcing data and (c) shows the normalized percent difference between part (a) and part (b). On the basis of this information computation of evapotranspiration becomes feasible.

Table 5.1. Modeled basin average energy fluxes in W/m^2 for June 1987 using ground based meteorological forcings and remotely-sensed forcings

	Net Radiation	Latent Heat	Sensible Heat
Ground Based Forcings	190	106	81
Remotely-Sensed Forcings	242	112	125

5.6 Future Directions

The promise of remote sensing for hydrological modeling is tremendous and is slowly being realized. There has been little effort to incorporate remotely-sensed observations in these models in a systematic fashion, even for those variables which we have reasonable confidence, such as solar radiation. There are several reasons for this. First, there is an enormous effort and expense required to assemble and process remotely-sensed data for large areas. The magnitude of this task, until it is attempted, is usually vastly underestimated, and after being discovered has probably deterred many a study. For example, purchasing a long-term record of GOES data, and then processing these data to derive solar radiation is a difficult task. Existing archives of solar radiation are at differing spatial and temporal resolutions and integration lengths. There is an urgent need for consistent, long-term solar radiation data that are suitable for macroscale hydrological modeling. The same is true for other types of remotely-sensed data. While the Pathfinder series of data sets have gone a long way towards filling these needs, there are gaps, most of which will be filled only with pressure from end-users in the hydrological community.

Secondly, frameworks for integrating observations with differing spatial and temporal resolutions into a modeling environment have been ad hoc, as are strategies for algorithm and model validation that involve scaling, such as comparisons between point observations and areal estimates. One promising approach is the use of data assimilation methodologies that merge observations with models. For example, TIROS Operational Vertical Sounder (TOVS) observations are used with a GCM in a data assimilation which then predicts fields of air temperature at coarse spatial resolution.

Lastly, the exploitation of distributed forcing fields generated from remotely-sensed data requires an evolution in hydrologic model structure and a rethinking of how such data are best used. Current structures are parameterized based on those data that are most frequently measured on the ground. This places a burden on remote sensing science by requiring derivation of variables such as air temperature that are difficult to estimate from space, while ignoring those that are easier, such as surface temperature. One possibility is to use satellite-measured skin temperature to update model skin temperatures, and thus keep the model from diverging from reality over time. The same method could be taken with other variables such as near-surface soil moisture state or snow cover, as measured with passive or active microwave sensors such as SSMR and SSM/I.

Other approaches may scrap parameterizations that involve several components, each of which have large uncertainties as derived from satellites, in favor of ones that have fewer elements. An example of the latter is the parameterization of latent heat flux in terms of net radiation. It has been shown that net radiation may be estimated empirically using solar radiation, which in turn is found relatively accurately from remote sensing. Whether or not this is an appropriate avenue of exploration for macroscale modeling remains to be decided. What is clear is that the capabilities of current and future remote sensing instruments, such as the EOS suite of sensors, must be studied carefully relative to hydrological modeling, and the models redesigned to fully take advantage of the data these instruments produce.

Acknowledgments

Thanks to Nancy Casey-McCabe for providing research assistance. This work is supported by NASA Grant NAGW-5194 (Dubayah) and NASA Contract NAS5-31719 (Wood).

References

Abdulla, F. A., D.P. Lettenmaier, E. F. Wood and J.A. Smith, Application of a Macroscale Hydrological Model to Estimate the Water Balance of the Arkansas-Red River Basin, J. Geophys. Res., 1996

Allord, G.J. and Scarpace, F.L. Improving Streamflow Estimates Through Use of Landsat. In Satellite Hydrology, 5th Annual William T. Pecora Memorial Symposium on Remote Sensing, Sioux Falls, SD. pp 284-291, 1979

Agbu, P.A. B. Vollmer and M.E. James, Pathfinder AVHRR Land Data Set. NASA Goddard Space Flight Center, Greenbelt. 1993

Arola, A., D.P. Lettenmaier, and E.F. Wood, Some preliminary results of GCIP modeling activities in the Arkansas-Red River basin, First International Scientific Conference on the Global Energy and Water Cycle Royal Society, London, 1994

Barrett, E.C., The estimation of monthly rainfall from satellite data. *Mon. Weather. Rev. 98*, pp. 322-327, 1970
Becker, F. and Li, Z.-L., Towards a local split window method over land surfaces, Inter. J. Remote Sens., 11, 369-393, 1990
Bishop, J.K. and W.B. Rossow, Spatial and temporal variability of global surface solar irradiance, J. Geophys. Res., 96, 16839-16858, 1991
Bondelid, T.R., T.J. Jackson and R.H. McCuen, Estimating runoff curve numbers using remote sensing data. Proc. Int. Symp. on Rainfall-Runoff Modeling. Applied Modeling in Catchment Hydrology, Water Resources Publications. Littleton, CO, pp. 519-528, 1982
Bras, R., Hydrology, Addison-Wesley, Reading, Mass., 643 pp., 1980
Brutsaert, W., On a derivable formula for long-wave radiation from clear skies, Water Resour. Res., 11, 742-744, 1975
Case, J.B., Report on the International Symposium on Topographic Applications of SPOT Data. *Photogrammetric Engineering and Remote Sensing,* Vol. 55, No. 1, pp. 94-98, 1989
Cess, R.D. and I.L. Vulis, Inferring surface solar absorption from broadband satellite measurements, J. Climate, 2, 974-985, 1989
Choudhury, B. J., S. B. Idso, and R. J. Reginato, Analysis of an empirical model for soil heat flux under a growing wheat crop for estimating evaporation be infrared-temperature based energy balance equation, Agric. For. Meteor., 39, 283-297, 1987
Czajkowski, K. P., T. Mulhern, S. N. Goward, J. Cihlar, R. O. Dubayah, and S. P. Prince, 1997b: Biospheric environmental monitoring at BOREAS with AVHRR observations. *Journal of Geophysical Research*, (in press)
Dubayah, R. and S. Loechel, Modeling topographic solar radiation using GOES data, J. Applied Meteor., 36, 141-154, 1997
Dubayah, R and D. Lettenmaier, Combining Remote Sensing and Hydrological Modeling for Applied Water and Energy Balance Studies, NASA EOS Interdisciplinary Working Group Meeting, San Diego, CA, March, 1997
Dümenil, L., and E. Todini, A rainfall-runoff scheme for use in the Hamburg climate model, in Advances in theoretical hydrology: A tribute to James Dooge, Eur. Geophys. Soc. Ser. on Hydrol. Sci., vol 1, ed. by J. P. O'Kane, 129-157, Elsevier, New York, 1992
Eck, T. and B. Holben, AVHRR split window temperature differences and total precipitable water over land surface, Inter. J. Remote Sens., 15, 567-582, 1994
Engman, E.T., G. Angus, and W.P. Kustas,. Relationship between the hydrologic balance of a small watershed and remotely sensed soil moisture. Proc. *IAHS Third Intl. Assembly, Baltimore, IAHS Publ. 186,* pp. 75-84, 1989
Engman, E. T. and R. J. Gurney, Remote sensing in hydrology, Chapman and Hall, London, 225 pp., 1991
Fortin, J.-P., and M. Bernier, Processing of remotely sensed data to derive useful input data for the hydrotel hydrological model. *IEEE.,* 1991
Frouin, R., D.W. Lingner, C. Gautier, K.S. Baker, and R.C. Smith, A simple analytical formula to compute clear sky total and photosynthetically available solar irradiance at the ocean surface, J. Geophys. Res., 94, 9731-9742, 1989
Gates, D. M., Biophysical Ecology, Springer-Verlag, New York, 611 pp., 1980
Gautier, C., and M. Landsfeld, Surface solar radiation flux and cloud radiative forcing for the Atmospheric Radiation Measurement (ARM) Southern Great Plains (SGP): A satellite and radiative transfer model study, J. Atmos. Sci., in press, 1997
Gautier, C., G. Diak, and S. Masse, A simple physical model to estimate incident solar radiation at the surface from GOES satellite data, J. Appl. Meteor.,19, 1005-1012, 1980
Goodrich, D. C., T.J. Schmugge, T.J. Jackson, C. L. Unkrich, T.O. Keefer, R.Parry, L.B. Bach, and S.A. Amer, Runoff simulation sensitivity to remotely sensed initial soil water content. *Water Resources Research, Vol. 30,* No. 5, pp. 1393-1405, 1994
Goward, S. N., R. H. Waring, D. G. Dye, and J. Yang, Ecological remote sensing at OTTER: satellite macroscale observations, Ecological Appl., 4, 332-343, 1994

Haralick, R.M., Wang, S., Shapiro, L.G. and Campbell, J.B. Extraction of Drainage Networks by Using a Consistent Labeling Technique. Remote Sensing of Environment, 18 pp. 163-175, 1985

Henderson-Sellers, A., A. J. Pitman, P. K. Love, P. Irannejad, and T. H. Chen, The project for intercomparison of land surface parameterization schemes (PILPS): phases 2 and 3, Bul. Amer. Meteor. Soc., 76, 489-503, 1996

Hollinger, J.P., J.L. Peirce, G.A. Poe, SSM/I instrument evaluation. IEEE Transactions on Geoscience and Remote Sensing, 28, 781-790, 1990

Huang, X. and T. J. Lyons, The simulation of surface heat fluxes in a land surface-atmosphere model, J. Appl. Meteorol., 34, 1099-1111, 1995. Johansen, O., Thermal conductivity of soils, Ph.D. Thesis, Trondheim, Norway (Corps of Engineers/CRREL Translation 637, 1977) ADA 044002, 1975

Jackson, T.J., Ragan, R.M. and Fitch, W.N. Test of Landsat-Based Urban Hydrologic Modeling. ASCE J. Water Resources Planning and Management Div. V 103. No. WR1, Proc. Papers 12950. pp. 141-158, 1977

Kneizys, F. X., E. P. Shettle, L. W. Abreu, J. H. Chetwynd, G. P. Anderson, W. O. Gallery, J. E. A. Selby, and S. A. Clough, Users Guide to LOWTRAN7, Report AFGL-TR-88-0177, Air Force Geophysics Laboratory, Bedford, MA, 1988

Kowuwen, N., E.D. Soulis, A. Pietroniro, J. Donald, and R.A. Harrington, Grouped response units for distributed hydrologic modeling. *Journal of Water Resources Planning and Management, Vol. 119*, No. 3, pp. 289-305, 1993

Liang, X., E. F. Wood, and D.P. Lettenmaier, A one-dimensional statistical-dynamic representation of subgrid spatial variability of precipitation in the two-layer VIC model, J. Geophys. Res., 101, 21403-21422, 1996

Liang, X., E. F. Wood and D.P. Lettenmaier, Surface soil moisture parameterization of the VIC-2L model: Evaluation and modifications, Global and Planetary Change, 13, 195-206, 1996

Liang, X., D.P. Lettenmaier, E. F. Wood and S.J. Burges, A simple hydrologically based model of land surface water and energy fluxes for general circulation models, J. of Geophys. Res., 99, 14,415-14,428, 1994

Lin, D.-S., E.F. Wood, J.S. Famiglietti and M. Mancini, Impact of microwave derived soil moisture on hydrologic simulations using a spatially distributed water balance model. Proceedings of the 6th International Symposium on Physical Measurements and signatures in Remote Sensing, Val d'Isere, France., 1994

Liston, G. E., Y. C. Sud, and E. F. Wood, Evaluating GCM land surface hydrology parameterizations by computing river discharges using a runoff routing model: application to the Mississippi basin, J. Appl. Meteor., 33, 394-405, 1994

Mauser, W. Modelling the Spatial Variability of Soil-Moisture and Evapotranspiration with Remote Sensing Data, Proc.Int. IAH Symp. Rem. Sens. and Water Resources, Enschede, Aug. 20-24, 1990, pp.249-260

McCumber, M. C. and R. A. Pielke, Simulation of the effect of surface fluxes of hat and moisture in a mesoscale numerical model, J. Geophys. Res., 86, 9929-9938, 1981

Meesen, B. W., F. E. Corprew, J. M. P. McManus, D. M. Myers, J. W. Closs, K. J. Sun, D. J. Sunday, and P. J. Sellers, ISLSCP Initiative I-Global Data Sets for Land Atmosphere Models, 1987-1988, Vols. 1-5, NASA, CD-ROM, 1995

Mettel, C., D. McGraw, and S. Strater, Money Saving Model, *Civil Engineering, 64(1)*, pp. 54-56. 1994

Nijssen, B., D.P. Lettenmaier, E. F. Wood, S.W. Wetzel and X. Liang), Simulation of runoff from continental-scale river basin using a grid-based land surface scheme, accepted Water Resour. Res., 1996

Neumann, P., W. Fett and G.A. Schultz, A Geographic Information System as data base for distributed hydrological models. Proc. International Symposium, Remote Sensing and Water Resources, Enschede, The Netherlands, August 1990, pp. 781-791

Ott, M. , Z. Su, A.H. Schumann, and G.A. Schultz, Development of a distributed hydrological model for flood forecasting and impact assessment of land use change in the international Mosel River Basins, Proceedings of the Vienna Symposium. IAHS Publ. No. 201, 1991

Ottle, C., D. Vodal-Madjar, and G. Girard, Remote sensing applications to hydrological modeling. J. of Hydrol. 105, 369-384, 1989

Papadakis, I., J Napiorkowski, and G.A. Schultz, Monthly runoff generation by non-linear model using multispectral and multitemporal satellite imagery. *Adv. Space Res., Vol 13*, No 5, pp. (5)181-(5)186, 1993

Peck, E.L., Keefer, T.N. and Johnson, E.R. Strategies for Using Remotely Sensed Data in Hydrologic Models. NASA Report No. CR-66729, Goddard Space Flight Center, Greenbelt, MD., 1981, 52 pp.

Peters-Lidard, C., E. Blackburn, X. Liang, E. F. Wood, The effect of soil thermal conductivity parameterization on surface energy fluxes and temperatures, J. Atmos. Sci., in press, 1997

Prabhakara, C. and G. Dalu, Remote sensing of the surface emissivity at 9 μ over the globe, J. Geophys. Res., 81, 3719-3724, 1976

Prihodko, L., Estimation of air temperature from remotely sensed observations, MA thesis, University of Maryland at College Park, 1992

Prihodko, L., and S. N. Goward, Estimation of air temperature from remotely sensed observations, Remote Sens. Environ., in review, 1997

Prince, S. D., and S. N. Goward, Global primary production: a remote sensing approach, J. Biogeography, 22, 2829-2849, 1995

Prince, S. D., Goetz, S. J., Dubayah, R., Czajkowski, K., and Thawley, M., Inference of surface and air temperature, atmospheric precipitable water and vapor pressure deficit using AVHRR satellite observations: validation of algorithms. Journal of Hydrology, in press, 1998

Pinker, R.T. and I. Laszlo, Modeling surface solar irradiance for satellite applications on a global scale, J. Appl. Meteor., 31, 194-211, 1992

Ragan, R.M and T.J. Jackson, Runoff synthesis using Landsat and SCS model. J. Hydraul. Div., ASCE 106, (HY5), pp 667-78, 1980

Rango, A., A. Feldman, T.S. George, III, and R.M. Ragan, Effective use of Landsat data in hydrologic models, Water Resour. Bull., 19, pp. 165-174, 1983

Rott, H., J. Aschbacher, and K.G. Lenhart, Study of River Runoff Prediction Based on Satellite Data. European Space Agency final Report, No. 5376, 1986

Rowntree, P.R. and J. Lean, Validation of hydrological schemes for climate models against catchment data, J. Hydrology, 155, 301-323, 1994

Schultz, G.A. and E.C. Barrett, Advances in remote sensing for hydrology and water resources management, Technical Documents in Hydrology, UNESCO, Paris, 1989, 102pp

Schultz, G.A. , Hydrological modeling based remote sensing information. *Adv. Space Res., Vol. 13*, No. 5, pp.(5)149-(5)166, 1993

Sellers, P.J., S.O. Los, C.J. Tucker, C.O. Justice, D.A. Dazlich, G.J. Collatz, and D.A. Randall, "A global 1 degree x 1 degree NDVI data set for climate studies. Part 2: The generation of global fields of terrestrial biophysical parameters from the NDVI," Inter. J. Remote Sens, in press

Smith, J.A., D.-J. Seo, M.L. Baeck, and M.D. Hudlow, An intercomparison study of NEXRAD precipitation estimates, Water Resour. Res., 32, 2035, 1996

Smith, W.L., Note on the relationship between total precipitable water and surface dew point, J. Appl. Meteor., 5, 726-727, 1966

Stamm, J.F., E.F. Wood, and D.P. Lettenmaier, Sensitivity of a GCM simulation of global climate to the representation of land surface hydrology, J. Climate, 7, 1218-1239, 1994

Strübing, G. and Schultz, A. Estimation of monthly river runoff data on the basis of satellite imagery. Proc. Hamburg Symposium. IAHS Publ. NO. 145, 1983, pp. 491-498

Tennessee Valley Authority (TVA), Heat and Mass Transfer Between a Water Surface and the Atmosphere, Laboratory Report no. 14, Norris, Tennessee, 1972

Thomas, D.M. and Benson, M.A. Generalized Streamflow Characteristics from Drainage Basin Characteristics. USGS Water Supply Paper. 1955. Washington, DC, 1970, 55 pp.

United Nations, Economic Commission for Asia and the Far East. Multipurpose river basin development, part 1, manual of river basin planning flood control series no. 7. United Nations Publication ST/ECAFE/SERF/7, New York, 1955

U.S. Army Corps of Engineers, Remote Sensing Technologies and spatial data applications. Res. Doc. No. 29, Hydrologic Engineering Center, Davis CA, 1987, pp. 1-5

U.S. Department of Agriculture. Soil conservation Service. National Engineering Handbook, Section 4, HYDROLOGY. U.S. Govt. Printing Office, Washington, DC., 1972, 544 pp.

Unsworth, M.H. and J.L. Monteith, Long wave radiation at the ground. I. Angular distribution of incoming radiation, Quarterly J. Royal Meteor. Soc., 101, 13-24, 1975

Wang, J.R., J.C. Shiue, T.J. Schmugge, and E.T. Engman, Mapping soil moisture with L-band radiometric measurements. *Remote Sens. Environ. 27*, pp. 305-312, 1989

Wood, E.F., D.-S. Lin, P. Troch, M. Mancini and T.Jackson, Soil moisture estimation: Comparisons between hydrologic model estimates and remotely sensed estimates. Proceedings of the ESA/NASA Workshop on Passive Microwave Remote Sensing, Saint-Lary, France, 11-15 January 1993

Wood, E. F., D. P. Lettenmaier, X. Liang, B. Nijssen and S. W. Wetzel Hydrological Modeling of Continental-Scale Basins, Annual Reviews of Earth and Planetary Sciences, 25, 1997

Wu, M.C. and C. Cheng, Surface downward flux computed by using geophysical parameters derived from HIRS 2/MSU soundings, Theoretical and Applied Climatology, 40, 37-51, 1989

Young, J.T., J.W. Hagens, and D.M. Wade, "GOES Pathfinder product generation system," 9th Conference on Applied Climatology, 9th, Dallas, American Meteorological Society, Boston, MA, 1995

Zebker, H.A. S.N. Madsen, J. Martin, K.B. Wheeler, T. Miller, Y. Lou, G.Alberti, S. Vetrella, and A. Cucci, The TOPSAR Interferometric Radar Topographic Mapping Instrument: *IEEE Trans. On Geoscience and Remote Sensing, V. 30*, pp.933-940, 1992

Zhao, R-J, Y-L Zhang, L-R Fang, X-R Liu and Q-S Zhang, The Xinanjiang model, Proc. Oxford Symp., Hydrol Forecast. IAHS Publ. 129, pp351-56, Wallingford, UK: Int. Assoc. Hydrol. Sci., 1980

Zhong, M., A. Weill, and O. Taconet, Estimation of net radiation and surface heat fluxes using NOAA-7 satellite infrared data during fair-weather cloudy situations of MESOGERS-84 experiment, Boundary Layer Meteor., 53, 353-370, 1990

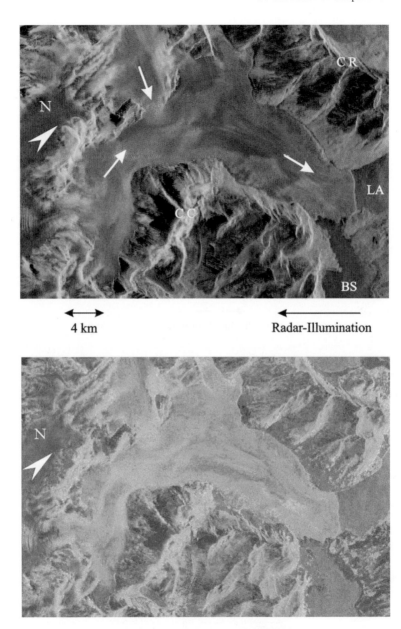

Colour Plate 2.A. Example for interferometric analysis, based on SIR-C data acquired on 9 and 10 October 1994 over the region of Moreno Glacier, Patagonia, Argentina. Top: Colour composite of C-band VV (green) and L-band VV (red) channels. The arrows show the flow direction of the glacier. Landmarks: LA – Lago (lake) Argentino, BS – Southern arm of LA, CR – Cordon Reichert (1600 m a.s.l.), CC – Cerro Cervantes (2400 m a.s.l.). Bottom: Degree of coherence ($0 \leq \rho \leq 1$) at L-band, colour coded

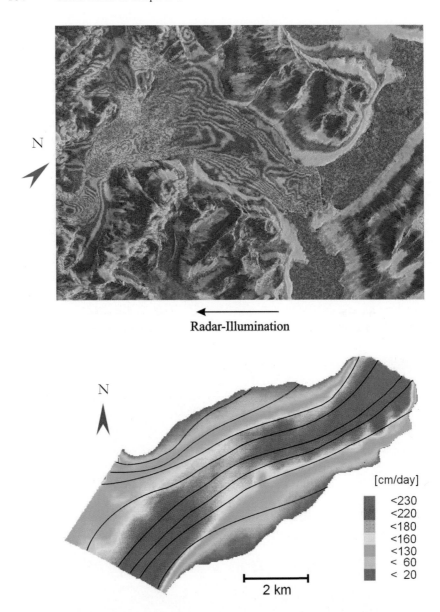

Colour Plate 2.B. Interferometric analysis based on the same data set as plate 2.A. Top: Interferogram, superimposed to the amplitude image. One colour cycle represents a phase shift of 2π. Bottom: Map of ice motion derived from the interferogram. The colours are related to the magnitude of the motion vector, the lines show the flow direction (Rott et al., 1998)

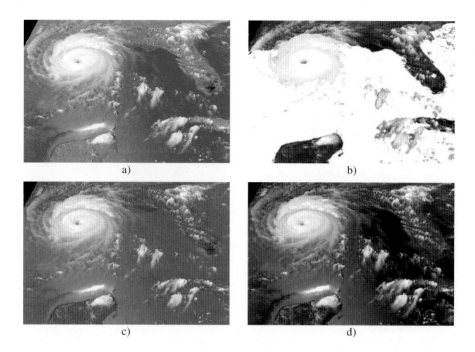

Colour Plate 3.A. a) Color composite of multi-band AVHRR imagery of Hurricane Andrew bearing down on Louisiana obtained on August 25, 1992 at 20:20 UT. The apparent height of the perspective rendering is inversely proportional to the cloud temperatures observed in band 5 (11.5 - 12.5 µm). **b)** AVHRR band 5 thermal infrared data were placed in the blue image processor memory plane. **c)** Band 2 (0.725 - 1.10 µm) near-infrared data were placed in the green memory plane. **d)** The visible band 1 (0.58 - 0.68 µm) data were placed in the red image plane (Source: F. Hasler, K. Palaniappan, M. Manyin and H. Pierce of NASA Goddard Space Flight Center)

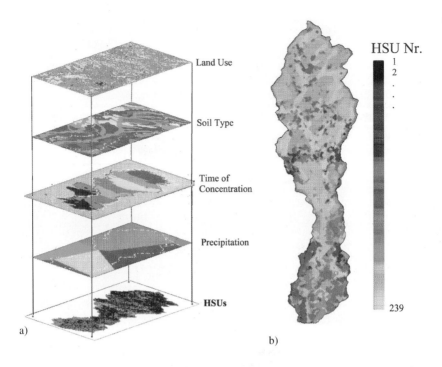

Colour Plate 4.A. Derivation of Hydrologically Similar Units (HSU's) by overlay operation: (a) the principle; (b) generated HSU's in the Nims river catchment (267 km², Mosel tributary), Germany (Su et al., 1992)

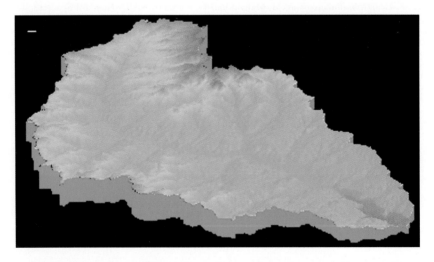

Colour Plate 4.B. Three-dimensional view of elevation in the Little Washita watershed, Oklahoma

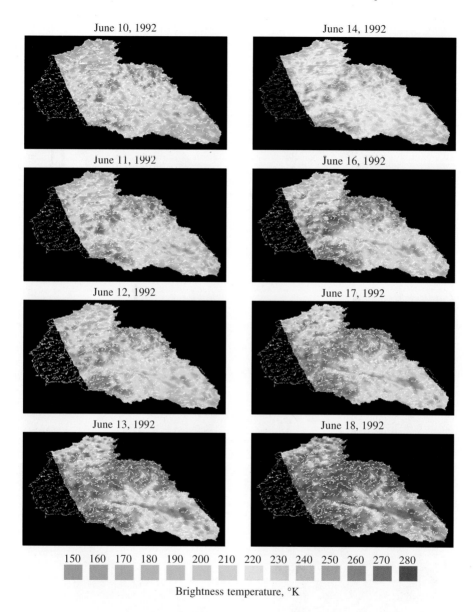

Colour Plate 4.C. 3D views of remotely sensed microwave brightness temperature data (0-5 cm depth) for the Little Washita watershed, Oklahoma, June 10-18, 1992. June 15 was the aircraft crew rest day. Contours derived from DEM have been overlaid on the 3D views for improved visualization (however the contours are not clearly visible in this diagram due to its size). Data were not collected for the southern portion of the watershed. Lower temperatures indicate wet soil and higher temperatures indicate dryer soil, and the sequence illustrates near-surface soil moisture dry down pattern over the eight days period.

108 Colour Plates of Chaps. 2–5

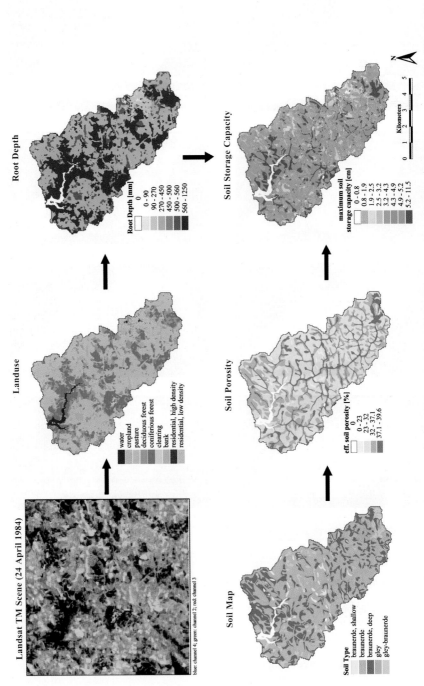

Color Plate 5.A. Generation of a soil storage capacity map of a catchment (Ennepe Dam, Germany) based on Landsat imagery and a digital soil map

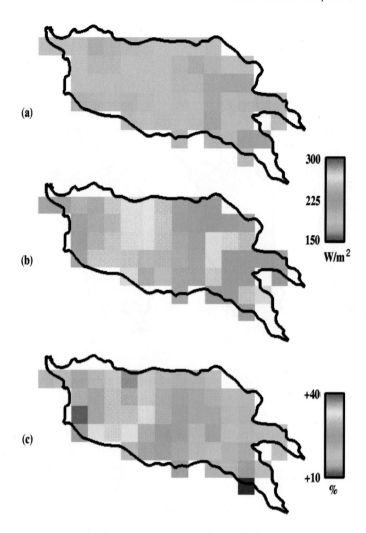

Color Plate 5.B. Net radiation from hydrologic model: (a) using forcings from ground based measurements; (b) using forcings derived from remote sensing data; (c) percent difference between (a) and (b)

110   Colour Plates of Chaps. 2–5

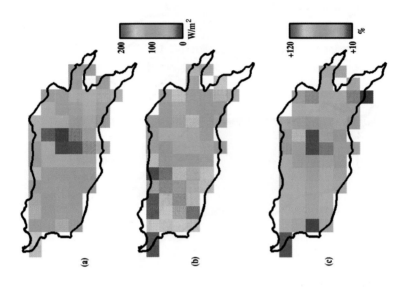

Color Plate 5.C. Latent heat flux from hydrologic model: (**a**) using forcings from ground based measurements; (**b**) using forcings derived from remote sensing data; (**c**) percent difference between (a) and (b)

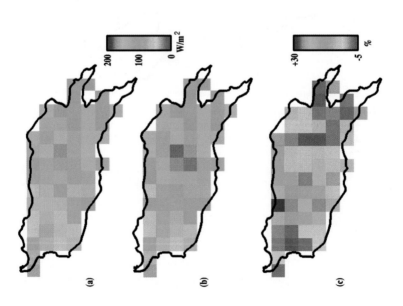

Color Plate 5.D. Sensible heat flux from hydrologic model: (**a**) using forcings from ground based measurements; (**b**) using forcings derived from remote sensing data; (**c**) percent difference between (a) and (b)

6 Precipitation

C. G. Collier

Telford Institute of Environmental Systems
Department of Civil and Environmental Engineering
University of Salford, United Kingdom

6.1 Introduction

Precipitation is the central component of the hydrological cycle, and as such is of primary importance in hydrology. It shows large and frequent spatial variations and usually exhibits rapid temporal variations. Table 6.1 illustrates the spectrum of significant precipitation events that can occur and emphasizes the important change of approach which is forced upon the analyst as he crosses the rain-day time averaging threshold. For time periods less than one day precipitation distributions are often analysed in terms of the 'cloud-scale' motions that cause them. For time periods greater than one day it becomes increasingly difficult, and ultimately impossible, to do this. Up to about 3 days, synoptic scale systems can provide a framework for analysis, although this can be difficult as such systems often move and develop significantly over this period.

The scales over which precipitation occurs are so large that methods of measurement are wide ranging, and usually complementary. Point measurements are insufficiently representative of sub-catchment scales, and hence it is difficult to understand and model hydrological processes using such data. This point is emphasised in Colour Plate 6.A which shows a rainfall situation for which point measurement from raingauges miss most of the rain as measured by ground-based radar. The basic measurement requirement of hydrology is for areal measurements albeit over sometimes very small, as well as very large, areas.

The hydrological requirement for measurements of precipitation is discussed by Collier (1996a) in terms of the maximum, minimum and most usual values for resolution, frequency of observations and accuracy (freedom from bias). Precision is also expressed qualitatively as the standard deviation of results from repeated trials under identified conditions. Hence, a measurement from a raingauge is spatially precise, but may be imprecise in time and quantity. On the other hand radar and satellite measurements are usually precise in time, less so in space, and much more imprecise in quantity. It is therefore important to regard all these types of measurement as complementary. In this chapter we discuss, in particular, radar and satellite measurements of precipitation. The reader is referred to other texts for detailed discussion of point measurements of rainfall and snowfall (see for example Collier, 1996b).

Table 6.1. Spectrum of precipitation-producing systems (after Mason, 1970)

HORIZONTAL SCALE

	10^4km	10^3km	10^2km	10km	5km	1km
	Planetary	Synoptic	Meso-scale	Convective or Small scale		Microscale Molecular
MIDDLE LATITUDES	Long Waves	Extra-Tropical Depressions	Fronts	Cumulo-nimbus		Boundary-Layer Eddies
	Sub-Tropical		Lee Waves	Showers		
	Anti-Cyclones	Anti-Cyclones	Squall Lines	Tornadoes		
TROPICS	ITCZ	Cloud Clusters				
	Easterly Waves	Tropical Cyclones	Meso-scale Convective Elements	Convective Cells		Boundary-layer eddies
		10^2HR	10 HR	1 HR		10^{-1}HR

Time Scale
SCALES OF MOTION

6.2 General approach

6.2.1 Ground-based radar

Ground-based radar offers areal measurements of precipitation from a single location, over large areas in near real-time. Both single and multi-polarisation radars have been used over a range of wavelengths. However, the most commonly used wavelengths lie in the region of the electromagnetic spectrum which includes wavelength of 3cm (x-band), 5 cm (c-band) and 10 cm (s-band).

As a radar beam rotates about a vertical axis, measurements are made at many ranges out to 100 km or more, and at difference azimuths, of the energy backscattered from precipitation particles in volumes above the ground. Given a radar wavelength λ, and considering a spherical raindrop with diameter D, we may define a back scattering cross section $\sigma_b(D)$ and a total attenuating cross section $\sigma_a(D)$ as proportional to D^6/λ^4. This is the justification for introducing a physical parameter called 'radar reflectivity factor', Z, defined as:

$$Z = \int_0^\infty N(D) D^6 dD \qquad (6.1)$$

where N(D) is the drop size distribution (DSD) within the resolution cell (Z in mm^6 m^{-3}, D in mm, N(D) in $mm^{-1}m^{-3}$). In the absence of attenuation along the radar path, and as long as the Rayleigh theory holds (targets very small compared to the radar wavelength), the back scattered power from a resolution cell is proportional to Z. However, when the ratio $\pi D/\lambda$ becomes larger than 0.1, then Mie theory should be used in place of Rayleigh theory. To take account of this effect, an 'equivalent' radar reflectivity fraction Z_e is generally considered. Z_e is the same as Z for light rain (mainly composed of small rain drops) but departs from it as the rainfall rate increases; the departure increases more rapidly as the radar wavelength decreases. It may be shown that if liquid precipitation uniformly fills the pulse volume, then the average power returned from precipitation at range r is proportional to Z/r^2, where Z is the so-called radar reflectivity factor given by the sum of the precipitation particle diameters raised to the sixth power.

$$Z = AR^B \qquad (6.2)$$

where A and B, empirical constants, depend upon the type of precipitation as shown in Table 6.2

Use of R:Z relationships to measure rain, modifying A and B as appropriate, would appear to be straightforward. There are a number of problems, however, arising from the characteristics of both the radar and the precipitation. The importance of these problems will depend upon the particular radar configuration in use and the meteorology of particular situations.

Table 6.2. Typical empirical relationships between reflectivity factor Z (mm^6 m^{-3}) and precipitation intensity, R (mm h^{-1}) (after Battan, 1973)

Equation	Precipitation Type	Reference
140 $R^{1.5}$	Drizzle	Joss et al. (1970)
250 $R^{1.5}$	Widespread rain	Joss et al. (1970)
200 $R^{1.6}$	Stratiform rain	Marshall and Palmer (1948)
31 $R^{1.71}$	Orographic rain	Blanchard (1953)
500 $R^{1.5}$	Thunderstorm rain	Joss et al. (1970)
486$R^{1.37}$	Thunderstorm rain	Jones (1956)
2000 $R^{2.0}$	Aggregate snowflakes	Gunn and Marshall (1958)
1780$R^{2.21}$	Snowflakes	Sekhan and Srivastava (1970)

Whilst weather radar does offer a means of making wide area measurements over specific important land and coastal areas, it clearly does not offer a practical method of making such measurements over the ocean, or over very large remote river catchments. The use of satellite data has consequently been investigated for this purpose ever since meteorological satellites have been flown.

6.2.2 Use of visible and infrared satellite data

Initially only visible and infrared channels were available but over recent years there has been increasing use made of passive (emitted from the Earth and clouds) microwave channels.

Visible/Infra-red techniques derive qualitative or quantitative estimates of rainfall from satellite imagery through indirect relationships between solar radiance reflected by clouds (cloud brightness temperatures) and precipitation. A number of methods have been developed and tested during the past fifteen years with a measured degree of success.

There are two basic approaches, namely the 'life-history' and the 'cloud-indexing' techniques. The first type makes use of data from geostationary satellites which produce images usually every half hour. It has been mostly applied to convective systems. The second type, also based on cloud classification, does not require a series of consecutive observations of the same cloud system.

It must be noted, however, that up to now none of these techniques has been shown to be 'transportable'. In other words, relationships derived for a given region and a given time period may not be valid for a different location and/or season. Other problems include difficulties in defining rain/no rain boundaries and inability to cope with the rainfall pasterns at the meso or local scales. Scientists working in this field are perfectly aware of these problems, and this is why it is current practice to speak of the derivation of 'Precipitation Indices' rather than rain rates.

6.2.3 Use of passive microwave satellite data

VIS-IR measurements represent observations of the upper surface of clouds only. In contrast it is often believed that microwave radiation is not affected by the presence of clouds. This statement is not generally true. Its degree of validity varies with the microwave frequency used as well as with the type of cloud being observed.

One major difference between infra-red and microwave radiation is the fact that whilst the ocean surface emissivity is nearly equal to one in the infra-red, its value (although variable) is much smaller in the microwave region (from 5 to 200 Ghz here). Therefore, the background brightness temperature (TBB) of the ocean surface appears much colder in the microwave. Over land the emissivity is close to one, but varies greatly depending on the soil moisture.

As far as microwaves are concerned several different effects are associated with the presence of clouds. They are highly frequency dependent. Absorption by cloud droplets is roughly proportional to the square of the frequency. A 2 km thick (non-precipitating) cumulus cloud transparent at 18 Ghz may be totally opaque (through absorption) at 160 Ghz, the resulting effect being an increase of brightness temperature (through emission) over the cold background. Hence, a 2 km thick precipitating cumulus cloud will be detected over the cold surface background at 18 Ghz, and the observed TBB will increase in proportion to the amount of 'vertically integrated' raindrops contained in the cloud.

Scattering by raindrops increases with drop size. The effect of scattering is to reduce the apparent TBB and is also frequency-dependent. The presence of ice in a precipitating cloud results in a decrease of TBB through scattering (ice particles scatter but do not absorb microwave radiation). This effect increases with frequency and particle size, and is used to indirectly delineate rain over the earth's surface.

Measurements of rainfall using passive microwaves falls into one or two classes depending on the particular effect used to detect precipitation (i.e. absorption or scattering). To exploit absorption a cold background is necessary. Thus techniques based on this effect can be applied only over the ocean using frequencies below 10 Ghz.

6.2.4 Space-borne radar

Measuring rain from a space-borne radar is very attractive since it can discriminate in range (altitude when operated from space), and radar reflectivity is directly related to rainfall rate. A number of approaches are currently being investigated. A particularly promising approach, known as the surface reference technique (Meneghini et al. 1983), exploits the fact that the scattering properties of the ocean surface vary on a much longer horizontal scale than does the rain rate field. Hence, the level of the signal received from the ocean surface through a rain cell relative to that outside the rain cell provides a means of monitoring the two way attenuation along a ray path which may then be related to the rainfall rate using known relationships.

A rain radar, operating at 13.786 and 13.802 Ghz is part of the instrument package on the joint US-Japan Tropical Rainfall Measurement Mission (TRMM) satellite which was launched on 28 November 1997 in a co-latitude orbit (see Theon et al., 1992). Since radiometers having visible (0.63, 1.6µm), infrared (3.75, 10.8, 12µm) and passive microwave (10.7, 19.4, 21.3, 37, 85.56 Ghz) channels also fly on this satellite, it is providing an ideal opportunity to combine a number of the techniques discussed in the next section. An example of the radar data being recorded is shown in Colour Plate 6.B.

6.3 Current Techniques

6.3.1 Single polarisation radar measurements of rainfall

In trying to relate radar measurements of reflectivity to rainfall rate, a number of problems arise from both the characteristics of the radar equipment and the radar site, and from the characteristics of the precipitation observed by the radar beam. These problems have been discussed extensively in the literature (see for example, Collier, 1996b).

Joss and Waldvogel (1990) concluded that errors in the radar measurement of surface precipitation are dominated by the effect of variations in the vertical profile of reflectivity (Fig. 6.1). Such variations occur on the scale of individual pixels (Kitchen and Jackson, 1993), and therefore it would seem that any correction method would have the greatest potential benefit by providing adjustment on a pixel-by-pixel basis. Kitchen et al. (1994) discuss three approaches to adjusting radar estimates of precipitation for both range and bright-band effects as follows.

Raingauge Adjustment. In attempting to compensate for the difficulties mentioned above, a range of procedures have been developed for adjusting radar measurements of precipitation using data from raingauges. These procedures have met with limited success, particularly for frontal rainfall (see for example Collier et al., 1983, Collier, 1986a,b). However, all gauge adjustment schemes suffer from the problem of random and bias error introduced by representativeness errors in the comparisons. Another weakness is that economic gauge networks cannot properly resolve the errors associated with a bright band close to the radar, or the spatial detail of the orographic enhancement. An advantage is that by relating the radar measurements to measurements of surface precipitation an attempt is made to deal with all sources of radar errors in a single process, including those due to the beam height above the ground, deviation of the Z-R relationship from that assumed, and imperfect radar calibration.

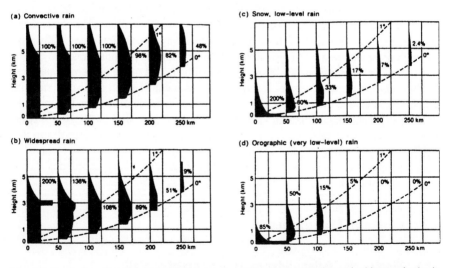

Fig. 6.1. Vertical profiles seen by the radar at various ranges in convective and widespread rain, in low-level rain or snow, and in orographic rain. The number in each figure gives the percentage (referred to the true melted water value which we would measure at ground level) in rain rated educed from the maximum reflectivity of the profile. A radar with a 1° beamwidth is assumed, in a flat country, which λ means that obstacles and radar horizon are of the same height as the radar itself (of the order of 100m). Of course, putting the radar on a high tower or on a mountain would change the situation (from Joss and Waldvogel. 1990, modified by Browning and Collier, 1989)

Recent use of radar-gauge comparisons has attempted to match the probability density functions of both radar estimates and raingauge measurements. The assumption made is that made by Calheiros and Zawadzki (1987) namely that the unconditional probability for any given rainfall intensity is constant everywhere in the radar field, and that this probability was sampled in a representative way by raingauges. The need for this assumption is removed by Rosenfeld et al. (1993) who assumed that the conditional probability density function of rainfall intensity, regardless of the integrated rain depth for any given rainfall regime, is the same everywhere. The Z_e:R relationship so derived varies greatly with the rainfall regime and with the radial distance from the radar

This work was developed further by Rosenfeld et al. (1994) and Rosenfeld et al. (1995a, b), and has become known as the Window Probability Matching Method (WPMM). Rainfall types are classified using the horizontal radial reflectivity gradients, the cloud depth as scaled by the effective efficiency, the bright band fraction within the radar field window and the height of the freezing level. The results of tests in Australia and Israel are very impressive, and comparisons of these results with radar data for convective cases in North West England (Fig. 6.2) suggest a more general applicability. It should be noted that the WPMM does not require that range effects be removed or that the bright-band be corrected in the radar imagery, but

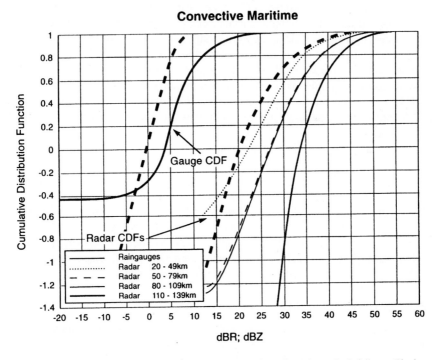

Fig. 6.2. General probability-matched relations between radar reflectivity and rainfall rate. The heavy dashed lines are for the Harrop Edge radar (NW England) data, and the other lines are as published by Rosenfeld et al. (1993)

rather that its presence and the fraction of enhanced echoes close to the 0°C or bright-band altitude be known.

Analytic methods based on using radar data alone. Procedures described by Harrold and Kitchingham (1975), Koistinen (1991), Gray (1991), Andrieu and Creutin (1995) and Andrieu et al. (1995), derive an average reflectivity profile by analysing data from several radar beam elevations at ranges up to a few tens of kilometres from the radar. The average profile is then used to correct data from longer ranges. Assumptions of spatial homogeneity are necessary, both in the derivation of the profile and in its application to data at a longer ranges.

Smith (1986) devised an analytic technique for reducing errors due to the bright band which proved very promising in tests. In routine operation, however, it was found to be susceptible to large errors when significant variations in the freezing level occurred over the area covered by a radar. Such variations are common in frontal precipitation in the UK (Colour Plate 6.C)

Physically-based methods using independent meteorological data. Austin (1987) (see also Dalezois and Kouwen, 1990, Fabry et al., 1992) recommended physically-based techniques. One such has recently been described by Kitchen et al. (1994). An idealised reflectivity factor profile was constructed from analysis of radar data. The heights of significant turning points in the profile are diagnosed from relevant meteorological data (for example surface temperatures) at each radar pixel. The parameterised profile is weighted by the radar-beam power profile and the surface precipitation rate found by an iterative method in real-time.

Since this technique can exploit a wide range of meteorological information from conventional observations and forecast models to derive the reflectivity factor profile Kitchen et al. (1994) felt that the technique is better than those using analyses of radar data along. This is undoubtedly true, although the strong dependence on other meteorological observations and model data makes the procedure vulnerable to the availability and accuracy of these data.

A somewhat different approach has been described by Hardaker et al. (1995). A microphysical model is used to calculate the reflectivity profile using as input lapse rate information from either radio-sondes or a mesoscale numerical model. Simple relationships between bright-band intensity and surface rainfall rate were derived which may be useful in the absence, in real-time, of extensive computing facilities or independent meteorological data. This approach enables improvements in the stylised bright band profile to be made, and tests using numerical model temperature profiles as input to a correction procedure showed encouraging results.

6.3.2 Measurement of snowfall and hail

Snowflakes make deviate significantly from the spherical shape assumed in the estimation of Eq. 6.1. It may be shown (Smith, 1984) that

$$Z_e = \frac{|K|_i^2}{|K|_w^2} \cdot Z \tag{6.3}$$

where $|K|_i^2$ and $|K|_w^2$ are the dielectric constants for ice and water respectively. There are two possible 'correct' values for $|K|_i^2$ depending upon how the particle sizes are determined. If the particle sizes used are melted drop diameters then $|K|_i^2$ is 0.208 and,

$$Z_e = 0.224\ Z \tag{6.4}$$

However, if the particle sizes are expressed as equivalent ice sphere diameters then $|K|_i^2$ is 0.176 and,

$$Z_e = 0.189Z \tag{6.5}$$

Normally radars use the "water equivalent" Z_e defined with $|K|_w^2 = 0.93$, and the dielectric factor is not changed when the precipitation form changes from liquid to solid. Table 6.3 compares equivalent radar reflectivity factors calculated for precipitation rates of 1 and 10 mmh^{-1} for rain, using the Marshall-Palmer relationships (Table 6.2) and for snow, using the Sekhan and Srivastava (1970) relationship (Table 6.2). At R = 1 mm h^{-1}, the value for snow is 3 dB higher than that for rain. Hence in general radar echoes from snow are not weaker than those from rain, although there is a tendency for the precipitation rates to be generally lower in snow than in rain.

Much work has been done on the accuracy of radar measurements of rainfall, but only a limited amount of data on the accuracy of radar measurements of snowfall has been obtained. In general, provided careful quality control and adjustment are exercised, radar measurements of snowfall can be as accurate as those for rainfall within about 50km of the radar site (see for example Collier and Larke, 1978, Boucher, 1981, Browning, 1983).

It should be noted that measurements of the size of, and energy associated with, hail are also possible using weather radar. Whilst quantitative measurements of hail are generally not important to hydrologists, hail can contaminate measurements of heavy

Table 6.3. Example values of R and Z_e for rain and snow (from Smith, 1984)

	Precipitation rate R (mm h^{-1})	
	1	10
Z_e (rain) - dBz	23	39
Z_e (snow) - dBz	26	48

convective rainfall important in flash flood prediction. Therefore techniques for recognising the presence of hail must be implemented in hydrological measurement systems. These techniques range from simple (and unreliable) threshold approaches to procedures involving the height of radar echoes (Waldvogel et al., 1979) and blending radar with satellite information (Hardaker and Auer, 1994). Multiparameter radar also offers reliable identification.

6.3.3 Multi-parameter radar

The departure of the shapes of precipitation particles from spherical gives rise to different radar reflectivity properties. Seliga and Bringi (1976) related signals in two orthogonal linear polarisation planes, horizontal (H) and vertical (V), to two-parametric dropsize distributions. Likewise, circular polarisation has also been used in this way by McCormich and Hendry (1972).

The obateness of raindrops, when falling at terminal velocity in air, increases with drop volume. Since models for the shape and minor-to-major axis ratio and fall speed data exist, it is possible to relate the dropsize distributions so measured to rainfall rates.

With a dual polarisation radar, both quantities Z_H and Z_V, can be measured and be used to calculate the so called differential reflectivity Z_{DR}:

$$Z_{DR} = \frac{\overline{Z_H}}{\overline{Z_V}} = \frac{\int_{D=0}^{D_{max}} \sigma_H(D) \cdot \exp(-3.67 D/D_0) \cdot dD}{\int_{D=0}^{D_{max}} \sigma_V(D) \cdot \exp(-3.67 D/D_0) \cdot dD} \tag{6.6}$$

Equation 6.6 shows that the Z_{DR} radar technique has the potential for accurately measuring rainfall rate without any need for raingauge adjustment. However, as Jameson et al. (1981) points out, single-point measurements of Z_{DR} may be associated with significantly diverse rainfall rates. The differential reflectivity technique, like other radar techniques, is adversely affected by the presence of reflectivity gradients below the radar beam, which may be significant in cases of isolated thunderstorms or orographic rainfall. In other words, even if the radar measures the rainfall rate accurately aloft within the beam, this measurement may still be unrepresentative of the rainfall rate at the surface.

Whilst differential reflectivity may not offer improved rainfall estimation, multi parameter radars do enable a range of other parameters to be derived (McCormich and Hendry, 1975). These parameters have been used to recognise precipitation type (Hendry and Antar, 1984). In addition, Holt (1988) showed that the difference phase parameter may be estimated from non-switched circularly polarised systems. Direct improvements in precipitation estimation may result from the application of these parameters. However, since C-band radars are implemented widely, particularly in Europe, which suffer for severe attenuation by heavy rainfall, the greatest operational impact of multiparameter technology may be in the area of attenuation recognition.

6.3.4 Satellite cloud indexing and life history methods of rainfall estimation

Cloud indexing was the first technique developed to estimate precipitation from space. It is based on the assumption that the probability of rainfall over a given area is related to the amount and type of cloudiness present over this area. Hence, one may postulate that precipitation can be characterised by the structure of the upper surface of the associated cloudiness.

A cloud structure analysis, either subjectively or objectively performed, is used as the basis of the definition of a criterion relating cloudiness to a co-efficient (or index) of precipitation. The characteristic may be, for instance, the number of image pixels above a given threshold level. Hence, the general approach for cloud indexing methods involving infra-red observations is to derive a relationship between a Precipitation Index (PI) and a function of the cloud surface area, S(TBB), associated with the brightness temperature (TBB) colder than a given threshold T_O. This relationship can be generally expressed as follows:

$$PI = A_0 + \sum_i A_i \cdot S_i[TBB_i] \tag{6.7}$$

for $TBB_i < T_O$. A_O and A_i are empirically determined constants. Examples of this approach have been reported by, for example, Barrett et al. (1986).

If desired, an additional term related to the visible image can be included on the right hand side of Equation 6.7. The next step is to associate PI to a physical quantity related in some way to rain. This is done by adjusting the coefficients A and the threshold level T_O by comparison with independent observations such as raingauge or radar data (see for example Arkin, 1979).

Richards and Arkin (1981) have shown that all cumuliform clouds with tops colder than a given threshold temperature T precipitate at a fixed rate G mm h^{-1}. It was found that T = 235°K and G = 3.3mm h^{-1} are typical of the eastern equatorial Atlantic. Arkin and Meisner (1987) developed this method into the GOES Precipitation Index (GPI).

One of the problems inherent to this technique is the bias created by the potential presence of high level non-precipitating clouds such as cirrus. Another limitation resides in the fact that the satellite measurement represents an instantaneous observation integrated over space, whilst raingauge observations are integrated over time at a given site.

Life-history methods, as indicated by their name, are based on the observations of a series of consecutive images obtained from a geostationary satellite. It has been observed that the amount of precipitation associated with a given cloud is also related to its stage of development, therefore two clouds presenting the same aspect (from the VIS-IR images point of view) may produce different quantities of rain depending on whether they are growing or decaying.

As with the cloud indexing technique, a relationship is derived between a Precipitation Index (PI) and a function of the cloud surface area, S(TBB), associated with a given brightness temperature (TBB) lying above a given threshold level, T_O. In addition the cloud evolution is taken into account and expressed in terms of rate of change of S(TBB) between two consecutive observations (see for example, Griffith et al. (1976), Stout et al. (1979), Scofield (1984)).

An equation as complex as desired, may be derived between PI and functions of S(TBB) and its derivative with respect to time:

$$PI = A_0 + A \cdot S(TBB) + A' \frac{d}{dt} S(TBB); \qquad (6.8)$$

for $TBB < T_O$

Here also, another step is necessary in order to relate the Precipitation Index defined by the equation to a physical quantity related to rain.

These approaches have now been underpinned by a more logical mathematical approach. Doneaud et al. (1981, 1984) derived the volumetric rainfall, V, from

$$V = \int_T \int_A R \, da \, dt = R_c \int_T \int_A da \, dt = R_c \sum_i A_i \Delta t_i \qquad (6.9)$$

where   R is the instantaneous local rain rate
da and dt are incremental elements of area and time respectively
R_C is the average rain rate.

The integrals are taken over the entire area A for duration T. The double integral is the area-time integral (ATI).

Following this work Chiu and Kedem (1990) developed a logistic regression model to estimate the fractional rainy area, which estimates the conditional probability that rain rate over an area exceeds a fixed threshold given the values of related covariates. Tests showed that this approach is superior to multiple regression. However, variabilities of meteorological parameters must be accounted for if this technique is to be applied to estimating rain rate from space.

Atlas et al. (1960) developed a unified theory for the estimation of both the total rainfall from an individual convective storm over its lifetime, and the area wide instantaneous rain rate from a multiplicity of such storms, by use of measurements of the areal coverage of the storms within a threshold rain intensity isopleth or the equivalent threshold radar reflectivity. Equation 6.6 was generalised to,

$$V = [\overline{A}(t) \cdot T] \cdot S(t) \qquad (6.10)$$

where the   ATI   = $A(\tau).T$
τ =   threshold
$S(\tau)$ = $R_c(\tau)/\phi$ and may be defined in terms of the probability density function (pdf)
$$\int_0^\infty RP(R) \, dR \bigg/ \int_0^\infty P(R) \, dR$$
ϕ =   the fraction of the total volumetric rain rate

Divide Eq. 6.10 by total area observed A_O then V/A_O is the average area wide rain rate $<R>$ and $A(\tau)/A_O$ is the fractional area, $F(\tau)$, covered by rain within the threshold, τ.

$$<R> = F(\tau) \, R_c(\tau)/\phi \qquad (6.11)$$

Errors over the whole observing area are 5-10%. The method is called the 'Height-Area Rainfall Threshold" or HART method. Morrisey (1994) examined the effect of data resolution on this and the ATI techniques. The results indicated that significant biases and random errors can arise when using data having spatial resolutions different from those used to calibrate the method.

6.3.5 Bispectral techniques

Radar systems can make measurements of precipitation over wide areas from a single location in real time, and are easier and more cost effective to operate than extensive raingauge networks. It is therefore attractive to design procedures to combine the radar data with satellite data in order to better understand the relationships between variations in visible and infrared or microwave brightness temperatures and rainfall.

The bispectral technique exploits information on the height of clouds derived from infrared measurements, and information on the depth of clouds derived from visible measurements. The visible and infrared data are each divided into a number of classes, and a table of 'probabilities' of rain for each class produced by comparison with radar data (Lovejoy and Austin, 1979). Examples of relationships for frontal rain are shown in Fig. 6.3. These are then applied to the satellite data beyond the area of radar coverage.

The precipitation formation process require the existence of large cloud droplets and/or ice particles in the cloud which often spread to the cloud top. These large

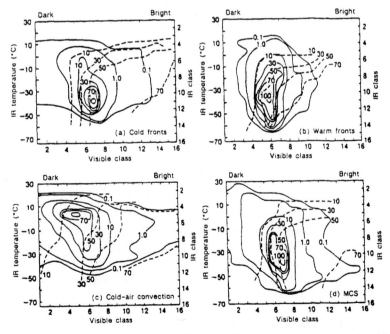

Fig. 6.3. The probability of precipitation as a function of IR temperature and visible class (dashed line). The percentage of precipitation pixels, relative to the total number of precipitation pixels, normalised to a maximum value of 100 (solid lines) from (Cheng et al., 1993)

particles absorb the 1.6μm and 3.7μm radiation much stronger than small cloud droplets. This effect makes it possible to calculate the effective radius (r_{eff} = integral volume divided by integral surface area) of particles (Arking and Childs, 1985). Rosenfeld and Gutman (1994) have shown that r_{eff} = 14μm can serve to delineate precipitating clouds regardless of the temperature of their tops.

A useful skill in estimating rain areas has been identified, although experience indicates that the probability of rain table should be changed seasonally. The radar verification shows that the main skill lies in identifying three classes, namely clear, cloudy with low probability of rain, and cloudy with a significant probability of rain. The accuracy of the results diminishes with increasing distance from the training radar. However, it was pointed out that use of a coarse resolution might allow the introduction of texture matrices. Rainfall over periods from ½ to 2 hours may be estimated with an accuracy of around 49% over areas of 10^5 km². However, more extensive assessments for different rainfall types are required to confirm this performance.

6.3.6 Passive microwave estimates of rainfall from space

Microwaves provide the measurements that are physically best related to the actual precipitation, especially in the longest wavebands. The interactions of passive MW with precipitation clouds and the surface are discussed by Rosenfeld and Collier (1998) using two wavebands, shorter (85 Ghz) and longer (19 Ghz).

(a)   Absorption-based measurements

>   Water drops have relatively large absorption/emission coefficient, increasing for the higher frequencies. The emission is proportional to the vertically integrated cloud and rain water in the low frequencies, but due to the increased emissivity for the higher frequencies the emission saturates for light rain intensities.

(b)   Scattering-based measurements

>   Ice particles have relatively small absorption/emission, but they are good scatterers of the MW radiation, especially at the higher frequencies. Therefore, at high frequencies (85 Ghz) the large scattering from the ice in the upper portions of the clouds makes the ice an effective insulator, because it reflects back down most of the radiation emitted from the surface and from the rain. The remaining radiation that reaches the MW sensor is interpreted as a colder brightness temperature. A major source of uncertainty for the scattering-based retrievals is the lack of a consistent relationship between the frozen hydrometers aloft and the rainfall reaching the surface.

Simple comparisons of the ESMR-5 (Electrically Scanning Microwave Radiometer) imagery operating at 19.33 Ghz (1.55 cm wavelength) with imagery from visible and infrared wavelength radiometers, ground-based radars and conventional meterological observations established, over water, an association of areas of relatively warm brightness temperature with areas of rainfall. Curves of 19.35 Ghz brightness temperatures and rainfall rate were published by Wilheit et al. (1977) using a model based upon the concept of many optically thin layers bounded on top by a variable freezing level. Improvements to the technique, including the use of both vertical and horizontal polarisation for measurements over land, have been proposed by Wilheit (1975).

Over land the passive MW algorithms can detect rain mainly by the ice scattering mechanism (b) above. This indirect rain estimation method is less accurate. Moreover, rainfall over land from clouds which do not contain significant amounts of ice aloft goes mostly undetected. A passive radiation method that is able to detect such rain is the 'effective radius' method of Rosenfeld and Gutman (1994).

Spencer (1986) proposed a technique based upon analysis of radiometric data gathered from the NIMBUS-7 SMMR (Scanning Multichannel Microwave Radiometer) at a frequency of 37 Ghz (0.81 cm wavelength). This radiometer system provides dual-polarized data at five different wavelengths (0.81, 1.42, 1.66, 2.80 and 4.54 cm) and can be used to detect a wide range of snow and ice features as well as rainfall. Spencer (1986) suggests that 37 Ghz can provide the best compromise for measuring rainfall over the ocean between an easily detectable precipitation signal (such as at 19.35 Ghz) and cloud penetration ability. However, it was noted that the scattering signal at 37 Ghz is not as strong as the emission signal of light precipitation at 19 Ghz. Hence the 37 Ghz technique may be better for observing heavy convective precipitation, and observations at 19 Ghz may relate better to measurements of light precipitation. Recently SMMR data have been used to derive rainfall fields over the Indian Ocean (Martin et al., 1993).

Much work is now being carried out to investigate the ability to measure precipitation of the Special Sensor Microwave Imager (SSM/I) 88.5 Ghz channels of the US Defense Meterological Satellite Programme (DMSP) satellite at present in polar orbit. The scattering-based technique is applicable at this frequency, and early results reported by Spencer et al. (1989) are very encouraging albeit for convective rainfall. It remains to be seen to what extent mid latitude frontal rainfall can be measured unambiguously. Recently, Ferraro et al. (1994) described how the identification of surfaces having signatures similar to that of rain can improve SSM/I estimation techniques. Figure 6.4 summarizes the relationship between rainfall rate and various passive microwave frequencies.

Measurements of precipitation have also been made using passive microwave imagery near 118 Ghz (Spencer et al. 1994, Schwartz et al., 1996). At these frequencies scattering from large, dense graupel in the tops of convective cells produces large decreases in radiances. However, liquid hydrometeors produce much smaller changes and overlying cloud may obscure precipitation at low levels. Finally, Kummerow and Giglio (1995) have described a technique for combining microwave and infrared observations, and Alder et al. (1991) have simulated microwave satellite observations using a three-dimensional cloud model.

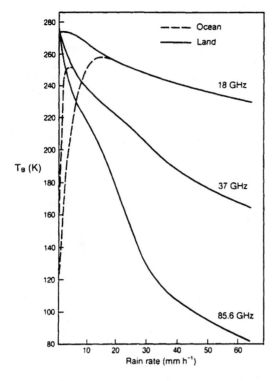

Fig. 6.4. Relationships between brightness temperature and rain rate at 18, 37 and 85.6 Ghz from the radiative transfer modelling of Wu and Weinman (1984). The vertical distribution of hydrometeors was based upon averaged radar results and assumed ice precipitation above and liquid precipitation below the freezing level. (From Spencer et al. 1989)

6.3.7 Sampling errors

Fabry et al. (1994) investigated the dependence of the accuracy of short-period radar rainfall acculuations on periodic sampling of the rain field. They noted that errors due to sampling can be greater than all the other errors if accumulations are improperly computed. If was found that the best accumulations are obtained with very high time resolution data.

In using satellite data to estimate precipitation sampling errors become even more important. In tropical regions there can be significant diurnal cycle in rainfall activity, and the phase and intensity of the cycle may vary from region to region (Albright et al., 1985). The low inclination orbit to be used for TRMM will precess in such a way as to sample a full diurnal cycle of equator crossing times over the course of a month. Unfortunately this is not the case for satellites in polar orbit, and using data from the satellites can result in large sampling errors.

The representation heterogeneity within global numerical weather prediction and climate models is important, and therefore sampling errors must be considered in both the disaggregation of model rainfall and in the aggregation of surface fluxes measured by satellites. Unfortunately global models use grids of characteristic length scale 50 km or larger and often fluxes from individual subgrid hydrological models which are coupled to large scale meteorological models are arithmetically averaged over the larger grid squares (see for example Mölders and Raabe, 1997). Whilst this is undoubtedly detrimental to both model predictions of convective rainfall and the resultant runoff, there is little else that can be done. Conversely it is necessary to ensure that model-derived rainfall is disaggregated over the hydrological model grid, as suggested by, for example, Eagleson (1984), in a physically meaningful way, so that subsequent hydrological model predictions of soil moisture and evaporation are realistic when inputted to the meteorological model. Recently Collier (1993) has used radar data to develop an improved method of disaggregation, but further work needs to be undertaken to develop techniques which match the characteristics of observing systems to numerical model grids.

6.4 The potential for improvement

6.4.1 Current performance levels

All satellite techniques suffer to a greater or lesser degree from errors arising from sampling. Indeed, these errors can be greater than all the other errors if accumulations are improperly computed. In tropical regions there can be a significant diurnal cycle in rainfall activity, and the phase and intensity of the cycle may increase the errors due to sampling. Nevertheless, since rain is a key component of the global hydrological cycle, the only way to obtain better measurements on a global scale is from the use of space-borne measurement techniques.

Satellite techniques are capable of estimating rainfall over areas in excess of 10^4 km^2 over periods of hours with an accuracy of around 50%. For much longer integration times of order a month, this accuracy may be improved to around 10 - 20%. However, for much smaller areas errors are much larger, and indeed the measurements are of little use for hydrological flow forecasting for catchment of size around a few hundred square kilometres.

Ground-based radar offers higher accuracy than satellite techniques over small areas ($\sim 10^2$km^2) and small time periods (minutes to an hour or so). Indeed, the most accurate accumulations are obtained with data collected every minute. Colour Plate 6.D shows a sequence of radar images at ten minute intervals. Note the highly variable nature of the rainfall generally, but the organization of a small bond in one location. Clearly then for most urban and many rural catchments radar offers the only realistic approach to real-time measurement of rainfall and snowfall (see also Chap. 16).

Unfortunately radar data require significant real-time quality control if this level of performance is to be consistently achieved. This has often lead to disappointment when implementing operational systems. It is imperative to understand the limitations of radar, and put in place the necessary processing system. Joss and Lee (1995) describe a procedure of this type, but other approaches as outlined in Section 6.3.1 may be adequate. Much will depend upon the operating environment, and the use to which the data are put.

There continues to be a need to test both radar and satellite measurement techniques. Routine assessment is necessary, both for these radar procedures which use independent meteorological and mesoscale model data such as the UK Nimrod System (Harrison et al., 1995), and for those such as the WPMM which do not use such data. Whatever techniques are used, it is generally recognised that radar data should be processed to remove spurious echoes such as ground clutter.

For satellite techniques, as part of the GPCP three (so far) Algorithm Intercomparison Projects (AIPs) have been organised over Japan (1989), over North West Europe (1991) and over the tropical Pacific Ocean as part of TOGA-COARE (1993). These AIPs should help us to understand a wide range of algorithms and new satellite instruments. So far they have tended to demonstrate that no one technique outperforms another on all occasions.

6.4.2 The future

Undoubtedly work will continue to improve individual measurement algorithms. It is necessary to ascertain whether the use of extensive independent meteorological observations and mesoscale model data are essential to achieve reliable radar estimates of rainfall, or whether the WPMM or some variant is acceptable.

Passive microwave data, particular at frequencies around 85.5 Ghz, seem to offer improved rainfall estimates from space. Sampling problems will remain, and one wonders whether it is sensible to keep refining individual algorithms. Perhaps the way forward is within the context of the data assimilation procedure of a numerical weather predication model. Numerical models are already capable of reproducing good rainfall distributions in situations where the rainfall is organised by orography. Models cannot however, as yet, reproduce convective rainfall, but the assimilation of satellite and radar data will undoubtedly lead to improvements in the quality of the model output.

There are dangers, of course, in seeking to improve numerical model rainfall output in this way. It is all to easy to regard a numerical model rainfall field as "observations". As work on data assimilation proceeds, there is an even greater need for independent observational datasets, particularly of precipitation which is so variable in space and time. It is heartening to note that international projects such as BALTEX strive to create the most accurate precipitation fields that can be achieved.

In the future there will be developments in both weather radar and satellite instrumentation. Work is already underway to exploit polarisation diversity techniques for operational ground-based radars, which promises to provide some improvement in precipitation measurement particularly for high rainfall rates. Undoubtedly the experience gained in TRMM will lead to improved spaceborne active radars, and it is anticipated that it will not be too long before such a system flies in a polar orbit providing global coverage.

References

Adler, R.F., Yeh, H.-Y.M., Prasad, N., Tao, W.-K and Simpson, J. (1991). Microwave simulations of a tropical rainfall system with a three-dimensional cloud model. *J. App. Met.*, **30**, 924-953

Andrieu, H. and Creutin, J.D. (1995). Identification of vertical profiles of radar reflectivity for hydrological applications using an inverse method. Part I: Formulations. *J. App. Met*, **34**, 225-239

Andrieu, H., Delrieu, G. and Creutin, J.D. (1995). Identification of vertical profiles of radar reflectivity for hydrological applications using an inverse method. Part II: Sensitivity analysis and case study. *J. App. Met*, **34**, 240-259

Arkin, P.A. (1979). The relationship between fractional coverage of high cloud and rainfall accumulations during GATE over the B-scale array. *Mon. Wea. Rev.*, **107**, 1382-1387

Arkin, P.A. and Meisner, B.N. (1987). The relationship between large-scale convective rainfall and cold cloud over the western hemisphere during 1982-84. *Mon. Wea. Rev.* **115**, 51-74

Arking, A. and Childs, J.D. (1985). Retrieval of cloud cover parameters from multispectral satellite images. *J. Climate App. Met.* **24**, 322-333

Atlas, D., Rosenfeld, D. and Short, D.A. (1990). The estimation of convective rainfall by area integrals, Part I: The theoretical and empirical basis. *J. Geophys. Res.*, **95**, No 03, 2153-2160.

Austin, P.M. (1987). Relation between measured radar reflectivity and surface rainfall. *Mon. Wea. Rev.* **115**, No. 5, 1053-1070

Barrett, E.C., D'Sousa, G. and Power, C.H. (1986). Bristol techniques for the use of satellite data in rain cloud and rainfall monitoring. *J. Br. Interplanet Soc.*, **39**, 517-526

Batton, L.J. (1973). Radar Observations of the Atmosphere, University of Chicago Press, Chicago. 324 pp.

Blanchard, D.C. (1953). Raindrop size distribution in Hawaiian rains. *J. Met*, **10**, 457-473

Boucher, R.J. (1981). Snowfall rates obtained from radar reflectivity within a 50km range, Tech. Rep. No. AFGL-TR-81-0265 (met. Div. Project 6670), U.S. Air Force Geophys,.Lab., 25 pp.

Browning, K.A. (1983). Air motion and precipitation growth in a major snowstorm. *Quart J.R. Met. Soc.*, **109**, 225-242

Browning, K.A. and Collier, C.G. (1989). Nowcasting of precipitating systems. *Rev. of Geophysics.*, **27**, No. 3, 345-370

Calheiros, R.V. and Zawadzki, I.I. 91987). Reflectivity-rain rate relationships for radar hydrology in Brazil. *J. App. Met.* **26**, 118-132

Cheng, M., Brown, R. and Collier, C.G. (1993). Delineation of rain areas using Meteosat IR-VIS data in the region of the United Kingdom. *J. App. Met*, **32**, 884-898

Chiu, L.S. and Kedem, B. (1990). Estimating the exceedance probability of rain rate by logistic regression. *J. Geophys. Res.*, **95**, D3, 2217-2227

Collier, C.G. (1986a) Accuracy of rainfall estimates by radar. Part I: Calibration by telemetering raingauges. *J. Hydrology*, **83**, 207-223

Collier, C.G. (1986b) Accuracy of rainfall estimates by radar. Part II: comparison with raingauge network. *J. Hydrology*, **83**, 225-235

Collier, C.G. (1993). The application of a continental-scale radar database to hydrological process parametrization within Atmopsheric General Circulation Models. *J. Hydrology*, **142**, 301-318

Collier, C.G. (1996a) Weather radar precipitation data and their use in hydrological modelling. Chapter 8 in Distributed Hydrological Modelling, editor J.C. Refsgaard, publ. Kluwer, 143-163

Collier, C.G. (1996b) Applications of Weather Radar Systems. A guide to uses of radar data in meteorology and hydrology, 2nd Edition, Praxis Publishing Ltd, Chichester and John Wiley, London, 390pp.

Collier, C.G. and Larke, P.R. (1978). A case study of the measurement of snowfall by radar: an assessment of accuracy. *Quart J.R. Met. Soc.*, **104**, 615-621

Collier, C.G., Larke, P.R. and May, B.R. (1983). A weather radar correction procedure for real-time estimation of surface rainfall. *Quart J.R. Met. Soc.* **104**, 509-608

Dalezois, N.R. and Kouwen, N. (1990). Radar signal interpretation in warm season rainstorms. *Nordic Hyd*, **21**, 47-64

Doneaud, A.A., Niscov, S.I., Priegrutz, D.L. and Smith, P.L. (1984). The area-time integral as an indicator for convective rain volumes. *J. Clim. App. Met*, **23**, 555-561

Eagleson, P.S. (1984). The distribution of catchement coverage by stationary rainstorms. *Water Resour. Res.*, **20**, 581-590

Fabry, F., Austin, G.L. and Tees, D. (1992). The accuracy of rainfall estimates by radar as a function of range. *Quart J.R. Met. Soc.* **118**, 435-453

Fernaro, R.R., Gody, N.C. and Marks, G.G.F. (1994). Effects of surface conditions on rain identification using the DMSP-SSM/I. *Remote Sensing Rev.* **11**, 195-209

Gray, W. (1991). Vertical profile corrections based on EOF analysis of operational data. Preprints 25th Int. Conf. on Radar Met., 24-28 June, Paris, ABS, Boston, 821-823

Griffith, C.G., Woodley, W.L., Grube, P.G., Martin, D.W., Stout, J. and Sidkar, D.N. (1976). Rain estimation from geosynchronous satellite imagery – visible and infrared studies. *Mon. Wea. Rev.* **106**, 1153-1171

Gunn, K. and Marshall, J. (1958). The distribution with size of aggregate snowflakes. *J. Met.* **15**, 452-461

Hardaker, P.J. and Auer, A.H. (1994) The separation of rain and hail using single polarisation radar echoes and IR cloud-top temperatures. *Met. Apps*, **1**, 201-204

Hardaker, P.J., Holt, A.R. and Collier, C.G. (1995). A melting layer model and its use in correcting for the bright-band in single polarisation radar echoes. *Quart. J.R. Met. Soc.*, **21**, 495-525

Harrold, T.W. and Kitchingman, P.G. (1975). Measurement of surface rainfall using radar when the beam intersects the melting layer. Preprints, 16th Radar Met. Conf. Houston, AMS, Boston, 473-478

Hendry, A. and Antar, Y.M.M. (1984) Precipitation particle identification with centimetre wavelength and dual-polarisation radars. *Radio Science*, **19**, No. 115-122

Holt, A.R. (1988). Extraction of differential propagation phase shift from data from S-band circularly polarised radars. *Electronics Letters*, **24**, 1241-1242

Jameson, A.R., Beard, K.V. and Bresch, J. (1981). Complications in deducing rain parameters from polarisation measurements. Preprints, 20th Conf. on Radar Met., 30 Nov. - 3 Dec., Boston, Mass, AMS, Boston, 586-589

Jones, D.M.A. (1956). Raindrop size distribution in Hawaiian rains. *J. Met*, **10**, 457-473

Joss, J. and Waldvogel, A. (1990). Precipitation measurement and hydrology: a review. in Battan Memorial and Radar Conference, Radar in Meteorology, editor D. Atlas, publ. AMS., Boston, chapter 29a, 577-606

Joss, J. and Lee, R. (1995). The application of radar-gauge comparisons to operational precipitation profile corrections. *J. App. Met*, **34**, 2612-2330

Joss, J., Schram, K., Thams, J.C. and Waldvogel, A. (1970). On the quantitative determination of precipitation by radar. Wissenschaftliche Mitteilung Nr 63, Zürich, Eidgenössische Kommission zum Studium der Hagelbildung und der Hagelabwehr

Kitchen, M. and Jackson, P.M. (1993). Weather radar performance at long range simulated and observed. J. App. Met, **32**, 975-985

Kitchen, M., Brown, R. and Davies, A.G. (1994). Real-time correction of weather radar data for the effects of bright-band, range and orographic growth in widespread precipitation. *Quart. J.R. Met. Soc.,* **120**, 1231-1254

Koistinen, J. (1991). Operational correction of radar rainfall errors due to the vertical reflectivity profile. Preprints, 25th Int. Conf. on Radar Met., 24-28 June, Paris, AMS, Boston, 91-94

Lovejoy, S. and Austin, G.L. (1979). The delineation of rain areas from visible and IR satellite data for GATE and mid-latitudes. *Atmos Ocean*, **17**, 77-92

Marshall, J.S. and (Mc)Palmer, W. (1948). The distribution of raindrops with size. *J. Met.*, **5**, 165-166

Martin, D.W., B.B. Hinton and B.A. Auvine. (1993). Three years of rainfall over the Indian Ocean. *Bull. Am. Met. Soc.*, **74**, No.4, 581-590

Mason, B.J. (1970). Future developments in meteorology: an outlook to the year 2000. *Quart. J.R. Met. Soc.*, **96**, No 409, 349-368

McCormich, G.C. and Hendry, A. (1972). Results of precipitation back-scatter measurements at 1.8cm with a polarisation diversity radar. Preprints 15th Radar Met. Conf., AMS, Boston, 35-38

McCormich, G.C. and Hendry, A. (1975) Principles for the radar determination of the polarisation properties of precipitation. *Radio Science*, **10**, 421-434

Meneghini, R., Echerman, J. and Atlas, D. (1983) Determination of rain rate from space borne radar using measurements of total attenuation. *IEEE Trans. Geoscience Remote Sensing*, GE-21, 34-43

Mölders, N. and Raabe, A. (1997) Testing the effect of a two-way-coupling of a meteorological and a hydrologic model on the predicted local weather. Atmos. Res., **45**, 81-107

Morrissey, M.L. (1994). The effect of data resolution on the area threshold method. *J. App. Met.*, **33**, 1263-1270

Richards, F. and Arkin, P.A. (1981). On the relationships between satellite observed cloud cover and preciptiation. *Mon. Wea. Rev.*, **109**, 1081-1093

Rosenfeld, D. and Gutman, G. (1994). Retrieving microphysical properties near the tops of potential rainclouds by multispectral analysis of AVHRR data. *J. Atmos. Res.*, **34**, 259-283

Rosenfeld, D. and Collier, C.G. (1998). Estimating surface precipitation. Chapter 4.1 in GEWEX, publ. Cambridge Univ. Press, editor K.A. Browning and R.J. Gurney, 124-133

Rosenfeld, D., Atlas, D., Wolff, D.B. and Amitai, E. (1994). The window probability matching method for rainfall measurements with radar. *J. App. Met*, **33**, 682-693

Rosenfield, D., Wolff, D.B., Atlas, D. (1993). General probability-matched relations between radar reflectivity and rain rate. *J. App. Met*, **32**, 50-72

Rosenfield, D., Amitai, E. and Wolff, D.B. (1995a) Classification of rain regimes by the three-dimensional properties of reflectivity fields. *J. App. Met.*, **34**, 198-211

Rosenfield, D., Amitai, E. and Wolff, D.B. (1995b) Improved accuracy of radar WPMM estimated rainfall upon application of objective classification criteria. *J. App. Met.*, **34**, 212-223

Schwartz, M.J., Barrett, J.W., Fieguth, P.W., Rosenkraz, P.W., Spina, M.S. and Staelin, D.H. (1996). Observations of thermal and precipitation structures in a Tropical Cyclone by means of passive microwave imagery near 118 Ghz. *J. App. Met*, **35**, 671-678

Scofield, R.A. (1984). A satellite-based estimate of heavy precipitation. In: Recent Advances in Civil Space Remote Sensing, SPIE Publication No. 481, 84-92

Seed, A.W. and Austin, G.L. (1990). Variability of summer Florida rainfall and its significance for the estimation of rainfall by gauges, radar and satellite. *J. Geophys, Res.*, **95**, D3, 2207-2215

Sekhon, R.S. and Srivastava, R.C. (1970) Snow size spectra and radar reflectivity. *J. Atm. Sci*, **27**, 229-267

Seliga, T.A. and Bringi, V.N. (1976). Potential use for radar differential reflectivity measurements at orthogonal polarization for measuring precipitation. *J. App. Met*, **15**, 69-95

Smith, C.S. (1986). The reduction of errors caused by bright-bands in quantitative rainfall measurement made using radar. *J. Atm. Ocean. Tech*, **3**, 129-141

Smith, P.L. (1984). Equivalent radar reflectivity factors for snow and ice particles. *J. Climate App. Met.*, **23**, No. 8, 1258-1260

Spencer, R.W. (1986) A satellite passive 37 Ghz scattering based method for measuring oceanic rain rates. *J. Climate App. Met.*, **25**, 754-766

Spencer, R.W., Goodman, H.M. and Hood, R.E. (1989). Precipitation retrieval over land and ocean with the SSM/I: Identification and characteristics of the scattering signal. *J. Atm. Ocean. Tech.*, **6**, 254-273

Spencer, R.W., Hood, R.E., La Fontiane, F.J., Smith, E.A., Platt, R., Galliano, J., Griffin, V.L. and Lobl, E. (1994). High-resolution imaging of rain systems with the advanced microwave precipitation radiometer. *J. Atm. Ocean Tech.*, **11**, 849-857

Stout, J.E., Martin, D.W. and Sikdar, D.N. (1979). Estimating GATE rainfall with gesynchronous satellite images. *Mon. Wea. Rev.* **107**, 585-598

Theon, J.S., Matsuno, T., Sakata, T. and Fuigano, N. (editors). (1992). The global role of tropical rainfall. publ. A. Deepak, 280 pp.

Waldvogel, A., Federer, B. and Grimm, P. (1979). Criteria for the detection of hail cells. *J. App. Met*, **18**, No. 12, 1521-1525

WCRP (1986). World Climate Research Programme. Report of the Workshop on Global Large-scale Precipitation Data Sets for the World Climate Research Programme, WCP-111, WMO/TD-No. 94, World Met. org., Geneva, 50 pp.

Wilheit, T.H. (1975). The electrically scanning microwave radiometer (ESMR) experiment. in Nimbus 6 User Guide, Goddard Space Flight Centre, NASA, **87**, 87-108

Wilheit, T.T., Chang, A.T.C., Rao, M.S.V., Rodgers, E.B. and Theon, J.S. (1977). A satellite technique for quantitatively mapping rainfall rates over the oceans. *J. App. Met.*, **16**, 551-560

Wu, R. and Weinman, J.A. (1984). Microwave radionces from precipitating clouds containing aspherical ice, combined phase, and liquid hydrometeors. *J. Geophys. Res.* **89**, 7170-7178

7 Land-use and Catchment Characteristics

B.G.H. Gorte
International Institute for Aerospace Survey and Earth Sciences (ITC),
Enschede, the Netherlands

7.1 Introduction

Vegetative cover or land cover influences hydrological processes in various ways. Interception and transpiration is a loss or sink term in the water balance of a catchment, and evapotranspiration losses have been shown to influence rainfall in downwind direction at regional scale (Savenije, 1995). The results of a great many paired catchment experiments have shown that evapotranspiration losses decrease in the following order; conifers, deciduous hardwoods/mixed hardwood, shrub (Bisch and Hewlett, 1982). Thus, spectral cover classifications allow the estimation of relative water losses.

A major step forward is the ability to calculate the actual evapotranspiration of each pixel over large areas, as related to the land cover, using multi-spectral and thermal data (Bastiaansen, 1998). See also Chap. 8.

The well known Runoff Curve Numbers method of the U.S. Soil Conservation Service (SCS) uses a hydrologic vegetation condition, combined with a soil description, for daily runoff estimation from rainfall. Soil units can be combined in a GIS with remotely sensed vegetation classes to map the curve numbers.

The rarity of overland flow under dense vegetation in temperate climates, often leading to saturation overland flow and piston flow has been described by many authors e.g. Ward and Robinson (1989). In contrast, poor vegetation on sloping lands usually could lead to rapid direct runoff. In smaller catchments vegetation delays runoff caused by high intensity rainfall bursts, resulting in lower peak runoff rates. The effects in larger catchments is still a matter of debate (Bruynzeel, 1990).

Soil erosion rates are strongly governed by vegetation. Dense vegetation prevents rain-splash, increases infiltration and provides vegetal retardance to overland flow. Hence, land cover classes, as determined by remote sensing, have an implicit hydrological significance in terms of water yield, peak flows and soil erosion. All available catchment data show that deforestation leads to highly increased sediment yields. Upstream deforestation can also lead to increased bedload, which may cause downstream damages in cultivated flood plains (Meijerink and Maathuis, 1997).

Vegetation can also influence water quality, as well as the observations of contaminated throughfall under pines in humid temperate regions with air pollution, which ultimately affects the water balance. As discussed in Chap. 14, nature of the vegetation and unsaturated zone conditions determine whether or not the vegetative cover induces net recharge. Examples are given of groundwater indicators and reference is made to the adaptation of vegetation to groundwater quality that may exist in certain regions. Finally, most wetlands are situated in a hydrological environment where vegetation reflects the variable inundation depths and durations, as influenced by the micro-topography and corresponding soil textures. In coastal regions, water salinity determines the vegetation associations.

7.2 Land cover Mapping with Remote Sensing

The various types of land cover govern much of the spectral reflection of the surface of the earth, which is measured by sensors on various remote platforms, such as multi-spectral and thermal scanners or push-broom arrays and active microwave (radar) imaging systems.

In most parts of the world, the land cover is highly dynamic. Apart from effects of seasonal rainfall, temperature and possible cyclicity of rainfall and droughts, man has influenced the vegetation by converting natural vegetation into agricultural lands where different crop rotations are practiced. Other dynamic aspects are the variable grazing pressure of range lands, forest fires, destruction of flood plain vegetation by floods, and so on. The frequent over-passes of the earth observation satellites allow one to monitor the changes. Since satellite observations are available since the early 1970's, it is possible to relate *e.g* trends in cover densities to stream flow.

The above brief review highlights the major role of vegetation in hydrology, but – implicitly – also points out that the effect of vegetation on hydrological processes has to be understood through field studies in order to apply transfer functions to convert remotely sensed cover data into (relative) hydrologic quantities or to parameter values for model input. The transfer functions are related e.g. to actual evapotranspiration, either through identification of crop types, followed by agro-ecologic models to predict transpiration or by functions relating evapotranspiration to remotely sensed vegetation indices. Some hydrologic models have in-build transfer functions, such as the SIMPLE model of Kouwen et al. (1990), as well as the models that use the curve number methods. Physically based remotely sensed parameters related to cover are perhaps limited to the actual evapotranspiration using surface energy balance methods.

The influence of scattering on the radar signal (images) is still in the research domain. Furthermore, classifications using remotely sensed data should aim at those classes which are meaningful in a hydrological sense. For example, dense vegetation near the surface strongly influences the surface hydrological processes, regardless of the botanical composition of that cover. Hence, spectral cover classes related to dense vegetation could be combined, unless experimental data is available to do otherwise. The coupling to field observations is essential; forest canopies are recorded by remote sensing, not the condition of the undergrowth, which determines much of the processes.

The remainder of this chapter is divided in three parts, dealing with (a) vegetation indices, (b) image classification and (c) the use of radar imagery. The separation between optical imagery in (a) and (b), *vs.* radar imagery in (c) is implicit. It was felt that integration of both kinds of imagery would be artificial in the context of vegetation indices and classification. Hydrologic model parameter assessment is particularly important for water management in changing environments (Chap. 19). Therefore, accurate assessment of parameters, as addressed in the current chapter, gains importance in multi-temporal cases. When comparing results of different dates, observed 'changes' due to errors and incompatibilities should not be mistaken as real changes in the parameters.

Case study area

Various techniques described in this chapter are illustrated using multi-temporal Landsat TM imagery of the Rio Verde do Mato Grasso area, located at the Eastern boundary of the Pantanal region, Mato Grosso do Sul, Brazil.

During the last two decades important transformations in land cover took place mainly in the Planalto, where the native vegetation (shrubs and forest) was mostly replaced by intensive cultivation methods. The systematic deforestation involved denudation of soils and caused rapid erosion, with consequences for the flood regime in the whole Pantanal, where also an increased sedimentation rate was detected (Hernandez Filho et al., 1995). The study area represents a good example where environmental dynamics linked to human activities have a strong impact on water management.

7.3 Vegetation Indices

The relevance of vegetation for hydrology was addressed in the introduction section. Quantified biophysical parameters from remote sensing are associated with irrigation management (Bastiaansen, 1998) (Table 7.1). The relation between remotely sensed measurements and vegetation parameters is captured in *vegetation indices*.

The interest in assessing vegetation growth and conditions from space dates back to 1972, when food crops were considered a strategic commodity, but the U.S. government was unaware of the disastrous crop situation in the Soviet Union (Calder, 1991). Crop yield prediction using satellites obtained political priority and funding for earth observation satellite programs was secured.

Field measurement of many crops revealed that a very specific reflectance characteristic occurs in the red/near-infrared part of the electro-magnetic spectrum. Visible light is mostly absorbed by vegetation. Even green reflectance is rather low, compared to most other materials (soils, rocks) that cover the earth's surface. At the same time, the near-infrared reflection of healthy vegetation is much higher than that of most other land covers. Reflectance curves of vegetation (for example Fig. 7.1) show a very steep ascent between visible red and near-infrared wavelengths (Tucker, 1979).

Table 7.1. Biophysical crop parameters retrievable from remote sensing measurements, and their association with irrigation management, from [Bastiaanssen, 1998]

Crop parameter	Process	Purpose
Fractional vegetation cover	Chlorophyll development, soil and canopy fluxes	Irrigated area
Leaf area index	Biomass, minimum canopy resistance, heat fluxes	Yield, water use, water needs
Photosynthetically active radiation	Photosynthesis	Yield
Surface roughness	Aerodynamic resistance	Water use, water needs
Broadband surface albedo	Net radiation	Water use, water needs
Thermal infrared surface emissivity	Net radiation	Water use, water needs
Surface temperature	Net radiation, surface resistance	Water use
Surface resistance	Soil moisture and salinity	Water use
Crop coefficients	Gross evapotranspiration	Water needs
Transpiration coefficients	Potential soil and crop evaporation	Water use, water needs
Crop yield	Accumulated biomass	Production

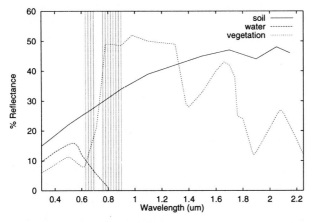

Fig. 7.1. Idealized spectral reflectance curves for soil, vegetation and water (from Mather, 1987), with spectral positions or red and near-infrared bands

7.3.1 Simple Vegetation Indices

An indication of the presence of vegetation in multi-spectral remotely sensed imagery is obtained by comparing reflection values in the near-infrared and red bands. One can choose between either the difference or the ratio between near-infrared and red — in both cases, high values indicate vegetation (Ray, 1994). This allows one to define two members of a large family of vegetation indices, called Difference Vegetation Index (DVI) and Ratio Vegetation Index (RVI):

$$DVI = NIR - R \qquad (7.1)$$

and

$$\text{RVI} = \frac{\text{NIR}}{\text{R}}, \tag{7.2}$$

where NIR and R are near-infrared (0.75 – 0.90 μm) and red (0.63 – 0-.70 μm) reflection measurements, respectively.

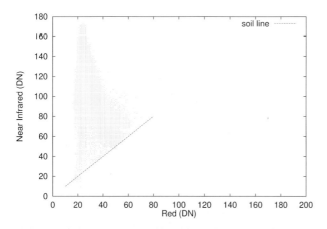

Fig. 7.2. Scattergram of TM bands 3 and 4 with soil line

To understand vegetation indices it is helpful to look at a scattergram, which is obtained by plotting the pixels of an image in the (R, NIR) space (Fig. 7.2), in this case based on Landsat TM. The triangular shape of the cluster of (R, NIR) vectors is typical for images of areas with non-uniform vegetation cover. The most densely vegetated pixels have (R, NIR) vectors at the top of the triangle: low R and high NIR. Various kinds of pixels without vegetation, notably bare soil pixels, are located on the *soil line* at the lower right edge of the triangle. The variation along this line is caused by differences in soil type and soil moisture. Dark and wet soils, as well as water, appear at the left end of the soil line.

Ratio indices assume that the soil line passes through the origin, and that the index of a (R, NIR) vector should depend on the angle between the vector (the line from the origin to (R, NIR)) and the soil line. The ratio vegetation index (RVI) shown above is the simplest one. Iso-vegetation lines pass through the origin. This is favorable when the illumination of the scene is not uniform, as for example in hilly terrain, where illumination depends on the orientation of slopes with respect to the sun. Assuming that this is a multiplicative effect and that it does not depend on the wavelength, ratio indices are are not affected by illumination differences. The results of RVI are between zero and infinity, although values below those on the soil line should not occur. Difference (or perpendicular) indices assume that the amount of vegetation in a pixel is related to the (Euclidean) distance between the (R, NIR) vector and its projection on the soil line. Iso-vegetation lines are parallel to the soil line. The difference index shown above is a perpendicular index. In flat terrain with uniform illumination the advantage might be that the results are less dependent on soil conditions than those of ratio indices. After normalizing the data into the the range of reflectances between 0 and 1, DVI values are between -1 and 1, although negative values (below the soil line) are not supposed to occur.

7.3.2 Normalized Difference Vegetation Index (NDVI)

A well-known index is the Normalized Difference Vegetation Index (NDVI), defined as

$$\text{NDVI} = \frac{\text{NIR} - \text{R}}{\text{NIR} + \text{R}} \tag{7.3}$$

Despite its name, NDVI is a ratio index, with isolines through the origin. NDVI results are independent of illumination. The range is theoretically between -1 and 1, but in practice the value of R is hardly ever larger that the one for NIR (Fig. 7.2), which reduces the range to [0 .. 1].

Figure 7.3 shows a near infrared band of a multi-spectral SPOT image, followed by three vegetation indices: DVI, RVI and NDVI. The near infrared band shows illumination differences due to the topography. These differences are still visible in the DVI, whereas they do not affect RVI and NDVI.

Many studies have shown a good correlation between NDVI and amount of vegetation. The results have proven to be useful in a large range of mapping scales, based on imagery with very different spatial resolutions, from SPOT-XS (20 m) to NOAA-AVHRR (1–4 km). The latter allows for the creation of global NDVI coverages on a regular basis, which enables scientists to study change in the biosphere in detail, whereas the former allow detailed studies on catchment scales.

It should be noted, that the information obtained from NDVI data is mostly used qualitatively, in relation to other NDVI data, e.g. to monitor vegetation over time, or to compare different regions with similar characteristics. Documentation of the FEWS (Famine Early Warning System) project of the U.S. Agency for International Development (US AID) (FEWS Bulletin, 1996), which is based on NOAA satellite imagery, states:

> The magnitude of NDVI is indicative of the level of photosynthetic activity in the vegetation being monitored. (...) FEWS receives an NDVI image for the entire continent of Africa three times a month. (...) The basic NDVI image gives an indication of where vegetation is present and active, but has limited value beyond that. More information is regularly gained by comparing a ten-day period (dekad) against 'normal' conditions for that time of year. In general terms, persistent worse-than-average conditions would be one piece of information in the FEWS "convergence of evidence" approach. However, it must be stressed that a satellite image alone does not give a definitive answer on food-security conditions on the ground.

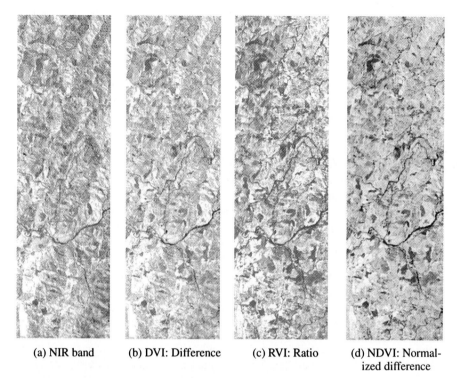

(a) NIR band   (b) DVI: Difference   (c) RVI: Ratio   (d) NDVI: Normalized difference

Fig. 7.3. Sensitivity of vegetation indices to terrain relief

7.3.3 Refined estimates

Despite the limitations of vegetation indices, estimation of parameters for hydrologic models based on remotely sensed imagery through vegetation indices is attractive. Research has resulted in various vegetation indices (Table 7.2) that have been proposed as improvements over NDVI. They are designed to reduce the effect of non-vegetation influences, such as illumination, soil type and soil moisture.

Lyon et al. (1998) reviewed a number of efforts that have evaluated the capabilities of vegetation indices, for example for rangeland productivity measurement (Richardson and Everitt, 1992), land-use change detection (Angelici et al., 1997) and forest clearcut location (Banner and Lynham, 1981). The introduction to this chapter mentioned the possibility to calculate the actual evapotranspiration per pixel, instead of treating it as the residual term in water balance equations. Which of the available vegetation indices is the most suitable for this purpose in different circumstances is yet to be investigated, for which the following considerations apply:

- Whereas the surface is characterized by *reflectances*, an image contains measurements of *reflection*, which are also influenced by illumination differences and atmosphere. Therefore, radiometric corrections must be performed.

- To correct for the influence of relief on illumination, a Digital Elevation Model is required with an accuracy that is compatible with the image resolution. When this is not available, ratio-based indices are to be preferred over perpendicular indices, because of their 'built-in' normalization. Ratio indices are implicitly trying to deal with an additional unknown variable, the illumination. Therefore, they cannot distinguish between poorly illuminated bright objects and well illuminated dark objects. This confusion is not necessary when illumination is known to be uniform, as in flat areas.
- Certain indices (e.g. NDVI) are known to work well in case of a high vegetation cover, but others perform better with less vegetation. In the second case, the influence of soil conditions is larger, for which *Soil Adjusted* indices (SAVI, SARVI, TSAVI) have been developed.
- The *Leaf Area Index* (LAI) is defined as the cumulative area of leaves per unit area of land at nadir orientation (Bastiaansen, 1998). It represents the total biomass and is indicative of crop yield, canopy resistance and heat fluxes. A non-linear relationship between LAI and various vegetation indices has been observed (Bunnik, 1978, Clevers, 1988) (Table 7.2).

7.3.4 Multi-temporal Vegetation Index

Of the study area, introduced in Sect. 7.2, Landsat TM imagery was available from 1985, 1990 and 1996 (Colour Plate 7.A). The images were acquired during the same season (September - October). Comparison of NDVI values clearly shows the development of the amount of vegetation during the period 1985 – 1996 and reflects the land-use changes in the area that were mentioned in the introduction.

Combined results of NDVI in three years are visualized in a color composite (Colour Plate 7.B), showing the NDVI of 1985 in blue, of 1990 in green and of 1996 in red. Red areas have little vegetation in 1985 and 1990, but vegetation increased between 1990 and 1996 — they are newly irrigated areas. White areas were densely vegetated all the time, whereas in blue and cyan areas vegetation has decreased, before or after 1990, respectively. The image shows quite some green areas. because 1990 was relatively wet.

7.4 Thematic Classification

To obtain thematic information from multi-spectral imagery, multi-dimensional, continuous reflection measurements from remote sensing images have to be transformed into discrete objects, which are distinguished from each other by a discrete thematic classification. Objects can be considered to consist of a unique type of land cover, such as wheat fields or conifer forests. The relationship is not one-to-one. Within different objects of a single class, and even within a single object, different reflections may occur. Conversely, different thematic classes cannot always be distinguished in a satellite image because they show (almost) the same reflection. In

Table 7.2. Overview of Vegetation Indices

Intrinsic Indices	
Difference Vegetation Index	$DVI = NIR - R$
Ratio Vegetation Index	$RVI = \dfrac{NIR}{R}$
Normalized Difference Veg. Index	$NDVI = \dfrac{NIR - R}{NIR + R}$
Normalized Difference Wetness Index	$NDWI = \dfrac{SWIR - MIR}{SWIR + MIR}$
Green Vegetation Index	$GVI = \dfrac{NIR + SWIR}{R + MIR}$
Soil-line Related Indices	
Perpendicular VI	$\begin{cases} PVI = \dfrac{NIR - aR - b}{\sqrt{1+a^2}}, \\ NIR_{soil} = a\, R_{soil} + b \end{cases}$
Weighted difference VI	$WDVI = NIR - \dfrac{NIR_{soil}}{R_{soil}} R$
Soil Adjusted VI	$SAVI = \dfrac{(1+L)(NIR - R)}{NIR + R + L}$
Soil Adjusted Ratio VI	$SARVI = \dfrac{NIR}{R + b/a}$
Transformed Soil Adjusted VI	$TSAVI = \dfrac{a * (NIR - a * R - b)}{R + a * NIR - a * b}$
Leaf Area Index	
Relation between Leaf Area Index and SAVI	$SAVI = c_1 - c_2 e^{-c_3 LAI}$

where $a = 0.96916, b = 0.084726, L = 0.5$ and $c_1 = 0.69, c_2 = 0.59, c_3 = 0.91$

such cases, deterministic methods are not sufficient. A probabilistic approach, however, may be able to describe the spectral variations within classes and to minimize the risk of erroneous class assignments.

Classification determines a thematic class from a user-defined set for each image pixel. The choice is made on the basis of reflection measurements stored in that pixel. The collection of measurements in one pixel is called measurement vector or feature vector. With M spectral bands the feature vector has M components and corresponds to a point in an M-dimensional feature space. The task of classification is to assign a class label to each feature vector, which means to subdivide the feature space into partitions that correspond to classes. This task can be achieved by *pattern recognition* (Ripley, 1996).

After the introduction, this section consists of six parts. First (a), image classification is introduced, followed by (b) an elaboration on maximum likelihood classification. Next, (c) some weaknesses are signalled and (d) impovement by refined probability estimates is suggested. Spatial image characteristics are considered in (e) image segmentation. Finally, (f) a case study is presented (Gorte, 1999).

7.4.1 Image Classification Methods

Reflections measured by satellite sensors depend on the local characteristics of the earth's surface. In order to extract information from the image data, we must find out this relationship.

Multi-spectral image classification can be an important step in the extraction of *thematic* information from satellite images. The aim is to automate this as much as possible by using suitable image processing and image analysis software. Due to the complexity of satelite images, automatic classification can at best be regarded complementary to visual image interpretation. Section 7.4.3 will discuss the advantages and disadvantages of computerized *vs.* visual methods.

In theory it is possible to base a classification on a single spectral band of a remote sensing image (for example, on SPOT pan-chromatic), but much better results can be obtained by using more bands (for example the three bands of a SPOT multi-spectral image) at the same time. Therefore, the focus will be on *multi-spectral* classification.

Looking at a certain image pixel in M bands simultaneously, M values are observed at the same time. In the example of multi-spectral SPOT, where $M = 3$ each pixel has three reflection values, in the green, red and infrared parts of the electro-magnetic spectrum, respectively. Such a group of M numbers is sometimes referred to as a *pattern* in statistics. Therefore, classification is an application of *statistical pattern recognition*.

Feature Vectors and the Feature Space. In classification jargon the image bands mentioned above are called *features*, because often transformations are applied to the spectral values of an image, prior to classification. They are called feature transformations, their results *derived* features. Examples are: principal components, HSI transformations, measures of texture etc.

In one pixel, the values in the M features can be regarded as components of a M-dimensional vector, the *feature vector*. Such a vector can be plotted in an M-dimensional space, called *feature space*.

Pixels with similar spectral characteristics, which are likely to belong to the same land cover class, are near to each other in the feature space, regardless how far they are from each other in the terrain and in the image. Pixels belonging to a certain class will hopefully form a so-called *cluster* in the feature space. Moreover, it is hoped that other pixels, belonging to other classes, fall outside this cluster, but into other clusters, belonging to those other classes.

A variety of classifications methods exists. A first distinction is between *unsupervised* and *supervised* classification.

Unsupervised Classification. A classification can be obtained by plotting all feature vectors of the image in a feature space, and then analyzing the feature space to group the feature vectors into clusters (Fig. 7.4). Software that does this automatically is called *clustering* software. The name for the process is *unsupervised classification*. Such software has no notion of thematic land cover class names, such as *town*, *road*, *wheat* etc. All it can do is find out that there seem to be (for example) 16 different *spectral classes* in the image and give them numbers (1 to 16). Subsequently, it can produce a raster map, in which each pixel has a value (from 1 to 16), according to the cluster to which the image feature vector of the corresponding pixel belongs.

After this process it is up to the user to invent the relationship between *spectral* and *thematic* classes. It is very well possible that he discovers that one thematic class is split into several spectral ones (classes 2, 3, 4 and 5 in Fig. 7.4), or (which is worse) that several thematic classes got caught in the same cluster (class 9 in Fig. 7.4).

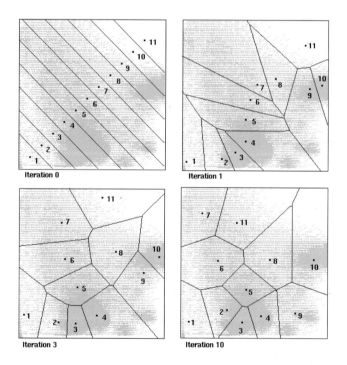

Fig. 7.4. Unsupervised classification (ISODATA) in a 2-dimensional feature space of principal components 1 and 2 of a Thematic Mapper image. Note that in the final iteration one apparent cluster is split into four classes (2, 3, 4 and 5), while class 9 contains two clusters

Various clustering algorithms exist. Usually, they are not completely automatic: The user must specify some parameters, such as the number of clusters that he approximately wants, the maximum cluster size (in the feature space!), the minimum distance (also in the feature space) that is allowed between different clusters etc. The software "builds" clusters as it is scanning through the image. Typically, when a cluster becomes larger than the maximum size, it is split into two clusters; on the other hand, when two clusters get nearer to each other than the minimum distance, they are merged into one.

Supervised Classification. In order to make a classifier work with thematic (instead of spectral) classes, some knowledge is needed about the relationship between classes and feature vectors.

Theoretically, this knowledge could come from a data base in which the relationships between (thematic) classes and feature vectors is stored. It is tempting to assume that in the past enough images of each kind of sensor have been analyzed as to know the spectral characteristics of all relevant classes.

Unfortunately, the observed feature vectors in a particular image are influenced by a large amount of other factors than land cover, such as: atmospheric conditions, sun angle (as function of latitude/time of day/date *and* as function of terrain relief), soil type, soil humidity, vegetation growing stage, wind, etc.

Trying to take all these influences into account is practically impossible, even if vast amounts of additionally required data (DEMs, soil maps etc.) were available.

Much more widely used are, therefore, classification methods where the relationship between reflections and cover classes is established for each separate image under consideration. Supervised classification is divided into two phases: a training phase, where the user trains the computer by telling for a limited number of pixels to what classes they belong **in this particular image**, followed by the decision phase, where the computer assigns a class label to all (other) image pixels, by looking for each pixel to which of the trained classes this pixel is most similar.

Training. During the training phase the user decides what classes to use. About each class he needs some ground truth, a number of places in the image area that are known to belong to that class. This knowledge must have been acquired beforehand, for instance as a result of fieldwork, or from an existing map, assuming that in some areas the class membership has not changed since the map was produced.

Suitable training software allows to display the image on the computer screen, to indicate the places where the ground truth is known and to enter the corresponding class names. Meanwhile the feature space is visible, where the training samples are plotted using distinct colors for the different classes, such that the user can judge whether the classes are spectrally distinguishable, whether each class corresponds to only one spectral cluster etc.

Decision Making. It is the task of the decision making algorithm to make a *partitioning* of the feature space, according to the training data. For every possible feature vector in the feature space (at least for those that actually occur in the image), the program decides to which of the sets of training pixels this feature vector is most similar. In addition, the program produces an output map where each image pixel is assigned a class label, according to the feature space partitioning.

Some algorithms are able to decide that feature vectors in certain parts of the feature space are not similar to any of the trained classes. They assign to those image pixels the class label *unknown*. In case the area indeed contains classes that were not included in the training phase, the result *unknown* may be better than assignemt to one of the classes in the user-defined set.

Simple classification algorithms are the *box* and the *minimum distance* classifiers.

During box classification, a box is created around the training feature vectors of each class. A box is rectangle in a two-dimensional feature space, a block in the three-dimensional case, or a hyper-block if there are more than three features. The position and size of the box of a class can be exactly around the feature vectors of the training samples (min-max method), or according to the mean vector and the standard deviations of these feature vectors. Each image pixel is then classified according to the box that contains its feature vector. In parts of the feature space where boxes overlap, it is usual to give priority to the smallest box. Feature vectors in the image that fall outside all boxes will be classified *unknown*.

A minimum distance-to-mean classifier first calculates for each class the mean vector of the training feature vectors. The feature space is partitioned by giving to each feature vector the class label of the nearest mean vector, according to Euclidian metric. Usually it is possible to specify a maximum distance threshold: if the nearest mean is still further away than that threshold, is is assumed that none of the classes is similar enough and the result will be *unknown*.

7.4.2 Maximum Likelihood Classification

Whereas the above-mentioned classifiers use rather heuristical decision rules, maximum likelihood classification has a statistical foundation.

Maximum likelihood aims at assigning a "most likely" class label C_i, from a set of N classes $C_1, \ldots, C_N$, to any feature vector **x** in an image.

The most likely class label C_i for a given feature vector **x** is the one with the highest *posterior probability* $P(C_i|\mathbf{x})$. Each $P(C_i|\mathbf{x})$, $i \in [1..N]$, is calculated, and the class C_i with the highest value is selected. The calculation of $P(C_i|\mathbf{x})$ is usually based on *Bayes formula*:

$$P(C_i|\mathbf{x}) = \frac{P(\mathbf{x}|C_i)\,P(C_i)}{P(\mathbf{x})}, \qquad (7.4)$$

with

$P(\mathbf{x}|C_i)$: class probability density
In Bayes formula, $P(\mathbf{x}|C_i)$ is the probability that some feature vector **x** occurs in a given class C_i. It tells us what kind of **x**s we can expect in a certain class C_i, and how often (relatively). It is the probability density of C_i, as a function of **x**. Supervised classification algorithms derive this information during the stage.

$P(C_i)$: prior probability

$P(C_i)$ is called the prior probability of class C_i, the probability for any pixel that it belongs to C_i, irrespective of its feature vector. It can be estimated on the basis of prior knowledge about the terrain, as the (relative) area that is expected to be covered by C_i.

$P(\mathbf{x})$: (class-independent) feature probability density

For a certain $\mathbf{x}$, in order to find the class C_i with the maximum posterior probability $P(C_i|\mathbf{x})$ the various $P(C_i|\mathbf{x})$ must be only compared. This comparison can be done with the same success without knowing $P(\mathbf{x})$ since it is the same for every C_i. Therefore, it is common to substitute $P(\mathbf{x})$ by a normalization factor

$$P(\mathbf{x}) = \sum_{j=1}^{N} P(\mathbf{x}|C_j) P(C_j) \tag{7.5}$$

This leaves, however, no room for an *unknown* class, since there is no way to find $P(\mathbf{x}|unknown)$ (by definition of *unknown*).

Gaussian Maximum Likelihood. To estimate class probability desities, it is commom to assume a multivariate normal (Gaussian) distribution of $P(\mathbf{x}|C_i)$ for each class C_i. . From the training data, for each class C_i the sample mean vector and the sample variance-covariance matrix are obtained, and they are used as estimates for the class mean vector $\mathbf{m}_i$ and the class variance- covariance matrix V_i. This gives the class probability density function for class C_i:

$$P(\mathbf{x}|C_i) = (2\pi)^{-M/2} |V_i|^{\frac{1}{2}} e^{-\frac{1}{2}(\mathbf{y}^T V_i^{-1} \mathbf{y})} \tag{7.6}$$

with:

M : the number of features
V_i : the $M \times M$ variance-covariance matrix of class C_i
$|V_i|$: the determinant of V_i
V_i^{-1} : the inverse of V_i
$\mathbf{y} : \mathbf{x} - \mathbf{m}_i$ ($\mathbf{m}_i$ is the class mean vector), as a column vector with M components
$\mathbf{y}^T$: the transposed of $\mathbf{y}$ (a row vector).

A classifier that does not take prior probabilities into account only needs probability densities $P(\mathbf{x}|C_i)$ for different C_i. By removing constants, taking a logarithms and multiplying the result by -2, equation 7.6 can be written as:

$$D_i(\mathbf{x}) = \ln|V_i| + \mathbf{y}^T V_i^{-1} \mathbf{y}. \tag{7.7}$$

D_i is a function of $\mathbf{x}$ and can be regarded as a measure of the dissimilarity between a feature vector $\mathbf{x}$ and a class mean $\mathbf{m}_i$ (because of $\mathbf{y}^T$ and $\mathbf{y}$) in the feature space. It is weighted by V_i^{-1} and compensated by $\ln|V_i|$ according to the within-class variability: From a given $\mathbf{x}$, a widely spread class seems more similar than a concentrated one. The second term of equation 7.7 is commonly referred to as Mahalanobis distance

$$M_i = \mathbf{y}^\mathrm{T} V_i^{-1} \mathbf{y}. \tag{7.8}$$

Most maximum likelihood implementations minimize on D_i or on M_i. The user is allowed to specify a threshold value: if all D_is or M_is are larger than this value, a pixel is classified *unknown*.

7.4.3 Discussion

Supervised classification can be used for thematic mapping, monitoring, agricultural acreage and yield estimation, forest inventories and the assessment of catchment characteristics.

Unlike human interpreters, however, classification only takes spectral characteristics into account. It does not consider shapes, patterns and other spatial associations. For example, the fact that a group of pixels is arranged in a long linear fashion does not help in classifying them as *road*. Likewise, the software cannot distinguish between a lake and a channel, or recognize a *town* based on its typical pattern that is caused by the road plan.

Classification relies on spectral separability of classes. When a pixel's feature vector falls in an overlapping area of two classes (for instance *grass* and *wheat*) in the feature space, this means that probably there are in the image some pixels with this feature vector belonging to *grass* and others, with the same feature vector, belonging to *wheat*. The classifier will treat all of them in the same way and classify them as one class, either *grass* or *wheat*. Similar problems occur with heterogeneous classes, such as *town*: a mixture of roads, roofs, trees, gardens, ponds etc. If these pixels are classified as *town*, then elsewhere some *forest* or *grass* pixels may be classified as *town* as well.

At boundaries between two distinct classes in the terrain we find pixels in the image where the reflection measurement is influenced by both classes. The feature vector of such a pixel may be outside both clusters (somewhere in-between them in the feature space). Therefore, such a "mixed' pixel will become *unknown*, or (worse!) be classified as something else.

Elongated terrain objects, such as roads, having a width less than the image resolution, usually influence the pixels that they intersect sufficiently to make them visible in the image. However, all these pixels are mixed. They are also influenced by the surrounding landscape. Therefore, such classes are difficult to classify. Classification is mainly suitable for classes of objects that form areas in the terrain.

Although usually classes with meaningful names are used, the system has no notion of class semantics. Classes are completely defined by a few distribution parameters, derived from training samples. Is is the users responsability to define separable classes and to select representative training samples.

7.4.4 Probability estimation refinements

Statistical classification methods do not always attempt to estimate various probabilities as accurately as possible. Perhaps it is assumed that the largest *a posteriori* probability occurs at the 'right' class also when estimated rather roughly. Even the most advanced of the methods described above suffers from two drawbacks.

- The probabilistics in Bayes' formula pertain to the entire image, while a decision is taken for each individual pixel. Having only one set of class *a priori* probabilities, the class assigned to a pixel depends on the situation in the entire image and is, therefore, influenced by areas that are far away from the pixel under consideration and have no relation with it.
- The assumption that reflections can be modeled as Normal distributions may be unrealistic for certain land-use classes, such as *built-up* and *agricultural* areas, which may consist of several land covers with different spectral signatures in different (unknown) proportions. Also signatures of land-cover classes are influenced by soil type, soil moisture, sun incidence angle (on slopes) etc. and may be inadequately modeled by Gaussian densities.

Local probabilities. Estimates of the various probabilities can be refined by making them local, i.e. pertaining to parts of the area, instead of to the entire area (Strahler, 1980, Middelkoop and Janssen, 1991).

If the user is able to subdivide the image into regions, such that different class mixing proportions occur in each region, and if these mixing proportions are known, then in each region the expected overall accuracy is higher with a region-specific set of priors (according to the mixing proportions) than with the 'global' set. Therefore, taking all regions together, the overall accuracy in the entire image also increases.

The necessary subdivision of the image can be made with additional (map) data, which may be stored, for example, in a geographic information system. The basic idea is that when deciding upon an class in a particular pixel, statistical data related to, for example, the soil type in that element are more relevant than statistics for the entire area. Until now, the objection against this approach was that such detailed statistics are generally not available. This information can be obtained from the distribution of reflections.

A posteriori probability values can be used to estimate class areas in an image region, by interpreting the class posterior probability at a pixel as the pixel's contribution to the total area of that class within its region (Duda and Hart, 1973).

This method was adapted by Gorte and Stein (1998) to estimate class areas in image subsets. Therefore, the area A_i^r, covered by class C_i in region r is the sum over all pixels p in r of the posterior probabilities $P(C_i|\mathbf{x}_p)$:

$$A_i^r = \sum_{p \in r} P(C_i|\mathbf{x}_p), \qquad (7.9)$$

where $\mathbf{x}_p$ is the feature vector of pixel p. Applying Bayes' formula gives:

$$A_i^r = \sum_{p \in r} \frac{P(\mathbf{x}_p|C_i) P(C_i)}{P(\mathbf{x}_p)}, \qquad (7.10)$$

in which $P(\mathbf{x}_p)$ can be obtained by normalization, according to

$$P(\mathbf{x}_p) = \sum_{j=1}^{N} P(\mathbf{x}_p|C_j) P(C_j), \tag{7.11}$$

where N is the number of classes.

The prior probability $P(C_i)$ for class C_i in region r can be estimated as the class area A_i^r divided by the total area A^r of r:

$$A_i^r = \sum_{p \subset r} \frac{P(\mathbf{x}_p|C_i)\frac{A_i^r}{A^r}}{P(\mathbf{x}_p)}. \tag{7.12}$$

From this equation, A_i^r can be solved iteratively, given the class probability densities $P(\mathbf{x}|C_i)$, as estimated from representative training samples.

Given an additional data set, in which the mapping units are expected to correlate with class occurrence (soil, elevation, geomorphology, etc.), class prior probabilities can be obtained per mapping unit. These are more specific for a pixel in such a unit than the ones that are estimated by the user for the entire image area. This, in turn, will increase the reliability of the posterior probability estimates.

When applying Bayes' formula per region, it must be noted that also the probability densities $P(\mathbf{x}_p|C_i)$ are region dependent and cannot be modeled by a single class probability density function. Therefore, non-parametric (k-Nearest Neighbor) estimation is applied, which estimates the probability $P(\mathbf{x}_p|C_i)$ that a class C_i pixel in region r has feature vector $\mathbf{x}_p$ as being proportional to the number k_i of class C_i samples in a neighborhood with k samples around x, divided by the total number N_i^r of training samples for class C_i that were involved in the classification of region r (Gorte, 1998).

The improved estimates of prior probabilities and probability densities increase the reliability of the posterior probabilities and, therefore, enhance subsequent decision processes, such as maximum posterior probability class selection (Strahler, 1980). Also the per-region class area estimates are more accurate than those obtained from standard Maximum Likelihood classifications.

7.4.5 Segmentation

Whereas classification only deals with spectral image characteristics, segmentation (Haralick and Shapiro, 1985) also attempts to take spatial characteristics (adjacency) into account. The purpose of image segmentation is to subdivide an image into regions that are homogeneous according to certain criteria, such that these regions correspond to area objects in the terrain (Fu and Mui, 1981). To combine the complementary information from segmentation and classification can be achieved in different ways (Cross et al, 1988), such as:

- Classify the unsegmented image and assign the majority class in each segment to the entire segment.
- Calculate the average feature vector per segment and find the 'most likely' class for this feature vector, using Maximum Likelihood

- Use the iterative procedure of Sect. 7.4.4 to combine pixel probability densities with segment prior probabilities. This approach will be applied in the case study of Sect. 7.4.6.

In Gorte (1998) an integrated segmentation and classification procedure is developed, which first gathers evidence concerning spatial and class-membership characteristics, and then combines these to delineate and identify relevant terrain objects. The procedure uses a pyramid of segmentations with different coarsenesses.

7.4.6 Case study in the Pantanal Area, Brazil

The purpose of the case study is to assess the rather drastic land use changes in the Pantanal region since 1985 (see Sect. 7.2).

Available were three Landat TM images of the Pantanal study area, from 1985, 1990 and 1996 (Colour Plate 7.A). A land-use survey from 1997 was available (Disperati et al., 1998) with classes according to level 3 of the CORINE legend (see also Chap. 19). From the survey map, training samples were selected in areas that were assumed to be the same in 1996 and 1997, allowing to establish spectral signatures for those classes using the 1996 image. Seven classes were selected, according to level 2 or the CORINE legend — in preliminary experiments, level 3 appeared too detailed for automatic classification of Thematic Mapper imagery. Two sets of samples were chosen, a training set and a test set.

Image segmentation (Gorte, 1996) (Sect. 7.4.5) was applied to the *multi-temporal* NDVI composite from Sect. 7.3.4 (Colour Plate 7.B). The result is a set of multi-temporal objects. These are regions of adjacent pixels showing the similar vegetation development (Colour Plate 7.C).

Iterative local prior probability estimation (Sect. 7.4.4) was applied to the 1996 image, using one set of priors in each segment of the multi-temporal segmentation. This gives the 1996 land-use map. Comparison with the test set shows that an overall accuracy of 71% was reached using local prior probabilities, as compared to 61% with maximum likelihood classification using the same band combination. Maximum likelihood classification with six bands gives 67% accuracy [1]. Regarding this rather low accuracy, it should be noted that test set pixels were chosen at random from the survey data, without considering their spectral values.

Since the 1996 and 1985 images are of the same season, it was assumed that the spectral signatures derived for the 1996 image are also valid for the 1985 image (Fig. 7.5). Under this assumption, also the 1985 land-use map could be made, despite the absence of ground truth for that year. Therefore, also a 1985 test set was not available.

To detect changes, a straightforward procedure is *post-classification comparison*, which involves overlaying the 1985 and 1996 land-use maps. The difficulty is that both classifications contain quite a large percentage of errors (which for 1985 is even unknown). Therefore, it is difficult to distinguish between real changes and those that are observed as a result of misclassification.

---

[1] Iterative local prior estimation is developed for SPOT-XS with 3 bands.

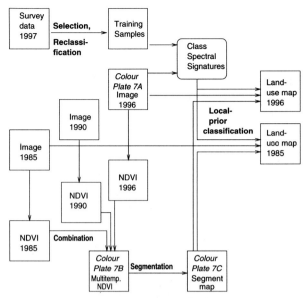

Fig. 7.5. Classification of 1985 and 1996 images with local prior probabilities according to segmentation of multi-temporal NDVI

Alternatively, *classification of a multi-temporal data set* was applied (Fig. 7.6). The 1985 and 1996 images were combined in a single 12 channel data set. From the two land-use classifications pixels were selected having high *a posteriori* probabilities. Moreover, only those pixels were chosen that were surrounded by pixels of the same class, to reduce mixed-pixel effects at boundaries. The selected pixels are the most reliable ones in the two land-use maps. The resulting maps have eight different pixel values: 1 – 7 for the seven classes, plus 0 for *not selected*. Overlaying these two maps gives 64 combinations, 15 of which contain a 0 (zero) and are no longer considered. From the remaining 49 combinations 31 were selected: the seven *no change classes*, having the same class in both years, plus 24 *change classes*, belonging to one class in 1985 and to another in 1996. The remaining 18 combinations did not occur in significant amounts of pixels.

The resulting change map has very many pixels with value 0. The remaining pixels (with a change-class number between 1 and 31) were used as training set for a maximum likelihood classification of the combined – multi-temporal – image. The assumption is that the 31 classes are distinguishable in the 12-dimensional feature space. The result is shown in Colour Plate 7.D, where the hatched areas indicate change: the narrow lines refer to 1985 and the wider ones to 1996. As far as this can be judged at this stage of the Pantanal project, the results are promising. Further evaluation will take place in the near future.

The problem of detection of landuse changes is also discussed in detail in Chap. 19 of this book.

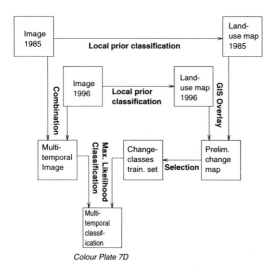

Colour Plate 7D

Fig. 7.6. Multi-temporal classification of combined 1985 - 1996 images with change classes

7.5 Radar

So far in this chapter only passive remote sensing data from various spectral bands have been used. But also active sensors are of relevance in the context of the chapter.

Active microwave images have been used in vegetation studies since the 1970's when aircraft radar coverages have been flown in many tropical (south and central America, Nigeria, Indonesia, etc.) and other regions. In fact, for many areas the tropical rain forests were mapped for the first time after the radar images became available.

Radars emit electro-magnetic pulses from the platform and the backscatter is received and transformed into an image. Imaging radars are therefore fundamentally different from sensors that use visible and infrared wavelengths. They operate in various parts of the microwave region of the spectral domain, denoted by letters – from shorter to longer wavelength: C, X, L, etc. (see also Chap. 6). In this chapter only a brief overview of the possibilities of cover classifications using active radar, or side looking airborne (or satellite) radar (SAR) is discussed, also in combination with multi-spectral data.

However, it may be useful to point out some limitations inherent to the system. In relief rich terrain there are problems known as lay-over and foreshortening, due to the geometry of the slant range in relation to the slope steepness or height difference of the top and base of the object; the relief appears displaced towards the platform. Correction is possible when a Digital Elevation Model is available, but it is not a routine matter. In mountainous terrain, radar shadows are present, depending on the slant range. Speckling or radar fading is another problem. Unlike for most multi-spectral systems, preprocessing of the data is required. Several speckle reduction algorithms are in use to attempt for correction. Often, the general solution is to average the number of pixels from the same cover type, or by multi-look averaging of the data, at the expense of loss of texture information.

An advantage of imaging radars is that they are insensitive to local weather conditions. The orbital SAR systems (ERS 1-2, JERS-1, Radarsat, SIR-C/X-SAR) provide multi-date data, which is valuable for monitoring the effects of cover on hydrology, particularly in cloudy weather types.

The backscatter signal of terrain, which is used to process an image, depends on the wavelength, cover species related (probably morphologic) features, surface roughness of the ground and moisture content, which leads to ambiguity of single radar data. Closed canopies of more or less uniform height cause attenuation of the signal, especially in the C and X band. The shorter wavelengths (C band and shorter) display a strong relation between the backscatter and species dependent properties and can be used to detect stress or diseases and the forest development (Hoekman, 1990). Natural forests with mixed species compositions may show a very limited dynamic range of radar backscatter. A high degree of canopy transparency was found for the longer wavelength L-band, which is useful for differentiation of broad forest classes and for detecting ground cover conditions, or flooding below the canopy (see also Figs. 10.8 and 10.9). It may be noted that multi-spectral data pertains to the canopy only.

An interesting specific application of radar was the detection of water below a woody forest canopy by the L and P bands. Presence of water increases the backscatter by ground-trunk two-bounce scattering(Hess et al., 1995). However, in non-woody wetlands, a decrease of the backscatter may occur because of the forward specular scattering of the pulse signal.

Many studies have used the maximum likelihood approaches or a decision tree method, based on training sets containing ground truth or local knowledge. It was found that single date, single frequency/polarization data has limitations, but multi-date and multichannel SAR leads to enhanced classification (Kasischke et al., 1997). As is the case with multi-spectral data, inclusion of textural features in the classification increases the accuracy of identification of cover types (Soares et al., 1997). Rignot et al. (1997) found that, at L band, multiple polarizations were required to obtain a reliable classification of deforestation and secondary re-growth in Rondonia, Brazil. The double reflection of emergent vegetation and water surface can produce a 3-6-dB increase in L band backscatter magnitude over non-flooded condition, and a simple model has been presented (Engheta and Elachi, 1982). Use was made of such differences for the monitoring of flooding of marsh lands using multi temporal, multi band and polarization radar data (Pope et al., 1997). The use of radar for cover identification was reviewed by Kasischke et al. (1997).

To decrease the dependency on empiricism, explicit relationships between radar backscatter and structural attributes have been studied (Hoekman, 1990), (Dobson et al., 1995), but more work is required before the results can be used at practical implementation level.

It was found that image fusion of micro-wave and optical sensors leads to the identification of more land cover classes and to improved overall classification accuracies (Schistad-Scholberg et al., 1994, Hussin and Shaker, 1996).

Acknowledgments

The author would like to thank the Earth Sciences Department of the University of Siena, Italy, for providing the image data of the Pantanal area (see Sect. 7.2). The analysis of the data was carried out in cooperation with Riccardo Salvini, Ph.D. student in this Department.

Thanks are also due to Prof. A.J.M. Meijerink for his contributions to the introduction and the radar section, and for his advise and comments.

References

Angelici G., N. Brynt and S. Friendman (1997). Techniques for land use change detection using Landsat imagery, Proc. of the American Society of Photogrammetry, Falls Church, Virginia, pp. 217–228

Banner A. and T. Lynham (1981). Multi-temporal analysis of Landsat data for forest cut over mapping — a trial of two procedures. Proc. of the 7th Canadian Symposium on Remote Sensing, Winnipeg, Manitoba, Canada, pp. 233–240

Bastiaanssen, W. (1998). Remote sensing in water resources management: the state of the art. International Water Management Institute (IWMI), Colombo, Sri Lanka

Bosch, J.M. and J.D. Hewlett (1982). A review of catchment experiments to determine the effects of vegetation changes on water yield and evapotranspiration. J. of Hydr. Vol 55, pp. 3-23

Bruynzeel, L.A. (1990). Hydrology of moist tropical forests and effects of conversion: a state of knowledge review. UNESCO, 224 p.

Bunnik, N.J.J. (1978). The multispectal refelctance of short wave radiation by agricultural crops in relation with their morphological and optical properties, PhD Thesis, Agricultural University of Wageningen, Papers 78-1, 175 pp.

Calder, N. (1991). Spaceship Earth, Viking Press, London

Clevers, J.G.P.W. (1988). The application of a weigted infrared-red vegetation index for extimating leaf area index by correcting for soil moisture, Remote Sens. Environ. 29, pp. 25-37

Cross, A.M., D.C. Mason and S.J. Dury (1988). Segmentation of remotely-sensed images by a split-and-merge process. IJRS 9(8), pp. 1329-1345

Disperati L., G. Righini, M. Bocci, P.L. Pantozzi, A.P. Fiori, S. Kozciac and A.C. Paranhos Filho (1998). Land cover changes in the Rio Verdo de Mato Grasso region (Pantanal, Mato Grasso do Sul, Brasil) as detected through remote sensing and GIS analysis. Technical report, Univ. of Siena (I)

Dobson, M.C., F.T. Ullaby and L.E. Pierce (1995). Land-cover classification and estimation of terrain attributes using synthetic aperture radar. Remote Sens. Environ. 51: 199-214

Duda, R.O. and P.E. Hart (1973). Pattern classification and scene analysis. John Wiley and Sons, New York, 465 pp.

Engheta, N. and C. Elachi (1982). Radar scattering from a diffuse vegetation layer over a smooth surface, IEEE Trans. Geosci. Remote Sens. 33: 848-857

FEWS Bulletin(1996). United States Agency for International Development (USAID), http://www.info.usaid.gov/fews/fews.html
Fu and Mui (1981). A survey on image segmentation, Pattern Recognition, 13:3-16, 1981
Gorte, B.G.H. (1996). Multi-spectral quadtree based image segmentation. IAPRS vol. XXXI B3, Vienna, pp. 251-256
Gorte, B.G.H. (1998). Probabilistic segmentation of remotely sensed images, Ph.D. thesis, Wageningen Agric. Univ., the Netherlands
Gorte, B.G.H. (1999), Change detection by classification of a multi-temporal image. Accepted for publication in: ISD'99 - Integrated Spatial Databases: Digital Images and GIS. Volume Editor(s): Peggy Agouris & Anthony Stefanidis, Springer Verlag
Gorte, B.G.H. and A. Stein (1998) Bayesian Classification and Class Area Estimation of Satellite Images using Stratification, IEEE Trans. Geosci. Remote Sens., Vol. 36, No. 3, May 1998, pp. 803 – 812
Haralick, R.M. and L.G. Shapiro (1985). Survey: Image segmentation techniques. Computer, Vision, Graphics and Image Processing, 29, pp. 100-132
Hernandez, Filho P., Ponzoni, F.J., Pereira, M.N., Pott, A., Pott, V.J. and Silva, M.P (1995). Mapeamento da vegetação e de uso da terra de parte a bacia do Alto Taquari (MS) considerando o procedimentode analise visual de imagens TM/LANDSAT e HRV/SPOT. In: Encontro sobre sensoriamento remoto aplicado a estudos no Pantanal - Livro de Resumas. Corumbà - MS, 9-12 de Outobro de 1995, 137-139
Hess L.L., J.M. Melack, S. Filoso and Y. Wang (1995). Delineation of inundated area and vegetation along the Amazon floodplain with the SIR-C synthetic aperture radar. IEEE Trans. Geosci. Remote Sens. 33, pp. 896–904
Hoekman, D.H. (1990). Radar remote sensing data for applications in forestry. Publ. Ph.D. thesis, Wageningen Agric. Univ., the Netherlands
Hussin, Y.A. and S.R. Shaker (1996). Optical and radar satellite image fusion techniques and their applications in monitoring natural resources and land use changes. Int. J. of Electronics and Communications. vol. 50, no.2, 169-176
Kasischke, E.S., J.M. Melack and M.C. Dobson (1997). The use of imaging radars for ecological applications - a review. Remote Sens. Environ. 59: 141–156
Kouwen, E.D., A. Soukis, A. Pietroniro, J, Donald, R.A. Harrington (1990). Flash flood forecasting with a rainfall runoff model designed for remote sensing inputs and geographic information system. Proc. Int. Symp. R.S. and Water Resources, IAS/Neth. Soc. R.S. pp. 805–814
Lyon, J.G., Ding Yuan, R.S. Lunetta and C.D. Elvidge (1998). A change detection experiment using vegetation indices. Photogrammetric Engineering & Remote Sensing, Vol. 64, No. 2, pp. 143–150
Mather, P.M. (1987). Computer processing of remotely-sensed images, an introduction, Wiley
Meijerink, A.M.J. and B.H.P. Maathuis (1997). Use of Remote Sensing to assess flood damage to rice lands caused by accelerated erosion and neo-tectonics, Komering River, Sumatra. River Flood Disasters. Proc. ICSU SC/IDNDR. IHP/OHP-Berichte, Sonderheft 10, Koblenz, pp. 97-105
Middelkoop, H. and L.L.F. Janssen (1991). Implementation of temporal relationships in knowledge based classification or satellite images. Photogrammetric Engineering & Remote Sensing, Vol. 57, No. 7, pp. 937 – 945
Pope, K.O., E. Rejmankova, J.F. Paris and R. Woodruff (1997). Detecting seasonal flooding cycles in marshes of the Yucatan peninsula with SIR-C polarimetric radar imagery. Remote Sens, Environ. 59: 157–166
Ray, T.W. (1994). A FAQ on Vegetation in Remote Sensing, Div. of Geological and Planetary Sciences, California Institute of Technology, ftp://kepler.gps.caltech.edu/pub/terrill/rsvegfaq.txt
Richardson, A. and J. Everitt (1992). Using spectral vegetation indices to estimate rangeland productivity, Geocarto International, 1:73–77

Rignot, E., W.A. Salas and D.L. Skole (1997). Mapping deforestation and secondary growth in Rondonia, Brazil, using imaging radar and Thematic Mapper data, Remote Sens. Environ. 59: 167–179

Ripley, B.D. (1996). Pattern recognition and neural networks, Cambridge University Press

Savenije, H.H.G. (1995). New definitions for moisture recycling and the relation with land use changes in the Sahel. J. of Hydr. 167, 57-78

Schistad-Scholberg, A.H., A.K. Jain and T. Taxt (1994). Multi-source classification of remotely sensed data; fusion of Landsat TM and SAR images. IEEE Trans. Geosci. Remote Sens. vol. 23, no.4, 768-778

Soares, J.V., C.D. Rennó, A.R. Formaggio, C. da Costa Freitas Yanasse and A.C. Frery (1997). An investigation of the selection of texture features for crop dissemination using SAR imagery. Remote Sens. Environ. 59: 234–247

Strahler, A.H. (1980). The use of prior probabilities in maximum likelihood classification of remotely sensed data. Remote Sensing of Environment, no. 10, pp. 135 – 163

Tucker, C.J. (1979). Red and Photographic Infrared Linear Combinations for Monitoring Vegetation, Remote Sensing of Environment, vol. 8, 127-150

Ward, R.C. and M. Robinson (1989). Principles of Hydrology McGraw-Hill (UK) 365 pp.

8 Evaporation

Massimo Menenti
The Winand Staring Centre, P.O.Box 125, 6700 AC, Wageningen, The Netherlands.

8.1 Introduction

8.1.1 General

Evaporation, the phase transition of liquid water to vapour, has attracted a rather considerable amount of attention and research work over the last two centuries (see Brutsaert, 1982 for a historic overview). This interest is due to the role played by evaporation in many processes in the earth-atmosphere system and the resulting practical relevance of observing and understanding evaporation in the context of hydrology, water management, meteorology, climatology, ecology and agriculture. The global role of evaporative processes is well established (Shukla and Mintz, 1982). Kustas and Norman (1996) referred to Wetherald and Manabe (1988) and Sato et al. (1989) while mentioning the impact of changes in available moisture released by evaporation on cloud formation, radiation budget and precipitation fields at global and continental scales. Many studies have documented the impact of evaporation on mesoscale atmospheric processes (e.g. Wang et al., 1995; Beljaars and Viterbo, 1994; Segal and Arritt,1992). Evaporation determines depletion of soil moisture and, therefore, crop water requirements of irrigated crops (e.g. Jensen et al., 1990) and partitioning of precipitation (Bruijnzeel, 1990).

The term *evaporation* is used in this chapter to mean the transition of water from the liquid to the vapor phase. The term *transpiration* is used in literature to mean that this process occurs in plant leaves. Evaporation may occur at or underneath the soil surface. The term *evapotranspiration* is widely used to mean evaporation from a mixture of vegetation and soil, such as from agricultural fields. Because of the clearer meaning, the term evaporation is preferred here.

Today's interest in evaporation spans a wide range of scientific disciplines and applications, although basic concepts and tools came to light in the context of agriculture and plant ecology. Thornthwaite (1948) proposed the concept of potential evaporation (E_0) to explain the relation of vegetation types with climate in terms of the difference between precipitation P and E_0. The equation proposed by Penman (1948) was meant to establish a general physical relation of climate with maximum water use by crops. The heritage of agricultural sciences was evident in the review of Eagleson (1982) on land surface processes at spatial scales from local to global. At the time (since the early 70's) there was already a widespread awareness of the sensitivity of atmospheric circulation to water available at land surfaces, as well as a growing body of work on detailed modeling of processes controlling water flow in the soil-plant-atmospheric continuum (Fritschen, 1982). As a matter of fact today's concept of Soil-Vegetation-Atmosphere-Transfer (SVAT) models had seen light already with Saxton et al. (1974) who described a Soil-Plant-Atmosphere-Water (SPAW) model.

Advances have been significant since those early stages. There seem to remain several outstanding issues, however. First and foremost the accuracy of both observations and models is moderate even under the well defined conditions of a field experiment. Absolute error in direct measurements of the instantaneous latent heat flux with eddy correlation devices is of the order of 30 Wm^{-2} (Lagouarde et al., 1996), while the error on daily evaporation calculated with numerical SVAT models is of the order of 1 mmd^{-1}, or a mean latent heat flux of 30 Wm^{-2} (Ottle et al., 1996). Moreover, what is required for improved understanding of environmental processes and for practical applications, such as irrigation water management, is spatial patterns of evaporation rather than local values. Direct measurements of evaporation require complex and costly devices such as eddy correlation systems or weighing lysimeters. Recent international experiments, such as EFEDA (Bolle et al., 1993) have seen an unprecedented deployment of eddy correlation devices. This notwithstanding, determination of spatial patterns of evaporation was only possible with additional remote sensing data (Pelgrum and Bastiaanssen, 1996). Numerical SVAT models have also been applied to produce spatial fields of evaporation (Harding et al., 1996) after calibration at five sites where eddy correlation measurements were available. No indication was given on the accuracy of the spatial patterns.

8.1.2 Remote sensing of land evaporation

Evaporation cannot be measured directly by means of spectral radiometric observations. The latter provide, however, information on atmosphere and land surface conditions useful to estimate evaporation, although concurrent ancillary data remain necessary. On the other hand, use of the spatial measurements of radiances, obtained with instruments on board satellites and aircrafts, is attractive precisely because of the need to determine spatial patterns of evaporation at heterogeneous land surfaces. The last few years have witnessed a renewed interest in this specialised area of remote sensing. Estimates of areal heat fluxes and of the land surface variables which control fluxes provide unique proxy data to study theoretical and modelling aspects of spatial variability (Beven and Fisher, 1996).

Two fundamental concepts need to be distinguished when dealing with evaporation of land surfaces (Fig. 8.1):

A. Potential evaporation E_0

B. Actual evaporation E_a

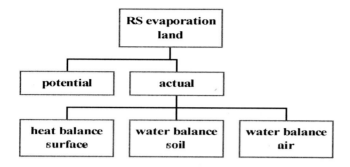

Fig. 8.1. Potential, actual evaporation and land surface processes

Ad A: Potential evaporation is meant to indicate the largest water loss rate from a vegetated land surface under given weather and climate conditions, but not limited by water supply. It is an obviously difficult concept to define precisely. Thornthwaite (1944) gave the definition "Potential evapotranspiration is the loss of water from a moist soil tract completely covered with vegetation and large enough for oasis effects to be negligible" as quoted by Monteith (1994). The term *evapotranspiration* indicates the combined contribution of evaporation from soil and open water with transpiration through the leaves.

Ad B: Water availability at land surfaces has a significant spatial and temporal variability so the actual water loss rate varies accordingly and is less than (or at most equal to) the potential rate. This is actual evaporation. It has been appreciated long ago (Thornthwaite, 1944) that water availability, evaporation, partitioning of net radiation and temperature are strongly interrelated. The latter is of particular relevance in the context of remote sensing as discussed in some detail later on.

The rate of water loss through the land surface relates to three different processes (Fig. 8.1) which all have been studied in the attempt to develop a method to obtain spatial patterns of actual evaporation. These are (Menenti, 1993):

A. heat transfer at the land – atmosphere interface

B. water flow in a soil column

C. water transport in the atmosphere

Most of the methods relying on space- or airborne instruments are based on simplified models of heat transfer at the land- atmosphere interface (land surface, case A). The processes A, B and C are further constrained to situations for which a balance equation must be fulfilled (see next section). Actual evaporation is then estimated from the balance equation after determining the other terms.

8.2 Evaporation and radiometric variables

8.2.1 Potential Evaporation

Choudhury (1997) used a combination (Penman, 1948) type equation to estimate and map global potential evaporation with satellite data:

$$E_0 = \frac{\Delta R_{ni} + \rho c_p D/(c r_e)}{\Delta + \gamma(r_s + r_H)/r_e} \qquad (\text{mm d}^{-1}) \qquad (8.1)$$

where:
Δ (hPa K^{-1}) is the differential of saturated vapour pressure with respect to temperature evaluated at the air temperature T_a (K);
γ (hPa K^{-1}) is the thermodynamic psychrometric constant;
R_{ni} (mm d^{-1}) is isothermal net radiation (evaporative water equivalent of daily total net radiation evaluated for isothermal conditions, i.e. with surface temperature equal to air temperature);
ρ and c_p are the density and specific heat of air ($\rho c_p = 1.2 \; 10^{-3}$ MJK^{-1} m^{-3} at the air temperature of 295 K and surface pressure of 1000 hPa);
D (hPa) is vapour pressure deficit;
r_s (s m^{-1}) is the surface resistance;
r_H (s m^{-1}) is the aerodynamic resistance for heat transfer from a virtual source height within the canopy to a reference level in the atmosphere;
r_e (s m^{-1}) is the effective resistance for heat transfer obtained by putting the aerodynamic resistance r_H and the resistance to long-wave radiative transfer, r_R, in parallel;
c is a constant equal to the product of latent heat of evaporation and density of water to give daily evaporation in units of (mm d^{-1}).

In terms of Eq. (8.1) weather and climate are defined by R_{ni} and D, while r_H depends both on weather (air temperature and humidity) and surface conditions (including water availability). The reduction of evaporation from potential to actual is parameterised by r_s: evaporation decreases with increasing r_s. This brings us to a somewhat more precise definition of potential land surface evaporation. With the exception of evaporation of water intercepted by leaves, the resistance to water flow through even a thin layer of soil or from open leaf stomata will not be zero. Potential evaporation can therefore be defined as the value given by Eq. (8.1) for a minimum value of r_s which depend on the land cover type, e.g. on the specific crop under consideration. The only truly climatological measure of potential evaporation applies to open water with $r_s = 0$.

To compute global potential evaporation Choudhury (1997) calculated R_{ni} using solar radiation and cloud cover data from the International Satellite Cloud Climatology Project (ISCCP) data set (Rossow et al., 1988), albedo was taken equal to 0.23 and net longwave radiation was estimated with a semi-empirical equation. The latter requires air temperature, vapour pressure and vapour pressure deficit.

Air temperature (near surface) was derived from TIROS Operational Vertical Sounder (TOVS) data (Susskind, 1993). Vapour pressure was also estimated with the TOVS data, using the TOVS precipitable water to obtain vapour pressure with a semi-empirical equation. The vapour pressure deficit was finally obtained calculating saturated vapour pressure from air temperature. The aerodynamic resistance was estimated using wind speed fields produced with a Four Dimensional Data Assimilation (4DDA) procedure (Schubert et al., 1993). Finally the minimum r_s was set at 70 (s m^{-1}). The calculated global evaporation (Table 8.1) was compared with lysimeter observations of evaporation from well watered grass at 35 locations. The most likely error at any location and month was 15 % and 5 % when taking all locations and months together. It should be noted that the values chosen for albedo and surface resistance may well be representative of well watered grass but not necessarily of other land cover types.

Table 8.1. Variables and radiometric observations used by Choudhury (1997) to compute global evaporation.

Variable	Radiometric observations	Source
Solar radiation	Reflected radiance	ISCCP Rossow et al. (1988)
Cloud cover	Reflected radiance	ISCCP Rossow et al. (1988)
Air temperature	Spectral emittance	TOVS Susskind (1993)
Vapour pressure	Spectral emittance	TOVS Susskind (1993)
Aerodynamic resistance	None	4DDA Schubert et al. (1993)

As defined by Penman (1948) potential evaporation applies to well watered grass. On the other hand, the maximum rate of evaporation for any land cover type is a useful agronomical and hydrological information. In terms of Eq. (8.1) differences in land cover type must be taken into account by using proper values of albedo, aerodynamic, effective and minimum surface resistance. In agronomic and irrigation practice so called crop coefficients are used to estimate the maximum rate of evaporation, E_m for a specific crop from E_0.

Two complementary methods to obtain crop coefficients using observations of spectral radiances were proposed by D'Urso and Menenti (1995 and 1996). The former method relies on Landsat Thematic Mapper images to translate estimates of the crop coefficients at a limited number of reference plots into a map. A two step numeric classification procedure is applied using performance indicators (calculated from image statistics) to optimize accuracy, separability and reliability.

The second method relies on an explicit equation to calculate crop coefficients using Eq. (8.1) in two different forms. Maximum evaporation, E_m, for different types of crops is calculated using the appropriate values of albedo, aerodynamic, effective and minimum surface resistance. Then Eq. (8.1) is used again with the values of albedo, aerodynamic, effective and minimum surface resistance applying to well watered grass to obtain E_0. The ratio (E_m / E_0) gives the crop coefficients which are defined as:

$$K_c = \frac{E_m}{E_0} \qquad (-) \qquad (8.2)$$

This will give an equation of the form:

$$K_c = f(R_n, T_a, e^*, e_a, u, r_i) \qquad (-) \qquad (8.3)$$

where:
R_n : net radiation (W m^{-2});
T_a : air temperature at reference height (K);
e^* : saturation vapour pressure at T_a (hPa);
e_a : actual vapour pressure (hPa);
u : wind speed (m s^{-1});
r_l : transport resistances (s m^{-1});

Note that the resistances r_H and r_s in Eq. 8.1 for E_0 are related to the Leaf Area Index, LAI, and to crop height. When applying this second method Landsat Thematic Mapper data are used in a quantitative manner to determine albedo and LAI. This procedure avoids the need for frequent acquisition of satellite data, since they are used to determine albedo and LAI which do not change very rapidly. Many authors noted the usefulness of spectral indices, as a measure of fractional vegetation cover, to estimate potential crop transpiration (Seevers and Ottmann, 1994; Inoue and Moran, 1997; Yang et al, 1997). Brasa- Ramos et al. (1996) described a method to map and monitor irrigation water requirements. Choudhury et al. (1994) established significant linear correlations between the K_c and spectral vegetation indices.

The high spatial resolution map (Colour Plate 8.A) of crop-specific E_m presented by D'Urso et al. (1999) demonstrates the practical scope of this information for irrigation water management (see also Chap. 17). A set of five Thematic Mapper images were used to determine the K_c, which change rather slowly throughout the irrigation season. This approach does not require very frequent acquisitions of satellite data and is feasible with the capabilities of current earth observation systems.

8.2.2 Actual Evaporation

Heat balance – surface. The heat balance at the land surface reads:

$$R_n + G + H + LE = 0 \qquad (W\ m^{-2}) \qquad (8.4)$$

where R_n is net radiation, G soil heat flux, H sensible heat flux and LE latent heat flux, i.e. the amount of energy L (J kg^{-1}) required in the liquid to vapour transition of E (kg m^{-2} s^{-1}); fluxes are counted positive when directed towards the surface.

This equation underlies most of the significant research effort dedicated to the development of a remote sensing approach to determine areal evaporation.

Imaging spectrometers onboard aircraft and satellites measure top-of-atmosphere radiances which are related to bottom-of-atmosphere radiances but cannot measure turbulent heat fluxes at the land atmosphere interface. Literature tells us another story, however, given the large numbers of scientific papers and technical reports dealing with the use of measurements of reflectance and emittance to obtain heat fluxes and particularly latent heat flux. Much of this tradition was initiated in the early 70-s by agricultural scientists studying the physical climate of crops, with special reference to irrigation.

The dependence of canopy temperature on solar radiation was studied by Stone et al. (1975) and early studies on the estimation of evaporation were presented by Heilman et al. (1976). The interrelation of heat balance with crop yield and soil water balance was addressed by Hatfield et al. (1978). The Heat Capacity Mapping Mission (HCMM), a small, low cost and focussed mission (1978-1980) boosted the interest of the scientific community for heat balance studies in relation with land surface hydrology. A detailed overview of the results of HCMM- and evaporation related results was presented by Reiniger and Seguin (1986). The relatively low spatial resolution of the HCMM radiometer led to consider large heterogeneous areas and the first conceptual and practical difficulties with the application of the early algorithms intended for homogeneous agricultural patches appeared. Moran and Jackson (1991) reviewed methods to estimate the spatial distribution of evaporation using remote measurements of surface temperature in combination with meteorological observations. The method assumes that observations of e.g. air temperature and humidity are available at the proper areal density, which is a relatively safe assumption when dealing with agricultural areas. In many other cases such as Alaska (Gurney and Hall, 1983) or the Libyan desert (Menenti, 1984) feasible approaches had to be based on variables observable with airborne or spaceborne radiometers only. In another review Menenti (1993) addressed the issue of the reference air temperature and of the inherent correlations in the spatial variability of land surface properties. Air temperature at some higher elevation in the atmospheric boundary layer has a limited spatial variability because of mixing and, therefore, is a suitable reference temperature for heat balance studies of heterogeneous land surfaces. This concept was demonstrated by Brutsaert et al. (1992) and Menenti and Choudhury (1993).

Water balance- soil. The water balance equation of a soil column reads:

$$P + I + Q + \delta m + \delta w + E = 0 \qquad (\text{kg m}^{-2}\,\text{t}^{-1}) \qquad (8.5)$$

where P is precipitation, I is inflow (capillary rise and lateral infiltration), Q is runoff, δm is change in soil water storage, δw is change in snow water equivalent and E is actual evaporation. It appears that the terms of Eq. (8.5) are hardly accessible with sensors on board satellites. Choudhury and DiGirolamo (1998) computed global evaporation at low spatial resolution (2.5° x 2.5°) with a biophysical process model based on Eq.(8.1) and a set of process- sub-models. Satellite observations were used to obtain radiative forcing (solar radiation, albedo; ISCCP see Table 8.1), canopy resistance as a function of Photosynthetically Active Radiation (PAR; ISCCP see Table 8.1), air temperature and vapour pressure (TOVS; see Table 8.1) and fractional vegetation cover (AVHRR).

Water balance – air. The water balance of an atmospheric column reads:

$$\frac{\delta v}{\delta t} + div\, q_v = E - P \qquad (kg\, m^{-2}\, t^{-1}) \qquad (8.6)$$

where δv is change in atmospheric water vapour and div q_v is horizontal divergence of vertically integrated vapour flux. Starr and Peixoto (1958) and Peixoto (1970) used this principle to study divergence of water vapour with world-wide atmospheric soundings. Oki et al. (1995) used vertical profiles of water vapour density produced with the 4DDA system of the European Centre for Medium range Weather Forecasts (ECMWF). These data, especially in tropical regions, depend heavily on the use of satellite sounders. Moreover the same approach can in principle be applied with the observations provided by satellite sounders only, although accuracy, vertical and horizontal resolutions might not be adequate. The same conceptual approach has been applied by Eichinger et al. (1991) using a Raman lidar to scan repeatedly a relatively small air mass close to the land surface to determine the vapour budget and evaporation.

The challenge to all remote-sensing methods to estimate land evaporation is how to determine model variables not directly related to feasible observations. The empty cells in the left column in Table 8.2 underscore this statement. The following section is meant to help readers through the maze of methods described in literature. Several reviews have been published on this subject, e.g Kustas et al. (1989 a, b), Kairu (1991) and Jha et al. (1996).

Table 8.2. Remote sensing of evaporation: feasible observations and required model variables

Feasible observations	Required model variables
1- Reflected directional radiances 0.4 - 2.5 μm	Surface albedo
2- Emitted radiances 8 - 14 μm	Surface temperature
3-Radar backscatter	Soil moisture (no robust algorithm)
4- Microwave emittance	Soil moisture
5- Raman lidar backscatter	Vapour concentration
6- 2. + spectral radiances	Surface emissivity
7- 1 + 2	Net radiation
Not observable close to land surface	Air temperature
Not observable close to land surface	Vapour pressure
9- 1 + modeling	Leaf Area Index
10- 1 + modeling ; Laser altimetry	Fractional vegetation cover
11- Laser altimetry	Aerodynamic roughness length
Not observable	Aerodynamic resistances for heat and vapour transfer

8.3 Remote Sensing of Land Evaporation: Applications and Modelling Approaches

8.3.1 General

Land surface temperature is controlled by the relative magnitude of heat flux densities at the land- atmosphere interface (Eq. 8.4). It is therefore related to the latent heat flux, even though this relationship is complex in a general case. Actual evaporation can be related to the heat balance using a general form of Eq. 8.1 (Menenti, 1984):

$$LE = \frac{\rho_a c_p [e^*(z) - e(z)] + s_a r_{ah}(R_n + G_E) + s_s \rho_a r_{sh} G_E}{\gamma(r_{av} + r_{sv}) + s_a r_{ah} + s_s \rho_a c_p r_{sh}} \quad (Wm^{-2}) \quad (8.7)$$

where:
s_a is the slope of the saturated soil vapour pressure curve in air;
s_s in soil;
$e^*(z)$ is saturated and e is actual vapour pressure at height z;
$r_{ah,v}$ is aerodynamic resistance to heat and vapour transfer in air;
$r_{s,h,v}$ is resistance to heat and vapour transfer between the liquid water- moist air interface and the physical boundary of the evaporating system (e.g. leaf or soil surface);
γ is psychrometric constant;
G_E is soil heat flux at liquid water – moist air interface.

This equation describes the influence of: available energy (i.e. R_n-G_E), vapour pressure deficit and the various flow resistances in soil and air on evaporation. The same latent heat flux density LE may be observed, therefore, for different (R_n-G_E)-values, given different combinations of the remaining variables such as vapour pressure and flow resistances.

A further consequence is that LE is not a one-value function of surface temperature. This can be easily shown by re-writing the previous equation as (Jackson et al., 1988):

$$T_0 - T_a = \frac{\frac{(r_i + r_e)}{\rho_a c_p}(R_n + G) - \frac{1}{\gamma}(e^* - e)}{1 + \frac{s}{\gamma} + \frac{r_i}{r_e}} \quad (K) \quad (8.8)$$

where the notation has been simplified by indicating as r_e resistances in air and as r_i resistances to transport in the evaporating system. This equation describes concisely how radiative, boundary layer and surface (r_i) conditions control surface temperature. The internal resistance r_i is higher for drier surfaces, but changes of surface temperature are related in a more complex way to r_i and to LE. For large r_i -values, for example, (T_0-T_a) does not depend on r_i anymore and increases with R_n

and/ or r_e. We may conclude that (T_0-T_a) is a straightforward indicator of evaporation when a limited range of environmental conditions is considered. A robust model of the dependence LE on (T_0-T_a) should account properly for all radiative, boundary layer and surface forcing factors.

With the availability of easy to use thermal infrared thermometers agricultural scientists undertook detailed experiments to study the dependence of leaf and canopy temperature on weather conditions and specifically on radiative forcing. Stone et al. (1975) documented the rapid response of canopy temperature (sorghum) to changes in net radiation due to passing clouds. Fluctuations in canopy temperature of up to 3 K over three minutes was observed. Heilman et al. (1976) presented estimates of evaporation using a simple resistance equation and the surface to air temperature difference. Remote sensing studies of the land surface heat balance, and of actual evaporation has been an active field of research during the last 25 years. A review of methods is presented in the following pages. In anticipation it may be mentioned that while no general and robust method is at hand, most methods described are applicable under specific conditions. The materials presented aim to help identifying the merits of different approaches. To help readers through the maze, the review has been organized as shown in Fig. 8.2. The sequence from [1] through [7] is closely related to the evolution of this research field over time.

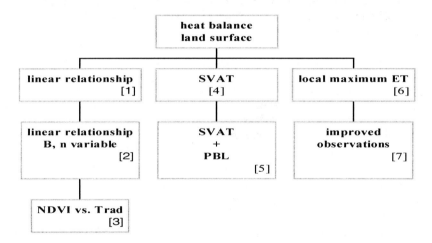

Fig. 8.2. Overview of methods based on the heat balance equation; numbers in brackets refer to text

8.3.2 Linear relationships between evaporation and land surface temperature [1]

The linear relationship given by Jackson et al. (1977):

$$(R_n - LE)_{day} = B(T_0 - T_a)^n \qquad \text{(mm d}^{-1}\text{)} \qquad (8.9)$$

where:
$(R_n - LE)_{day}$: daily integral of the difference $(R_n - LE)$;
B = constant
T_0 : instantaneous surface temperature at midday;
T_a : instantaneous air temperature at midday;
n = 1
was a milestone in the development of remote sensing methods to determine actual evaporation through the surface heat balance approach. This equation indicates that the excess available energy (i.e. Rn-LE) is proportional to the difference between surface and air temperature. This empirical relationship implies that the ratio of sensible heat flux to (T_0-T_a), i.e. the aerodynamic resistance to sensible heat transfer, is a constant. Net radiation, as well as LE, was observed at one location over a homogenous crop of wheat. Both R_n and LE are daily values, while (T_0-T_a) is instantaneous. Interest for this approach developed rapidly and diverse experimental, theoretical and modelling studies were undertaken. Land heterogeneity was given scarce attention initially. At heterogeneous land changes in net radiation are significant (e.g. Menenti, 1984) and force changes in surface temperature comparable with the observations of Stone et al. (1975) on temporal changes. This explains why investigations dealing with regions where albedo was rather variable used measurements of both surface temperature and albedo to estimate evaporation (Gurney and Hall, 1983; Menenti, 1984). Studies have been published, however, using this linear relationship in modified forms (e.g. Nieuwenhuis et al., 1985) to map actual evaporation. It should be noted that the spatial variability of other land surface properties such as aerodynamic roughness determines spatial patterns of surface temperature related to evaporation in a more complex manner.

Initially, improvements were sought along two avenues:

— use of different values of B in the linear relationship (see 8.3.3) to account for differences in land surface properties (Seguin and Itier, 1983; Sandholt and Andersen, 1993; Carlson and Buffum, 1989; Thunnissen and Nieuwenhuis, 1990);

— detailed modelling of heat and momentum transfer at the land – atmosphere interface (e.g. Carlson et al., 1981; see 8.3.4);

8.3.3 Improved linear relationships [2]

The linear relationship rests on many assumptions, one of which is the uniqueness of the radiometric surface temperature T_{rad}. The latter is not unique as a matter of principle due to the dependence on view angle when a target is observed with a radiometer. While changes with view and azimuth angles may be negligible for a complete canopy, changes are large for partial canopies (e.g. Kimes et al., 1983). Caselles et al. (1992) demonstrated that large deviations from linearity were due to the view and azimuth angles and that different values of the parameters in the linear relationship had to be calculated for each observation geometry and target. Building upon the results of Wetzel et al. (1984), Carlson and Buffum (1989) proposed a different type of linear relationship by relating $(R_n - LE)_{day}$ to the rate of increase in surface temperature during the morning. This approach avoids the need for ground observations of air temperature. As with the original linear relationship, the parameters depend on meteorological and surface variables and a planetary boundary layer model was used to study this dependence. Carlson and Buffum (1989) observed that the parameters were highly sensitive to wind speed and surface roughness.

8.3.4 Relationships between evaporation, surface temperature and spectral indices [3]

Price (1990) explicitly related observed variability of surface temperature and a spectral index to the range of actual evaporation. The spectral index provides a measure of the green vegetation cover. Potential evaporation may be estimated independently with e.g. the methods mentioned in the preceding pages. Maximum evaporation for a given amount of green vegetation may be estimated as described by Inoue and Moran (1997). Minimum and maximum surface temperatures change as a function of the amount of green vegetation: the range is largest for bare surfaces and decreases with increasing green vegetation.

Carlson et al. (1995) proposed to modify the simplified linear relationship to account for the spatial variability of vegetation cover through variable B (the constant in the linear relationship) and n (an exponent accounting for deviations from linearity). To determine the required parameters a scatter plot of NDVI vs. radiometric surface temperature was used. To explain this pattern and to estimate the B and n parameters a SVAT model was used. The results indicated that deviations from linearity were significant, although a family of linear relationships (i.e. for different combinations of the two parameters B and n) could still be a practical and useful approximation. To estimate the near surface air temperature, the surface radiometric temperature corresponding with the apex of the scatter plot, i.e. at the highest observed NDVI was taken. The air temperature at 50 m elevation was used as a reference and estimated from the near surface air temperature as $(Ta - 1)$. Carlson et al. (1995) noted that the estimation and mapping of evaporation requires accurate values of net radiation.

The slope of the NDVI vs. T_{rad} relationship has been interpreted as a proxy of canopy resistance by Nemani and Running (1989). Smith and Choudhury (1991) analyzed in detail the relation of surface radiant temperature and NDVI for agricultural land and native evergreen forests in New South Wales. Surface radiant temperature increased with decreasing NDVI for agricultural land, but not for forests. They also noted that changes in soil water availability had opposite effects on the slope of the NDVI vs. T_{rad} relationship according to whether the reduction in evaporation was from soil or from vegetation. Another evaluation of the NDVI vs. T_{rad} relationship was presented by Friedl and Davis (1994) who used observations of a tall-grass prairie in NE Kansas. While the negative correlation was observed consistently through time, the slope was highly date- and time-specific. The authors concluded that these findings were likely due to the large differences in T_{rad} between soil and vegetation, rather than to the land surface energy balance.

8.3.5 Soil Vegetation Atmosphere Transfer (SVAT) models [4]

This approach was rather popular in the early years of thermal infrared remote sensing and heat-balance studies. Use of SVAT models requires a detailed and realistic description of land surface – boundary layer interactions. Initially numerical models were used to establish case-specific look-up tables to estimate evaporation from a pair of day-night observations of surface radiometric temperature (e.g. Soer, 1977 and 1980; Rosema and Bijleveld, 1977; Carlson and Boland, 1978). Camillo et al. (1986) described the use of a SVAT model to estimate soil hydrologic properties by calibrating the calculated soil brightness temperature against remote measurements. Using this method, values were estimated for hydraulic conductivity, matric potential and soil moisture at saturation, and a soil texture parameter. An inversion algorithm to infer heat fluxes at the land surface was validated by Taconet and Vidal Madjar (1988). Their SVAT model was driven by satellite (METEOSAT and AVHRR) thermal infrared measurements and weather observations while ground based measurements at the same spatial resolution as the satellite data were used for validation. The performance of the inversion was improved by using a boundary layer model to characterize the surface layer instead of ground-based weather observations.

Difficulties with the determination of all model variables led to a trade-off. Area estimates of evaporation were obtained (e.g. Moran and Jackson, 1991) with measurements of surface reflectance, temperature and normalized difference vegetation index (a measure for the amount of green vegetation) and ground measurements of all the remaining land surface variables.

The SVAT used did not describe flow processes in the atmospheric boundary layer. The near surface air layer was characterised using ground measurements of air temperature, humidity and wind at a reference height (e.g. 2m) to solve the system of heat transfer equations. The difficulty with this approach is that both air temperature and humidity are strongly coupled with the surface. Vieira and Hatfield (1984) for example proposed to estimate surface temperature from air temperature.

Almost twenty years of research in this direction seem to indicate that estimation of evaporation by model inversion is severely limited by the significant requirements for ancillary data and by the relatively poor accuracy of remote measurements of radiometric surface temperature (Feddes et al., 1993; Bastiaanssen et al., 1994). Majumdar (1991) concluded that field experiments were needed to obtain satisfactory, site-specific estimates of model variables for further use in evaporation mapping. Taconet et al. (1995) used an advanced model to estimate evaporation from a single measurement of surface radiometric temperature at midday. Independent measurements of soil water content, of vegetation density and of canopy height were required. Once the accuracy of estimates is improved, however, SVAT models may become again an essential tool to obtain daily evaporation from instantaneous radiometric observations.

8.3.6 Integrated SVAT and Planetary Boundary Layer (PBL) models [5]

Integrated land surface – PBL models are attractive since upper boundary conditions can be defined using large-scale variables, i.e. truly constant over large distances These models use some height in the atmospheric boundary layer as a reference. Wieringa (1986), for example, proposed to use a blending height as a reference height. Horizontal gradients in PBL variables vanish at the blending height due to horizontal mixing. In other words, the effect of spatial heterogeneity on the PBL vanishes at the blending height. Brutsaert et al. (1992) chose the PBL as a reference.

The advantage of choosing a higher reference height is two-fold. Coupling with surface conditions is weaker, so air temperature at the top of the PBL changes over larger spatial scales in comparison with near surface air temperature.

Carlson et al. (1981) developed a detailed land surface – atmospheric boundary layer model to study sensible and latent heat fluxes, soil moisture availability (thermal inertia) in a urban – rural environment. The function of the model was to partition net radiation into sensible and latent heat fluxes and to determine the daily amplitude of surface temperature. Effective land surface variables were defined to match the observed surface temperatures with simulated ones through heat balance modeling. Four different sub-systems were considered: a mixing layer, a 50 m surface turbulent layer, a thin transition layer which contains many surface obstacles and a 1 m ground layer. The model was applied with observations provided by the imaging radiometer on-board the Heat Capacity Mapping Mission (HCMM).

The necessity of using observations of the atmospheric boundary layer to study heat fluxes at the land surface was further demonstrated by Diak and Whipple (1993). Diak et al. (1994) linked top-of the-atmosphere radiances, simulated to reproduce the observations of the High resolution Interferometer Sounder (HIS), to heat fluxes and skin temperature at the surface using a planetary boundary layer (PBL) model. The interest of this approach is that HIS spectral radiances relate to both the PBL (at some wavelengths) and the surface (at other wavelenghts). A further refinement of this approach was presented by Anderson et al. (1997) who coupled a dual- source model for sensible heat transfer at the surface with a parameterization of the development of the PBL in response to surface heating. A sequence of observations of radiometric surface temperature T_{rad} gives the rate of change of T_{rad} from which heat fluxes at the surface are estimated by model inversion. The coupling with PBL growth eliminates the need for observations of near surface air temperature. Soil and vegetation temperatures within a mixed target are estimated using a parameterization of the angular dependence of T_{rad}.

8.4 Current trends: improved observations and improved parameterizations.

8.4.1 Local maximum evaporation and land surface temperature [6]

When dealing with heterogeneous land surfaces two issues have to be dealt with separately:

A. Surface temperature provides information on actual evaporation in a relative sense only: given the energy available (R_n-G) and Planetary Boundary Layer (PBL) conditions, including the aerodynamic properties of the surface, a higher surface temperature indicates lower evaporation.

B. Accurate estimates of evaporation require accurate determination of land surface variables, particularly of the maximum evaporation and minimum surface temperature attainable under the given conditions (available energy and PBL conditions).

In other words the maximum evaporation and the corresponding range from minimum to maximum surface temperature changes spatially for a given radiative and PBL forcing. To use surface temperature as a measure of evaporation, i.e. of water availability, both maximum evaporation and the range from minimum to maximum surface temperature must be determined first.

Implicitly, both the linear relationship and the look-up table methods attempt to address this issue, although evaporation mapping would require mapping of $(Rn - G)$, B and n. On the other hand the early results (e.g. Stone et al., 1975; Jackson et al., 1977) indicated that observations of surface temperature were highly correlated with actual evaporation under precisely defined experimental conditions. Analysis of multi-spectral data sets, i.e. concurrent measurements of surface temperature, albedo and spectral indices brought evidence of well identified clusters of data points related to the range of actual evaporation. Goward et al. (1985) provided evidence of the correlation of NDVI with T_{rad}. Menenti et al. (1986) studied patterns of temperature and albedo in the oasis-playa-desert environment of Tunisia. Menenti et al. (1989) exploited the correlation of surface albedo with surface temperature to map actual evaporation.

Observations of surface temperature give a measure of evaporation, relative to maximum evaporation, at constant green vegetation. This concept, i.e. the vegetation index/ temperature trapezoid (VITT; Fig. 8.3) was described in detail by Moran et al. (1994a). Carlson et al. (1995) used a SVAT model to determine linear isopleths (at constant soil moisture availability and relative evaporation) relating NDVI to surface radiant temperature. Once the range of surface temperature, the maximum evaporation and the isopleths have been determined, this method makes straightforward use of observed surface temperatures possible.

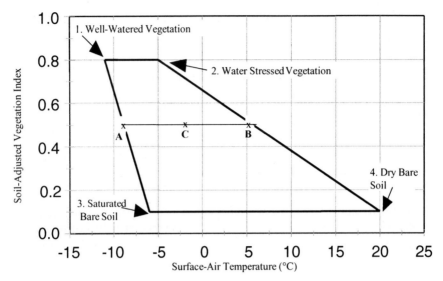

Fig. 8.3. The Vegetation Index Temperature Trapezoid (VITT); the stress index is given by the ratio (AC/AB) (Moran et al., 1994a)

Detailed experimental evidence to support the occurrence of trapezoidal patterns in scatter plots of NDVI vs. T_{rad} was presented by Carlson et al. (1990). Segments relating to bare soil and vegetation were clearly identified and had significantly different slopes when observations collected on one date were considered, while significantly larger scatter was observed on other dates. The minimum (vegetation) and maximum (bare soil) T_{rad} were estimated as the mean values of image samples having minimum variance. In contrast with the approach of Price (1990) both the envelop and the minimum and maximum temperature were determined from image statistics. These extreme conditions may not be always observable. The latter is necessary to estimate heat fluxes at the surface using the asymptotic surface temperatures of sunlit bare soil and sunlit vegetation and the relationship between NDVI and surface temperature (see 8.3.4). The NDVI vs. T_{rad} approach has been applied and evaluated by Yang et al. (1997) to determine actual evaporation of sugarcane with Thematic Mapper data.

The range in surface temperature and maximum evaporation can also be calculated as a function of surface albedo and PBL conditions as described by Menenti and Choudhury (1993). The maximum value of (T_0-T_a) can be determined by taking the limit of Eq. (8.8) for $r_i \to \infty$:

$$(T_0 - T_a)_u = \frac{r_e}{\rho_a c_p}(R_n + G) \qquad (K) \qquad (8.10)$$

And the minimum value for $r_i \to 0$:

$$(T_0 - T_a)_l = \frac{\frac{r_e}{\rho_a c_p}(R_n + G) - \frac{1}{\gamma}(e^* - e)}{1 + \frac{s}{\gamma}} \quad (K) \tag{8.11}$$

The ratio of actual to net evaporation, E_a / E_{max}, becomes then:

$$\frac{E_a}{E_{max}} = 1 - SEBI \quad (-) \tag{8.12}$$

where the Surface Energy Balance Index (SEBI) is:

$$SEBI = \frac{\dfrac{(T_0 - T_a)}{r_e} - \dfrac{(T_0 - T_a)_l}{r_{e,l}}}{\dfrac{(T_0 - T_a)_u}{r_{e,u}} - \dfrac{(T_0 - T_a)_l}{r_{e,l}}} \quad (-) \tag{8.13}$$

where:
T_a is air potential temperature at the top of the PBL.
r_e, $r_{e,l}$, $r_{e,u}$ = external resistances

Net radiation depends on surface albedo α as $R_n = (1 - \alpha) R_{sw}$ where R_{sw} is the solar irradiance flux density. A trapezoid can now be constructed by estimating the range of α, e.g. as expected in the region under study. This results in the diagram presented in Fig. 8.4. The upper line indicates limiting conditions for $r_i \to \infty$ and the lower line for for $r_i \to 0$. The ratio E_a/E_m is easily determined as indicated by the position of T_0 relative to the upper and lower bound. The agreement with observations was good (Fig. 8.5).

Practical use of the SEBI diagram requires meteorological observations of air temperature, vapour pressure and net radiation. Moreover the resistance r_e has to be determined. As indicated above in relation with SVAT models, processes at the land –atmosphere interface cannot be studied independently of the Planetary Boundary Layer. Accordingly, Menenti and Choudhury (1993) proposed to use SEBI with PBL instead of near surface values of air temperature, vapour pressure, friction velocity and net radiation. Once the upper and lower bounds (Fig. 8.4) have been calculated, values of SEBI are obtained by plotting paired observations of [T_0, α] in this diagram. The value of E_a / E_{max} is then easily determined as indicated in Fig. 8.4.

The use of PBL variables implies that PBL and radiative forcing is constant over large areas and observations of [T_0, α] describe how elements of the land surface react to forcing.

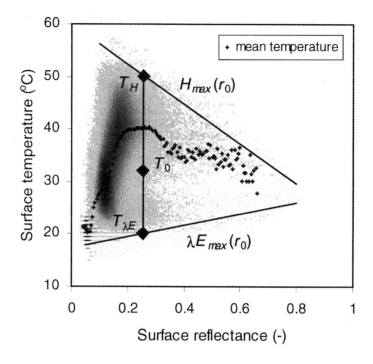

Fig. 8.4. The Surface Energy Balance Index: diagram of reference land surface states; data points obtained with TM radiometric observations; Piano di Rosia, Italy; August 23rd 1997 (courtesy of G. Roerink, SC-DLO)

8.4.2 Improved observations of land surface variables [7]

The quest for simple operational methods to estimate and map actual evaporation (Fig. 8.2) has come to highlight the need for accurate observations of a wider set of land surface variables. The methods described in the previous paragraph require relatively complex models to construct the case-specific algorithms which make direct use of remote measurements of spectral radiances. The Local Surface Energy Balance (LSEB) approach of Moran et al. (1989) is based on straightforward pixel-wise calculation of the heat balance and transfer equations with emphasis on accurate determination of all variables involved. The LSEB approach relies on the combination of meteorological observations at a single location with airborne or satellite measurements of reflected and emitted radiances. The values obtained with Thematic Mapper data differed from the ground-based measurements by less than 12%.

This work proves that remote sensing estimates of evaporation are feasible and sufficiently accurate when use of airborne and satellite measurements is limited to the time of acquisition, to a relatively homogeneous land surface and when all nec-

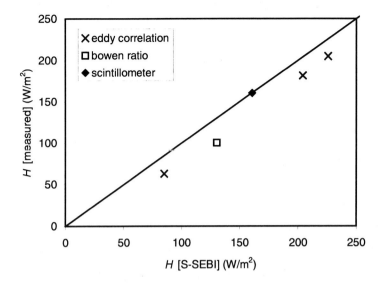

Fig. 8.5. Sensible heat flux H: comparison of SEBI estimates with ground measurements; scintillometer value is a path-length average of H; Piano di Rosia, Italy; August 23rd 1997; (courtesy of G. Roerink, SC-DLO)

essary meteorological variables are measured. Evaporation mapping becomes challenging when other surface types are considered such as open canopies in arid environments and when appropriate meteorological data are not available. Estimation of sensible heat flux with a single source one-dimensional equation becomes cumbersome and other parameterizations taking into account the large differences in T_{rad} between soil and vegetation are necessary (Moran and Jackson, 1991).

Kustas and Norman (1996) reviewed the state of the art on remote sensing of evaporation. Errors on radiant fluxes have a large impact on estimated evaporation, so more attention has to be dedicated to the estimation of radiant fluxes. Shortwave solar radiation, R_{sw}, and surface albedo can be determined with e.g. GOES observations (Pinker et al., 1995). This gives shortwave net radiation. Longwave net radiation has been estimated using space-borne sounders (Darnell et al., 1992). Other methods are based on estimates of surface albedo and temperature to obtain the upwelling components and on meteorological data to obtain the downwelling components (Moran et al., 1989). Soil heat flux can be obtained by estimating the ratio G/R_n through spectral indices (Choudhury et al., 1994). Errors: $R_{sw} = 10\%$ daily, 20-30% hourly; $R_n = 10\%$ hourly (when using meteorological data).

Assessment of error of estimate on turbulent fluxes is an issue on itself: when combining the error on at-surface radiances, surface albedo and T_{rad} with formal error propagation, absolute errors of 100 to 150 Wm^{-2} are obtained (Kustas and Norman, 1996). On the other hand, many authors have compared remote sensing estimates of evaporation with measurements to find differences significantly smaller than 100 Wm^{-2}.

Two land surface variables appear in many parameterizations of evaporation and turbulent heat fluxes:

— the Leaf Area Index (LAI);

— the aerodynamic resistances for turbulent transport of momentum and of sensible heat fluxes

Accurate determination of these variables made the difference between good and poor agreement with observations in many studies of latent and sensible heat fluxes.

Leaf Area Index. Wollenweber (1995) used a one-dimensional model to study the impact of spatial variability of land surface variables on heat fluxes. The LAI was used to parameterize roughness length, zero plan displacement and stomatal resistance. Since these parameterizations are not linear, spatial aggregation of either LAI or of the calculated heat fluxes will give different results. Kaneko and Hino (1996) proposed a method to estimate evaporation based on concurrent estimates of LAI (inferred from NDVI), of vapour pressure deficit at leaf temperature (which requires an estimate of the latter) and of stomatal resistance. On the other hand Nemani et al. (1993) developed a procedure to estimate LAI using a corrected NDVI by including spectral reflectance in the mid-infrared. They concluded that relation of NDVI with LAI was poor in their forest area and that the proposed method significantly improved the LAI estimates. The impact of the correction was significant on estimates of evaporation.

Aerodynamic resistances. The most widely used parameterizations of momentum and heat fluxes are derived from flux-profile relationships and relate mean (along the vertical direction) flux to the difference in either wind speed or temperature between the surface and a reference height. The accuracy of such parameterizations depends on the accuracy of the aerodynamic resistance (r_a, the ratio of the flux density to the difference in the state variable).

Kalma (1989) evaluated several parameterizations to conclude that the estimation of r_a is difficult. Under conditions of large spatial variability of bare and vegetation patches, accurate values of the resistance for momentum require accurate values of the aerodynamic roughness length (Humes et al., 1994). The use of thermal infrared radiometers to estimate surface temperature and sensible heat flux density brings an additional difficulty. The proper reference for the flux-profile relationship is an aerodynamic surface temperature, T_{aer}, while T_{rad} relates to directional measurements of radiance. The difference between T_{aer}, and T_{rad} is significant and depends on environmental conditions (Choudhury et al., 1986).

To determine the resistance for heat fluxes the difference in temperature between foliage and bare soil must, therefore, be taken into account. Norman et al. (1995) described an explicit dual source heat transfer model. A different approach to estimate sensible heat flux using observations of T_{rad} was proposed by Lhomme et al.

(1994). They used an empirical relationship between ($T_{foliage} - T_{soil}$) and ($T_{rad} - T_{air}$) to estimate the resistance for sensible heat transport. Kustas (1990) compared a single source model, using two different procedures (to estimate the kB^{-1} parameter) to correct for the difference ($T_{rad} - T_{aer}$) with a dual-source model. Better agreement with observations was obtained for the single source model using a theoretical equation to estimate the kB^{-1} parameter.

These two examples lead to a preliminary conclusion: relatively simple models of heat transfer at the land atmosphere interface can be sufficiently accurate if a few key-variables can be determined with sufficient accuracy and capturing the fundamental physics of the process, as in the case of the ($T_{rad} - T_{aer}$) difference.

8.5 Spatial variability

The work reviewed in the preceding pages indicates that spatial variability of evaporation, and of the land surface variables determining it, is at the same time a problem and an asset. The methods exploiting image context (e.g Price, 1990; Bastiaanssen, 1995) take advantage of range of variability to derive additional equations. On the other hand, spatial variability makes it impossible to use methods such as the linear relationships, which apply to small changes in T_{rad} around the reference value. Even more to the point, spatial variability implies that the spatial resolution of observations has to be selected on the basis of the spatial structure of the landscape.

Kustas et al. (1990) applied the LSEB approach of Moran et al. (1989) to different agricultural fields and found that errors were of the order of 2 mmd^{-1}, which is significant. Better accuracy was achieved in another study in a semiarid rangeland basin by Kustas et al. (1994 a,b). To compute sensible heat flux the factor kB^{-1} was calculated with an empirical equation determined earlier with observations collected in the same basin. Spatial extrapolation was done with a scheme similar to the one described by Menenti (1979, 1984) based on the estimate of first order differentials relative to the values of state variables and fluxes observed at a reference site. In the study of Kustas et al. (1990) spatial extrapolation was done using the differentials with respect to surface temperature and albedo. Changes in roughness lengths for momentum and heat transfer were neglected and the values at the reference site kept constant over the watershed. Analysis of observations did show that this assumption was not correct, since the empirical coefficient used to estimate the roughness length for heat transfer had to be changed significantly on each day used for the study.

Most of the equations necessary to model exchanges of momentum and heat at heterogeneous land surfaces are non-linear. Spatial averaging of the independent variables in these equations has therefore a potentially large impact on the variables calculated with the non-linear equations. This is also the case when using observations at a spatial resolution insufficient to capture the spatial structure of the landscape. More precisely, if the spatial resolution is larger than the smaller homogeneous element in the landscape, spatial averaging will always contribute to the error on estimated evaporation. Given the spatial resolution and the equations to be computed, weighing coefficients can be estimated to minimize the error due to spatial aggregation. Examples of this approach were presented by Pelgrum and Bastiaanssen (1996), Bastiaanssen et al. (1996) and Hipps et al. (1996). The underlying problem remains the definition and determination of a measure of the spatial structure of landscapes. For this purpose Menenti et al. (1996) explored the use of spatial statistics of very high resolution laser measurements.

The issue of spatial aggregation (up-scaling) is of significant theoretical and practical complexity and will require significant attention in the next few years. The underlying cause of complexity is that the two fundamental concepts of homogeneity (of each landscape element) and heterogeneity (of a landscape comprising different elements) are relative rather than absolute concepts. It is therefore difficult to translate these two concepts into a measure of homogeneity and heterogeneity. As a consequence, it is difficult to measure the spatial structure of a specific landscape and to assess the error due to spatial aggregation of observations at a given resolution.

8.6 Accuracy

The issue of accuracy has haunted remote sensing studies of land evaporation since the early years. Design of a validation experiment for maps of actual evaporation computed with nearly instantaneous image data is challenging because of fundamental and practical reasons. The very nature of direct measurements of latent heat flux density by e.g. eddy correlation systems differs from estimates obtained with measurements of spectral radiances. The latter is local in space and instantaneous in time, while the former integrates turbulent fluxes over some distance upwind of the observation station. Due to the intermittent nature of turbulent heat transfer, measurements of surface radiant temperature may reveal changes too rapid to be seen in the time integrated eddy correlation measurements. In homogeneous areas these difficulties may be limited, while in drier, heterogeneous landscapes it becomes challenging to compare correctly local direct measurements of evaporation with the estimates obtained with image data.

Literature does provide useful, albeit somewhat conflicting, indications on accuracy. Moran et al. (1989) reported errors on LE of 1%, -3% and -3% for different dates and crops at the Maricopa Agricultural Research Centre in Arizona. Two studies on the Walnut Gulch rangeland basin, also in Arizona, gave slightly different results: Moran et al. (1994b) obtained a mean absolute difference of 40 Wm^{-2} with a LE-range of 0 to 340 Wm^{-2} observed at the eight observation stations in the watershed, while Kustas et al. (1995) obtained an average error of 15% over three days of observations. Other results support the general approach of combining a resistance model with measurements of T_{rad} to estimate heat flux densities. Choudhury (1986) achieved an accuracy of <10% on daily LE and <5% on 10 days cumulative LE, while Zhang et al. (1995) obtained 50 Wm^{-2} when comparing estimates with measurements at individual stations within a homogeneous surface and 28 Wm^{-2} when comparing spatial averages of both measurements and estimates.

Error propagation and Monte Carlo simulation studies for specific heat balance models give significantly larger error bounds (e.g. Bastiaanssen et al., 1994). The precise reason for this discrepancy is unclear.

Errors due to spatial averaging were studied by Bresnahan et al. (1997) who evaluated the error of estimate for decreasing spatial resolutions. The error in the mean increased erratically as resolution coarsens due to the non-linearity of the model. Absolute error increased asymptotically. These results seem to confirm that spatial patterns in independent variables are important in determining the correct averaging procedure and scale for a regional process model.

Airborne eddy correlation measurements have the potential of providing a better standard for remote sensing of evaporation, although this author is not aware of any successful study in this direction. These observations should also be useful to understand how to select spatial samples of evaporation maps for comparison with local measurements (see e.g. Schuepp et al., 1989)

The issue of accuracy deserves further attention by recalling the results of Allen et al. (1989) on the comparison of different equations to estimate potential evaporation against accurate measurements with lysimeters. Deviations of monthly values from measurements were 10 % to 15 % for the best formula and 30 % for the worst one. This provides a measure of the challenge to meet when estimating instantaneous actual evaporation. On the other hand large changes in evaporation are strongly correlated with differences in land cover, so it is expected that useful estimates of actual evaporation can be obtained with observations capturing such differences in land cover.

8.7 Applications

A significant amount of evidence has been reviewed in the previous pages to document the progress made in remote sensing of land evaporation. Not surprisingly the accuracy of estimates could be documented for relatively simple situations such as agricultural fields or small, adequately equipped watersheds. On the other hand several applications to larger and more complex areas have been documented in literature.

Seguin (1981 and 1989) described the use of thermal infrared observations to assess and monitor agricultural production. The relevance of heat balance studies for land degradation studies was mentioned by Myers et al. (1980). Sandholt and Andersen (1993) used AVHRR and meteorological data to estimate and map evaporation in northern Sahel. Nemani et al. (1995) used spectral vegetation indices and surface temperature to assess the impact of land cover changes on climatic processes through changes in net radiation and evaporation. Ottle and Vidal Madjar (1994) demonstrated that the performance of a hydrological model of the Adour basin could be improved with remote sensing estimates of evaporation. The independent estimates of groundwater losses by evaporation in the Libyan desert led Menenti (1984) to new insights in the hydrology of groundwater reservoirs in the Sahara, confirmed by follow up investigations in the Western Desert of Egypt (Bastiaanssen and Menenti, 1990). The SEBAL algorithm described by Bastiaanssen (1995) was applied to assess irrigation performance in the Nile Delta by Bastiaanssen et al. (1992). Seguin et al. (1991) applied the linear relationship approach to produce an indicator of drought conditions in agricultural areas in France over a three years period.

Examples and additional references on applications are given in Chap. 17.

8.8 Current and Future Observations

The techniques and applications reviewed in the previous chapter span a period of time during which there has been a significant evolution in sensor technology. Changes in the type of measurements obtained with earth observation satellites have been limited. A major technological evolution of airborne sensors has taken place however. The driving force of such evolution has been the development of the new International Earth Observing System (IEOS), a very diverse suite of instruments developed by space agencies: NASA, ESA, NASDA and CNES. In many cases a series of forerunners of the sensors scheduled for flight in the time frame 1999 - 2005 has been developed to test new technological solutions. The extent and direction of technological evolution can be documented by looking at three categories: current satellite sensors, current airborne sensors and future satellite sensors. For accurate and detailed information on missions and satellites for earth observation the reader is referred to Kramer (1994). See also appendices to Chap. 20.

Remote sensing studies of land evaporation have been hampered by the lack of:

— simultaneous and accurate measurements of surface albedo and temperature at spatial resolutions consistent with the spatial variability of terrestrial landscapes;

— measurements of the geometry of canopies to determine surface roughness and aerodynamic resistances;

— measurements of PBL conditions.

The near future bears promises of significant improvements on all three aspects:

— The ASTER sensor system on-board Terra-1 (NASA EOS AM-1) will provide sufficient spectral sampling in the VNIR (Visible – Near Infrared) region to determine albedo, while the multi-spectral TIR measurements will provide for the first time a real opportunity for simultaneous determination of surface temperature and emissivity.

— The Vegetation Canopy Lidar (NASA/ ESSP/ VCL) will for the first time provide measurements of the height and possibly gap-fraction of vegetation canopies. This advancements in the characterization of the geometry of vegetation canopies should improve significantly the accuracy of current estimates of turbulent heat and momentum fluxes at land surfaces.

— The advanced satellite sensors and, in the long run, DIfferential Absorption Lidars (DIAL) should provide measurements of PBL height, temperature and humidity to specify better boundary conditions when modeling land evaporation.

8.9 Summary and Conclusions

In this chapter the use of remote sensing to address all aspects of land evaporation has been reviewed. Although most literature on land evaporation relates to the use of thermal infrared observations and the land surface heat balance, a broad spectrum of multi-spectral measurements by space- and airborne instruments has been used to estimate potential evaporation and crop water requirements. The chart (Fig. 8.2) illustrates the evolution of methods over time: the vertical links are meant to indicate how each type of method evolved, while as regards the heat balance approach an overall shift has occurred from left (linear relationships) to right (improved observations and parameterizations). The reader is also referred to the synoptic table on methods and outstanding issues of Kustas and Norman (1996).

One may say that the early expectations for a simple and general algorithm have been replaced by the awareness of the need for a stepwise approach to address outstanding observational and modelling issues.

Among the likely developments in the coming years the author would like to mention:

The documented impact of the difference ($T_{rad} - T_{aer}$) on estimates of latent heat flux will stimulate work on new observational methods. This difference depends on the difference ($T_{foliage} - T_{soil}$) and directional thermal infrared measurements have the potential of capturing foliage and soil temperatures, so we should see significant developments in this direction.

Canopy architecture and local topography play a significant role in land –atmosphere interactions. Very high resolution measurements by means of laser systems provide nearly ideal observations to capture the geometry of land surface and to estimate related land surface variables (e.g. Menenti and Ritchie, 1994).

Estimation of land surface evaporation requires some knowledge on the conditions of the atmospheric boundary layer. It appear that two complementary approaches will have to be explored:

a) use of back-scatter and differential absorption lidars;

b) assimilation of land surface variables in high resolution models of land – atmospheric processes.

The determination of actual evaporation through the water balance of an atmospheric column should become significantly more accurate with the launch of advanced sounders and the development of new 4DDA systems.

References

Allen, R.G., M.E. Jensen, J.L. Wright and R.D. Burman, 1989. Operational estimates of reference evapotranspiration. Agronomy Journal vol. 81: 650-662

Anderson, MC; Norman, JM; Diak, GR; Kustas, WP; Mecikalski, JR, 1997. A two source time integrated model for estimating surface fluxes using thermal infrared remote sensing. Rem. Sens. Environ. vol. 60 (2): 195-216

Bastiaanssen, W.G.M. and M. Menenti, 1990. Mapping groundwater losses in the western desert of Egypt with satellite measurements of surface reflectance and surface temperature. Proceedings and Information. No. 42. TNO Committee on Hydrological Research, The Netherlands: 61-90

Bastiaanssen, W.G.M., C.W.J. Roest, M.A. Abdel-Khalek and H. Pelgrum, 1992. Monitoring crop growth in large irrigation schemes on the basis of actual evapotranspiration: Comparisons of remote sensing algorithm and simulation model results. Proc. Intern. Conf. on Advances in Planning, Design and Management of Irrigation Systems.: 473-483

Bastiaanssen, W.G.M., M. Menenti, P.J. van Oevelen and R.A. Feddes, 1994. Spatial variability and accuracy of remote sensing estimates of evaporation and soil moisture within EFEDA '91. In: T. Keane and E. Daly (eds), The balance of water: present and future. AGMET Group, Dublin, Ireland: 175-192.

Bastiaanssen, W.G.M., 1995. Regionalization of surface flux densities and moisture indicators in composite terrain. Ph.D. Thesis Agricultural University of Wageningen and Report 109, DLO- Winand Staring Centre, Wageningen, The Netherlands.: 195 p

Bastiaanssen, W.G.M., H. Pelgrum, M. Menenti and R.A. Feddes, 1996. Estimation of surface resistance and Priestly-Taylor alpha parameter at different scales. In: J.B. Stewart et al. (eds), Scaling up in hydrology using remote sensing. Wiley, Chichester, UK: 93-111.

Beljaars, A.C.M. and P. Viterbo, 1994. The sensitivity of winter evaporation to the formulation of aerodynamic resistance in the ECMWF model. Boundary Layer Meteorol. vol. 71: 135-149

Beven, K.J. and J.Fisher, 1996. Remote sensing and scaling in hydrology. in: J.B. Stewart et al. (eds). Scaling in Hydrology using Remote Sensing. Wiley, Chichester U.K.: 1-18

Bolle, H.J. et al. (30 co-authors), 1993. EFEDA: European field experiments in a desertification-threatened area. Annales Geophysicae vol. 11: 173-189

Brasa Ramos, A., F.M. de Santa Olalla and V. Caselles, 1996. Maximum and actual evapotranspiration for barley (Hordeum vulgare L.) through NOAA satellite images in Castilla-La Mancha, Spain. J. Agr. Eng. Res. Vol. 63(4): 283-293

Bresnahan, P.A., D.R. Miller and A. Robertson, 1997. Choice of data scale: predicting resolution error in a regional evapotranspiration model. . Agric. For. Meteorol. vol. 84 (1-2): 97-113

Bruijnzeel, L.A. 1990. Hydrology of moist tropical forests and effects of conversion: a state of knowledge review. UNESCO, Paris

Brutsaert, W., 1982. Evaporation into the atmosphere. Reidel, Dordrecht, The Netherlands.

Brutsaert, W., A.Y. Hsu and T.J. Schmugge, 1992. Parametrization of surface heat fluxes above forest with satellite thermal sensing and boundary layer soundings. Proc. IGARSS '92: 1505 – 1507

Camillo, P.J., P.E. O'Neill and R.J. Gurney, 1986. Estimating soil hydraulic parameters using passive microwave data. IEEE Trans. Geosc. Rem. Sens. Vol. 24 (6): 930 - 936

Carlson, T.N. and F.E. Boland, 1978. Analysis of urban-rural canopy using a surface heat flux/temperature model. J. Appl. Meteorol. vol. 17: 998-1013

Carlson, T.N., J.K. Dodd, S.T. Benjamin and J.N. Cooper, 1981. Satellite estimates of the surface energy balance, moisture availability and thermal inertia. J. Appl. Meteorol. vol. 20: 67-87

Carlson, T.N. and M.J. Buffum, 1989. On estimating total daily evapotranspiration from remote surface temperature measurements. Rem. Sens. Environ. vol. 29: 197-207

Carlson, T.N., E.M.Perry and T.J.Schmugge, 1990. Remote estimation of soil moisture availability and fractional vegetation cover for agricultural fields. Agric. For. Meteorol. vol. 52: 45-69

Carlson, T.N., W.J. Capehart, R.R. Gillies, 1995. A new look at the simplified method for remote sensing of daily evapotranspiration. Rem. Sens. Environ. vol. 54: 161-167

Caselles, V., J.A. Sobrino, and C.Coll, 1992. On the use of satellite thermal data for determining evapotranspiration in partially vegetated areas. Int. J. Rem. Sens. Vol.13 (14): 2669-2682

Choudhury, B.J., 1986. Analysis of a resistance-energy balance method for estimating daily evaporation from wheat plots using one time of day infrared temperature observations. Rem. Sens. Environ. vol. 19 (3): 253-268

Choudhury, B.J., R.J. Reginato and S.B. Idso, 1986. An analysis of infrared temperature observations over wheat and calculation of latent heat flux. . Agric. For. Met. Vol. 37 (1): 75-88

Choudhury, B.J., N.U. Ahmed, S.B. Idso, R.J. Reginato and C.S.T. Daughtry, 1994. Relations between evaporation coefficients and vegetation indices studied by model simulations. Rem. Sens. Environ. vol.50 (1): 1-17

Choudhury, B.J., 1997. Global pattern of potential evaporation calculated from the Penman-Monteith equation using satellite and assimilated data. Rem. Sens. Environ. vol.61: 64-81

Choudhury, B.J. and N.E. DiGirolamo, 1998. A biophysical process-based estimate of global land surface evaporation using satellite and ancillary data: I. Model description and comparison with observations. J. Hydrol. Vol. 205: 164-185

Darnell, W.L., W.F. Staylor, S.K. Gupta, N.A. Ritchey and A.C. Wilber, 1992. Seasonal variation of surface radiation budget derived from International Satellite Cloud Climatology Project C1 data. J. Geophys. Res. Vol. 97 (15): 741-760

Diak, G.R. and M.A. Whipple, 1993. Improvements to models and methods for evaluating the land-surface energy balance and effective roughness using radiosonde reports and satellite measured skin temperatures. Agric. For. Met. Vol. 63: 189-218

Diak, G.R., C.J. Scheuer, M.S.Whipple and W.L. Smith, 1994. Remote sensing of land-surface energy balance using data from the High-resolution Interferometer Sounder (HIS): A simulation study. Rem. Sens. Environ. Vol. 48 (1): 106-118

D'Urso, G. and M.Menenti, 1995. Mapping crop coefficients in irrigated areas from Landsat TM images. SPIE Int. Soc. Optical Engineering, Bellingham, USA. Vol 2585: 41-47

D'Urso, G. and M.Menenti, 1996. Performance indicators for the statistical evaluation of digital image classifications. ISPRS J. Photogr.& Rem. Sens. Vol. 51(2): 78-90

D'Urso G., M.Menenti M. and A. Santini, 1999. Regional application of one-dimensional water flow models for irrigation management. Agric. Water Manag., 40: 291-302

Eagleson, P.S. (ed.), 1982. Land surface processes in atmospheric general circulation models. Cambridge Univ. Press: 560 pp.

Eichinger, W., D. Cooper, G. Katul, M.Parlange, D. Holtkamp, C. Quick, R. Karl, R. Shaw, K. Paw-U, T. Hsia, L. Hipps and W. Dugas, 1991. Derivation of evaporative fluxes with a Raman scanning Lidar. Eos, Trans. Amer. Gephys. Union vol. 72(44): 149

Feddes, R.A., M. Menenti, P. Kabat and W.G.M. Bastiaanssen, 1993. Is large-scale inverse modelling of unsaturated flow with areal average evaporation and surface soil moisture as estimated from remote sensing feasible? J. Hydrol. Vol. 143 (1-2): 125-152

Friedl, M.A. and F.W. Davis, 1994. Sources of variation in radiometric surface temperature over a tallgrass prairie. Rem. Sens. Environ. vol.48 (1): 1-17

Fritschen, L.J., 1982. The vertical fluxes of heat and moisture at a vegetated land surface. In: P.S. Eagleson, (ed.), 1982. Land surface processes in atmospheric general circulation models. Cambridge Univ. Press: 169-226

Goward, S.N., G.D. Cruikshanks and A. Hope, 1985. Observed relations between thermal emission and reflected spectral radiance of a complex vegetated landscape. Rem. Sens. Environ. vol.18: 137-146

Gurney, R.J. and D.K. Hall, 1983. Satellite derived surface energy balance estimates in the Alaskan sub-artic. J. Clim. Appl. Meteorol. vol. 22: 115-125

Harding, R.J., C.M. Taylor and J.W. Finch, 1996. Areal average surface fluxes from mesoscale meteorological models: the application of remote sensing. In: J.B. Stewart et al. (eds.). Scaling up in hydrology using remote sensing. J. Wiley & Sons. Ltd., Chichester UK: 59-76

Hatfield J.L., R.J. Reginato, R.D. Jackson, S.B. Idso and P.J. Pinter, 1978. Remote sensing of surface temperature for soil moisture, evapotranspiration and yield estimation. Proc. 5th Canadian Symposium on Remote Sensing: 460-465

Heilman J.L., E.T. Kanemasu, N.J. Rosenberg and B.L. Blad, 1976. Thermal scanner measurement of canopy temperatures to estimate evapotranspiration. Rem. Sens. Envir. vol. 5 (2): 137-145

Hipps, L.E., D. Or and C.M.U. Neale, 1996. Spatial structure and scaling of surface fluxes in a Great Basin ecosystem. In: J.B. Stewart et al. (eds), Scaling up in hydrology using remote sensing. Wiley, Chichester, UK: 113-125.

Humes, K.S., W.P. Kustas and M.S. Moran, 1994. Use of remote sensing and reference site measurements to estimate instantaneous surface energy balance components over a semiarid rangeland watershed. Water Resour. Res. Vol. 30 (5): 1363-1373

Inoue, Y. and M.S. Moran, 1997. A simplified method for remote sensing of daily canopy transpiration - a case study with direct measurements of canopy transpiration in soybean canopies. Int. J. Rem. Sens. Vol.18 (1): 139-152

Jackson, R.D., R.J. Reginato and S.B. Idso, 1977. Wheat canopy temperatures: a practical tool for evaluating water requirements. Water Resour. Res. vol. 13: 651-656

Jackson, R.D., 1985. Evaluating evapotranspiration at local and regional scales. Proc. IEEE vol. 73: 1086-1096

Jackson, R.D., W.P. Kustas and B.J. Choudhury, 1988. A re-examination of the crop-water stress index. Irrig. Sci. vol. 9: 309-317

Jensen, M.E., R.D. Burman and R.G. Allen, 1990. Evaporation and irrigation water requirement. ASCE Manual No. 70. American Society of Civil Engineers, New York: 332 pp.

Jha, C.S., B. Gharai, M.S.R. Murthy and C.B.S. Dutt, 1996. Estimation of evapotranspiration at larger scales using remotely sensed data. Tropical Ecol. vol.37 (1): 93-100

Kairu, E.N., 1991. A review of methods for estimating evapotranspiration, particularly those that utilize remote sensing. GeoJournal vol.25 (4): 371-376

Kalma, J.D., 1989. A comparison of expressions for the aerodynamic resistance to sensible heat transfer. Technical Memorandum No. 89-6. Division of Water Resources, Institute of Natural Resources and Environment, CSIRO, Australia. 17 pp

Kaneko, D. and M. Hino, 1996. Proposal and investigation of a method for estimating surface energy balance in regional forests using TM derived vegetation index and observatory data. Int. J. Rem. Sens. Vol.17 (6): 1129-1148

Kimes, D.S., 1983. Remote sensing of row crop structure and component temperatures using directional radiometric temperatures and inversion techniques. Rem. Sens. Envir. Vol. 13: 33-55

Kustas, W.P., B.J. Choudhury, M.S. Moran, R.J. Reginato, R.D. Jackson, L.W. Gay and H.L. Weaver, 1989 a. Determination of sensible heat flux over sparse canopy using thermal infrared data. Agric. For. Met. Vol. 44: 197-216

Kustas, W.P, R.D. Jackson-and G. Asrar, 1989 b. Estimating surface energy-balance components from remotely sensed data. in: G. Asrar (ed.). Theory and applications of optical remote sensing. John Wiley & Sons; New York; USA. 604-627

Kustas, W.P., 1990. Estimates of evapotranspiration with a one- and two-layer model of heat transfer over a partial canopy cover. J. Appl.Meteor. vol. 29 (8):704-715

Kustas, W.P., M.S. Moran, R.D. Jackson, L.W. Gay, L.F.W. Duell, K.E. Kunkel and A.D. Matthias, 1990. Instantaneous and daily values of the surface energy balance over agricultural fields using remote sensing and a reference field in an arid environment. Rem. Sens. Environ. vol. 32 (2-3): 125-141

Kustas, W.P., E.M. Perry, P.C. Doraiswamy and M.S. Moran, 1994a. Using satellite remote sensing to extrapolate evapotranspiration estimates in time and space over a semiarid rangeland basin. Rem. Sens. Environ. vol. 49(3): 275-286

Kustas, W.P., M.S. Moran, K.S. Humes, D.I. Stannard, P.J. Pinter Jr., L.E. Hipps, E. Swiatek, and D.C. Goodrich, 1994b. Surface energy balance estimates at local and regional scales using optical remote sensing from an aircraft platform and atmospheric data collected over semiarid rangelands. . Water Resources Res. Vol. 30(5):1241 – 1259

Kustas, W.P., L.E. Hipps and K.S. Humes, 1995. Calculation of basin-scale surface fluxes by combining remotely sensed data and atmospheric properties in a semiarid landscape. Boundary-Layer Meteorol. vol. 75: 105-124

Kustas, W.P. and J.M. Norman, 1996. Use of remote sensing for evapotranspiration monitoring over land surfaces. In: A. Rango and J.C. Ritchie (eds.). Special issue: remote sensing applications to hydrology. Hydrological Sciences J. vol. 41 (4): 495-516

Lagouarde, J.P., K.J. McAneney and A.E. Green, 1996. Scintillometer measurements of sensible heat flux over heterogeneous surfaces. In: J.B. Stewart et al. (eds), Scaling up in hydrology using remote sensing. Wiley, Chichester, UK: 147-160

Lhomme-JP; Monteny-B; Chehbouni-A; Troufleau-D Determination of sensible heat flux over Sahelian fallow savannah using infra-red thermometry. Agric. For. Met. Vol. 68 (1-2): 93-105

Majumdar, T.J.,1991. Regional mapping of evapotranspiration rates using MOS-I VTIR data. ITC Journal No. 2: 86-89

Menenti, M., 1979. Defining relationships between surface characteristics and actual evaporation rate. TELLUS Newsl. 15. JRC, Ispra: 21 p

Menenti, M., 1984. Physical aspects and determination of evaporation in deserts applying remote sensing techniques. Report 10 (special issue). Inst. Land Water Manag. Res., Wageningen, The Netherlands: 202 p

Menenti, M., A. Lorkeers and M. Visser, 1986. An application of thematic mapper data in Tunisia ITC Journal vol. 1: 35-42

Menenti, M., W.G.M. Bastiaanssen, D. van Eick and M.H. Abd El Karim, 1989. Linear relationships between surface reflectance and temperature and their application to map actual evaporation of groundwater. Adv. Space Res. 9(1): 165-176.

Menenti, M., 1993. Understanding land surface evapotranspiration with satellite multispectral measurements. Adv. Space Res. vol. 13 (5): 89-100

Menenti, M. and B.J. Choudhury, 1993. Parametrization of land surface evapotranspiration using a location dependent potential evapotranspiration and surface temperature range. in: H.J. Bolle et al. (eds.) Exchange Processes at the Land Surface for a Range of Space and Time Scales. IAHS Publication 212: 561-568

Menenti, M. and J.C. Ritchie, 1994. Estimation of effective aerodynamic roughness of Walnut Gulch watershed with laser altimeter measurements. Water Res. Research (special issue MONSOON '90) 30(5): 1329-1337.

Menenti, M., J.C. Ritchie, K.S. Humes, R.Parry, Y. Pachepsky, D. Gimenez and S. Leguizamon, 1996. Estimation of aerodynamic roughness at various spatial scales. In: J.B. Stewart et al. (eds), Scaling up in hydrology using remote sensing. Wiley, Chichester, UK: 39-58.

Monteith, J.L. 1994. Fifty years of potential evaporation. Proc. Conf. Trinity College. Dublin, 7-9 sept.: 29-45

Moran, S.M., R.D. Jackson, L.H. Raymond, L.W. Gay and P.N. Slater, 1989. Mapping surface energy balance components by combining Landsat Thematic Mapper and ground based meteorological data. Rem. Sens. Environ. vol. 30: 77-87

Moran, S.M. and R.D.Jackson, 1991. Assessing the spatial distribution of evapotranspiration using remotely sensed inputs. J. Environ Qual. vol. 20: 725-737

Moran, M.S., T.R.Clarke, Y. Inoue and A.Vidal, 1994a. Estimating crop water deficit using the relation between surface-air temperature and spectral vegetation index. Rem. Sens. Environ. vol. 49(3): 246-263

Moran, M.S., W.P. Kustas, A.Vidal, D.I. Stannard, J.H. Blanford and W.D. Nichols, 1994b. Use of ground-based remotely sensed data for surface energy balance evaluation of a semiarid rangeland. Water Resources Res. Vol. 30(5):1339 – 1349

Myers, V.I., H.S. Mann, D. Moore, M. DeVries and M. Abdel-Hady, 1980. Remote sensing for monitoring resources for development and conservation of desert and semi-desert areas. In: H.S. Mann (ed.) Arid-Zone-Research-and-Development: 505-513

Nemani, R.R. and S.W. Running, 1989. Estimation of regional surface resistance to evapotranspiration from NDVI and thermal-IR AVHRR data. J. Appl. Meteorol. vol.28 (4): 276-284

Nemani, R.R., L. Pierce, S.W. Running and L. Band, 1993. Forest ecosystem processes at the watershed scale: sensitivity to remotely-sensed Leaf Area Index estimates. Int. J. Rem. Sens. Vol.14(13): 2519-2534

Nemani, R.R., S.W. Running and T.R. Karl, 1995. Satellite monitoring of global land cover changes and their impact on climate. Climatic Change vol. 31(2-4): 395-413

Nieuwenhuis, G.J.A., E.H. Smidt and H.A.M. Thunnissen, 1985. Estimation of regional evapotranspiration of arable crops from thermal infrared images. Int J. Rem. Sens. vol. 6: 1319-1334

Norman, J.M., W.P. Kustas, K.S. Humes and M.S. Moran, 1995. A two-source approach for estimating soil and vegetation energy fluxes from observations of directional radiometric surface temperature. Agric. For. Met. Vol. 77: 263-293

Oki, T., K. Musiake, H. Matsuyama and K. Masuda, 1995. Global atmospheric water balance and runoff from large river basins. Hydrol. Processes vol. 9: 655-678

Ottle,C. and D. Vidal-Madjar, 1994. Assimilation of soil moisture inferred from infrared remote sensing in a hydrological model over the HAPEX-MOBILHY region. J. Hydrol. vol.158(3-4): 241-264

Ottle, C., D. Vidal-Madjar, A.L. Cognard, C. Loumagne and M. Normand, 1996. Radar and optical remote sensing to infer evapotranspiration and soil moisture. In: J.B. Stewart et al. (eds), Scaling up in hydrology using remote sensing. Wiley, Chichester, UK: 221-233

Peixoto, J.P., 1970. Pole to pole divergence of water vapour. Tellus vol. 22: 17-25

Pelgrum, H. and W.G.M. Bastiaanssen, 1996. An intercomparison of techniques to determine the area averaged latent heat flux from individual in situ observations: a remote sensing approach using EFEDA data.Water Resources Res. Vol. 32(9): 2775-2786

Penman, H.L., 1948. Natural evaporation from open water, bare soil and grass. Proc. Roy. Soc. (London) vol. A 193: 120-145

Pinker, R.T., R. Frowin and Z. Li, 1995. A review of satellite methods to derive surface shortwave irradiance. Rem. Sens. Environ. vol. 51(3): 108-124

Price, J.C., 1990. Using spatial context in satellite data to infer regional scale evapotranspiration. IEEE Trans. Geosc. Rem. Sens. Vol. 28 (5): 940-948

Reiniger, P. and B.Seguin, 1986. Surface temperature as an indicator of evapotranspiration and soil moisture. Remote Sensing Reviews vol. 1: 277-310

Rosema, A. and J.H. Bijleveld, 1978. Test of algorithm for the determination of soil moisture and evaporation from remotely sensed surface temperatures. Report to the Commission of the European Communities, Joint Research Centre. Contr. Nr. 752-77-08: 33 p

Rossow, W.B., L.C. Gardner and P.J. Lu, 1988. International Satellite Cloud Climatology Project (ISCCP) documentation of cloud data. WMO-TD No. 266. World Meteorological Organization, Geneva: 75 pp

Sandholt, I. and H.S. Andersen, 1993. Derivation of actual evapotranspiration in the Senegalese Sahel, using NOAA-AVHRR data during the 1987 growing season. Rem. Sens. Envir. Vol. 46 (2): 164-172

Sato, N., P.J. Sellers, E.K. Randall, J. Schneider, J. Shukla, J.L. Kinter III, Y. T. Hou and E. Albertazzi, 1989. Effects of implementing the simple biosphere model (SiB) in a general circulation model. J. Atmos. Sci. vol. 46 (18): 2757-2782

Saxton, K.E., H.P.Johnson and R.H. Shaw, 1974. Modeling evapotranspiration and soil moisture. Trans. ASAE vol.17: 673-677

Schuepp, P.H., R.L. Desjardins, J.I. Macpherson, J.B. Boisvert and L.B. Austin, 1989. Interpretation of airborne estimates of evapotranspiration. IAHS Publication. No. 177: 185-196

Schubert, S.D., J. Pfaendtner and R. Rood, 1993. An assimilated data set for Earth science applications. Bull. Am. Meteorol. Soc. Vol. 74: 2331-2342

Seevers, P.M., and R.W. Ottmann, 1994. Evapotranspiration estimation using a normalized difference vegetation index transformation of satellite data. Hydrol. Sci. J. vol. 39 (4): 333-345

Segal, M. and R.W. Arritt, 1992. Non-classical mesoscale circulations caused by surface sensible heat-flux gradients. Bull. Amer. Meteorol. Soc. Vol. 73: 1593-1604

Seguin, B., 1981. Bioclimatological aspects of crop production. In: A.Berg (ed.) Application of remote sensing to agricultural production forecasting. A.A. Balkema, Rotterdam, The Netherlands: 33-45

Seguin, B. and B. Itier, 1983. Using midday surface temperature to estimate daily evaporation from satellite thermal IR data. Int. J. Rem. Sens. vol. 4: 371-383

Seguin, B.,1989.Use of surface temperature in agrometeorology. in: F. Toselli (ed.) Applications of remote sensing to agrometeorology. Kluwer Academic Publishers, Dordrecht, The Netherlands: 221-240

Seguin, B., J.P. Lagouarde and M. Savane, 1991. The assessment of regional crop water conditions from meteorological satellite thermal infrared data. Rem. Sens. Envir. Vol. 35 (2-3): 141-148

Shukla, J. and Y. Mintz, 1982. Influence of land surface evapotranspiration on the Earth's climate. Science vol. 215: 1498-1501

Smith, R.C.G. and B.J. Choudhury, 1991. Analysis of normalized difference and surface temperature observations over southeastern Australia. Int. J. Rem. Sens. Vol. 12: 2021-2044

Soer, G.J.R., 1977. The TERGRA model – A mathematical model for the simulation of the daily behaviour of crop surface temperature and actual evapotranspiration. NIWARS Publ. 46. Delft, The Netherlands: 44 p

Soer, G.J.R., 1980. Estimation of regional evapotranspiration and soil moisture conditions using remotely sensed crop surface temperature. Rem. Sens. Envir. Vol. 9: 27-45

Starr, V.P. and J.P. Peixoto, 1958. On the global balance of water vapour and the hydrology of deserts. Tellus vol. 10: 189-194

Stone, L.R., E.T. Kanemasu and M.L. Horton, 1975. Grain sorghum canopy temperature as influenced by clouds. Rem. Sens. Envir. vol. 4 (2): 177-181

Susskind, J., 1993. Water vapour and temperature. In: R.J. Gurney et al. (Eds.). Atlas of satellite observations related to global change. Cambridge Univ. Press, New York N.Y.: 89-128

Taconet, O. and D. Vidal-Madjar, 1988. Application of a flux algorithm to a field-satellite campaign over vegetated area. Rem. Sens. Envir. Vol.26 (3): 227-239.

Taconet O., A. Olioso, M. Ben-Mehrez and N. Brisson, 1995. Seasonal estimation of evaporation and stomatal conductance over a soybean field using surface IR temperatures. Agric. For. Met. Vol. 73 (3-4): 321-337

Thornthwaite, C.W. 1944. Report of the Committee on Transpiration and Evaporation. Trans. AGU vol. 25: 683-693

Thornthwaite, C.W., 1948. An approach towards a rational classification of climate. Geogr. Rev. vol. 38: 55-94

Thunnissen, H.A.M. and G.J.A. Nieuwenhuis, 1990. A simplified method to estimate 24-h evapotranspiration from thermal infrared data. Rem. Sens. Envir. Vol. 31: 211-225

Vieira, S.R and J.L. Hatfield, 1984. Temporal variability of air temperature and remotely sensed surface temperature for bare soil. Int. J. Rem. Sens. Vol.5(3): 587-596

Wang, J., Y. Ma, M. Menenti and W.G.M. Bastiaanssen, 1995. The scaling-up of processes in the heterogeneous landscape of HEIFE with the aid of satellite remote sensing. Journal Meteorological Society of Japan vol. 73(6): 1235-1244.

Wetherald, R.T. and S. Manabe, 1988. Cloud feedback processes in a general circulation model. J. Atmos. Sci. vol. 45: 1397-1415

Wetzel, P.J., D. Atlas and R. Woodward, 1984. Determining soil moisture from geosynchronous satellite infrared data. A feasibility study. J. Clim. Appl. Met. Vol. 23: 375-391

Wieringa, J., 1986. Roughness- dependent gepgraphical interpolation of surface wind speed averages. Q.J. R. Met. Soc. Vol. 112: 867-889

Wollenweber, G.C., 1995. Influence of fine scale vegetation distribution on surface energy partition. Agric. For. Met. Vol. 77 (3-4): 225-240

Yang, X.H., Q.M. Zhou and M. Melville, 1997. Estimating local sugarcane evapotranspiration using Landsat TM image and a VITT concept. . Int. J. Rem. Sens. Vol.18(2): 453-459

Zhang,L., R. Lemeur, J.P. Goutorbe and M.S. Moran, 1995. A one-layer resistance model for estimating regional evapotranspiration using remote sensing data. . Agric. For. Met. Vol. 77 (3-4): 241-261

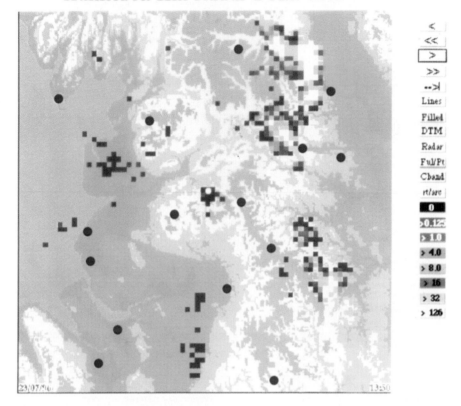

Colour Plate 6.A. Rainfall observed by the Hameldon Hill radar located in north west England. Raingauge observations available in near real-time are shown by dots. Note that the full distribution of the rainfall is not observed by the raingauges

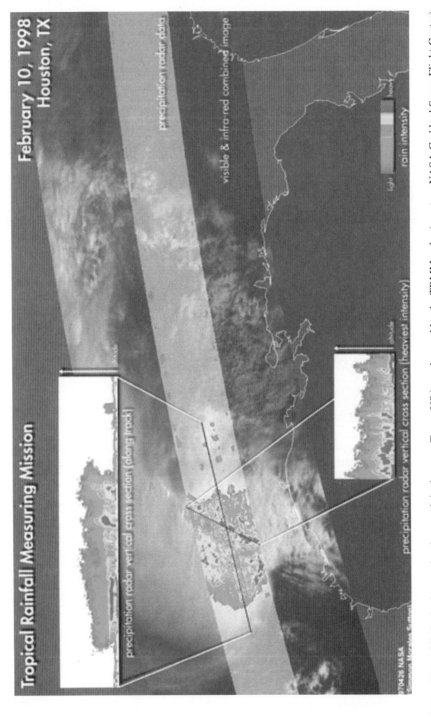

Colour Plate 6.B. Radar image showing precipitation over Texas, USA as observed by the TRMM radar (courtesy NASA Goddard Space Flight Centre)

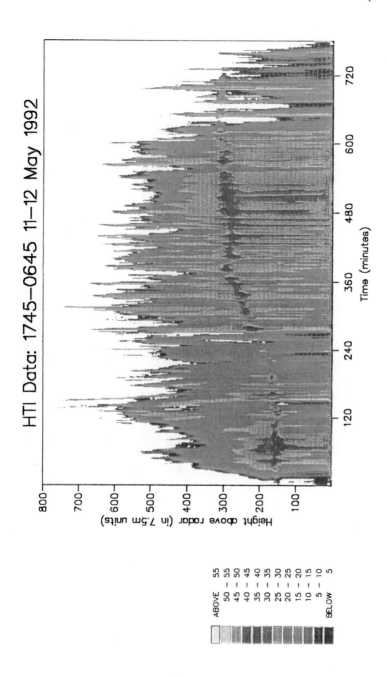

Colour Plate 6.C. Example of the variation of radar reflectivity with time on 11 May 1992 as observed with a vertical pointing X-band radar located at Salford in NW England (courtesy Dr K A Tilford). Note the change in the height of the bright-band

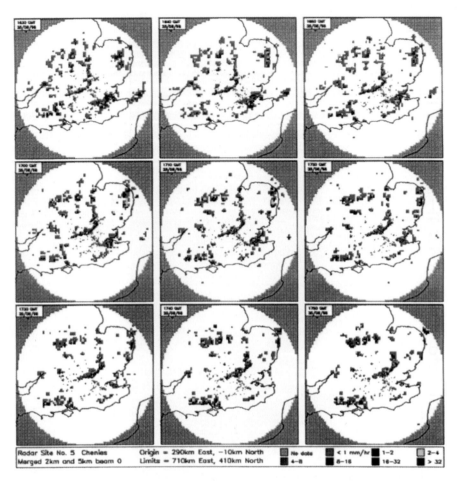

Colour Plate 6.D. Sequence of radar images at ten minute intervals observed by the Chenies radar on a 2 km by 2 km grid over south east England. Note the general variability of the convective rainfall together with the narrow band of more organized severe convection (rainfall rates > 32 mm h^{-1}). The radius of each circle is 75 km (courtesy of Dr P J Hardaker, Crown Copyright)

Colour Plate 7.A. Colour composite Thematic Mapper bands 4,3,7, Pantanal, Brazil, 28 Sept 1996

Colour Plate 7.B. Multi-temporal NDVI, Pantanal, Brazil, Red: 1996, Green: 1990, Blue: 1985

Colour Plate 7.C. Multi-temporal segmented NDVI, Pantanal, Brazil, Red: 1996, Green: 1990, Blue: 1985

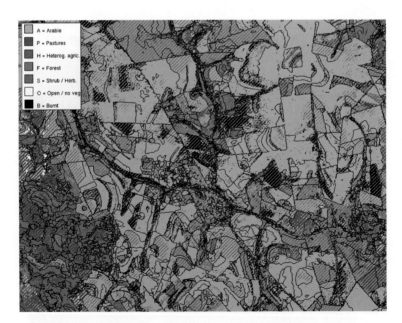

Colour Plate 7.D. Multi-temporal classification (1985-1996) with stable classes (uniform colors), change classes (hatched) and segmentation boundaries, Pantanal, Brazil. In the hatched areas the thinner diagonal stripes refer to 1985 and the wider ones to 1996.

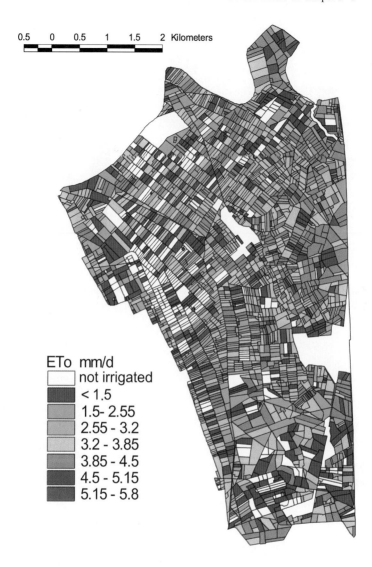

Colour Plate 8.A. Maximum crop evaporation: values for each cadastral unit have been calculated using the average within the unit of canopy variables determined with TM observations at 30 m spatial resolution; Sinistra Sele irrigation scheme in Southern Italy; July 25th 1994

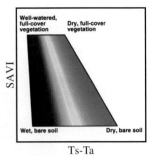

Colour Plate 8.B. Map of estimated water deficit index; scale as indicated by the VITT diagram on the right side; July 6th 1994 (courtesy of S. Moran USDA-ARS)

9 Soil Moisture

Edwin T. Engman
Head, Hydrological Sciences Branch,
Laboratory for Hydrospheric Processes,
NASA/Goddard Space Flight Center

9.1 Introduction

Soil moisture is an environmental descriptor that integrates much of the land surface hydrology and is the interface between the solid earth surface and the atmosphere. By far the most important role for soil moisture is in controlling and regulating the interaction between the atmosphere and the land surface. The redistribution of solar energy over the globe is central to studies in climate and weather. Water serves a fundamental role in this redistribution through the energy associated with evapotranspiration, the transport of atmospheric water vapor, and precipitation. Residence times for atmospheric water is on the order of a week and for soil moisture from a couple of days to months, emphasizing the active nature of the hydrologic cycle. Understanding the importance of the land-surface hydrology to climate has emerged as an important research area since the mid 1960s when researchers at the Geophysical Fluid Dynamics Laboratory placed a land hydrology component into their general circulation model (GCM) (see Manabe et al., 1965; Manabe, 1969).

The role of soil moisture is equally important at smaller scales. Recent studies with mesoscale atmospheric models have similarly demonstrated a sensitivity to spatial gradients of soil moisture. Chang and Wetzel (1991) have concluded that the spatial variations of vegetation and soil moisture affect the surface baroclinic structures through differential heating which in turn indicate the location and intensity of surface dynamic and thermodynamic discontinuities necessary to develop severe storms. An analysis of the 1988 drought in the mid-western United States by Atlas et al. (1993) has shown that these conditions could be modeled accurately only when the soil moisture values were realistic. A reanalysis (Beljaars et al., 1996) of the conditions leading to the 1993 floods in the United States illustrated improved precipitation forecasts, both in quantity and location, when realistic soil moisture values were used in the model.

Soils, their properties and their spatial distribution all reflect the historical hydroclimate processes that formed the soils and control the current land surface response to meteorological inputs, i.e., precipitation and potential ET. Soil moisture patterns, both spatial and temporal, may be the key to understanding the spatial variability and scale problems that are paramount in scientific hydrology. Operational flood forecasting is based on limited measurements of rainfall and river stage and in some cases some type of soil moisture description usually in the form of an antecedent precipitation index. In many cases poor forecasts are attributed to lack of information about the initial conditions, i.e., soil moisture. For example, hydrologists at the NOAA Kansas City River Forecast Center believe that soil moisture has been their most troublesome parameter affecting forecasts (Wiesnet, 1976) and Georgakakos et al. (1996) have identified soil moisture as the most sensitive variable controlling runoff in the development of an operational flash flood prediction system.

As important as this seems to our understanding of hydrology, soil moisture has not had widespread application in the modeling of these processes. The main reason for this is that it is a very difficult variable to measure, not at a point in time, but at a consistent and spatially comprehensive basis. The large spatial and temporal variability that soil moisture exhibits in the natural environment is precisely the characteristic that makes it difficult to measure and use in Earth science applications. For the most part our understanding of the role of soil moisture in hydrology has been developed from point studies where the emphasis has been on the variability of soil moisture with depth. Much of our failure to translate this point understanding to natural landscapes can be traced to a realization that soil moisture varies greatly in space but with no obvious means to measure the spatial variability. As a parallel consequence, most models have been designed around the available point data and do not reflect he spatial variability that is known to exist.

9.2 General Approach

It has been shown that the soil moisture can be measured to some extent by a variety of techniques using all parts of the electromagnetic spectrum. Successful measurement of soil moisture by remote sensing depends upon the type of reflected or emitted electromagnetic radiation. Table 9.1 summarizes the advantages and disadvantages of each approach. However, it will be seen that only the microwave region of the spectrum can provide a quantitative approach to estimate soil moisture under a variety of topographic and vegetation cover conditions.

Gamma radiation techniques. Airborne soil moisture measurement by gamma radiation is based on detecting the difference between the natural terrestrial gamma radiation flux for wet and dry soils. The presence of water in the upper soil layers increases the attenuation of the gamma radiation from below; thus the flux is less for wet soils then for dry soils. Quantitative estimates of soil moisture require calibration flight lines to determine the background soil moisture value, Mo, and the background gamma count rate, Co. The current soil moisture, M can be estimated according to

$$M = \frac{C/Co(100 + 1.11 Mo) - 100}{1.11} \quad (9.1)$$

where C is the measured gamma count rate. A more complete description can be found in Carroll (1981). Because the atmosphere also attenuates the gamma radia-

Table 9.1 Summary of remote sensing techniques for measuring soil moisture (after Engman, 1991)

Wavelength region	Property observed	Advantages	Disadvantages
Gamma radiation	Attenuation of naturally emitted radiation	Existing airborne program. Averages over a line.	Limited spatial resolution. Limited to low elevation flights. Empirical calibration.
Reflected solar	Albedo. Index of refraction	Data available	No unique relationship between spectral reflectance and soil moisture. Surface only. Clouds.
Thermal Infrared	Surface temperature. Measured diurnal range of surface or crop temperature.	High spatial resolution, wide swath. Relationship between temperature and soil water pressure independent of soil type.	Bare soil only. Clouds. Surface topography and local meteorologic conditions can cause noise. Surface layer only.
Active micro-wave	Backscatter coefficient, dielectric constant.	All weather, high resolution.	Surface roughness, vegetation, topography, limited swath width. Radio frequency interference
Passive micro-wave	Brightness temperature, soil temperature, emissivity, dielectric constant.	All weather, wide swath, good sensitivity can compensate for moderate vegetation	Limited spatial resolution. Radio frequency interference dense vegetation

tion flux from the soil, this approach is limited to low elevation aircraft flights, less than 300m above the land surface.

Visible/near-infrared techniques. Reflected solar radiation is not a particularly useful approach to estimating soil moisture because it is very difficult to quantify the estimate. While it is true that wet soil will generally have a lower albedo than dry soil (Crist and Cicone, 1984), and this difference can theoretically be measured, confusion factors such as organic matter, roughness, texture, angle of incidence, color, plant cover, and the fact that it is a transient phenomenon, all make this approach impractical (Jackson, et al., 1978).

Thermal infrared techniques. The thermal infrared portion of the spectrum offers a theoretically sound approach to measuring soil moisture. After meteorological inputs to the soil surface have been accounted for, surface temperature is primarily dependent upon the thermal inertia or the soil. The thermal inertia, in turn, is dependent upon both the thermal conductivity and heat capacity which increases with soil moisture according to the following relationship (Price, 1982)

$$DT_s = T_s(PM) - T_s(AM) = f(1/D) \tag{9.2}$$

where DTs is the diurnal temperature difference between the afternoon surface temperature Ts(PM) and the early morning surface temperature, Ts(AM), and D is the diurnal thermal inertia given by

$$D = w \cdot Qc \cdot k \tag{9.3}$$

where w corresponds to the day length, Qc is the volumetric heat capacity and k is the thermal conductivity. The diurnal thermal inertia D describes the ability for the soil to resist temperature change. For example, a dry sand has a relatively low value of D compared to a wet clay because the thermal conductivity of the sand is lower than that for the wet clay and the volumetric heat capacity for dry soil is lower than that for wet soils. Thus, by measuring the amplitude of the diurnal temperature change, one can develop a relationship between the temperature change and soil moisture. However, the relationship between the diurnal temperature and soil moisture depends upon soil type and is largely limited to bare soil conditions (van de Griend et al., 1985).

Microwave Techniques. Microwave techniques for measuring soil moisture include both the passive and active microwave approaches, with each having distinct advantages. The theoretical basis for measuring soil moisture by microwave techniques is based on the large contrast between the dielectric properties of liquid water and of dry soil. The large dielectric constant for water is the result of the water molecule's alignment of the electric dipole in response to an applied electromagnetic field. For example, at L-band frequency the dielectric constant of water is approximately 80 compared to that of dry soils which is on the order of 3-5. Thus, as the soil moisture increases, the dielectric constant can increase to a value of 20, or greater (Schmugge, 1983). Figure 9.1 illustrates the change in dielectric constant for soil at several microwave frequencies.

For passive microwave remote sensing of soil moisture from a bare surface, a radiometer measures the intensity of emission from the soil surface. This emission is proportional to the product of the surface temperature and the surface emissivity which is commonly referred to as the microwave brightness temperature (T_B) and can be expressed as follows (Schmugge, 1990):

$$T_B = t(H) \cdot \left[rT_{sky} + (1-r)T_{Soil} \right] + T_{atm}, \tag{9.4}$$

where t(H) is the atmospheric transmissivity for a radiometer at height H above the soil, r is the smooth surface reflectivity, T_{soil} is the thermometric temperature of the soil, T_{atm} is the average thermometric temperature of the atmosphere, and T_{sky} is the contribution from the reflected sky brightness. For typical remote sensing applications using longer microwave wavelengths (greater than 5 cm, which are better for soil moisture), the atmospheric transmission will approach 99%. The atmospheric, T_{atm}, and sky, T_{sky}, contributions are both on the order of 5°K, each of which are small compared to the soil contribution. Thus neglecting these two terms, Eq. 9.4 can be simplified to

$$TB = (1-r)T_{Soil} = eTsoil \tag{9.5}$$

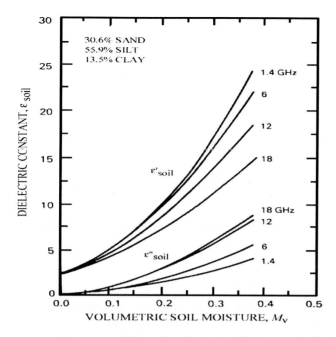

Fig. 9.1. The real and imaginary parts of the dielectric constant as a function of volumetric soil moisture for a loam soil measured at four frequencies (after Ulaby et al. 1986)

where $e = (1-r)$ is the emissivity and is dependent upon dielectric constant of the soil and the surface roughness. Thus over the normal rage of soil moisture, a decrease in the emissivity from about 0.95 to 0.60 or lower can be expected. This translates to a change in brightness temperature on the order of 80 degrees K. Though the relationship between emissivity and brightness temperature is linear (see Eq. 9.5), the soil moisture has a non-linear dependence on reflectivity because the reflection coefficient R of the ground is related in a non-linear way to the dielectric constant of the soil (ε). For horizontal polarization, the reflection coefficient is given by

$$R = \frac{\cos q - b}{\cos q + b} \tag{9.6}$$

where $b = \sqrt{\varepsilon - \sin^2 q}$ and q is the angle of incidence. The expression for vertical polarization can be written in a similar way. The dielectric constant, ε, is a complex quantity and the empirical relationships between dielectric constant and soil moisture derived by Dobson et al. (1985) show that dielectric constant has a non-linear dependence on soil moisture. However, even though the brightness temperature -soil moisture relation has a strong theoretical basis, most algorithms are empirical in that they depend upon ground data for the relationship.

For the active microwave approach over a bare soil, the measured radar backscatter, s_s, can be related directly to soil moisture by

$$s_s = f(R, A, M_V) \tag{9.7}$$

where R is a surface roughness term, a is a soil moisture sensitivity term, and M_V is the volumetric soil moisture. Although R and a are known to vary with wavelength, polarization, and incidence angle, there is no satisfactory theoretical model suitable for estimating these terms independently. Thus, as is the case for the passive microwave approach, the relationship between measured backscatter and soil moisture requires an empirical relationship with ground data, even for bare soils.

An additional approach for using soil moisture data derived with microwave approaches is through change detection. This approach can be used for both passive or active microwave data. The change detection method minimizes the impact of target variables such as soil texture, roughness, and vegetation because these tend to change slowly, if at all, with time.

9.3 Sensor-Target Interactions

As discussed above, microwave techniques for measuring soil moisture have a strong theoretical basis. In addition, they are not limited to cloud-free and bare-soil conditions because the microwave approach can sense through cloud cover and, in many cases, through a vegetation canopy. Each of the two basic approaches, passive and active, offer different but distinct advantages. The differences being in their instrument characteristics and their interaction with the characteristics of the target.

There are a number of target and target-sensor characteristics that affect the measurement of soil moisture. These include the effects of soil texture, the depth of measurement, surface roughness, vegetation effects, and instrument parameters such as incidence angle and frequency. Each of these are discussed in more detail below.

Soil Texture. Soil texture affects the microwave sensing of soil moisture in the way that the dielectric constant changes with the relative amounts of sand, silt, and clay in the soil. Figure 9.2 shows this effect with laboratory data and an empirical model developed by Wang and Schmugge (1980). However, it can be seen that this effect is relatively small and given the overall accuracy of the methods and uncertainty in other factors, texture effects can be neglected for practical purposes.

Measurement Depth. The same principles control the depth of soil that is being measured by the microwave technique, whether it is passive, as discussed above, or active. In a series of careful field experiments with a C-band, HH polarization radar, Bruckler et al. (1988) showed experimental results of penetration depth compared with soil moisture that followed very closely to the theoretical curve for a uniform profile.

The relationship between emissivity and soil moisture depends upon the dielectric contrast across the air-soil interface. Consequently, this results in some uncertainty as to exactly how thick the soil layer is for determining the dielectric con-

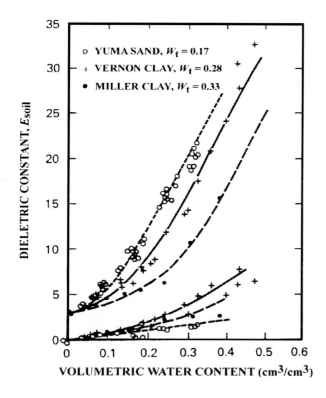

Fig. 9.2. A comparison of laboratory measurements of the real and imaginary parts of the dielectric constant and model predictions (smooth curves) for three soils as functions of volumetric soil moisture at 1.4 GHz (after Wang and Schmugge, 1980). Wt is the transition moisture content where the dielectric constant approaches that for a liquid

stant. According to Wilheit (1975), the layer of soil would be on the order of a tenth of a wavelength or less. Mo et al. (1980) determined that the radiometric sampling depth is between 0.06 and 0.1 times the wavelength. In an experiment comparing dry-down measurements of soil layers at three frequencies, Newton et al. (1982) found that for L-band (21 cm wavelength) the sampling depth was about two-tenths of the wavelength. The fact of the matter is that measurement depth is not a constant but related to the total amount of water in the soil layer, and thus to the moisture content, and to the operational frequency of the sensor. A reasonably good idea about the measurement depth can be obtained from penetration depth, dp, inside the soil. This is given by

$$dp = k \cdot \left| \mathrm{Im} \sqrt{\varepsilon} \right| \tag{9.8}$$

where $k = 2p/l$ is free space propagation constant, ε is dielectric constant of soil, Im denote the imaginary part and the vertical bars refer to the absolute value. The plot of penetration depth versus soil moisture for various frequencies is shown in Fig. 9.3 (Ulaby et al., 1982).

Surface Roughness. Microwave emission from the soil is related to the reflectivity of the surface, which if smooth can be calculated by the Fresnel equations. Smoothness in microwave terms is a relative term, being dependent upon the wavelength. That is, a surface that is smooth for one wavelength, say 21 cm L-band (1.4 GHz), may not be smooth for 6 cm C-band (4.9 GHz), and 2.8 cm K-band (10.7 GHz). The effect of a rough surface is to increase the surface emissivity and thus to decrease the sensitivity to soil moisture (Newton and Rouse, 1980).

Choudhury et al. (1979) have shown that surface roughness can affect the soil reflectivity r' in the following way:

$$r' = r \cdot \exp(-h \cos^2 q) \tag{9.9}$$

where r is the smooth surface reflectivity, h is a roughness parameter ($= 4 \sigma^2 k^2$) proportional to the root mean square height variations of the soil surface, σ, and q is the incidence angle. Here the roughness parameter, σ, appears as squared in the exponential. Therefore, the value of reflectivity will decrease rapidly with the slight increase of σ. This is what makes reflectivity very sensitive to s compared to soil moisture. It should be mentioned here that the roughness-reflectivity depend-

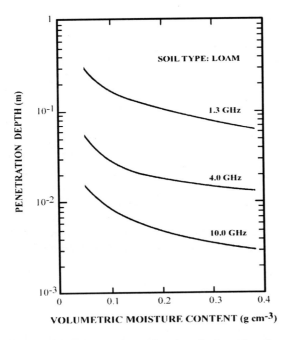

Fig. 9.3. Penetration depth as a function of volumetric soil moisture and frequency (after Ulaby et al. 1986)

ence issue is under a great deal of investigation, because relation 9.6 is violated under certain conditions of roughnesses and incidence angles. Wang et al. (1980) assumed the random roughness was independent of incidence angle and simplified Eq. 9.6 to

$$R = R_0 \exp(-h) \tag{9.10}$$

More recent work by Promes et al. (1988) has shown that, for L-band, assuming smooth field emissivity for most dry land agricultural conditions in which the row height to row spacing is less than 2 will result in an error of less than 3%.

Theis et al. (1986) have demonstrated the possibility of using a multisensor approach for improving the estimates of soil moisture under field conditions. In this case, the effects of surface roughness were accounted for with scatterometer measurements. These were then used in a soil moisture equation which included terms related to the emissivity measured by the radiometer and to the scatterometer roughness term. Inclusion of the roughness term improved the r'^2 values from 0.22 to 0.65 for C-band and from 0.69 to 0.95 for L-band.

Although roughness may not be a serious limitation for passive sensors, at least for most natural surfaces, it is a major factor for radar. In many cases the effects of roughness may be equal or greater than the effects of soil moisture on the backscatter. Thus the soil moisture problem becomes one of determining the roughness effect independently so that a model can be inverted to yield a measure of soil moisture.

The role of surface roughness in soil moisture estimation for the active case needs to be understood through surface scattering processes. The theoretical work on surface scattering can be divided into three categories: The small perturbation model (SPM), the physical optics model (POM) and geometrical optics model (GOM). In a broad sense, the geometrical optics model is best suited for a very rough surface, the physical optics model is suitable for surfaces with intermediate scales of roughness, and the small perturbation model is suitable for surfaces with short correlation lengths. Figure 9.4 describes the region of validity for the three models in terms of Kl and Kσ, where K is the wave number, l is the correlation length, and σ is the standard deviation of the surface roughness heights. The mathematical expressions to calculate surface backscatter using these models and their regions of validity in terms of RMS height, correlation length and wavelength can be found in Ulaby et al. (1982). An examination of these surface backscattering expressions employing different scattering models shows that even though the backscatter increases due to the increase of surface roughness, the soil moisture sensitivity to backscatter diminishes due to sharp rate of decrease in the value of reflectivity. As a result of two competing effects, the roughness effects overshadow the soil moisture effects.

Unfortunately, even if roughness data are available, Oh et al. (1992) have shown that the typical values of Kσ and Kl found in the natural environment result fall in the area outside of the various models regions of validity (Fig. 9.4). Consequently, most people have had little success using these models.

However, based on the scattering behavior in limiting cases and experimental data, Oh et al. (1992) have developed an empirical model in terms of the rms

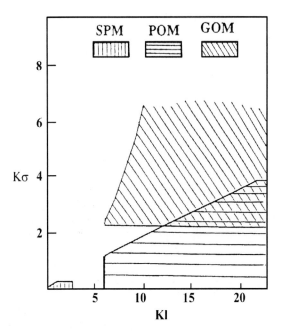

Fig. 9.4. Illustration of the regions of validity for the Small Perturbation Model (SPM), the Physical Optics Model (POM) and the Geometric Optics Model (GOM) as functions of the roughness and correlation lengths normalized by the wave number, K (after Oh et al. 1992)

roughness height, the wave number and the relative dielectric constant. By using this model with multipolarized radar data the soil moisture content and the surface roughness can be determined. The key to this approach is the copolarization ratios (hh/vv) and cross-polarization ratios (hv/vv) are given explicitly in the terms of the roughness and the soil dielectric constant. Results from this model look very good and if further testing proves as valid, this approach will be a major step forward in determining soil moisture from radar backscatter. Furthermore, this model appears to work well in the roughness domains that the more classical methods have failed in the past.

Vegetation Cover. The effect of vegetation is to attenuate the microwave emission from the soil; it also adds to the total radiative flux with its own emission. The degree to which vegetation affects the determination of soil moisture depends upon the mass of vegetation and the wavelength. Barton (1978) used an aircraft mounted 2.8 cm radiometer to measure soil moisture over bare soils and uniform grass cover. Although he demonstrated a strong relationship between brightness temperature and moisture for the bare fields, no relationship for the grass sites could be perceived.

In studies over bare soil and sorghum, Newton and Rouse (1980) found no sensitivity to soil moisture with the 2.8 cm measurements over the sorghum, but with the 21 cm data the radiometer was sensitive to soil moisture even under the tallest sorghum.

Basharinov and Shutko (1975) and Kirdiashev et al. (1979) studied a variety of crops in the USSR with wavelengths varying from 3 cm to 30 cm. For wavelengths greater than 10 cm, their results indicate that one can expect a decrease in sensitivity of about 10-20% for small grains over what would be expected for bare soil. With broad leaf crops such as corn, the sensitivity could decrease by as much as 80% for wavelengths shorter than 10 cm, and 40% for a 30 cm wavelength. Thus, from these studies the wavelength effect can be seen, that is, a vegetation canopy is more transparent for longer wavelengths than for shorter wavelengths.

Jackson et al. (1982) developed a parametric approach based on a theoretical model proposed by Basharinov and Shutko (1975). This model treats the vegetation as an absorbing layer that can be quantified in terms of the water content of the vegetation by the following relationship:

$$M_V = 78.9 - 78.4[1 + (e-1)\exp(0.22W)] \qquad (9.11)$$

where M_V is the volumetric soil moisture (0-2.5 cm), *e is* the measured emissivity, and W is the water content of the vegetation (kg/m^2). Figure 9.5 illustrates the effect of vegetation on soil moisture. Jackson and Schmugge (1991) have analyzed a large amount of published data to verify previous findings and they have defined a vegetation parameter that is based on the optical depth of the canopy. This parameter appears to be inversely related to the wavelength and can represent four types of vegetation classes (leaf dominated, stem dominated, grasses, and trees and shrubs). However, at longer wavelengths, a single value of the parameter might be

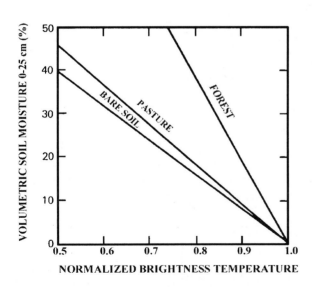

Fig. 9.5. The relationships between normalized brightness temperature and volumetric soil moisture for different types of vegetation, L-band, H polarization (after Jackson et al. 1982)

used for any cover type. Furthermore they speculate on how this parameter could be estimated using visible and near infrared satellite data in an operational sense. These studies point out the possibility of a total satellite remote sensing approach for soil moisture without any ground sampling.

Dead vegetation can also have an attenuating effect on the microwave emissions from the soil as was demonstrated by Schmugge et al. (1988). Aircraft experiments with an L-band pushbroom microwave radiometer over the Konza prairie grasslands showed that, for areas that had not been burned, a buildup of a thatch layer serves as a highly emissive layer above the soil, thus masking the emission of the soil itself. Where there was an absence of this thatch layer because of burning or grazing, the microwave sensitivity to soil moisture was as expected for bare soil.

As with the roughness case, the effect of vegetation on the active microwave sensing of soil moisture is greatly dependent upon the instrument incidence angle, frequency, and polarization. These effects are illustrated in Fig. 9.6 for a corn canopy. In the top half of Fig. 9.6, it can be seen that the attenuation for the horizontal polarization is relatively weak, but the vertically polarized data are attenuated to a much greater degree because of their relationship to the canopy structure,

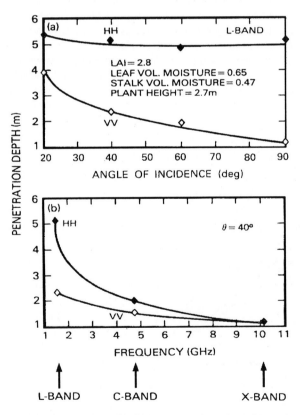

Fig. 9.6. The effect of polarization, incidence angle and frequency on the penetration depth of a corn canopy (after NASA, 1988)

which consists primarily of vertical stalks. The effect of frequency on penetration depth can be seen in the lower part of Fig. 9.6. It is readily apparent that the penetration depth increases with a decrease in frequency or an increase in wavelength.

With radar the effect of the vegetation canopy adds more complexity to the problem. Now to determine soil moisture, one must determine the soil roughness effects and the effects of the vegetation canopy, which is not a trivial exercise.

All kinds of vegetation contain water along with some plant structure. Both of these vegetation parameters are responsive to microwaves. Therefore, the radar response from vegetated areas will have the integrated effect of vegetation and soil. Most of the microwave models have been constructed by replacing the vegetated regions with a random medium whose statistical characteristics are related to physical quantities of the medium.

9.4 Hydrologic Examples

Within the past few years there have been a number of very successful demonstrations of measuring soil moisture with both passive and active microwave systems. The successes with the passive systems have been limited to aircraft campaigns but the SAR results have been obtained from aircraft, the Space Shuttle and satellites.

The examples of successful use of passive microwave radiometers for measuring soil moisture have been associated with large scale field campaigns designed to examine the role of soil moisture in hydrology and land-atmosphere interactions. These field campaigns have been conducted in a fairly typical way as far as the remote sensing was concerned. During the experiment period, extensive ground data and hydrologic data were collected. Soil moisture measurements consisted of neutron meter, TDR, and a large number of gravimetric samples. In addition, a number of more traditional hydrologic measurements were made such as rainfall, streamflow, meteorological variables and ground based energy and flux stations. Two AMES Research Center based NASA aircraft have typically participated in the experiments, the DC-8 and C-130. The DC-8 carried the three frequency polarimetric Synthetic Aperture Radar (AIRSAR) and the C-130 carried the NS001 thematic mapper simulator, the Thermal Imaging Mapper, the Electronically Steered Thinned Array Radiometer (ESTAR) or its predecessor the Push Broom Microwave Radiometer (PBMR)and a USDA laser profiler.

Some of the more successful campaigns are briefly described below:

The First ISLSCP Field Experiment (FIFE) was carried out over a 15 x 15 km grassland site near Manhattan, Kansas. This area became the focus for an extended monitoring program of satellite, meteorological, biophysical, and hydrological data acquisition from early 1987 through October 1989. During FIFE the airborne L-band, four-beam Push Broom Microwave Radiometer (PBMR) was used to map the spatial distribution of soil moisture over a small (37.7 ha) drainage basin. By using low elevation, overlapping flight lines during a drying period, the spatial patterns of soil moisture were mapped as shown in Fig. 9.7 (Wang et al., 1989). These patterns resembled those developed from a simple draining slab model and

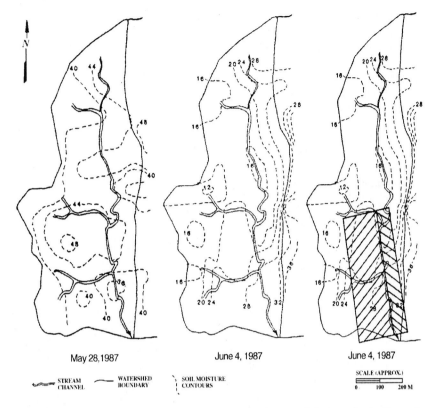

Fig. 9.7. The temporal and spatial patterns of soil moisture in a small drainage basin illustrating the drying pattern (after Wang et al., 1989) and the area of drainage (cross hatched area) necessary to sustain base flow which over lays the high soil moisture region (after Engman et al., 1989)

illustrated the capability to identify the active regions of a basin (Engman et al., 1989).

Monsoon 90 and 91 were campaigns to investigate the effects of changmoisture on the surface radiation balance, the hydrologic cycle, and the feedbacks to the atmosphere. The experiments were conducted in a hydrologically instrumented semiarid rangeland watershed, the Walnut Gulch Experimental Watershed operated by the USDA-Agricultural Research Service, which is located in the southwestern United States about 120 km southeast of Tucson, Arizona. One of the really first examples of using passive microwave measurements in a hydrologic context was reported by Jackson, et al. (1993). Using both the PBMR and the ESTAR during Monsoon 90 and 91, they showed how soil moisture patterns were related to the extent and amount of rainfall resulting from small, isolated convective storms. Goodrich et al. (1994) compared the measured rainfall, modeled soil moisture and the microwave mapped soil moisture. In studies of prestorm rainfall, they concluded that both the measured passive microwave soil moisture and the model produced estimates (from measured rainfall) produce adequate definition of the antecedant rainfall or soil moisture condition.

A Multisensor Aircraft Campaign (MAC) was conducted in July of 1990 with a focus on microwave sensors, soil moisture and humid region hydrology. The experiment was conducted at the U.S. Department of Agriculture, Agricultural Research Service's Northeast Watershed Research Center experimental watershed near Klingerstown PA. The passive microwave measurements showed a good correlation with the ground data and may yield a reliable technique for calibrating hydrologic models (Wood et al., 1993) and improve the water budget calculations of a basin (Lin et al., 1994).

Multisensor Aircraft Campaigns for hydrology (WASHITA 92 and 94) were conducted in June of 1992 and April and October of 1994 over the Little Washita Watershed, a U.S. Department of Agriculture (USDA) research facility near Chickasha, Oklahoma. The Little Washita Watershed is a 610 sq. km drainage basin situated in the southern part of the Great Plains in southwest Oklahoma. This MAC focused on both passive microwave and Synthetic Aperture Radar (SAR). The Passive microwave instrument was theL-band Electronically Steered Thinned Array Radiometer (ESTAR) while the SARs were the JPL AIRSAR (polarimetric C, L and P band) and the Shuttle Imaging Radar (SIR-C/X-SAR) which had X, C and L bands. An even more dramatic example of hydrologically significant soil moisture was derived with the ESTAR during the WASHITA 92 (Jackson et al., 1995). From a hydrologic perspective we were able to follow a drying period from very wet to dry over a period of ten days. It had rained for 26 consecutive days in Oklahoma when we arrived and initial conditions were very wet. The drying pattern as well as the spatial variability reflected by different soil properties is shown in Colour Plate 9.A. Colour Plate 9.A illustrates the spatial and temporal changes in soil moisture measured with ESTAR during 1992 in which we were able to follow a dry down from an extremely wet condition. Colour Plate 9.B illustrates the changes in soil moisture observed in1994 with the shuttle imaging radar.

The patterns illustrated in Colour Plate 9.A and their similarity to the soil texture map has inspired research to investigate if certain physical and hydrologic properties might be determined from a temporal series of the rates of change in soil moisture. The 1992 ESTAR data have been analyzed to produce estimates of soil texture (represented by the percent sand to clay ratio, and the saturated hydraulic conductivity (Mattikalli et al., 1998). Examples of these results are shown in Colour Plate 9.C.

Verhoest et al., (1998) used a principal components analysis to separate the effects of vegetation, topography and soil moisture from each other using eight ERS-1&2 scenes over the Zwalm catchment in Belgium. The resulting product (Colour Plate 9.D) is an excellent rendition of what one would expect from a partial contributing area process. The dark blue areas represent the storm runoff contributing areas and the orange and yellow represent regions that could be considered to be recharge areas and not areas that produce storm runoff.

9.5 Future Microwave Remote Sensing of Soil Moisture

The previous sections have discussed the basis of microwave remote sensing for soil moisture and presented a discussion of various target-sensor interactions. As promising as microwave remote sensing for soil moisture appears to be, the future for using microwave data for operational use is somewhat uncertain. For the next few years, researchers will be limited by the lack of suitable data. Currently the ERS-1&2 SAR from the European Space Agency, the JERS-1 SAR from the Japanese and the Canadian RADARSAT are the only operational active microwave satellite sensors with frequencies suitable for soil moisture. Although these instruments should be valuable data sources for extending our knowledge of SAR for measuring soil moisture, to date very little data have been available for this purpose. Fortunately there have been some intensive and science-driven aircraft experiments conducted in the last several years and these data are beginning to become available to the scientific community. These should be invaluable for providing sample data for developing and testing algorithms as well as answering some of the target-sensor questions.

Looking ahead to when there may be more microwave sensors on orbiting platforms, one confronts the basic differences between passive and active instruments and the intended use of the data. Comparing the instruments simplistically, the active sensors have the capability to provide high spatial resolution data (on the order of tens of meters) but their sensitivity to soil moisture may be confused more by roughness, topographic features, and vegetation than the passive systems. On the other had, the passive systems, although less sensitive to target features, can provide spatial resolutions only on the order of tens of kilometers. One then must consider how the data will be used. Meteorological and climate models currently use computational cells on the order of 10-100 km which may be well within the capacity of future passive systems. However, if one is interested in more detailed hydrologic process studies and partial area hydrology, the passive data would appear to be of little use. It is in this context that the active systems appear promising. For example, existing and planned SAR's can provide at least 20-30 m resolution over a swath width of 100 km. RADARSAT has the capability of a scanning mode (SCANSAR) to cover a much wider swath (300-500 km) at a reduced resolution (250 m) as well as a high precision mode with a spatial resolution approaching 8 m. There are a number of soil moisture missions on the drawing boards or in the proposal stage that would provide a major boost for the applications of soil moisture in hydrology. These include two proposals based on passive microwave aperture synthesis, the NASA HYDROSTAR and the CNES/ESA SMOS and the NASA-JPL LIGHTSAR with a polarametric L-band sensor. More certain is the AMSR passive microwave (C-band) instrument that will fly on the Japanese ADEOS II and the EOS-PM and the CMIS instrument that will be on future US weather satellites.

Research Needed. There are going to be more and more orbiting microwave sensors suitable for measuring soil moisture in both the near and distant future. However, before the hydrologic community makes good use of these data there are a number of research questions that must be addressed. Perhaps the foremost research need is to have the hydrology community try using the available data with hydrologic models. This will undoubtedly lead to an improved understanding of the value of these data band and the modification or development of new models to effectively use these data.

There is a need to develop and improve algorithms to extract volumetric soil moisture directly from the microwave measurement (backscatter coefficient or brightness temperature). To do this, the other target characteristics of vegetation and surface roughness will have to be parameterized. As discussed previously, there is great progress being made in these areas but much more needs to be done. Connected directly to this need is a need to better understand the effects of surface roughness on the measured microwave response with respect to incidence angle, azimuth angle, wavelength, and polarization. New methods to measure surface roughness need to be explored. The dual frequency or Dk radar technique has been quite effective to measure sea-surface roughness (Schuler et al., 1991). Implementation of this technique to soil surfaces can provide accurate measure of roughness over a very large scale. Also, there is a need to understand the effect of the vegetation canopy on the microwave response. Progress has been made in this area with both the modeling of radar (Ulaby et al. 1990) and a dual frequency, dual polarization approach for passive microwave (Njoku, 1999). Vegetation variables include the geometry for the individual plant as well as the canopy as a whole, the water content (and perhaps the biochemical makeup) of the plant, and its stage of growth. Microwave variables would include the incidence angle, the azimuth angle, wavelength, and polarization. Some difficult problems such as soil moisture estimation from rocky soil, effects of discontinuous canopy or vegetation clumps on soil moisture estimation, etc., also need to be addressed.

There is a need to investigate the use of change detection algorithms or statistical techniques like the principal components analysis (Verhoest et al. 1998) for determining the relative soil moisture of an area and whether or not this information can be useful for hydrologists. Change detection should minimize the influence of target variables such as roughness and vegetation, at least over short time intervals. With change detection it is assumed that the only target change occurring is the soil moisture. Thus, any measured changes in brightness temperature or backscatter can be related directly to changes in soil moisture. Fortunately, both the brightness temperature and backscatter relationships with soil moisture are approximately linear. There is also a reasonable basis for expecting change detection methods to provide adequate data for agricultural and hydrologic applications if the data are collected from a long term orbiting platform. Long term (multiseason or year) data will establish the upper (wet) and lower (dry) limits for the change algorithm.

There is a need to develop software procedures for correcting the effects of terrain on the microwave response. Active microwave (SAR) is especially sensitive to this. This includes foreshortening, layover, and local incidence angle effects. Also, a potential issue is the relative accuracy of the DEM data with respect to the spatial resolution of the microwave data and the potential effect of subpixel variability on the measured signal.

There is also a need to investigate the potential for polarimetric SAR and its potential for abstracting target information such as the surface roughness and vegetation characteristics. Studies of this technique need to be carried out with carefully conceived ground data collection programs.

References

Atlas, R., N. Wolfson, and J. Terry, "The effect of SST and soil moisture anomalies on GLA model simulations of the1988 US summer drought". J. Climate, 6:2034-2048, 1993

Barton, I.J., "A case study of microwave radiometer measurements over bare and vegetated surfaces," J. Geophys. Res. 83:3515-3517, 1978

Basharinov, A.Y., and Shutko, A.M. "Simulation studies of the SHF radiation characteristics of soils under moist conditions, NASA Tech. Trans., TTF-16, Greenbelt, MD, 1975

Beljaars, A., P. Viterbo, M. Miller, and A. Betts, "The anomalous rainfall over the US during July 1993 – sensitivity to land-surface parameterization and soil moisture". Monthly Weather Review, 124(3): 362-383, 1996

Bruckler, L., Witono, H., and Stengel, P. "Near surface soil moisture estimation from microwave measurements". Remote Sens. Environ, 26:101-121, 1988

Carroll, T.R., "Airborne soil moisture measurements using natural terrestrial gamma radiation". Soil Sci. 132(5):358-366, 1981

Chang, J.-T. and P.J. Wetzel, "Effects of spatial variations of soil moisture and vegetation on the evolution of a pre-storm environment: a numerical case study." Mon. Wea. Rev., 119: 1368-1390, 1991

Choudhury, B.J., Schmugge, T.J., Newton, R.W., and Chang, A., "Effect of surface roughness on the microwave emission from soils, J. Geophys. Res. 81:3660-3666, 1979

Crist, E.P., and R.C. Cicone, "Physically based transformation of Thematic Mapper data: VTM tasseled cap". IEEE Trans. Geosci. Remote Sensing, GE-22:256-263, 1984

Dobson, M., F. Ulaby, M. Hallikainen and M. El-Rayes, "Microwave dielectric behavior of wet soil-part II: dielectric mixing models," 23, pp. 3546, 1985

Engman, E.T., G.Angus, and W.P.Kustas, "Relationship between the hydrologic balance of a small watershed and remotely sensed soil moisture." Proc. IAHS Third International Assembly, Baltimore. IAHS Publ. No. 186: 75-84, 1989

Engman, E.T., "Applications of microwave remote sensing of soil moisture for water resources and agriculture". Remote Sens. Environ., 35:213-226, 1991

Fast, J.D. and M.D. McCorcle, "The effect of heterogeneous soil moisture on a summer baroclinic circulation in the central United States". Mon. Wea. Rev., 119: 2140- 2167, 1991

Georgakakos, K., J. Sperfslage, and A. Guetter, "Operational GIS-based models for NEXRAD radar data in the U.S.". Proc. Of the Int. Conf. On Water Resources and Environmental Res., Oct. 29-31, Kyoto, Japan, Vol. I: 603-609, 1996

Goodrich, D.C., T.J.Schmugge, T.J.Jackson, C.L.Unkrich, T.O.Keefer, R.Parry, L.B.Bach and S.A.Amer, "Runoff simulation sensitivity to remotely sensed initial soil water content". Water Resources Res., 30(5): 1393-1405, 1994

Jackson, T.J., Schmugge, T.J., and Wang, J.R., "Passive microwave sensing of soil moisture under vegetation canopies, Water Resour. Res. 18(4):1137-1142, 1982

Jackson, T.J. and T.J. Schmugge, "Correction for the Effects of Vegetation on the Microwave Emission of Soils." IEEE Int. Geosci. Remote Sensing Symp. (IGARSS) Digest. pp. 753-756, 1991.

Jackson, T.J. and T.J. Schmugge, "Passive microwave remote sensing system for soil moisture: Some supporting research". IEEE Trans. Geosci. Remote Sensing GE-27: 225-235, 1989

Jackson, T.J., D.M. Le Vine, A.J. Griffis, D.C. Goodrich, T.J. Schmugge, C.T. Swift, and P. O'Neill, "Soil moisture and rainfall estimation over a semiarid environment with ESTAR microwave radiometer". IEEE Trans. Geoscience and Remote Sensing, 31(4): 836-841, 1993

Jackson, T.J., D. Le Vine, C.T. Swift, T.Schmugge, and F. Schiebe, "Large area mapping of soil moisture using ESTAR passive microwave radiometer in WASHITA 92". Remote Sens. Environ., 53:27-37, 1995

Jackson, T.J., P.E. OíNeill, and C.T. Swift, " Passive microwave observation of diurnal surface soil moisture". IEEE Trans. on Geoscience and Remote Sensing, 35(5): 1210-1222, 1997

Jackson, R.D., J. Ahler, J.E. Estes, J.L. Heilman, A. Kakle, E.T. Kanemasu, J. Millard, J.C. Price and C. Wiegand, "Soil moisture estimation using reflected solar and emitted thermal radiation". In Soil Moisture Workshop, NASA Conf. Publ. 2073, Chap. 7, 219pp, 1978

Kirdiashev, K.P., Chukhlantsev, A.A. and Shutko, A.M., "Microwave radiation of the earth's surface in the presence of vegetation cover, Radio Eng. Electron. (Engl. Transl.) 24:256-264, 1979

Lin, D.-S., E.F.Wood, J.S.Famiglietti and M. Mancini, "Impact of microwave derived soil moisture on hydrologic simulations using a spatially distributed water balance model". Proc. of the 6th Inter. Symp. on Physical Measurements and Signatures in Remote Sensing, Val d'Isere, France, 1994

Manabe, S., J. Smagorinsky, and R.J. Strickler, "Simulated climatology of a general circulation model with a hydrological cycle," Mon. Weather Rev., 93, pp. 769-798 1965

Manabe, S., "Climate and ocean circulation, I, The atmospheric circulation and the hydrology of the Earth's surface," Mon. Weather Rev., 91, pp. 739-774, 1969

Mattikalli, N.M., E.T. Engman, T.J. Jackson, and L.R. Ahuja, "Microwave remote sensing of temporal variations of brightness temperature and near surface soil water content during watershed-scale field experiment, and its application to estimation of soil physical properties".Water Resour. Res., 34(9) 2289-2299, 1998

Mo. T., Schmugge, T.J., and Choudhury, B.J., "Calculations of the spectral nature of the microwave emission from soils, NASA Tech Memo 82002, Greenbelt, MD, 1980

Newton, R.W., and Rouse, J.W., "Microwave radiometer measurements of moisture content". IEEE Trans. Antennas Propagat. AP-28:680-686, 1980

Newton, R.W., Black, Q.R., Makanvand, S., Blanchard, A.J., and Jean, B.R., "Soil moisture information and thermal microwave emission". IEEE Trans. Geosci. Remote Sens. GE-21:300-307, 1982

Njoku, E., and J. Kong, "Theory of Passive Microwave Remote Sensing of Near-Surface Soil Moisture," J. Geophy. Res., 82, pp. 3108-3118, 1977

Njoku, E. G. and L. Li, "Retrieval of land surface parameters using passive microwave measurements at 6-18 GHz". IEEE Trans. Geosci. Rem. Sens., in press, 1999

NASA (1988), "SAR Instrument Panel Report," Earth Observing System

Oh, Y., K. Sarabandi, F.T. Ulaby, "An Empirical Model and an Inversion Technique for Radar Scattering from Bare Soil Surfaces." IEEE Trans. on Geosci. and Remote Sensing GE-30 (2): 370-381, 1992

Price, J.C., "On the use of satellite data to infer surface fluxes at meteorological scales". J. Appl. Meteorol. 21: 1111-1122, 1982

Promes, P.M., Jackson, T.J., and O'Neill, P.E., "Significance of agricultural row structure on the microwave emissivity of soils". IEEE Trans. Geosci. Remote Sens., 26(5):580-589, 1988

Schmugge, T.J., "Remote Sensing of soil moisture: Recent advances". IEEE Trans. Geosci. Remote Sens. GE-21(3):336-344, 1983

Schmugge, T., "Measurements of surface soil moisture and temperature". in Remote Sensing of Biosphere Functioning (R.J. Hobbs and H.A. Mooney, Eds.), Springer-Verlag, New York, pp. 31-62, 1990

Schmugge, T.J., Wang. J.R., and Asrar, A., "Results from the pushbroom microwave radiometer flights over the Konza Prairie in 1985". IEEE Trans. Geosci. Remote Sens. GE-26:590-596, 1988

Schuler, D.L., W.C. Keller and W.J. Plant, "A three frequency scatterometer technique for measurement of ocean wave spectra," IEEE J. of Oceanic Engg., 16, pp. 244, 1991

Theis, S.W., Blanchard, B.J., and Newton, R.W., "Utilization of vegetation indices to improve microwave soil moisture estimates over agricultural lands". IEEE Trans. Geosci. Remote Sens. GE-22(6):490-496, 1984

Theis, S.W., Blanchard, B.J., and Blanchard, A.J., "Utilization of active microwave roughness measurements to improve passive microwave soil moisture estimates over bare soils". IEEE Trans. Geosci. Remote Sens. GE-24(3):334-339, 1986

Ulaby, F., R. Moore and A. Fung, "Microwave Remote Sensing: Active and Passive", Addison-Wesley Publishing Company, Reading, MA, 1064 pp, 1982

Ulaby, F., R. Moore and A. Fung, "Microwave Remote Sensing: Active and Passive", VOL III, Addison-Wesley Publishing Company, Reading, MA, 1986

Ulaby, F., K. Sarabandi, K. McDonald, M. Whitt, and C. Dobson, "Michigan Microwave Scattering Model". Int. J. Remote Sensing, 11:12223-1253, 1990

van de Griend, A., P.J. Camillo, and R.J. Gurney, "Discrimination of soil physical parameters, thermal inertia and soil moisture from diurnal surface temperature flucuations". Water Resour. Res. 21(7):997-1009, 1985

Verhoest N. E. C., P. A. Troch, C. Paniconi, F. P. De Troch, 1998. "Mapping basin-scale variable source areas from multitemporal remotely sensed observations of soil moisture behavior", Water Resour. Res., Vol. 34, p. 3235-3244

Wang, J.R., and Schmugge, T.J., "An empirical model for the complex dielectric permittivity of soils as a function of water content, IEEE Trans. Geosci. Remote Sens. GE-18:288-295, 1980

Wang, J.R., Newton, R.W., and Rouse, J.W., "Passive microwave remote sensing of soil moisture: The effect of tilled row structure". IEEE Trans. Geosci. Remote Sens. GE-18:296-302, 1980

Wang, J.R., J.C. Shiue, T.J.Schmugge and E.T.Engman, "Mapping soil moisture with L-band radiometric measurements". Remote Sens. Environ., 27: 305-312, 1989

Wiesnet, D.R., "Remote sensing and its applications to hydrology". In J.C. Rodda (ed.) Facets of Hydrology, John Wiley & Sons, London, 368pp, 1976

Wilheit, T.T., Jr., "Radiative Transfer in a plane stratified dielectric, NASA Report X-911-75-66, Goddard Space Flight Center, Greenbelt, MD, 19 pp., 1975

Wood, E. F., D.-S. Lin, M. Mancini, D. Thongs, P.A. Troch, T.J. Jackson, J.S. Famiglietti and E.T. Engman, "Intercomparisons between passive and active microwave remote sensing, and hydrological modeling for soil moisture". Adv. Space Res., 13(5): 167-176, 1993

10 Remote Sensing of Surface Water

Geoff Kite[1] and Alain Pietroniro[2]

[1]International Water Management Institute, THAEM,
35660 Menemen, Izmir, Turkey
[2]National Water Research Institute, NHRC, Environment Canada,
11 Innovation Blvd, Saskatoon, Saskatchewan, Canada S7N 3H5

10.1 Introduction

The rapid growth of population in many countries together with a generally increasing standard of living is increasing demands on surface water for irrigation, industry and urban water supply and is decreasing the quality of the available surface water (Schultz and Barrett, 1989). Surface water may occur in liquid form as lakes, reservoirs and rivers and in its solid form as snow, glacier and river and lake ice. Remote sensing has a major role to play in estimating the areal extent and water content of both these phases. This chapter concentrates on the use of remote sensing to detect and measure surface water in its liquid phase with brief diversions from this topic to describe remotely-sensed measurements of related properties of surface water such as sediment load and water quality.

Studies in the U.S.A. (e.g. Castruccio et al., 1980; Rango, 1980; Carroll, 1985) have suggested possible benefit/cost ratios ranging from 75:1 to 100:1 for using remotely sensed data in hydrology and water resources. These estimates are based on savings from flood prevention and improved planning of irrigation and hydro-electric production, all of which require information on location and amounts of surface water. Measurements of surface water extent taken at intervals of many years are also useful for determining long term changes in water regimes (e.g. Foster and Parkinson, 1993; Birkett and Mason, 1995).

Unfortunately, as Schultz (1988) pointed out, satellite sensors do not measure hydrological data directly; the hydrology is obtained only after interpretation of the measured electromagnetic radiation. In many cases the analysis of remotely sensed data consists only of developing a regression equation between the desired information and the available pixel intensities. Such regressions are generally not applicable beyond the time and space constraints of the original data and add little if anything to our understanding of what is actually being measured. Apart from the direct estimation of surface water characteristics from remotely sensed data, such data may be used as inputs to hydrological models that may simulate river flows and lake levels (Schultz, 1994, Kite and Pietroniro, 1996, and Schultz, 1996).

This chapter first describes the location and measurement of surface water-bodies such as lakes, while briefly introducing some quality aspects of these water bodies (sedimentation and water quality are described in more detail in Chaps. 12 and 13).

The chapter then examines surface water levels (lakes and rivers) measured directly from satellites and the possibility of converting these levels into flows. Remote sensing is increasingly used for flood monitoring and the chapter describes the various satellite sensors used for this purpose. Direct estimation of water levels and the subsequent estimate of flow are not the only means of using remotely sensed data for surface water applications. Data from satellite may also be used as input to hydrological models to simulate surface water levels and flows. The use of such models is covered in more detail in Chap. 5; only a brief mention is included here.

There are many other excellent reports and conference proceedings, which include descriptions of the use of remotely sensed data for measurement of surface water. Examples of such publication include Hydrologic Applications of Space Technology (Johnson 1996), Remote Sensing in Hydrology (Engman and Gurney, 1991), Remote Sensing and Geographic Information Systems for Design and Operation of Water Resources Systems (Baumgartner et al., 1997) and a series of workshop proceedings on the application of remote sensing to hydrology (Kite et al., 1994; 1996). NERAC (1993) contains a bibliography of citations on remote sensing of wetlands and swamps.

10.2 Surface Water Detection

Locating and delineating surface waters are most easily done using remotely sensed data in the near-infrared and visible wavelengths (Swain and Davis, 1978). Since water absorbs most energy in the near- and middle-infrared wavelengths (> 0.8µm) there is little energy available for reflection at these wavelengths. Vegetation and soil, on the other hand, have lower reflectance in the visible bands (0.4 - 0.8µm) and higher reflectance in the near- and mid-infrared wavelengths (Fig. 10.1). Thus, on grayscale infrared images or on multi-spectral scanner images in the reflective infrared portion of the spectrum, water bodies appear dark and stand out in stark contrast to surrounding vegetative and soil features (Swain and Davis, 1978).

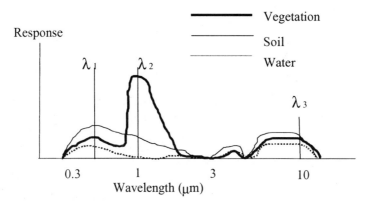

Fig.10.1. Reflectance vs. wavelength for soil, vegetation and water (adapted from Swain and Davis, 1978)

Water has a low reflectivity in the wavebands between 0.7 and 3.0 μm (Engman and Gurney, 1991). This region of the spectrum aligns itself best with Band 7 (0.8-1.1 μm) on the Landsat MSS sensor, Band 4 of the Landsat TM sensor (0.76-0.90 μm), Band 3 of the SPOT-HRV sensor (0.79-0.89 μm) and Band 2 (0.72-1.1 μm) of the NOAA AVHRR series. All of these sensors have been shown to be quite effective at these wavelengths for mapping open water regions. Weisnet. (1979) noted that surface water inventories are readily detected at these wavelengths and point to some of the earliest work for flood delineation using the near-infrared channel of both the ERTS-1 (Landsat MSS) satellite (Rango and Salomonson, 1974) and the VHRR-IR channel of the early NOAA satellites (Weisnet et al., 1974).

Microwave remote sensing platforms are also sensitive to water discrimination and have the distinct advantage of nearly all-weather viewing. Active sensors such as ERS-1 and 2, JERS-1 and Radarsat have all shown potential for estimating open water boundaries because of the specular reflection of the incident wave and very low return at the operating angles of these satellites (Crevier and Pultz, 1997; Hall, 1996, Yamagata and Yasuoka, 1993). The surrounding land surface will usually behave as a diffuse reflector, providing a return signal to the satellite as is depicted in Fig. 10.2.

Water surfaces, can also be subject to Bragg resonance effects in the radar image due to surface waves. Radar backscatter at incident angles of 20 degrees to 70 degrees is generated mostly through the Bragg resonance effect where the incident pulse responds to the short ripples or waves. These surface waves generate a backscatter response at the small wavelength of the imaging radar. Bragg resonance is a useful notion in oceanographic applications where mapping fronts, eddies or ocean currents is desired. This resonance effect can however cause confusion when discriminating between land and open water, and in some cases, the open-water radar backscatter gray levels can make the coastline obscure. In general X and C-band receivers are sensitive to centimeter surface waves while L-band radars are sensitive to decimeter surface wave heights. Low incidence angles will also increase the backscatter response from an open water target and large incidence angles are often recommended for surface water delineation. Polarization can also have a large effect on radar

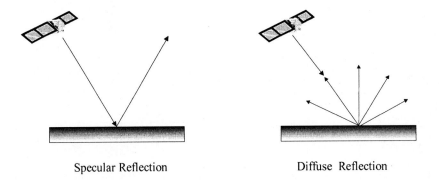

Specular Reflection　　　　　　　　Diffuse Reflection

Fig. 10.2. Active microwave response to a smooth surface and a rough surface

backscatter. The vertically polarized transmit and receive signal on the ERS series of satellites is much more sensitive to surface winds and waves than the same frequency C-HH SAR onboard the Radarsat satellite. This was clearly shown by Barber et al., (1996) for flood delineation along the Assiniboine River, Manitoba, Canada. They concluded that C-HH data acquired at large incidence angles (greater than 45 degrees) maximized their ability to detect flooded areas through specular reflection of the incident wave.

10.3 Lake and Reservoir Area Estimates

There are many examples of lake and reservoir area estimates that rely on the contrast of the near-infrared and visible wavelengths for coastline detection. This marked contrast was demonstrated with two images of the Aral Sea in Usbekistan/Kasachstan as shown by Foster and Parkinson (1993). Using a DMSP daytime visible image from 1973 and a NOAA 11 acquired visible image from 1990, the dramatic reduction in area of the inland sea (due to irrigation diversions from the in-flowing rivers) is clearly demonstrated (Fig. 10.3). El-Baz (1989) used stereo photographs from space shuttle flights in 1981 and 1984 to show a difference of more than 20 metres in the level of Lake Nasser between these dates. Such information was no doubt available from water level gauges but the technique would be useful for lakes that are not routinely measured.

Mouchot et al. (1991) used Landsat TM for a morphological analysis of water bodies in the Mackenzie Delta, northwestern Canada. The Mackenzie Delta has an area of 12,000 km^2 and contains a complex interconnected network of many thousands of lakes, ponds, channels and waterways (Roberts et al., 1994). Subtle temporal changes in the water regime result in a limnological network that is sporadically

 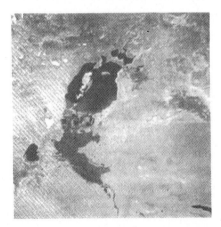

Fig. 10.3. DMSP daytime images of the Aral Sea acquired in August, 1973 (left) and September, 1990 (right) showing the shrinkage of the sea and enlarged islands. An irrigated floodplain at the bottom centre of both images is also clearly visible (after Foster and Parkinson, 1990)

reconfigured. Such changes have effects on water quality, transportation, fishing and vegetation distribution (Mouchot et al., 1991). The brightness values of Landsat band 5 (mid-infrared) were initially set to a threshold to separate water from land features and morphological analysis was then used to define further the shapes of the water bodies and the channel connections. In the same area, Roberts et al. (1994) used airborne multi-spectral video imagery to provide data for understanding the frequency, timing and duration of flooding events in the more than 25,000 lakes of the delta during the annual ice breakup. In the study transect, the flooding times of 3,216 lakes were determined from lake areas.

Pietroniro and Prowse (1997) used a time series of Landsat MSS bands 4-7, Landsat TM bands 1-7 and SPOT panchromatic data for the Peace-Athabasca Delta (PAD) in northern Alberta, Canada to document changes in areal water extent over a 15 year period within one of the world's largest freshwater deltas. This area is particularly well suited to remote sensing studies because of its remoteness, size and lack of accessibility. This dynamic ecosystem is affected by seasonal, annual and longer-term changes in water levels and extents that are rarely documented. In their study, a time series of satellite images was geometrically corrected to a UTM projection and then classified as water/no water using a parallelepiped scheme with training areas of known water bodies. The time-series of images highlights the extent of surface water change that occurred in this ecologically sensitive region during a significant drying trend (Colour Plate 10.A). This trend has been attributed to a combination of anthropogenic forcing through the construction of an upstream reservoir and climatic factors, both of which are believed to have affected the frequency of dynamic ice-jam flooding (Prowse and Lalonde, 1996). To define hydrologically connected areas, an algorithm was applied to find all pixels satisfying a four-point connectivity. The images were then converted to vector polygons and overlain on the original scenes. Colour Plate 10.B shows an example of the resulting image. The vectors could then be compared to existing lake level estimates and a simple relationship between lake areal extent and lake elevation determined. Both these applications provided vital hydrologic information that was otherwise unavailable for this remote site.

In a study of the Great Lakes of the Mackenzie Basin, Birkett and Kite (1997) used NOAA AVHRR images in the visible, near- and thermal-infrared bands. The data were obtained from the NOAA Satellite Active Archive System (SAA) using 2048 pixel-wide swath width LAC images. It was found that channel 2 (near-infrared) showed maximum contrast between water and land, and channels 1 (visible) and 2 were used to detect ice. Due to their location and size, the Mackenzie lakes are often cloud-covered; for example, during the month of August 1994, only six images out of fifty were cloud-free with another fourteen having partial cloud cover. In winter, the situation was even worse; Great Slave Lake had no cloud-free images from October to December 1994. Estimates of lake surface area were made using a local iso-luminance contour (LIC) routine (see next paragraph). This semi-automatic routine had been previously used for small lakes (Harris, 1994) but it had difficulty dealing with the complex coastlines of the Mackenzie lakes and often failed to close the lake perimeter. Instead of LIC, a manual system of identifying the lake perimeter was used. Lake surface areas in pixels were converted to square kilometres using satellite and lake altitudes, lake latitudes and locations of the centre-lake pixels. Errors in measured areas were determined to be in the order of 3-4% when compared to ground measurements.

Time series of lake areas can provide a record of climatic change; in particular, the areas of closed lakes (without river outflows) display considerable variation. The areas of lakes greater than 100 km^2 may easily be measured from Landsat MSS or TM to an accuracy of 1% (Harris, 1994); however, there is a need for frequent and regular coverage to obtain sufficient cloud-free passes. The cost of such numbers of Landsat images is high and the use of NOAA AVHRR is preferable provided that sufficient accuracy can be maintained. Harris (1994) describes the development and use of a technique, which uses sub-pixel edge detection, based on local iso-luminance contours (LIC). Edge detection is done within a moving circle of pixels. For each neighborhood, the brightest and darkest pixel intensities are determined and, if the difference exceeds a specified threshold, an edge with iso-luminance equal to one half the sum of the brightest and darkest pixel intensities is plotted. The resulting disjointed edge plot is transformed to a continuous boundary using a line-joining algorithm. Using the LIC technique on a degraded Landsat image showed comparable accuracy to using simpler techniques on the full image. In order to use the LIC technique for more areas, Birkett and Mason (1995) constructed a global database of large (>100km^2) lakes (1,403) and evaluated the availability of satellite passes for each lake.

Remotely sensed data may also be used to discover lakes. In 1974-75 an airborne radio-echo sounding survey indicated the presence of a large previously-unknown lake beneath about 4 km of ice in central East Antarctica. Analysis of changes in surface slope derived from waveform products of ERS-1 radar altimeter data confirmed the presence of this lake and allowed estimations of the lake area (10,000 km^2) and mean depth (125 m); comparable in size to Lake Ontario (Kapitsa et al., 1996).

Previously-known relationships between water depth and some other measurable parameter can be used to convert remotely sensed lake areas into lake volumes (Higer and Anderson, 1985). In a study of the Florida Everglades, Higer and Anderson (1985) used channels 5 and 7 to classify a Landsat MSS scene into water and vegetation. A two-band false-colour technique was used with ground measurements of the variation of vegetation density with depth to classify the MSS image into 10 water-depth classes. The volume of water in the area was determined by summing the products of class area and depth.

The properties of natural microwave emissions from the earth can be used to distinguish between open water and lake ice. As a development of earlier work using the DMSP SSM/I passive microwave sensor to measure snow water equivalent, Walker has used SSM/I 85 Ghz brightness temperatures to discriminate between areas of ice cover and open water on large lakes in Canada (Great Slave, Great Bear, Great Lakes) and to monitor spatial and temporal patterns of ice freeze-up and decay. Low brightness temperatures over known lake areas indicate the presence of open water (Walker, 1997).

Quite apart from the measurement of existing lakes and reservoirs, remotely sensed data may be used to select future reservoir locations (Schumann and Geyer, 1997).

Land use data classified from a Landsat TM scene of the Prüm basin, Germany were used with a digital elevation model in a GIS to identify more than 9000 potential reservoir locations and to determine the characteristics of selected sites.

10.4 Wetlands

While the contrast between water and vegetation makes the delineation of lake areas relatively easy, the similarity in reflectance between vegetation and the surrounding ground (Fig. 10.1) complicates the measurement of vegetation laden swamps and wetlands. Remote sensing techniques have increasingly been used for wetland assessment and since the 1980's, remotely sensed data from satellites has been considered the most important tool for the identification and monitoring of wetlands (Wang et al., 1998). Klemas et al.(1993) note that the advantages of using satellite imagery for mapping wetlands over conventional aerial photography include timeliness and reduced costs. Jensen et al., (1986) showed that Landsat MSS data was useful in mapping the location and area of wetlands, and that Landsat MSS was well suited for regional assessment of inland wetlands.

For larger areas, Mason et al. (1992) have shown that in near-infrared (channel 2) data from an AVHRR image of East Africa the clear-water Lakes Turkana and Albert are distinct but the 10-50,000 km^2 Sudd swamp is barely distinguishable. However, using channel 4 (thermal infrared) data, the Sudd shows very clearly because of the higher thermal inertia of the water/vegetation compared to the land. Using daily sequences of 30-minute Meteosat images, Mason et al. (1992) found that the greatest contrast between water, vegetation and land occurred during early afternoon. A simple thresholding technique was used on the brightness histograms of early afternoon images for the year 1988 to show inter-annual variations in Sudd area. A fundamental limitation to the technique was found during the rainy season (May to October) when the increased atmospheric humidity and the wet ground reduced contrast between the swamp and its surroundings. Improvements to the thermal imaging were made by using AVHRR channel 3 (3.7μm) with its higher resolution and higher contrast.

The location and measurement of water hyacinth in Lake Kyoga, Uganda has been addressed by classifying SPOT images into 6 categories of concentration (RCSSMRS, 1995). The contrast between water, vegetation and land in the infrared has also been used to monitor the development of lake deltas. Haack (1996) describes the use of the Landsat MSS near-infrared (0.8-1.1μm) channel with images from 1973 to 1989 to study the increasing size of the Omo River delta in Lake Turkana, Kenya, using the fact that the newly-formed delta was covered with active vegetation.

Radar sensors can also be utilized in wetland identification and have been used successfully for mapping. However, space-borne SAR platforms are limited to single frequency and single polarization, limiting their usefulness for vegetation discrimination. None the less, JERS-1 L-band radar imagery has been applied to mangrove vegetation mapping (Aschbacher et al., 1995). Adam et al., 1998 successfully used Radarsat C-HH data to map open and flooded vegetation in a large wetland region. In their study, they identified floodwater distribution by segmenting the images into three distinct classes: open water, flooded willow, and non-flooded areas. Supervised classification results of the original images were not acceptable due to the large local variance introduced by speckle. To improve the classification results they utilized image tone and texture. Initially, speckle was minimized with a 7x7 Gamma MAP (Maximum *A Posteriori*) filter applied to each image. The resulting classifications were significantly improved with a Kappa coefficient of over 90%. Textural information is further added to the classification, via a coefficient of variation scene. With this additional information Kappa was slightly reduced, however, visual inspection showed that some channel and lake edges were more accurately classified as flooded willows.

There have been some studies exploring using multi-temporal radar imagery for wetland mapping and monitoring. Kasischke and Bourgeau-Chavez (1997) evaluated the utility of SAR imagery for monitoring the hydrology of wetland ecosystems using two-dates ERS-1 imagery. Wang et al., 1998 also investigated the use of multidate ERS-1 data for wetland classification in Southern Ontario, Canada and used scenes of ERS-1 imagery from nine different months. The results are compared with Landsat TM imagery of the same area. In their case, accuracy was measured by comparing the results derived from digital image classification with the ground truth obtained from airphoto interpretation, map data, and field investigation. They found that although a cloud free Landsat TM performed better for wetland identification, an accuracy of over 80% could be achieved using more than 3-date ERS-1 data.

10.5 Lake Levels

The advent of satellites with onboard radar altimeters such as Geosat, Seasat, ERS-1 and the recent U.S.A/France TOPEX/Poseidon has made possible the measurement of lake and (large) river levels to centimetre accuracy with frequent repeat cycles. Remote measurement of lake levels from space using radar altimetry has been demonstrated experimentally using data from the Geosat and Seasat missions (Birkett, 1994). Birkett (1994) has described the use of Geosat data to measure levels of Lake Ontario and the Caspian Sea to within 10cm accuracy. These satellites are no longer in service but the ERS-1 and TOPEX/Poseidon satellites are operational and carry advanced radar altimeters capable of accurately measuring surface height changes, with the most advanced being the TOPEX/Poseidon altimeter. The TOPEX/Poseidon mission is designed to obtain a global view of earth's ocean topography with sufficient accuracy to improve models designed to forecast global ocean circulation. The payload consists of a dual frequency radar operating at the C-band (5.3 GHz) and Ku-band (13.6 GHz). The Ku-band secondary channel allows for correction of propagation delays in the ionosphere, reducing a significant error source in the measurement. The satellite sends out radar pulses and measures the time of the return reflection from the nearest surface directly below the satellite (Fig. 10.4). A 10 km footprint contributes to the measured return, and the accuracy of an individual height measurement is estimated to be about 13 cm. Using the onboard satellite navigation and knowledge of the orbit, the average ocean height is then estimated (Zieger et al. 1991). Sources of error from altimeter measurements can be attributed to instrument errors, atmospheric corrections, geoid undulation, temporal variation and spacecraft location errors. Atmospheric effects are primarily related to ionosphere electron content and water vapour attenuation in the troposphere. These errors are also outlined in Fig. 10.4. All radar altimeters estimate surface height by bouncing a microwave pulse off of the ocean surface. By estimating the time of the echo return to the satellite, distance is easily resolved.

Stuttard et al. (1994) used data from the ERS-1 radar altimeter to estimate levels of Lake Nakuru, Kenya, for April and May 1993. None of the altimeter tracks crossed the lake and so an off-ranging correction was derived by modelling the altimeter response within the footprint using an ocean-like backscatter with a polar response of 6 degrees half-power width. The error estimate of +/- 80cm was dominated by the error in the off-ranging correction. Results of other work (Cudlip et al., 1994) showed that for lakes directly crossed by the altimeter track, RMS lake level errors of +/-15cm are obtainable. Of this error, +/-10cm is orbit error.

Similarly Cudlip et al. (1992) derived a 20m interval contour map for the Sudd swamp in southern Sudan from Seasat radar altimeter data. This swamp covers an area of up to 50,000 km^2 of seasonally inundated floodplain for which it is very difficult to obtain ground-based data and for which visible and near-infrared sensors are ineffective during long rainy seasons. In a concurrent study Mason et al. (1992) used data from Geosat for four ground tracks across the Sudd for 1987-88 and developed water slopes across the Sudd for different seasons. Radar altimeter data can

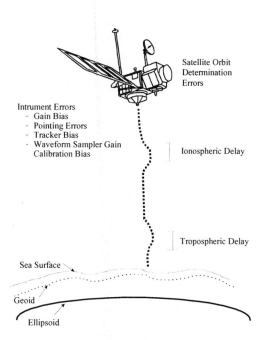

Fig. 10.4. A schematic representation of altimeter function and sources of error for the TOPEX/POSEIDON satellite (adapted from Zieger et al., 1991)

also be used to identify ground features since the height and shape of the waveform are highly dependent on the surface roughness.

Birkett and Kite (1997) used data from the NRA and SSALT radar altimeters onboard the joint NASA/CNES TOPEX/Poseidon satellite to measure levels of Lake Athabasca, Great Slave lake and Great Bear Lake in the Mackenzie Basin for September 1992 to December 1995. Results showed that the altimeters agreed well with the recorded lake levels except for periods when the lakes were ice covered. At such times, the altimeters record a dip in altimetric "lake surface" during January to April and then a recovery to normal in May, suggesting that the altimeter is recording the effect of an increasing ice and snow thickness during the winter decreasing to zero during melt. The raw altimeter waveform data were extracted for pass 095 across Lake Athabasca for 1994/95. At the end of April the echoes are broad-peaked but during May, when the ice begins to melt, the waveforms change shape, becoming multi-peaked, and reverting to normal broad-peaked during July, when the lake is completely ice-free.

Lake Victoria in East Africa has an interesting history of major changes in lake level (Kite, 1982) and it is important that accurate measurements of the lake level be obtained. Figure 10.5 plots water level vs. time for a position in Lake Victoria about 100km west of Kisumu for a series of TOPEX/Poseidon satellite passes from June 1992 to December 1993. The data show a plausible pattern of lake level changes but no surface-based measurements are currently available for comparison.

Stuttard et al. (1994) conclude that at the moment, the application of radar altimetry for routine lake level monitoring of a specific lake is not appropriate on the grounds of cost, complexity and accuracy. The role of radar altimetry is seen as monitoring a large series of closed lakes worldwide for long-term climatic change studies.

So far, this section has described methods of estimating lake levels directly from remotely sensed data. However, there is also a less direct method; many hydrological models use remotely sensed data as inputs, along with ground-based data, to simulate river flows and lake levels. Such models are described in more detail in Chap. 5 of this publication as well as in other publications such as Schultz (1994), Kite and Pietroniro (1996) and Schultz (1996). As an example, the SLURP model (Kite, 1995) has been applied to simulate the inter-annual and seasonal variation in levels of a prairie wetland near Saskatoon, Saskatchewan, Canada (Su et al., 1997). This model uses NOAA AVHRR or Landsat data to divide a basin into sub-areas of different land cover. For each vegetation type the model then carries out a vertical water balance and accumulates water from each land class and sub-area. The NDVI vegetation activity index is used to compute evapotranspiration for each land class. Figure 10.6 shows an example of the simulated and observed levels of the prairie slough.

10.6 River Levels and Flows

While streamflow may be measured in the field or may be computed in a hydrological model, many watersheds are ungauged and flow estimates from satellite data would be extremely useful. While streamflow cannot be measured directly from satellite it may, in some circumstances, be computed from parameters estimated from remotely

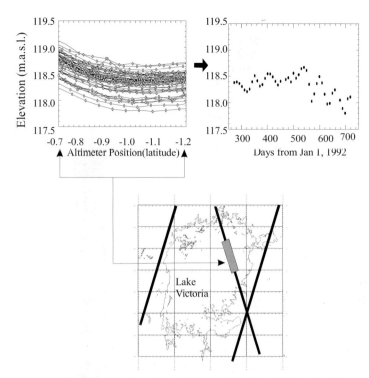

Fig. 10.5. TOPEX/Poseidon time series of lake levels for Lake Victoria. Dark lines on map indicate swath path of the satellite. The highlighted gray area indicated sampling region for time series analysis

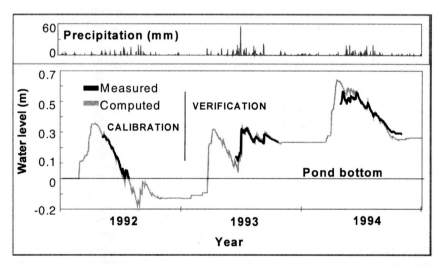

Fig. 10.6. Comparison of recorded wetland water levels with those computed using the SLURP hydrologic model for a small prairie wetland (slough) in Saskatchewan, Canada. Measured values (in black) are for ice-free periods only

sensed data. At a simple level, river flows at specific return periods may be related to geomorphological characteristics. For example, Inglis (1947) suggested that the 100-year flood is a linear function of the river's meander length; if the meander length can be measured from a satellite image then the flood flow can be estimated. Engman and Gurney (1991) recommend the use of the visible red band (MSS band 5, TM band 3) for discerning stream channel networks. Solomon (1993) recommends the use of ERS-1 synthetic aperture radar (SAR) for delineating river networks in densely forested areas, taking advantage of the SAR's all-weather day/night capability and its ability to detect water beneath forest canopies.

Similarly, the flow in a wide shallow river may be estimated from the channel width measured by satellite. Leopold and Maddock (1953) suggested that width, W, is related to discharge, Q, as:

$$W = a Q^b$$

where a and b are constants. Smith et al. (1995) used the C-band (5.3 Ghz) ERS-1 SAR to measure the width of a heavily-braided section of the Iskut River in British Columbia twenty-eight times in the period April 1992 to December 1993. Each image was radiometrically calibrated and fitted to the Goddard Earth model ellipsoid and an "effective width", W_e, of the river was computed as a simple count of those pixels classified as water within a 10km x 3km control area divided by the length of the river section. A regression equation of the form $W_e = aQ^b$ was then developed between the "effective width" and the corresponding river discharges, Q, measured by Water Survey of Canada at each overpass. The processed SAR pixel resolution of ERS-1 is 12.5m and the "effective width" of the Iskut River varied between 100 and 1100 m during the measured period. The corresponding discharges varied between 240 and 6350 m^3/s. Smith et al. (1996) extended this study to a total of 41 SAR images of the Iskut, Taku and Tanana Rivers with similar types of relationship and suggest that discharges of ungauged rivers (with strongly braided channels) can be estimated within a factor of 2.

If a river is wide enough and has a stable stage-discharge relationship, then the discharge may also be estimated from measurements of river level using a radar altimeter. For example, Cudlip et al. (1992) used Seasat data to produce a river surface elevation profile for the Amazon River from 32 altimeter crossings of the river over a 17 day period in July 1978. The accuracy of the river elevations is estimated at +/− 10 cm to +/− 20cm. If a rating curve exists for a location along this profile, then the river discharge could be computed. Koblinsky et al. (1993) concluded, however, that Geosat radar does not provide sufficient accuracy or coverage to estimate river levels in the Amazon Basin and recommended that TOPEX/Poseidon could be used with better results.

For runoff over shorter time periods it is necessary to develop more complex models. For example, to simulate monthly runoff in a basin in southwestern France, Strubing and Schultz (1985) used NOAA AVHRR infrared images to estimate different densities of cloud cover and to develop a "mean daily temperature-weighted cloud cover index". The mean monthly runoff for a particular month was then a linear function of the sum of the daily product of the cloud index and a system response function. The response function was derived from calibration of estimated and recorded runoff. A similar model using Meteosat data was developed by Papadakis (see Chap. 18). Development of this concept has led to extremely advanced distributed hydrological models such as the NOAA Nile forecasting system, which estimates inflows to Lake Nasser based on satellite estimates of precipitation over Ethiopia (Barrett, 1993). In this model, METEOSAT data are used to estimate precipitation because of the almost total lack of recorded precipitation data in the source area of the Blue Nile. The precipitation data are input to a distributed (11,000 grid squares at 30 km^2 each) rainfall-runoff model consisting of a vertical water balance, a hillslope routing component and a channel routing component. Another model estimating flows of the Blue Nile with the aid of remote sensing data is described in Chap. 18.

Other hydrological models use remotely sensed data for land cover classifications and vegetation activity indices. The SLURP semi-distributed model (Kite, 1995) divides a basin into hydrological sub-basins and into land classes using either NOAA AVHRR or Landsat data depending on the size of the basin. The model then carries out a vertical water balance for each land cover area within each sub-basin and accumulates water down the basin. Evapotranspiration is computed for each vegetation type using a vegetation activity index derived from NOAA AVHRR data. Figure 10.7 shows an example of river flow derived in this way for the Kootenay River at Skookumchuck, British Columbia for 1986-1990.

Remote sensing may be used not only to locate and monitor existing river channels but also to locate palaeo rivers. Landsat MSS and TM were used by Philip and Gupta (1993) to map three distinct stages in the migration of the Burhi-Gandak River in north-eastern India. Visible images, when enhanced by contrast stretching, revealed palaeo-features such as abandoned channels, meander scars and oxbow lakes; such information has practical importance for siting of water supply wells. Radar remote sensing allows the identification of geological formations and deep basement struc-

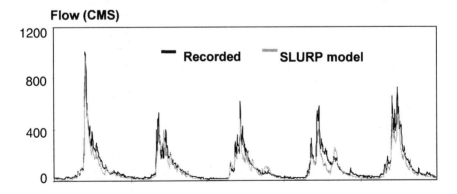

Fig. 10.7. Daily observed and computed hydrographs of the Kootenay River, British Columbia, Canada at the Skookumchuck gauge using the SLURP hydrologic model

tures. Analysis of SIR-C/X-SAR radar images obtained during two space shuttle flights in 1994 was used to reveal the bedrock structures of different ages that have controlled the course of the River Nile (Stern and Abdelsalam, 1996). Palaeochannels of the Nile have been identified and show the relatively recent deflection of the river to form the great bend in northern Sudan.

10.7 Flood Extent

Flooding in North America causes an average of $1 billion damages and several dozen deaths each year (Paterson et al., 1996). Accurate and timely information on flood extent can help emergency personnel make better decisions about where to deploy resources and how to plan evacuations, and can allow more accurate damage assessments.

Many of the techniques described in the previous section on measuring lake areas may also be used to measure flood extent. Satellite sensors in the visible and infrared bands such as NOAA-AVHRR and the Landsat TM and MSS have long been used to provide estimates of flood extent and flood hazard areas (e.g. Rango and Anderson, 1974). However, such sensors are dependent on cloud-free conditions which are rare during major floods and cannot be reliably used for under-canopy flooding. For example, Mertes (1994) used Landsat TM data to map flood events on the Amazon River and found that the water surface on the flood plain was frequently masked by vegetation.

Over the last 15 years the use of airborne and satellite synthetic aperture radar (SAR) with its cloud penetrating day and night capability has developed considerable potential for measuring flood extent (Pultz and Crevier, 1997). Ormsby and Blanchard (1985) carried out some of the earliest experimental work on SAR response to flooded vegetation using X-band, C-band and L-band imagery. They concluded that response depended on wavelength, plant volume and the geometry of the inundated vegetation (Fig. 10.8). The study notes that at X-band wavelengths, the incoming energy is scattered within the forest canopy, yet provides enhanced return within saturated sedges and grasslands. At L-band frequencies, the response from short vegetation was minimal if any, as noted in Fig. 10.9.

Wang et al. (1995) have also studied radar backscatter from flooded forests in Brazil. They concluded that the ratio of C band (such as on the ERS-1 and RADARSAT satellites) backscatter from flooded forest to non-flooded forest is about 1.8 at an incidence angle of 20° but decreases to about 1.0 at 60°. Brakenridge et al. (1994) used ERS-1 data to monitor the Mississippi flood of 1993. SAR images are often combined with optical or infrared images and with data from other sources such as maps to produce images containing both flood extent and geographic locators such as roads, and railways. Puyou-Lascassies et al. (1997) describe the use of ERS-1 images before, during and after the 1994 flood event in the Camargue (south-eastern France) as part of a space-based European risk management system.

Based on the use of C-band airborne experiments, Pultz et al. (1991) suggested that larger incidence angles and shorter revisit times than available on ERS-1 would give better results for flood events. Pultz and Crevier (1997) describe the use of

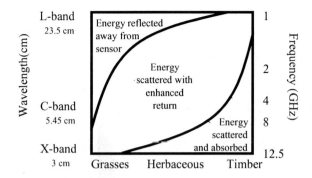

Fig. 10.8. Radar response of flooded vegetation at a variety of wavelengths and for different inundated surface vegetation (adapted from Ormbsy and Blanchard, 1985)

RADARSAT to monitor the severe flooding in the Red River of North Dakota and Manitoba in spring 1996. Standard beam mode images at varying incidence angles showed clear distinctions between dark flooded areas and multi-toned non-flooded agricultural areas (Colour Plate 10.C). This approach was later used operationally to map the areal extent of the "Flood of the Century" during the Spring of 1997 in North Dakota, USA and Southern Manitoba, Canada (see Fig. 10.10).

Pietroniro and Prowse (1997) used RADARSAT images to study the advent of the 1996 flood in the Peace-Athabasca Delta (PAD). This was the first flood in the delta

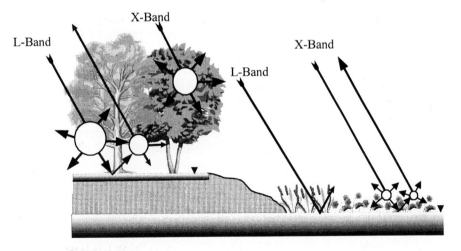

Fig. 10.9. Schematic of the effects of flooded vegetation on X-band and L-band Radar (adapted from Ormbsy and Blanchard, 1985)

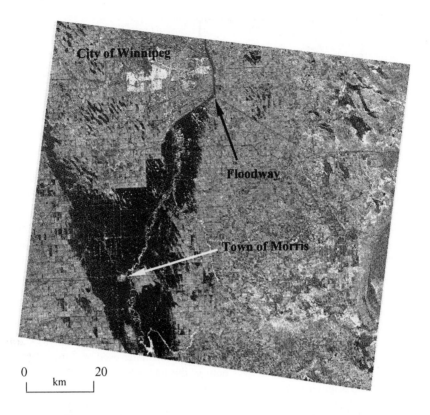

Fig. 10.10 – Radarsat standard mode 6 image of the 1997 "Flood of the Century" near Winnipeg, Manitoba acquired on May 1, 1997. The flood extent is clearly evident as is the ring dike around the town of Morris and the floodway providing protection for the City of Winnipeg (RADARSAT data © Canadian Space Agency 1997. Received by the Canada Centre for Remote Sensing (CCRS). Processed and distributed by RADARSAT International. Image provided by T.J.Pultz, CCRS)

since 1974 and was caused by ice jams along the Peace River. The size and remote nature of this flood necessitates the use of remote sensing to assist in mapping those regions inundated. Initially, aerial photography was acquired; however, analysis of the aerial photos shows that flooded areas dominated by willow trees can be very difficult to delineate. The ability to obtain synthetic aperture radar (SAR) imagery at any time of day and nearly all weather conditions led to the acquisition of a RADARSAT image immediately following the 1996 spring flood. The scene (Colour Plate 10.D) displays two distinctive inundated features. Open water exhibits its typical low return while the occurrence of regions with very high backscatter is typical of flooded willow stands. The high reflectance is a result of a double bounce between the water surface and emerging tree structures.

If the flooded area is large enough, then passive microwave emissions from the Earth's surface may also be useful. Measurements are expressed as brightness temperatures in degrees Kelvin and are proportional to the product of the surface temperature and the emissivity of the medium. Using the differences between vertically and horizontally polarised brightness temperatures (ΔT) minimizes the effects of atmospheric interference. For 37 Ghz data from the Nimbus-7 satellite scanning multichannel microwave radiometer (SMMR), heavily vegetated land shows the lowest ΔT (< 4 K) grading up to about 30 K as vegetation cover decreases. Calm surface water is easily distinguishable with a ΔT of about 60 K.

Sippel et al. (1994) used daytime 37 Ghz SMMR brightness temperatures at 6-day intervals for 1979-1985 to estimate seasonal changes in inundation area for a 34,500 km^2 area along the Amazon River near Manaus. The resolution of SMMR at the 37 Ghz frequency is about 25 km and Sippel et al. (1994) used a linear mixing model incorporating the 4 K and 60 K endpoints to determine inundation fractions of mixed pixels. Hamilton et al. (1996) extended the fractional analysis of SMMR data to study seasonal and inter-annual inundation patterns in the Pantanal, an area which varies between 11 and 110 thousand square kilometres located mainly in Brazil. By developing a regression between flooded area and river stage Hamilton et al. (1996) were able to reconstruct a 95-year flood record using recorded historic river levels

It is also possible to estimate sediment concentration and total sediment transport during major flood events. Mertes (1994) used Landsat TM data to measure sediment concentration during the 1977 and 1989 flood events on the Amazon River. Field data were combined with depth averaged velocities from a 2D hydrodynamic model and the Landsat distribution of suspended sediment concentration to give vertical deposition rates and mass balances of flood deposits.

10.8 Conclusion

Studies have suggested that the use of remotely sensed data in hydrology and water resources could yield benefit/cost ratios in the order of 100:1 from savings in flood damage and from improved planning. However, the current use of remotely sensed data in operational monitoring of surface waters is limited. This is partly because visible data from satellite are limited to relatively cloud-free and daylight conditions. Even for relatively simple tasks such as estimating the areal extent of water bodies such as large lakes, clouds obscure the analysis. Microwave satellites offer the potential of (almost) all-weather application and the synergistic application of combined electro-optical remote sensing for surface water estimates needs to be examined more closely.

Efforts should also be made by the data suppliers to educate potential customers in the advantages of remotely sensed data. Certainly, the real strength of satellite remote sensing applications is in alleviating some of the hydrometric data collection and management problems facing many of the developing nations. Developed nations with large expanses of low populated or remote areas can also benefit greatly from some of the observational techniques presented. In both cases, the lack of appropriate education and training of operational agencies and consultants in remote sensing techniques limits the thinking to traditional techniques.

Historic satellite data are often difficult to obtain and there is a need to archive more of the data than is currently done. In order to achieve this, potential users should be made aware of the usefulness of these data. Before the advent of satellites, remotely sensed data came mainly from aircraft sensors flown on specific missions. Now, remotely sensed data for hydrological modelling generally means data from non-mission specific satellites. Such satellites have disadvantages for hydrology; the sensors are not ideal for hydrological purposes, the resolutions are often inadequate and the return periods are often too long. Also, many of the current satellite systems need direct user requests in order for scenes to be acquired for a specific area of interest. This requires operational agencies to be pro-active and committed for many years of data acquisitions to monitor long-term trends.

The successful applications of remote sensing to surface water observations have been summarized in this chapter. The potential of remote sensing for such applications is now well understood, although new technology and research will continually be presented. The operational use of remote sensing for surface water applications is slowly gaining acceptance. As an example, Radarsat data were used to track the movement of the flood wave during the "Flood of the Century" in Southern Manitoba and North Dakota during the spring of 1997 in the USA and Canada. This is testament to a shift in thinking within operational agencies, fostered by hydrologic and remote sensing research in surface water applications.

References

Adam, S, J. Weibe, M. Collins and A. Pietroniro, 1998. Evaluation of Multi-date ERS-1 and Multispectral landsat Imagery for Wetland Detection in Southern Ontario, *Can. J. Rem. Sens.*,24,1, 69-79

Aschbacher, , J.R., J.P. Delsol, T.B. Binarko Seselo, S. Vibulsresth, 1995, An Integrated Comparitive Approach to Mangrove Vegetation Mapping using Avanced Remote Sensing and GIS Technologies: Preliminary Results, *Hydrobiologia* 295, 285-294

Barber, D.G., K.P. Hocheim, R. Dixon, D.R. Mosscrop and M.J. McMullan, 1996, The Role of Earth Observation technologies in Flood Mapping: A Manitoba Case Study, *Can. J. Rem. Sens.*,22,1, 137-143

Barrett, C.B., 1993. The development of the Nile hydrometeorological forecast system. *Wat. Res. Bull.*, 29, 6, 933-938

Baumgartner, M.F., G.A. Schultz and A. Ivan Johnson, 1997. Remote sensing and geographic information systems for design and operation of water resources systems (Proc. Rabat Symposium, April, 1997). *IASH* Publ. 242

Birkett, C.M., 1994. Radar altimetry - A new concept in monitoring global lake level changes. EOS Trans. *AGU*, 75, 24, 273-275

Birkett, C.M. and I.A. Mason, 1995. A new global lakes database for a remote sensing program studying climatically sensitive large lakes. *J. Great Lakes Res.*, 21, 3, 307-318

Birkett, C.M. and G.W. Kite, 1997. Derivation of lake areas and elevations for the Mackenzie Basin using satellite remote sensing. In: G.W.Kite, A. Pietroniro and T.J. Pultz (eds.), Application of Remote Sensing in Hydrology, Third International Workshop, Goddard Space Flight Centre, Washington, D.C., USA, 16-18 October, 1996, *NHRI* Symposium No.17, 19-32

Carroll, T.R., 1985. Snow surveying. Yearbook of Science and Technology, McGraw-Hill, New York, 386-398

Castruccio, P.A., H.L. Loats Jr., D. Lloyd and P.A.B. Newman, 1980. Cost/benefit analysis for the Operational Application of Satellite Snowcover Observations (OASSO). Proc. Final Workshop on OASSO, Washington, NASA CP-211b, 201-222

Crevier, Y. and T.J. Pultz, 1997. Analysis of C-Band SIR-C/X SAR Radar Backscatter over a Flooded Environment, Red River, Manitoba.. in Kite, G.W., A. Pietroniro and T.J. Pultz (eds.), 1997. Application of Remote Sensing in Hydrology, Third International Workshop, Goddard Space Flight Centre, Washington, D.C., USA, 16-18 October, 1996, *NHRI* Symposium No.17, 47-60

Cudlip, W., D.R. Mantripp, C.L. Wrench, 1994. Corrections for Altimeter low-level processing ath the Earth Observation data Centre. *Int. J. Remote Sens*, ,March, 15, 5, 889-914

Cudlip, W., J.K. Ridley and C.G. Rapley, 1992. The use of satellite radar altimetry for monitoring wetlands. In: Remote Sensing and Global Change, Proc. 16th Annual Conference, *Remote Sensing Society*, London, 207-216

El-Baz, Farouk, 1989. Monitoring lake Nasser by space photography. In: Remote Sensing and Large-Scale Global Processes (Proc. IAHS Third Int. Assembly, Baltimore, MD, May, 1989). *IAHS* Publ. 186, 177-182

Engman, E.T. and R.J. Gurney, 1991. Remote sensing in hydrology. *Chapman and Hall*, London

Foster, J.L. and C.L. Parkinson, 1993. Indications and effects of human activities. In: Gurney, R.J., J.L. Foster and C.L. Parkinson (eds.) Atlas of satellite observations related to global change. *Cambridge University Press*, Cambridge, 425-443

Haack, B., 1996. Monitoring wetland changes with remote sensing: An East African example, *Environ. Manag.*, 20, 3, 411-419

Hall, D. K. 1996. Remote Sensing Applications to Hydrology: Imaging Radar. *Hydrological Sciences*, 41,4, 609-624

Hamilton, S.K., S.J. Sippel, S.J. and J.M. Melack, 1996. Inundation patterns in the Pantanal wetland of South America determined from passive microwave remote sensing. *Arch. Hydrobiol.*, 137, 1, 1-23

Harris, A.R., 1994. Time series remote sensing of a climatically sensitive lake. *Remote Sens. Environ.*, 50, 2, 83-94

Higer, A.L. and D.G. Anderson, 1985, 1985. Water volume estimates by Landsat data. In: Hydrological Applications of Remote Sensing and Remote Data Transmission (Proc. Hamburg Symposium, August 1983). *IAHS* Publ. 145, 569-575

Inglis, Sir Claude, 1947. Meanders and their bearing on river training. Maritime and Waterways Paper No. 7, Inst. Civ. Eng., London

Johnson, A.I. (ed.), 1986. Hydrologic Applications of Space Technology. Proc. Coca Beach Workshop, *IAHS Publ.* 160

Jensen, J.R., M.E. Hodgson, E. Christensen, H.E. , Mackey, H.E.Jr., Tinney, R.L. and R. Sharitz, 1986. Remote Sensing Inland Wetlands: A Multi-spectral Approach, *Photogrammetric Engineering and Remote Sensing,* 52,1,87-100

Kapitsa, A.P., J.K. Ridley, G.de Q. Robin, M.J. Siegert and I.A. Zotikov, 1996. A large deep freshwater lake beneath the ice of central East Antarctica, *Nature*, 381, 6584, 684-686

Kasischke, E.S. and Bourgeau-Chavez, L.L., 1997. Monitoring South Florida Wetlands Using ERS-1 SAR imagery, *Photogrammetric Engineering and Remote Sensing, 63,3,281-291*

Kite, G.W. and A. Pietroniro, 1996, "Remote Sensing Applications in Hydrological Modelling", *Hydro. Sci. J.,* 41,9, 563-591

Kite, G.W., 1982. Analysis of Lake Victoria levels. *Hydro. Sci. J.*, 27,2,99-110.

Kite, G.W., 1995. Use of remotely sensed data in the hydrological modelling of the Upper Columbia Watershed. *Can. J. Rem. Sensing*, 22, 1, 14-22

Kite, G.W. and A. Pietroniro, 1996. Remote sensing applications in hydrological modelling. *Hydro. Sci. J*, 41, 4, 563-591

Kite, G.W., A. Pietroniro and T.J. Pultz (eds.), 1997. Application of Remote Sensing in Hydrology, Third International Workshop, Goddard Space Flight Centre, Washington, D.C., USA, 16-18 October, 1996, *NHRI* Symposium No.17

Kite, G.W., A. Pietroniro and T.J. Pultz (eds.), 1995. Application of Remote Sensing in Hydrology, Second International Workshop, National Hydrology Research Institute, Saskatoon, SK., Canada, 18-20 October, 1994, *NHRI* Symposium No.14

Klemas, V. Dobson, J.E., R.L. Ferguson and K.D. Haddad, 1993. A Coastal Land Cover Classification System for the NOAA Coastwatch Change Analysis Project. *Journal of Coastal Research,* 9,3,862-872

Koblinsky, C.J., R.T. Clarke, A.C. Brenner and H. Frey, 1993. Measurement of river level variations with satellite altimetry. *Wat. Resour. Res.*, 29, 6, 1839-1848

Leopold, L.B., and T. Maddock, 1953. The hydraulic geometry of stream channels and some physiographic implications, *U.S. Geol. Surv.* Prof. Paper 252

Mason, I.M., A.R. Harris, J.N. Moody, C.M. Birkett, W. Cudlip and D. Vlachogiannis, 1992. Monitoring wetland hydrology by remote sensing: A case study of the Sudd using infra-red imagery and radar altimetry. Proc. Central Symposium of the International Space Year, Munich, *ESA* SP-341, 79-84

Mertes, L.A.K., 1994. Rates of flood-plain sedimentation on the central Amazon River. *Geology*, 22, 171-174

Mouchot, M-C, T. Alfoldi, D. de Lisle and G. McCullough, 1991. Monitoring the water bodies of the Mackenzie Delta by remote sensing methods. *Arctic,* 44, Supp. 1, 21-28

NERAC, 1993. Wetlands: Remote sensing. *NERAC, Inc.*, Tolland, CT, USA

Ormsby, J.P. and B.J. Blanchard, 1985. Detection of Lowland Flooding Using Active Mircowave Systems. *Photogram. Eng. and Rem. Sens.*, 51(3): 317-328

Philip, G. and R.P. Gupta, 1993. Channel pattern transformation in the Burhi-Gandak River, Bihar, India: A study based on multidata sets. *Geocarta Int.*, 8, 3, 47-51

Pietroniro, A. and T.D. Prowse, 1997. Environmental monitoring of the Peace-Athabasca Delta using multiple satellite data sources. In: G.W.Kite, A. Pietroniro and T.J. Pultz (eds.), Application of Remote Sensing in Hydrology, Third International Workshop, Goddard Space Flight Centre, Washington, D.C., USA, 16-18 October, 1996, *NHRI* Symposium No.17, 237-252

Pietroniro, A., T.D. Prowse, and M. Demuth, 1996. Mapping areal extent of lakes and perched basins in the Peace-Athabasca Delta, *18th Canadian Symposium on Remote Sensing / 26th International Symposium on Remote Sensing of Environment*. Vancouver, British Columbia, Canada, March 26, 468-471

Prowse, T.D. and V. Lalonde, 1996. Open-water and Ice-Jam Flooding of a Northern Delta, *Nord. Hydrol.*, 27, 1/2, 85-100

Pultz, T.J., R. Leconte, L. St-Laurent and L. Peters, 1991. Flood mapping with airborne SAR imagery. *Can. Wat. Resour. J.*, 16, 2, 173-189

Pultz, T.J. and Y. Crevier, 1997. Early demonstrations of RADARSAT for applications in hydrology. In: G.W.Kite, A. Pietroniro and T. Pultz (eds.), Application of Remote Sensing in Hydrology, Third International Workshop, Goddard Space Flight Centre, Washington, D.C., USA, 16-18 October, 1996, *NHRI* Symposium No.17, 271-281

Puyou-Lascassies, Ph., J. Harms, H. Lemonnier and J.P. Cauzac ,1997. Use of space technologies to manage flood disasters towards a space based European risk management system. In: G.W. Kite, A. Pietroniro and T.J. Pultz (eds.), Application of Remote Sensing in Hydrology, Third International Workshop, Goddard Space Flight Centre, Washington, D.C., USA, 16-18 October, 1996, *NHRI* Symposium No.17, 283-294

Rango, A. and A.T. Anderson, 1974. Flood hazard studies in the Mississippi River basin using remote sensing. *Water Resour. Bull.*, 10, 5, 1060-1081

Rango, A. (1980) Operational applications of satellite snowcover observations. *Water Resour. Bull.*, 16, 6, 1066-1073

RCSSMRS, 1995. Mapping of the distribution of water hyacinth using satellite imagery, pilot study in Uganda. Publication of RCSSMRS/French Technical Assistance, Nairobi, Kenya

Roberts, A., K. Bach, C. Kirman, L. Lesack and P. Marsh, 1995. Multitemporal classification of flood parameters in the Mackenzie Delta from airborne MSV and SAP imagery. In: Kite, G.W., A. Pietroniro and T.J. Pultz (eds.), 1994. Application of Remote Sensing in Hydrology, Second International Workshop, National Hydrology Research Institute, Saskatoon, SK., Canada, 18-20 October, 1994, *NHRI* Symposium No.14, 127-135

Schultz, G.A., 1988. Remote sensing in hydrology. *J. Hydrol.*, 100,1/3, 239-265

Schultz, G.A., 1994. Meso-scale modelling of runoff and water balances using remote sensing ad other GIS data. *J. Hydrol.*, 39, 2, 121-141

Schultz, G.A., 1996. Remote sensing applications to hydrology: runoff. *J. Hydrol.*, 41, 4, 453-475.

Schultz, G.A. and E.C. Barrett, 1989. Advances in remote sensing for hydrology and water resources management. *UNESCO* IHP-III Project 5.1, Technical Documents in Hydrology, Paris

Schumann, A.H. and J. Geyer, 1997. Hydrological design of flood reservoirs by utilization of GIS and remote sensing, 1997. In: Remote sensing and geographic information systems for design and operation of water resources systems (Proc. Rabat Symposium, April, 1997). *IAHS* Publ. 242, 173-180

Sippel, S.J., S.K. Hamilton, J.M. Melack and B.J. Choudrey, 1994. Determination of inundation area in the Amazon River floodplain using the SMMR 37 Ghz polarization difference. *Remote Sens. Environ.*, 48, 1, 70-76

Smith, L.C., B.L. Isacks, R.R. Forster, A.L. Bloom and I. Preuss, 1995. Estimation of discharge from braided glacial rivers using ERS-1 synthetic aperture radar: First results. *Wat. Resour. Res.*, 31, 5, 1325-1329

Smith, L.C., B.L. Isacks and A.L. Bloom, 1996. Estimation of discharge from three braided rivers using ERS-1 synthetic aperture radar: Potential application to ungaged basins. *Wat. Resour. Res.*, 32, 7, 2021-2034

Smith, S.E., K.H. Mancy, A.F., D.A. Latif and E.A. Fosnight, 1985. Assessment and monitoring of sedimentation in the Aswan High Dam Reservoir using Landsat imagery. In: Hydrological Applications of Remote Sensing and Remote Data Transmission (Proc. Hamburg Symposium, August 1983). *IAHS* Publ. 145, 499-508

Solomon, S.I., 1993. Methodological considerations for use of ERS-1 imagery for the delineation of river networks in tropical forest areas. In: Proceedings of the First ERS-1 Symposium, Eur. Space Agency. *ESA* SP-359, 595-600

Stern, R.J., M.G. Abdelsalam, 1996. The Origin of the Great Bend of the Nile from SIR-C/X SAR Imagery. *Science*, 24, 5293, 1696-1698

Strübing, G. and G.A. Schultz, 1985. Estimation of monthly river runoff data on the basis of satellite imagery. In: Hydrological Applications of Remote Sensing and Remote Data Transmission (Proc. Hamburg Symposium, August 1983). *IAHS* Publ. 145, 491-498

Stuttard, M.J., J.B. Hayball, G. Narciso, M. Suppo and A. Oroda, 1994. Use of GIS to assist hydrological modelling of lake basins in the Kenyan Rift Valley. *Proc. Agi 94*, Birmingham, U.K., 13.4.1-13.4.8

Su, M., W.J. Stolte and G. van der Kamp, 1997. Modelling wetland hydrology using SLURP. Proc. Scientific Meeting, Can. Geophys. Union, Banff, Alberta, p. 198 (abstract)

Swain, P.H. and S.M. Davis, 1978. Remote Sensing: The quantitative approach. *McGraw-Hill*, New York

Walker, A., 1997. Personal communication

Wang, J. J.Shang, B. Brisco and R.J. Brown, 1998. Evaluation of Multi-date ERS-1 and Multispectral landsat Imagery for Wetland Detection in Southern Ontario, *Can. J. Rem. Sens.*,24,1, 60-68

Wang, Y., L.L. Hess. S. Filoso and J.M. Melack, 1995. Understanding the radar backscattering from flooded and non-flooded Amazonian forests: Results from canopy backscatter modeling. *Remote Sens. Environ.*, 54, 324-332

Weisnet, D.R., 1979. *Applications of Remote Sensing to Hydrology*, World Meterological Organization, Operational Hydrology Report no. 12

Weisnet, D.R., McGinnis, D.F and J.A. Pritchard, 1974. Mapping of the 1973 Mississippi River Floods with the NOAA-2 Satellite, *Water Resources Bulletin, 10, 1040-1049*

Yamagata, Y. and Yasuoka, Y., 1993. Classification of Wetland Vegetation by Texture Analysis Methods Using ERS-1 and JERS-1 Images. International Geoscience and Remote Sensing Symposium (IGARSS '93). 4, 1014-1016

Zieger, A.R, Hancock, D.W, Hayne, G.S, and Purdy, C., 1991. NASA Radar Altimeter for the TOPEX/Poseidon Project, *Proc. of the IEEE*, 79,6, 810-825

11 Snow and Ice

A. Rango[1], A.E. Walker[2] and B.E. Goodison[2]

[1]USDA Hydrology Laboratory, Agricultural Research Service,
Beltsville, Maryland 20705, USA
[2]Climate Research Branch, Atmospheric Environment Service, 4905
Dufferin Street, Downsview, Ontario M3H 5T4, Canada

11.1 Role of Snow and Ice

The occurrence of precipitation in the form of snow as opposed to rain typically causes a change in how a drainage basin responds to the input of water. The reason for the modified hydrological response is that snow is held in cold storage on a basin for an extended period of time before it enters the runoff process. There is such a vast difference in the physical properties of snow and other natural surfaces that the occurrence of snow on a drainage basin can cause significant changes in the energy and water budgets. As an example, the relatively high albedo of snow reflects a much higher percentage of incoming solar shortwave radiation than snow-free surfaces (80% for relatively new snow as opposed to roughly 15% for snow-free vegetation). Snow may cover up to 53% of the land surface in the northern hemisphere (Foster and Rango, 1982) and up to 44% of the world's land areas at any one time. On a drainage basin basis, the snow cover can vary significantly by elevation, time of year, or from year to year. The Rio Grande basin at Del Norte, Colorado is 3419 km^2 in area and ranges from 2432 m a.s.l. at the streamgage up to 4215 m a.s.l. at the highest point in the basin. Figure 11.1 compares the snow cover depletion curves obtained from Landsat data in 1977 and 1979 in elevation zones A (780 km^2; 2432-2926 m), B (1284 km^2; 2926-3353 m), and C (1355 km^2; 3353-4215 m) in the Rio Grande basin. In a period of two years from April 10, 1977 to April 10, 1979, a great difference in seasonal snow cover extent was experienced. Landsat data show that 49.5% or 1693 km^2 were covered by snow on April 10, 1977. Only two years later in 1979, 100% or 3419 km^2 were snow covered on April 10.

Snow cover and the equivalent amount of water volume stored supplies at least one-third of the water that is used for irrigation and the growth of crops worldwide (Steppuhn, 1981). In high mountain snowmelt basins of the Rocky Mountains, USA, as much as 75% of the total annual precipitation is in the form of snow (Storr, 1967), and 90% of the annual runoff is from snowmelt (Goodell, 1966). The variation of runoff depth can vary greatly from year to year for mountain basins because of differences in the seasonal snow cover. In the Rio Grande basin at Del Norte, Colorado, the seasonal snow cover in the two years cited above, 1977 and 1979, produced vastly different annual runoff depths of 7.8 cm and 35.36 cm for 1977 and 1979, respectively (Rango and Martinec, 1997).

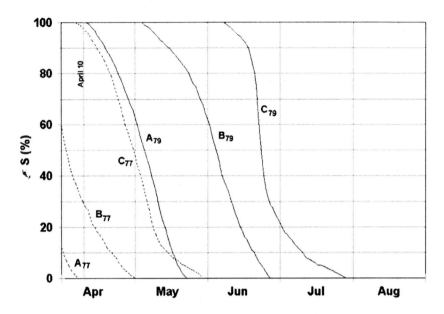

Fig. 11.1. Comparison of snow cover depletion curves for the Rio Grande basin at Del Norte, Colorado derived from Landsat data for elevation zones A, B, and C for 1977 and 1979. On 10 April, the snow cover is 100% in all zones for 1979 and 87.5%, 29%, and 0% in zones A, B, and C for 1977

11.2 General Approach

11.2.1 Gamma Radiation

Snow Water Equivalent Methodology. The use of gamma radiation to measure snow water equivalent (SWE) is based on the attenuation of natural terrestrial radiation by the mass of water in the overlying snow cover. Gamma radiation is emmitted from potassium (40K), uranium (238U), thorium (208Ti) radioisotopes in the soil, with the majority of the emission coming from the top 20cm. The intensity of gamma radiation is measured using a gamma radiation spectrometer which is generally flown on an airplane for snow survey measurements. Fritzsche (1982) provides a description of the physics behind the airborne gamma radiation spectrometer flow by the U.S. National Weather Service (NWS) and methods of calibration.

Airborne gamma measurements of snow water equivalent in a basin are generally acquired along an established network of flight lines. A background (no-snow) flight is conducted in the fall before snow accumulation and then the same lines are flown during the winter when snow covered. When an airborne gamma snow survey is conducted, several flight lines are usually flown within a basin, providing information on the spatial variation in SWE over the basin.

Airborne snow water equivalent measurements using gamma radiation are made using the relationship given in Eq. 11.1 (Carroll, 1990):

$$\text{SWE} = \frac{1}{A}\left(\ln\frac{C_0}{C} - \ln\frac{100+1.11M}{100+1.11M_0}\right) \tag{11.1}$$

Where:
    SWE = snow water equivalent, g/cm^2
    C and C_0 = uncollided terrestrial gamma counts over snow and bare ground, respectively
    M and M_0 = percent soil moisture in snow covered and bare soil areas, respectively
    A = Radiation alternation coefficient in water, cm^2/g

Advantages. The use of airborne gamma measurements for snow water equivalent retrieval is a reliable technique that has been proven effective for operational hydrological monitoring in many countries. Basin surveys can be completed in a matter of hours producing information on the spatial distribution of snow water equivalent and a representative estimate of the volume of snow contained in the basin. The method is applicable to all sizes of basins. As demonstrated by the NWS operational airborne gamma snow survey program (Carroll, 1990), the derived SWE information can be made available to user agencies within several hours of the airborne survey, thus providing timely information for hydrological forecasts. Carroll (1986) presents a cost-benefit analysis of airborne gamma SWE data acquired in support of snowmelt flood forecasting for a major flood event that occurred in the U.S. in February 1985.

Limitations. Since gamma radiation is attenuated by water in all phases, the radiation measurements will include the effects of water in the soil as well as the snow cover mass and thus the contribution of soil moisture has to be taken into account. Inaccurate estimates of soil moisture will yield over- or under-estimations of snow water equivalent. By conducting a background airborne gamma flight in the fall before snow accumulation, the soil moisture background is measured, but it will not be representative if subsequent rainfall adds moisture to the soil or if moisture migrates upward from the soil to the snow during winter. Ground-based soil moisture measurements conducted along the flight line during the winter over-snow flight can help contribute to a more accurate snow water equivalent (Carroll et al., 1983).

Data Availability and Platforms Used. As discussed earlier, the use of airborne gamma surveys for snow water equivalent retrieval in support of hydrological monitoring have been conducted in several northern countries. One of the largest operational programs is maintained by the U.S. National Weather Service. Over 1500 flight lines have been established in the U.S. and Canada, and each winter from January to April snow water equivalent measurements are collected continuously over many of these lines (Carroll, 1990). After each survey, data are transmitted to the NWS National Operational Hydrologic Remote Sensing Center (NOHRSC) for archiving, processing and distribution to forecast offices, other government agencies, and the general public. The NOHRSC web site (http://www.nohrsc.nws.gov) provides a description of the NWS airborne gamma survey program, locations of flight lines, and access to data products. Historical data are also available from the U.S. National Snow and Ice Data Center in Boulder, Colorado. Low-flying aircraft are generally the platforms used to acquire airborne gamma measurements. The NWS airborne gamma survey program uses Aero Commander and Turbo Commander twin engine aircraft which fly at 150m above the ground during the gamma surveys (Carroll, 1990), which prevents airborne gamma measurement in rugged topographic areas.

11.2.2 Visible Imagery

Snow and Ice Cover Extent Methodology. Snow cover can be detected and monitored with a variety of remote sensing devices. The greatest application has been found in the visible and near infrared region of the electromagnetic spectrum. The red band (0.6-0.7µm) of the multispectral scanner subsystem (MSS) on the Landsat satellite was used extensively for snow cover mapping because of its strong contrast with snow-free areas. Originally, snow extent mapping was performed manually using photointerpretive devices and MSS photographs (Bowley et al., 1981). More recently, digital mapping of the snow cover has been the preferred approach (Baumgartner et al., 1986; 1987; Dozier and Marks, 1987; Baumgartner and Rango, 1995).

The greatest problem hindering Landsat (and SPOT) snow mapping in the past was a poor observational frequency. Depending on the Landsat satellite being used, each study area or drainage basin was only revisited every 16-18 days. In areas with minimal cloud cover during the snowmelt season, this was a sufficient frequency of observation. In most mountain snow areas, however, this observational frequency is inadequate because cloud cover will often hide the underlying snow from the satellite sensors.

A spectral channel (1.55-1.75µm) currently only available regularly on the TM instrument can be used to assist in mapping snow cover when clouds partially cover a drainage basin. In this band, clouds are usually more reflective than snow (Dozier, 1989). As a result, automatic discrimination between snow and clouds is possible. Although useful, this capability does not overcome the problem of a complete cloud cover or an inadequate frequency of observation.

When a basin is partially snow covered, a method has been developed to estimate the snow cover in the cloud-obscured parts of the basin (Lichtenegger et al., 1981; Baumgartner et al., 1986; Ehrler et al., 1997). The method uses digital topographic data and assumes that pixels with equal elevation, aspect and slope have the same relative snow coverage over all the basin. With this extrapolation method, information from the cloud-free portion of the basin can be used to estimate the snow cover in the cloud-covered parts of the basin.

As a result of the Landsat (and SPOT) frequency of observation problem, many users have turned to the NOAA polar orbiting satellite with the AVHRR, which has a resolution of about 1 km in the 0.58-0.68µm red band. The frequency of coverage is twice every 24 hours (one daytime pass and one nighttime pass). The major problem with the NOAA-AVHRR data is that the resolution of 1 km may be insufficient for snow mapping on small basins. As the nighttime pass of NOAA-AVHRR cannot be used for snow mapping in the visible spectrum, only one-half of the overflights can be used. However, several NOAA satellites may pass overhead at different times on a given day in specific locations.

Despite the various problems mentioned, visible aircraft and satellite imagery have been found to be very useful for monitoring both the buildup of snow cover in a drainage basin and, even more importantly, the disappearance of the snow covered area in the spring. This disappearance or depletion of the snow cover is important to monitor for snowmelt runoff forecasting purposes. It has been recommended (Rango, 1985) that the optimum frequency of observation of the snow cover during depletion would be once a week. Depending on the remote sensing data used, it could be very difficult to obtain this frequency. Certain snowmelt-runoff applications have been possible with as few as two to three observations during the entire snowmelt season (Rango, 1985).

In cryospheric applications other than snow cover, Landsat images have proved to be very useful for collecting certain basic data from glaciers including long-term surface velocities determined by comparison of displacement on different image dates and short-term high flow rates of surging glaciers. Perhaps the easiest remote observation to make, if the data are available, is the location of the snow line on a glacier at the end of the melt season which can be related to the annual net mass balance (Braslau & Bussom, 1979).

Formation and dissipation of river ice on the Ottawa River was monitored daily using visible imagery from NOAA satellites. The break-up of 14 ice-covered reaches was observed during the melt period (McGinnis & Schneider, 1978). Both Landsat and NOAA satellite data were used to study the ice break-up on the Mackenzie River in the Canadian Arctic (Dey et al., 1977). These satellite observations are adequate to detect patterns of river ice break-up in these remote regions at much less the cost of conventionally collecting the necessary hydrological data in the Arctic environment.

Lake ice studies using visible and infrared remote sensing data have primarily been made for increasing the length of the navigation season on major lakes. Borodulin and Prokacheva (1985) combined aircraft and satellite observations of ice to provide a 35-year data set. They were able to develop regression relations between remote sensing ice-covered area on 15 April and the ice cover thickness and the duration of the ice melting period.

Advantages. There are several advantages to using visible satellite data for snow and ice applications. First, the data are relatively easy to interpret, and it is also relatively easy to distinguish snow from snow-free areas (as well as ice from ice-free areas). If necessary, the analysis of visible satellite data for snow mapping can be accomplished on microcomputer-based systems (Baumgartner and Rango, 1995). The visible satellite data is available in a range of resolutions from 20 m resolution (SPOT) to 8 km resolution (NOAA satellites) which allows applications on small basins up to applications on continental size areas. Because these data have progressed to an operational status in some cases, much snow extent data are readily available to many users on the world wide web free of charge (see section on Current Applications).

Limitations. There are also several disadvantages to using visible satellite data for snow and ice mapping. Nothing can directly be learned about snow water equivalent or ice thickness from the visible data. Snow cover extent is valuable in forecasting of snowmelt runoff, but clouds can be a significant problem in restricting observations, especially when the frequency of observation is every 16 days.

Data Availability and Platforms Used. There is no restriction on acquiring visible or near infrared data for a particular area; however, you must have financial resources available to purchase the more expensive, high resolution data such as Landsat-TM and SPOT. The primary satellites and sensors employed are SPOT, Landsat-TM, DMSP, NOAA-AVHRR, and GOES. Airplanes are infrequently, but effectively, used to provide areal snow cover data.

11.2.3 Thermal Infrared

Thermal infrared remote sensing currently only has minor potential for snow mapping and snow hydrology. Thermal infrared mapping, like visible mapping, is hindered by cloud cover. In addition, the surface temperature of snow is not always that much different from the surface temperatures of other surfaces in rugged mountain terrain where elevation and aspect differences can cause major changes in temperature. This makes it extremely difficult to distinguish snow cover from other features, especially during the snowmelt period. When clouds permit, the big advantage of thermal snow cover mapping is that the mapping can be done during the nighttime overpasses.

In hydrological forecasting, monitoring the surface temperature of the snowpack may have some direct application to delineating the areas of a basin where snowmelt may be occurring. If the snow surface temperature is always below 0°C, no melt is occurring. If the snow surface is at 0°C during the daytime and below 0°C at night, the melt-freeze cycle is occurring but it is uncertain whether melt water is being released at the bottom of the snowpack. If the snow surface stays at 0°C both day and night, then it is highly likely that the snowpack is isothermal and that melt is occurring and being released at the base of the snowpack. Much research needs to be done to determine whether thermal infrared remote sensing can play a useful role in assisting in snowmelt runoff modelling and forecasting.

11.2.4 Passive and Active Microwave

Snow Water Equivalent and Associated Applications. Snow on the earth's surface is, in simple terms, an accumulation of ice crystals or grains, resulting in a snowpack which over an area may cover the ground either completely or partly. The physical characteristics of the snowpack determine its microwave properties; microwave radiation emitted from the underlying ground is scattered in many different directions by the snow grains within the snow layer, resulting in a microwave emission at the top of the snow surface being less than the ground emission. Properties affecting microwave response from a snowpack include: depth and water equivalent, liquid water content, density, grain size and shape, temperature and stratification as well as snow state and land cover. The sensitivity of the microwave radiation to a snow layer on the ground makes it possible to monitor snow cover using passive microwave remote sensing techniques to derive information on snow extent, snow depth, snow water equivalent and snow state (wet/dry). Because the number of scatterers within a snowpack is proportional to the thickness and density, SWE can be related to the brightness temperature of the observed scene (Hallikainen and Jolma, 1986); deeper snowpacks generally result in lower brightness temperatures.

The general approach used to derive SWE and snow depth from passive microwave satellite data relates back to those presented by Rango et al. (1979) and Kunzi et al. (1982) using empirical approaches and Chang et al. (1987) using a theoretical basis from radiative transfer calculations to estimate snow depth from SMMR data. As discussed in Rott (1993), the most generally applied algorithms for deriving depth or snow water equivalent (SWE) are based on the generalized relation given in Eq. 11.2

$$SWE = A + B\ ((T_B(f1) - T_B(f2))/(f2 - f1))\ \text{in mm, for SWE} > 0 \qquad (11.2)$$

where A and B are the offset and slope of the regression of the brightness temperature difference between a high scattering channel ($f2$, commonly 37GHz) and a low scattering one ($f1$, commonly 18 or 19 Ghz) of vertical or horizontal polarization. No single global algorithm will estimate snow depth or water equivalent under all snowpack and land cover conditions. The coefficients are generally determined for different climate and land covered regions and for different snow cover conditions; algorithms used in regions other than for which they were developed and tested usually provide inaccurate estimates of snow cover. Also, accurate retrieval of information on snow extent, depth, and water equivalent requires dry snow conditions, because the presence of liquid water within the snowpack drastically alters the emissivity of the snow, resulting in brightness temperatures significantly higher than if that snowpack were dry. Therefore, an early morning overpass (local time) is the preferred orbit for retrieval of snow cover information to minimize wet snow conditions. It is also recognized that knowledge of snowpack state is useful for hydrological applications. Regular monitoring allows detection of the onset of melt or wet snow conditions (Goodison and Walker, 1995).

The accuracy of the retrieval algorithms is a function of the quality of both the satellite data and the snow cover measurements used in their development. Empirical methods to develop algorithms involve the correlation of observed T_B with coincident conventional depth measurements (as from meteorological stations) or ground SWE data (such as from areally representative snow courses). Goodison et al. (1986) used coincident airborne microwave data and airborne gamma data collected over the Canadian prairie area, supplemented by special ground surveys, to derive their SWE algorithm which has now been used for over 10 years in operational hydrological forecast operations. Non-forested open prairie areas have generally shown the best correlation between areal SWE and brightness temperature (e.g. Kunzi et al., 1982; Goodison et al., 1986; Chang et al., 1987). Hallikainen and Jolma (1986) and Hallikainen (1989) report on Finnish studies over various landscapes. Rott and Nagler (1993) report on European algorithm development which incorporates information from 19, 37, and 85 GHz channels for snow cover mapping; the 85GHz data were of use in mapping very shallow snowcovers (<5cm depth). They use a decision tree for snow classification, including the separation of snow/no snow and the calculation of depth. On a global scale, Grody and Basist (1996) use a decision tree to produce an objective algorithm to monitor the global distribution of snowcover, which includes steps to separate snow cover from precipitation, cold deserts and frozen ground. With time, algorithms can be expected to become more sophisticated as they incorporate filters to eliminate or minimize errors or biases.

Several investigators have attempted to use SAR data to map snow cover area and to infer the snow water equivalent. In the snow cover aspects, Nagler and Rott (1997) indicate that the European Remote Sensing (ERS) satellite SAR (C-band) cannot distinguish dry snow and snow free areas. When the snow becomes wet, the backscattering coefficient is significantly reduced and the wet snow area can be detected. In order to derive a snow map for mountain regions, both ascending and descending orbit images must be compared with reference images acquired from the same positions. This technique was used to successfully derive snow cover for input to the Snowmelt Runoff Model (SRM) for two drainage basins in the Austrian Alps (Nagler and Rott, 1997). Haefner and Piesberger (1997) also used ERS SAR data to map wet snow cover in the Swiss Alps. Shi and Dozier (1998) are developing multiband-multipolarization methods that have some promise for obtaining snow water equivalent with SAR data, however, the required satellite sensors are not likely to be available in the near future.

Ice and Glaciers. Microwave data are used to derive other cryospheric information. Although the resolution of passive microwave satellite data prevents its use in deriving information on mountain glaciers and river ice, satellite SAR data has been shown to be useful (Rott and Nagler, 1993; Leconte and Klassen, 1991; Rott and Matzler, 1987). Glacier snowline mapping is an important input in hydrological models and in the computation of glacier mass balance. Adam et al. (1997) used ERS-1 C-band, VV SAR data to map the glacier snowline within 50-75m of ground based measurements. The technique could separate wet, melting snow from glacier ice and bedrock, but was not applicable when the snow was dry, since dry snow was transparent to C-band SAR. Some research has been done on freeze-thaw applications in permafrost areas (England, 1990; Zuerndorfer et al., 1989) but other sensors with higher resolution (e.g. Landsat) may be more suited to mapping permafrost areas (Leverington and Duguay, 1997; Duguay and Lewowicz, 1995).

The spatial and temporal patterns of lake ice freeze-up and break-up can be monitored using passive microwave measurements over large lakes because of the large difference between the microwave emissivity of the lake ice and fresh water. Due to the coarse resolution of satellite passive microwave radiometers, the technique is limited to large lakes (>100km^2 in size). The use of passive microwave remote sensing for lake ice thickness determination is limited to longer wavelengths (e.g., 1-5GHz frequencies), which can be sensed through the entire ice thickness, as opposed to shorter wavelengths (e.g., 37GHz) which are emitted from the near surface and are sensitive to the overlying snow cover (Hall and Martinec, 1985). The potential application of higher frequency data for ice thickness determination has been investigated by Chang et al. (1997) using airborne microwave data and a layer radiative transfer model to calculate ice thickness, however, the small sample size limited statistically significant results.

Active microwave remote sensing has an advantage over passive techniques for lake ice information retrieval in that the higher spatial resolution means that lakes of all sizes can be monitored. In addition to providing spatial and temporal information on lake ice freeze-up and break-up and ice thickness, radar backscatter is related to the internal structure of the lake ice. Hall (1993) provides an overview of the capability of active microwave remote sensing for retrieval of lake ice information. The application of satellite SAR for lake ice remote sensing has been investigated and demonstrated in recent years using data from the ERS-1 satellite (e.g., Hall et al., 1994; Jeffries et al., 1994; Leshkevich et al., 1994).

Advantages. Passive microwave data provides several advantages not offered by other satellite sensors. Studies have shown that passive microwave data offer the potential to extract meaningful snowcover information, such as SWE, depth, extent and snow state. SSM/I is a part of an operational satellite system, providing daily coverage of most snow areas, with multiple passes at high latitudes, hence allowing the study of diurnal variability. The technique has generally all-weather capability (although affected by precipitation at 85GHz), and can provide data during darkness. The data are available in near-real time, and hence can be used for hydrological forecasting. SAR has an additional advantage of having resolution of about 25 m making it very useful for mountain snowpacks.

Limitations. There are limitations and challenges in using microwave data for deriving snow cover information for hydrology. The coarse resolution of passive microwave satellite sensors such as SMMR and SSM/I (~25km) is more suited to regional and large basin studies, although Rango et al. (1989) did find that reasonable SWE estimates could be made for basins of less than 10,000km^2. Heterogeneity of the surface and the snowcover within the microwave footprint results in a mixed signature, which is ultimately represented by a single brightness temperature that is an areally weighted mean of the microwave emission from each surface type within the footprint. Hence, an understanding of the relationship between snow cover and surface terrain and land cover (e.g., Goodison et al., 1981) is as important for developing remote sensing applications in hydrology as for conventional hydrology analyses in snow covered areas.

Another challenge is to incorporate the effect of changing snowpack conditions throughout the winter season. Seasonal aging, or metamorphism, results in a change in the grain size and shape, and this will affect the microwave emission from the snowpack. In very cold regions, depth hoar characterized by its large crystal structure enhance the scattering effect on the microwave radiation, resulting in lower surface emission producing an overestimate of SWE or snow depth (Hall, 1987 and Armstrong et al., 1993). The increase in T_B associated with wet snow conditions currently prevents the quantitative determination of depth or water equivalent since algorithms will tend to produce zero values under these conditions. The best way to view the seasonal variability in microwave emission from the snowpack is to compile a time series of satellite data spanning the entire season which can then be related to changes in the pack over the season (Walker et al., 1995).

SAR data has a major disadvantage of not being able to detect the dry snowpack. Additionally, the SAR data is difficult to process with backscattering from rough surfaces beneath the snow causing interpretation problems. The optimum bands around 1 cm wavelength are not represented in any existing or planned SAR instrument.

Data Availability and Platforms Used. Passive microwave data for hydrological and climatological studies are available from sensors operated onboard NASA's Nimbus 7 (SMMR) and the US Defense Meteorological Satellites (DMSP). Massom (1991) provides comprehensive details of the individual satellites. Data are archived and available in a variety of formats at NSIDC; the EASE-Grid brightness temperature product (Armstrong and Brodzik, 1995), a gridded 25km resolution global data set (12.5 km at 85.5GHz) is particularly suited to historical hydrological analyses where the user may wish to run different regional snow cover algorithms or integrate other geophysical information. SSM/I data are available in near real-time for operational hydrological applications from the NOAA National Environmental Satellite Data and Information Services (NESDIS).

11.2.5 Related Applications

Hand-Drawn Snow Maps. It is possible to use a base map, such as a snow-free Landsat image, and map the location of the snowline in a basin during an aircraft overflight. This approach is even possible from the ground in a very small basin (say less than several km^2 in area) by climbing to the highest point and visually transferring the area covered by snow to a base map. These techniques are most useful where access to the basin or aircraft flights are easily arranged.

Photointerpretation. Where labor is inexpensive in certain developing countries, it is still possible to take the photos from satellite overpasses and manually map the area covered by snow and save a considerable amount of money. Computer systems have become so reasonable in price, however, that the photointerpretive approach is no longer used very much.

Frequency Modulated-Continuous Wave (FM-CW) Radar. In recent years FM-CW radars of various types have found applications in snow and ice. Sturm et al. (1996) used an X-band FM-CW radar mounted on a towed sled to make dry snow depth measurements in Alaska. Much care has to be taken in the data collection with coincident field verification of the snow depths. With frequent field calibration, the radar determined snow depths were accurate to about 2 cm. A K_a-band FM-CW radar was mounted on a helicopter to continuously measure the thickness profile of freshwater ice (Yankielun, 1992). The ice thickness resolution of this system is about 3 cm. There appear to be significant safety and trafficability applications for ponds, lakes, and rivers using FM-CW radars.

11.3 Current Applications

11.3.1 NOHRSC - Snow Cover and Snow Water Equivalent Products

Because areal snow cover extent data have been available since the 1960s, various investigators have found many useful applications. A team of scientists from a variety of US government agencies developed plans in the early 1980s for operational snow mapping by the US National Weather Service (NWS) for hydrological purposes. In 1986, NWS adopted these plans and proceeded to develop operational remote sensing products, mostly for snow hydrology. The most widely distributed products of the NWS National Operational Hydrologic Remote Sensing Center (NOHRSC) are periodic river basin snow cover extent maps from NOAA-AVHRR and the Geostationary Operational Environmental Satellite (GOES). Figure 11.2 shows a snow cover map from NOAA satellites for the United States and parts of Canada for the period 27-31 January 1997. Digital maps for about 4000 basins in North America are produced about once per week and are used by a large group of users including the NWS River Forecast Centers and individual water authorities. On about 10% of these basins, the mapping is done by elevation zone (Carroll, 1995). Data distribution is possible in real time through a variety of electronic methods such as the Internet and with the assistance of Geographic Information Systems. The 1 km resolution of the

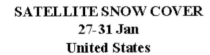

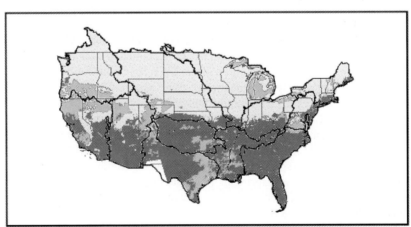

Fig. 11.2. NOAA-NOHRSC satellite snow cover map for the United States and parts of Canada 27-31 January 1997

product makes it useful on basins or sub-areas greater than 200 km^2 in area (Rango et al., 1985), and various users employ the data to assist in hydrological forecasting using models. NOHRSC products are continually under development, and the latest products and services can be accessed by visiting their World Wide Web homepage at http://www.nohrsc.nws.gov.

In addition to producing operational snow cover extent data, NOHRSC also produces operational airborne gamma radiation snow water equivalent data. The difference between the NOHRSC airborne radiation measurements over bare ground and snow covered ground is used to calculate a mean areal snow water equivalent value with a root mean square error of less than one cm (Carroll, 1995). Immediately after each airborne snow survey, the snow water equivalent derived for each flight line are used in a GIS to generate a contoured surface of snow water equivalent for the region of the survey. After each survey, users of the data are able to operationally obtain by electronic means a contour map of snow water equivalent in the region and the mean areal snow water equivalent for each basin in the region.

11.3.2 Canadian Prairie Snow Water Equivalent Mapping

In Canada, a federal government program (Climate Research Branch, Atmospheric Environment Service) has been ongoing since the early 1980's to develop, validate and apply passive microwave satellite data to determine snow extent, snow water equivalent and snowpack state (wet/dry) in Canadian regions for near real-time and operational use in hydrological and climatological applications. Goodison and Walker (1995) provide a summary of the program, its algorithm research and development, and future thrusts. For the prairie region a snow water equivalent algorithm was empirically derived using airborne microwave radiometer data (Goodison et al., 1986), and tested and validated using Nimbus-7 SMMR and DMSP SSM/I satellite data (Goodison, 1989).

With the launch of the first SSM/I on the DMSP F-8 satellite in 1987, came the ability to access passive microwave data in near real-time and generate snow cover products for users within several hours after data acquisition. Since 1989, the Climate Research Branch prairie SWE algorithm has been applied to near-real time SSM/I data to generate weekly maps depicting current SWE conditions for the provinces of Alberta, Saskatchewan and Manitoba in western Canada. Thirkettle et al. (1991) describe the procedures for data acquisition, processing and mapping of the SWE information. The prairie maps are disseminated by fax machine (see Fig. 11.3) to water resource agencies and meteorological offices throughout the prairie region where they are used to monitor snow cover conditions, plan for field surveys, and make forecasts regarding spring water supply conditions including potential flooding or drought. In the winter and spring of 1994, the maps were particularly useful for monitoring the high SWE conditions in southern Manitoba and North Dakota preceding the devastating Red River flood (Warkentin, 1997).

After 10 winter seasons in operation, the Canadian prairie SWE mapping program has successfully demonstrated a useful application of SSM/I derived snow cover information for operational hydrological analyses. It is also a cooperative program in

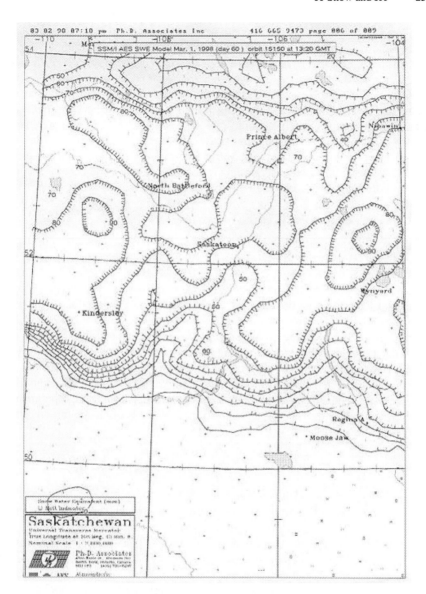

Fig. 11.3. Example of an operational snow water equivalent distribution map sent out to water resources agencies in Canada by facsimile transmission

that user feedback has served to enhance the validation and the refinement of the SSM/I SWE algorithm (Goodison and Walker, 1995). One enhancement has been the development of a wet snow indicator (Walker and Goodison, 1993), which overcomes a major limitation of the passive microwave technique described in Sect. 11.2.3 by providing the capability to discriminate wet snow areas from snow-free areas and hence a more accurate retrieval of snow extent during melting conditions.

The Climate Research Branch prairie SWE algorithm has been applied to Nimbus-7 SMMR and DMSP SSM/I data to create a 20 year time series of maps depicting winter SWE conditions over the Canadian prairie region, for the purpose of investigating seasonal and interannual variability in support of the Branch's climate research activities in assessing climate variability and change. Walker et al. (1995) presents the SMMR time series in the form of an atlas. Colour Plates 11.A and 11.B are two winter snowpack extremes from the SSM/I time series, which contributed to drought conditions in summer 1988 and flooding in areas of Manitoba and Alberta in spring 1997, illustrating the extremes in SWE that can be experienced in this region.

11.3.3 Snowmelt Runoff Forecast Operations

Very few hydrological models have been developed to be compatible with remote sensing data. One of the few models that was developed requiring direct remote sensing input is the Snowmelt Runoff Model (SRM) (Martinec et al., 1998). SRM requires remote sensing measurements of the snow covered area in a basin. Although aircraft observations can be used, satellite-derived snow cover extent is the most common.

Two versions of SRM are now available. The most commonly-used version employs the degree day approach to melting the snow cover in a basin (Martinec et al., 1998). To date, this version of SRM has been tested on over 80 basins in 25 countries worldwide. This version of SRM has also been frequently used for forecasting. Each day, the water produced from snowmelt and from rainfall is computed, superimposed on the calculated recession flow, and transformed into daily discharge from the basin according to Eq. 11.3:

$$Q_{n+1} = (c_{Sn} \cdot a_n (T_n + \Delta T_n) \cdot S_n + c_{Rn} P_n) \frac{A \cdot 10000}{86400} (1 - k_{n+1}) + Q_n k_{n+1} \qquad (11.3)$$

where Q = average daily discharge [m^{3}s^{-1}]
  c = runoff coefficient expressing the losses as a ratio (runoff/precipitation), with c_S referring to snowmelt and c_R to rain
  a = degree-day factor [cm·°C^{-1}·d^{-1}] indicating the snowmelt depth resulting from 1 degree-day
  T = number of degree-days [°C·d]
  ΔT = the adjustment by temperature lapse rate when extrapolating the temperature from the station to the average hypsometric elevation of the basin or zone [°C·d]
  S = ratio of the snow covered area to the total area
  P = precipitation contributing to runoff [cm]. A preselected threshold temperature, T_{CRIT}, determines whether this contribution is rainfall and immediate. If precipitation is determined by T_{CRIT} to be new snow, it is kept on storage over the hitherto snow free area until melting conditions occur.

A = area of the basin or zone [km^2]

$k = \dfrac{Q_{m+1}}{Q_m}$ (m, m+1 are the sequences of days during a true recession flow period).

k = recession coefficient indicating the decline of discharge in a period without snowmelt or rainfall

n = sequence of days during the discharge computation period. Equation (1) is written for a time lag between the daily temperature cycle and the resulting discharge cycle of 18 hours. In this case, the number of degree-days measured on the nth day corresponds to the discharge on the n + 1 day. Various lag times can be introduced by a subroutine.

$\dfrac{10000}{86400}$ = conversion from cm km^{2}d^{-1} to m^{3}s^{-1}

T, S and P are variables to be measured or determined each day. c_R, c_S, lapse rate to determine ΔT, T_{CRIT}, k and the lag time are parameters which are characteristic for a given basin or, more generally, for a given climate.

Figures 11.4 and 11.5 show SRM simulations for 1977 and 1979, respectively, for the Rio Grande basin at Del Norte, Colorado that were obtained using the snow cover data in Fig. 11.2. It is apparent that the SRM simulations adequately approximate the measured flows in two vastly different snowmelt runoff seasons.

A second version of the model has recently been developed (Brubaker et al., 1996) that adds a net radiation index to the degree day index to melt snow from a basin's hydrologic response units (based on elevation and aspect). This version of SRM is more physically based, and thus requires more data than the original version of SRM.

The degree day version of SRM can be obtained by visiting the SRM Worldwide Web homepage (http://hydrolab.arsusda.gov/cgi-bin/srmhome).

Norway Power Applications. Norway has been using satellite snow cover data for planning of hydroelectric power generation since 1980. This approach is a simple digital ratioing which is converted to a percentage of snow cover by pixel in the study basin (Andersen, 1991; 1995). NOAA-AVHRR has again been used as the data source for this system because of its daily coverage. Snow cover maps are produced for the various basins, and snow cover by elevation zone in a basin is also a product. The data are now input to snowmelt runoff models for the prediction of streamflow (Andersen, 1995).

Spain Power Applications. Spain is also using NOAA-AVHRR snow cover data for the forecasting of snowmelt runoff volume during the spring and summer months in the Pyrenees. Development of subpixel analysis techniques (Gomez-Landesa, 1997) has allowed snow cover mapping on basins as small as 10km^2 using the AVHRR data. Colour Plate 11.C shows the NOAA-AVHRR derived snow cover for the Cinca basin (798.1km^2) in the Central Pyrenees of Spain on 10 March 1998. The different gray levels correspond to different percents of snow cover in each NOAA-AVHRR pixel (Gomez-Landesa and Rango, 1998). This approach could make NOAA-AVHRR data

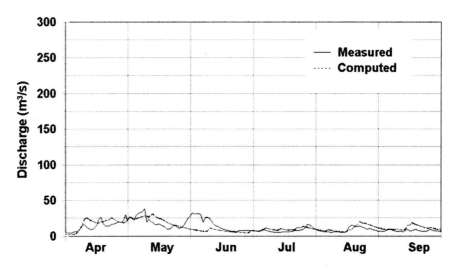

Fig. 11.4. SRM computed versus measured snowmelt runoff for 1977 on the Rio Grande basin at Del Norte, Colorado

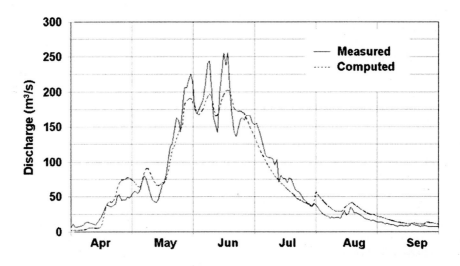

Fig. 11.5. SRM computed versus measured snowmelt runoff for 1979 on the Rio Grande basin at Del Norte, Colorado

more widely useable for hydrological applications after it is tested in different geographic regions. The snow cover data for each basin or zone are input into the Snowmelt Runoff Model (SRM) for use in forecasting the seasonal snowmelt runoff volume in the Pyrenees to assist in planning hydropower production.

India Runoff Forecasts. Some good examples of operational application of snow extent data are found in India. Initially Ramamoorthi (1983; 1987) started to use NOAA-AVHRR data in a simple regression approach for empirical forecasts of seasonal snowmelt runoff in the Sutlej River Basin (43,230 km^2) and the forecasts were extended to other basins. Ramamoorthi (1987) also decided to use satellite data as input to a snowmelt runoff model for shorter term forecasts. This idea was developed by Kumar et al. (1991) and satellite data were input to SRM for use in operational forecasts of daily and weekly snowmelt runoff on the Beas (5144 km^2) and Parbati (1154 km^2) Rivers in India. Figure 11.6 shows the SRM results of Kumar et al. (1991) who produced very accurate simulations on the Beas river at Thalot, India with very little historical data. Kumar et al. (1991) conclude that short term runoff models, such as SRM, can be effectively put to use in hydro-electric schemes already in operation or can be developed for projects being planned or under construction. Apparently, because of the isolation of these remote basins, the only way forecasts can be made in this region is with the use of remote sensing data.

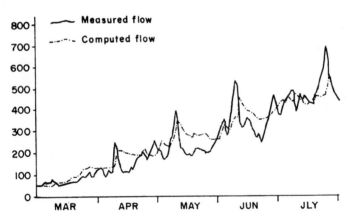

Fig. 11.6. SRM computed versus measured snowmelt runoff on the Beas River at Thalot, India for 1987 (Kumar et al., 1991)

11.4 Future Directions

11.4.1 Improved Resolution in the Passive Microwave

The most significant problem standing in the way of effective snow water equivalent mapping from space is a poor resolution at present in the passive microwave. This is about to change as some new passive microwave instruments such as the Advanced Microwave Scanning Radiometer (AMSR) on EOS will be flown in space that should improve resolution at about 0.8 cm wavelength from the existing 25 km to a projected 8 km. It will give investigators currently experienced with the 25 km capabilities the chance to compare the improved resolution advantages over study areas previously used.

11.4.2 Improved Algorithms in the Passive Microwave

It is generally agreed that no single algorithm will produce representative values on a global basis due to spatial variations in land cover, terrain and snow cover characteristics. Hence, a regional approach to snow cover algorithm development has been adopted by several research groups (e.g., Goodison and Walker, 1995; Solberg et al., 1998).

In 1992, a strategy for future snow cover algorithm development was outlined by the passive microwave research community at an international workshop – "Passive Microwave Remote Sensing of Land-Atmosphere Interactions". This strategy is outlined in Choudhury et al. (1995). The priorities identified at this meeting included: (i) the need for signature research using ground-based microwave radiometry to understand the influence of vegetation cover and physical variations in snowpack structure; (ii) the development of theoretical models to characterize microwave interactions with snowpack properties; (iii) the incorporation of land cover information (e.g., type and density) into snow cover retrieval algorithms; (iv) the identification of target areas with good in-situ snow cover measurement for algorithm development and validation; and (v) investigation of "mixed pixels", characterized by high spatial variations in snow distributions and/or land use, and the associated variations in microwave emission that contribute to the brightness temperature measured for the pixel. Since that meeting, research has focussed on these priority areas with a common goal of developing algorithms that will produce consistently representative information on snow cover parameters over as much of the globe as possible. Examples of recent passive microwave snow cover research developments in relation to the above priorities include Chang et al. (1997), Matzler (1994), Woo et al. (1995), Foster et al. (1996), De Sève et al. (1997), Sun et al. (1997), and Kurvonen and Hallikainen (1997).

11.4.3 Outlook for Radar Applications

The application of current satellite active microwave sensors for snow cover information retrieval is generally limited to wet snow detection and mapping (Rott, 1993), mainly due to the single frequency (typically C-band) and single transmit/receive polarization characteristics of these sensors. Thus, the outlook for improved snow cover information retrieval from active microwave sensors has focussed on multifrequency and multi-polarization systems (Rott, 1993).

In 1994, a multi-frequency and multi-polarization SAR system (SIR-C/X-SAR) was flown on two NASA Space Shuttle missions, providing research data sets for investigating the potential of an advanced SAR system for retrieval of snow cover information. The SIR-C (Spaceborne Imaging Radar-C) operated at L- and C-bands, and the X-SAR (X-Band Synthetic Aperture Radar) operated at X-band, thus providing simultaneous data for 3 radar frequencies (1.25, 5.3 and 9.6 GHz). The SIR-C also operated at 4 different transmit/receive polarizations to provide polarimetric data. The results of snow cover investigations using SIR-C/X-SAR data demonstrated a potential for using multi-frequency polarimetric SAR data to retrieve information such as the discrimination of wet snow areas (Li and Shi, 1996), snow cover wetness or liquid water content (Shi and Dozier, 1995; Matzler et al., 1997), and the water equivalent of dry snow (Shi and Dozier, 1996).

Although currently the availability of routine SAR data from satellites is limited to single frequency and single polarization, two polarimetric SAR sensors are planned for launch in the near future. The Advanced Synthetic Aperture Radar (ASAR), scheduled to be launched on the European Space Agency's Envisat-1 platform in 2000, will provide C-band in a number of alternating polarization modes. Radarsat-2 is the Canadian Space Agency's C-band SAR follow-on to the current Radarsat, which is planned for launch in late 2001, and will have new advanced capabilities, including dual or quad polarization options on selected beam modes. The availability of polarimetric SAR data from these satellite platforms should enhance the use of active microwave sensors for routine snow cover monitoring in support of hydrological applications, especially in small, alpine basins. The appropriate frequency and polarization combinations for operational monitoring by a single space platform, however, will be unlikely in the future.

11.4.4 Integration of Various Data Types

Combinations of two different remote sensors can at certain times increase the information available about the snowpack. As an example, a method has been developed in Finland which combines several different approaches to measure or estimate snow water equivalent (Kuittinen, 1989). Ground-based point measurements of water equivalent were made as usual about twice a month. One airborne gamma ray flight was made at the beginning of the snowmelt season to give line transect values of snow water equivalent. All available NOAA-AVHRR satellite images in the Spring are used to provide areal snow water equivalent estimates based on a relationship between the percentage of bare spots in the snow cover and snow water equivalent (Kuittinen, 1989). The point, line and areal snow water equivalent values were used with a method of correlation functions and weighting factors, as suggested by Peck et al. (1985), to determine an areal value based on all data. Evaluations in the Finland study have shown that the error of the estimate of areal snow water equivalent is less than 3.5 cm (Kuittinen, 1989). Carroll (1995) also developed a system which combines ground-based snow observations and airborne gamma ray snow water equivalent observations to yield gridded snow water equivalent values. Satellite snow cover extent data are used to constrain the snow water equivalent interpolations to regions where snow cover is observed to be present. Colour Plate 11.D presents the NOHRSC snow water equivalent data (1 inch = 2.54 cm) resulting from using a geographic information system to integrate satellite, airborne, and ground snow data (9-12 April 1997) for Eastern North Dakota during the 1997 Red River of the North snowmelt flood.

With technological advances in data processing and transmission, data and derived snow and ice products from many of the current sensors are available to the hydrological community in near real-time (e.g. within 6-24 hours of satellite overpass). The development of the Internet and World Wide Web has facilitated the availability of many remote sensing derived snow and ice products to users right from their computers. As the technology related to data access continues to advance, the hydrological community can expect to see an expanded variety of satellite data products available to them for use in hydrological monitoring and modeling. When both the AM-1 and PM-1 EOS platforms are launched, NASA will make a variety of snow products available to users from MODIS (daily snow cover and 8-day composite maximum snow cover at 500m resolution; daily climate modeling grid (CMG) snow cover and 8-day composite CMG maximum snow cover at ¼° x ¼° resolution) and AMSR (daily global snow-storage index map; pentab (5-day) composite snow-storage index map). The cost associated with satellite-derived snow and ice products will vary depending on the data policy associated with the satellite sensor. For commercial satellite platforms, such as Radarsat-1 and 2 SAR, derived products may come with a price tag in the thousands of dollar range, whereas products from federally funded satellite platforms (e.g. NOAA AVHRR, DMSP SSM/I) are generally available at no cost via the Internet.

References

Adam, S., Pietroniro, A. and Brugman, M.M.: Glacier snowline mapping using ERS-1 SAR imagery. Remote Sensing of Environment, 61, 46-54 (1997)

Andersen, T.: SNOWSAT-Operational snow mapping in Norway. Proc. First Moderate Resolution Imaging Spectroradiometer (MODIS) Snow and Ice Workshop, NASA Conf. Publ. CP-3318, NASA/Goddard Space Flight Center, Greenbelt, MD 1995, pp. 101-102

Andersen, T.: AVHRR data for snow mapping in Norway. Proc. 5th AVHRR Data Users Meeting, Tromsoe, Norway 1991

Armstrong, R.L. and Brodzik, M.J.: An earth-gridded SSM/I data set for cryospheric studies and global change monitoring. Advances in Space Research, 16(10), 155-163 (1995)

Armstrong, R. L., Chang, A., Rango, A., and Josberger, E.: Snow depths and grain-size relationships with relevance for passive microwave studies. Annals of Glaciology, 17, 171-176 (1993)

Baumgartner, M. F. and Rango, A.: A microcomputer-based alpine snowcover and analysis system (ASCAS). Photogrammetric Engineering & Remote Sensing, 61 (12), 1475-1486 (1995)

Baumgartner, M. F., Seidel, K., and Martinec, J.: Toward snowmelt runoff forecast based on multisensor remote-sensing information, IEEE Trans. Geosci. Remote Sens. 25, 746-750 (1987)

Baumgartner, M.F., Seidel, K., Haefner, H., Itten, K.I., and Martinec, J.: Snow cover mapping for runoff simulations based on Landsat-MSS data in an alpine basin. Proc. Hydrological Applications of Space Technology, Cocoa Beach Workshop, IAHS Publ. No. 160, 1986, pp. 191-199

Borodulin, V. V. and Prokacheva, V. G.: Studying lake ice regimes by remote sensing methods. In: Hydrological Applications of Remote Sensing and Remote Data Transmission. Proc. Hamburg Symp., IAHS Publ. No. 145, 1985, pp. 445-450

Bowley, C. J., Barnes, J. C., and Rango, A.: Satellite snow mapping and runoff prediction handbook, NASA Technical Paper 1829, National Aeronautics and Space Administration, Washington, D. C. 1981, 87 pp.

Braslau, D. and Bussom, D. E.: Landsat sensing of glaciers with application to mass-balance and runoff. In: Proc. Modeling Snow Cover Runoff, S. C. Colbeck and M. Ray (eds.), Hanover, New Hampshire: U.S. Army Cold Regions Res. and Eng. Lab 1979, pp. 77-82

Brubaker, K., Rango, A., and Kustas, W.: Incorporating radiation inputs into the Snowmelt Runoff Model. Hydrological Processes, 10, 1329-1343 (1996)

Carroll, T. R.: Remote sensing of snow in the cold regions. In: Proc. First Moderate Resolution ImagingSpectroradiometer (MODIS) Snow and Ice Workshop, Nasa Conf. Publ. CP-3318, NASA/Goddard Space Flight Center, Greenbelt, MD 1995, pp. 3-14

Carroll, T.R.: Airborne and satellite data used to map snow cover operationally in the U.S. and Canada. Proc. International Symposium on Remote Sensing and Water Resources, Enschede, The Netherlands, 1990, pp. 147-155

Carroll, T.R., Glynn, J.E. and Goodison, B.E.: A comparison of U.S. and Canadian airborne gamma radiation snow water equivalent measurements. Proc. 51st Annual Western Snow Conference, Vancouver, Washington, 1983, pp. 27-37

Carroll, T.R.: Cost-benefit analysis of airborne gamma radiation snow water equivalent data used in snowmelt flood forecasting. Proc. 54th Annual Meeting of the Western Snow Conference, Phoenix, Arizona, 1986, pp. 1-11

Chang, A.T.C., Foster, J.L., Hall, D.K., Goodison, B.E., Walker, A.E., Metcalfe, J.R. and Harby, A.: Snow parameters derived from microwave measurements during the BOREAS winter field campaign. Journal of Geophysical Research 102(D24), 29, 663-29, 671 (1997).

Chang, A.T.C., Foster, J.L. and Hall, D.K.: Nimbus-7 SMMR derived global snow cover parameters. Annals of Glaciology, 9, 39-44 (1987)

Choudhury, B.J., Kerr, Y.H., Njoku, E.G. and Pampaloni, P. (eds.): Working group A1: Snow. In: Passive Microwave Remote Sensing of Land-Atmosphere Interactions, VSP, Utrecht, The Netherlands, 1995, pp. 651-656

De Sève, D., Bernier, M., Fortin, J.-P. and Walker, A.: Preliminary analysis of snow microwave radiometry using the SSM/I passive-microwave data: the case of La Grande River watershed (Quebec). Annals of Glaciology, 25, 353-361 (1997)

Dey, B., Moore, H. And Gregory, A. F.: The use of satellite imagery for monitoring ice break-up along the Mackenzie River, NWT. Arctic 30(4), 234-242 (1977)

Dozier, J. and Marks, D.: Snow mapping and classification from Landsat Thematic Mapper data. Annals of Glaciolology 9, 1-7 (1987)

Dozier, J.: Spectral signature of alpine snow cover from the Landsat Thematic Mapper. Remote Sens. Environ. 28, 9-22 (1989)

Duguay, C. R., and Lewkowicz, A. G.: Assessment of SPOT panchromatic imagery in the detection and identification of permafrost features, Fosheim Peninsula, Ellesmere Island, N.W.T. Proceedings of the 17th Canadian Symposium on Remote Sensing, Saskatoon, Saskatchewan, 1995, pp. 8-14

England, A. W.: Radiobrightness of diurnally heated, freezing soil. IEEE Transactions on Geoscience and Remote Sensing, 28(4), 464-476 (1990)

Erhler, C., Seidel, K., and Martinec, J.: Advanced analysis of snow cover based on satellite remote sensing for the assessment of water resources. In: Remote Sensing and Geographic Information systems for Design and Operation of Water Resources Systems, IAHS Publication No. 242, 1997, pp.93-101

Foster, J. L. and Rango, A.: Snow cover conditions in the northern hemisphere during the winter of 1981, Jour. Clim. 20, 171-183 (1982)

Foster, J., Chang, A. and Hall, D.: Improved passive microwave algorithms for North America and Eurasia. Proc. of the Third International Workshop on Applications of Remote Sensing in Hydrology, Greenbelt, Maryland, U.S.A., 1996, pp. 63-70

Fritzsche, A.E.: The National Weather Service Gamma Snow System Physics and Calibration. Publication No. NWS-8201, EG&G, Inc., Las Vegas, Nevada (1982)

Gomez-Landesa, E.: Evaluacion de Recursos de Agua en Forma de Nieve mediante Teledeteccion usando satelites de la sine NOAA (Evaluation of water resources in the form of snow by remote sensing using NOAA satellites). Universidad Politenica de Madrid, Madrid, Spain, Ph.D. Thesis 1997

Gomez-Landesa, E. and Rango, A.: Snow cover remote sensing and snowmelt runoff forecasts in the Spanish Pyrenees using the SRM model. Proceedings of the Fourth International Workshop on Applications of Remote Sensing in Hydrology, NHRI Symposium Report, Santa Fe, NM, 1998, 12pp

Goodell, B. C.: Snowpack management for optimum water benefits. ASCE Water Resources Engineering Conference, Denver, Colorado: Conference Preprint 379, 1966

Goodison, B.E.: Determination of areal snow water equivalent on the Canadian prairies using passive microwave satellite data. Proc. 1989 International Geoscience and Remote Sensing Symposium (IGARSS '89), Vancouver, Canada, 1989, pp. 1243-1246

Goodison, B.E. and Walker, A.E.: Canadian development and use of snow cover information from passive microwave satellite data. In: Passive Microwave Remote Sensing of Land-Atmosphere Interactions (Choudhury, B.J., Kerr, Y.H., Njoku, E.G. and Pampaloni, P., eds.), VSP, Utrecht, The Netherlands, 1995, pp. 245-262

Goodison, B.E., Rubinstein, I., Thirkettle, F.W. and Langham, E.J.: Determination of snow water equivalent on the Canadian prairies using microwave radiometry. In: Modelling Snowmelt Induced Processes, IAHS Publication No.155, 1986, pp. 163-173

Goodison, B.E., Ferguson, H.L. and McKay, G.A.: Measurement and Data Analysis. In, Handbook of Snow (ed. D.M. Gray and Male, D. H.), Pergamon Press Canada Ltd., 1981, pp. 191-274

Grody, N.C. and Basist, A.N.: Global identification of snowcover using SSM/I measurements. IEEE Transactions on Geoscience and Remote Sensing, 34 (1), 237-249 (1996)

Haefner, H. and Piesberger, J.: High alpine snow cover monitoring using ERS-1 SAR and Landsat TM data. In: Remote Sensing and Geographic Information Systems for Design and Operation of Water Resources Systems (Proc. Rabat Symp.), IAHS Publication No. 242, 1997, pp. 113-118

Hall, D. K., Fagre, D. B., Klasner, F., Linebaugh, G., and Liston, G.: Analysis of ERS-1 synthetic aperture radar data of frozen lakes in northern Montana and implications for climate studies. Journal of Geophysical Research, 99(C11), 22, 473-22, 482 (1994)

Hall, D.K. and Martinec, J.: Remote sensing of ice and snow. New York: Chapman and Hall, 189 pp, 1985

Hall, D. K.: Influence of depth hoar on microwave emission from snow in northern Alaska. Cold Regions Science and Technology, 13, 225-231 (1987)

Hall, D. K.: Active and passive microwave remote sensing of frozen lakes for regional climate studies. In, Snow Watch ?92 - Detection Strategies for Snow and Ice (eds. R. G. Barry, B. E. Goodison, and E.F. LeDrew), Glaciological Data Report GD-25, World Data Center A for Glaciology (Snow and Ice), Boulder, Colorado, 1993, pp. 80-85

Hallikainen, M. and Jolma, P.: Development of algorithms to retrieve the water equivalent of snow cover from satellite microwave radiometer data. In: Proc. 1986 International Geoscience and Remote Sensing Symposium (IGARSS '86), Zurich, Switzerland, 1986, pp. 611-616

Hallikainen, M. Microwave radiometry of snow. Advances in Space Research, 9(1), (1)267-(1)275 (1989)

Jeffries, M. O., Morris, K., Weeks, W.F., and Wakabayashi, H.: Structural and stratigraphic features and ERS 1 synthetic aperture data backscatter characteristics of ice growing on shallow lakes in NW Alaska, winter, 1991-1992, Journal of Geophysical Research, 99(C11):22, 459-22, 471 (1994)

Kuittinen, R.: Determination of snow water equivalents by using NOAA-satellite images, gamma ray spectrometry and field measurements. In: Remote Sensing and Large-Scale Global Processes, IAHS Publication No. 186, 1989, pp. 151-159

Kumar, V. S., Haefner, H. and Seidel, K.: Satellite snow cover mapping and snowmelt-runoff modelling in Beas Basin. In: Snow Hydrology and Forests in High Alpine Areas. Proc. Vienna Symp. IAHS Publ. No. 205, 1991, pp. 101-109

Kunzi, K.F., Patil, S., and Rott, H.: Snow cover parameters retrieved from Nimbus-7 Scanning Mutlichannel Microwave Radiometers (SMMR) data, IEEE Transactions on Geoscience and Remote Sensing, GE-20(4), 452- 467 (1982)

Kurvonen, L. and Hallikainen, M.: Influence of land-cover category on brightness temperature of snow. IEEE Transactions on Geoscience and Remote Sensing, 35 (2), 367-377 (1997)

Leconte, R. and Klassen, P.D.: Lake and river ice investigations in northern Manitoba using airborne SAR imagery. Arctic, 44 (Supp. 1), 153-163 (1991)

Leshkevich, G., Piche, W. And Clemente-Colon, P.: Great Lakes ice research applications demonstration. In: Proc. Second ERS-1 Symposium: Space at the Service of our Environment, Hamburg, Germany, ESA SP-361, 1994, pp. 675-679

Leverington, D. W., and Duguay, C. R.: A neural network method to determine the presence or absence of permafrost near May, Yukon Territory, Canada. Permafrost and Periglacial Processes, 8, 205-215 (1997)

Li, Z. and Shi, J.: Snow mapping with SIR-C multipolarization SAR in Tienshen Mountain. In: Proc. 1996 International Geoscience and Remote Sensing Symposium (IGARSS '96), Lincoln, Nebraska, U.S.A., 1996, pp. 136-138

Lichtenegger, J., Seidel, K., Keller, M., and Haefner H.: Snow surface measurements from digital Landsat MSS data. Nordic Hydrol. 12, 275-288 (1981)

Martinec, J., Rango, A., and Roberts, R.: Snowmelt Runoff Model (SRM) user's manual. Geographica Bernensia P35, Department of Geography, University of Berne, 1998, 84pp

Massom, R.: Satellite remote sensing of polar regions: Applications, limitations, and data availability. London, Belhaven Press, 1991

Matzler, C.: Passive microwave signatures of landscapes in winter. Meteorol. Atmos. Phys., 54, 241-260 (1994)

Matzler, C., Strozzi, T., Weise, T., Floricioiu D.-M. and Rott, H.: Microwave snowpack studies made in the Austrian Alps during the SIR-C/X-SAR experiment. International Journal of Remote Sensing, 18 (12), 2505-2530 (1997)

McGinnis, D. F. And Schneider, S. R.: Monitoring river ice break-up from space. Photogrammetic Engineering & Remote Sensing, 44 (1), 57-68 (1978)

Nagler, T. and Rott, H.: The application of ERS-1 SAR for snowmelt runoff modeling. In: Remote Sensing and Geographic Information Systems for Design and Operation of Water Resources Systems (Proc. Rabat Symp.), IAHS Publication No. 242, 1997, pp. 119-126

Peck, E. L., Johnson, E.R., Keefer, T.N., and Rango, A.: Combining measurements of hydrological variables of various sampling geometries and measurement accuracies. In: Hydrological Applications of Remote Sensing and Remote Data Transmission (Proc. Hamburg Symp.), IAHS Publ. No. 145, 1985, pp. 591-599

Ramamoorthi, A. S.: Snow cover area (SCA) is the main factor in forecasting snowmelt runoff from major basins. In: Large Scale Effects of Seasonal Snow Cover. Proc. Vancouver Symp. IAHS Publ. No. 166, 1987, pp. 279-286

Ramamoorthi, A. S.: Snow-melt run-off studies using remote sensing data. In: Proc. Indian Acad. Sci. 6 (3), 279-286 (1983)

Rango, A., Martinec, J., Chang, A.T.C., Foster, J., and van Katwijk, V.: Average areal water equivalent of snow in a mountain basin using microwave and visible satellite data. IEEE Transactions on Geoscience and Remote Sensing GE-27(6), 740-745 (1989)

Rango, A.: The snowmelt-runoff model. Proc. ARS Natural Res. Modeling symp. Pingree Park, CO, USDA-ARS-30, 1985, pp. 321-325

Rango, A., Martinec, J., Foster, J., and Marks, D.: Resolution in operational remote sensing of snow cover. In: Hydrological Applications of Remote Sensing and Remote Data Transmission. (Proc. Hamburg Symp.), IAHS Publ. No. 145, 1985, pp. 371-382

Rango, A., Chang, A. T. C. and Foster, J. L.: The utilization of spaceborne microwave radiometers for monitoring snowpack properties. Nordic Hydrol. 10, 25-40 (1979)

Rango, A. and Martinec, J.: Water storage in mountain basins from satellite snow cover monitoring, In: Remote Sensing and Geographic Information Systems for Design and Operation of Water Resources Systems (Proc. Rabat Symp.), IAHS Publication No. 242, 1997, pp. 83-91

Rott, H. and Matzler, C.: Possibilities and limits of synthetic aperture radar for snow and glacier surveying. Annals of Glaciology, 9, 195-199 (1987)

Rott, H.: Capabilities of microwave sensors for monitoring areal extent and physical properties of the snowpack. In: Proc. NATO Advanced Res. Workshop on Global Environmental Change and Land Surface Processes in Hydrology, Tucson, U.S., 1993

Rott, H. and Nagler, T.: Snow and glacier investigations by ERS-1 SAR - First results. Proceedings, First ERS-1 Symposium: Space at the Service of Our Environment, Cannes, France, 1993, pp. 577-582

Shi, J. and Dozier, J.: Estimation of snow water equivalence using SIR-C/X-SAR. In: Proc. 1996 International Geoscience and Remote Sensing Symposium (IGARSS '96), Lincoln, Nebraska, U.S.A., 1996, pp. 2002-2004

Shi, J. and Dozier, J.: Inferring snow wetness using C-band data from SIR-C's polarimetric synthetic aperture radar. IEEE Transactions on Geoscience and Remote Sensing, 33 (4), 905-914 (1995)

Shi, J. and Dozier, J.: On estimation of snow water equivalence. In: Proc. Of the Fourth International Workshop on Applications of Remote Sensing in Hydrology, National Water Research Institute, Saskatoon, Saskatchewan, 1998 (In Press), 10pp.

Solberg, R., Hiltbrunner, D., Koskinen, J., Guneriussen, T., Rautiainen, K. and Hallikainen, M.: SNOWTOOLS: Research and development of methods supporting new snow products. In: Proc. XX Nordic Hydrology Conference, Helsinki, Finland, 1998

Steppuhn, H.: Snow and Agriculture. In: Gray, D. M. and Male, D. N. (eds.) Handbook of Snow: Principles, Processes, Management and Use. Toronto: Pergamon Press 1981, pp. 60-125.

Storr, D.: Precipitation variations in a small forested watershed. Proc. 35th Annual Western Snow Conference. 1967, pp. 11-16

Sturm, M., Holmgren, J. A., and Yankielun, N. E.: Using FM-CW radar to make extensive measurements of arctic snow depth: problems, promises, and successes. EOS Trans., AGU 77 (46) F196 (1996)

Sun, C., Neale, C.M.U., McDonnell, J.J. and Cheng, H.-D.: Monitoring land-surface snow conditions from SSM/I data using an artificial neural network classifier. IEEE Transactions on Geoscience and Remote Sensing, 35 (4), 801-809 (1997)

Thirkettle, F., Walker, A., Goodison, B. and Graham, D.: Canadian prairie snow cover maps from near real-time passive microwave data: from satellite data to user information. In: Proc. 14th Canadian Symposium on Remote Sensing, Calgary, Canada, 1991, pp. 172-177

Walker, A.E. and Goodison, B.E.: Discrimination of a wet snow cover using passive microwave satellite data. Annals of Glaciology, 17, 307-311 (1993)

Walker, A., Goodison, B., Davey, M., and Olson, D.: Atlas of Southern Canadian Prairies Winter Snow Cover from Satellite Passive Microwave Data: November 1978 to March 1986. Atmospheric Environment Service, Environment Canada. 1995

Warkentin, A.A.: The Red River flood of 1997: An overview of the causes, predictions, characteristics and effects of the flood of the century. CMOS Bulletin, 25 (5), (1997)

Woo, M.K., Walker, A., Yang, D. and Goodison, B.: Pixel-scale ground snow survey for passive microwave study of the Arctic snow cover. In: Proc. of the 52nd Annual Meeting of the Eastern Snow Conference, Toronto, Ontario, Canada, 1995, pp. 51-57

Yankielun, N. E.: An airborne millimeter-wave FM-CW radar for thickness profiling of freshwater ice. CRREL Report 92-20, US Army Corps of Engineers, Cold Regions Research and Engineering Laboratory, Hanover, New Hampshire, 1992, 77 pp.

Zuerndorfer, B., England, A. W., Wakefield, G. H.: The radiobrightness of freezing terrain. Proceedings, International Geoscience and Remote Sensing Symposium (IGARSS ?89), Vancouver, Canada, 1989, pp. 2748-2751

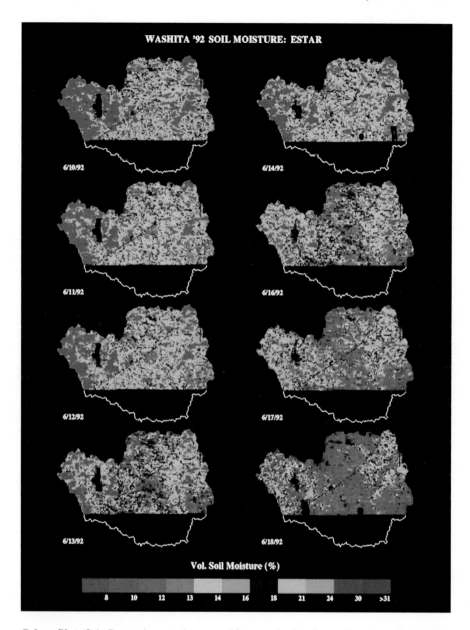

Colour Plate 9.A. Remotely sensed patterns (signatures) of surface soil moisture for the little Washita Watershed near Chickasha, Oklahoma, USA. The spatial and temporal changes in soil moisture from wet (blue and green) to dry (orange and red) resulting from ET and drainage are shown by the surface few cm of measured soil moisture from the airborne ESTAR instrument (Courtesy of Peggy O'Neill, NASA, Goddard Space Flight Center)

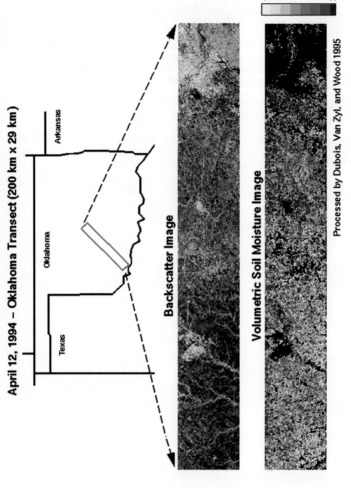

Colour Plate 9.B. Soil moisture mapped from the SIR-C/X-SAR mission over a 200 km strip in Oklahoma, USA. The moisture can be seen to increase from southwest to northeast in response to recent rain in the north east portion of the image (Courtesy of Eric Wood and Wade Crow, Princeton University)

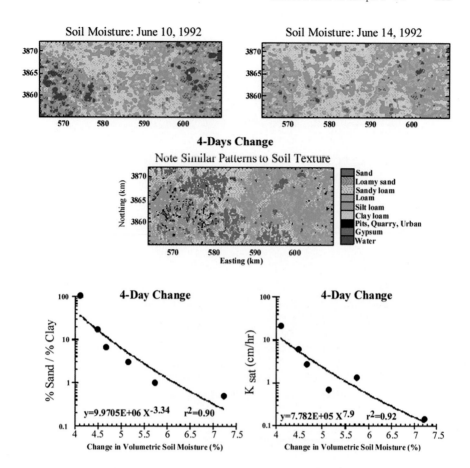

Colour Plate 9.C. Illustration of how soil hydrologic properties (texture and Ksat) can be estimated from space-time changes in remotely sensed soil moisture (Mattikalli et al., 1998)

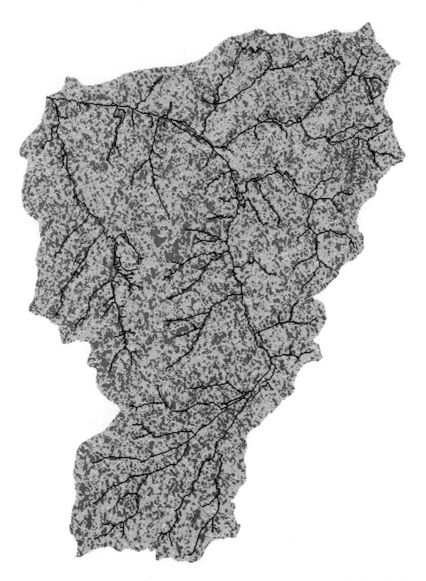

Colour Plate 9.D. An illustration of using the ERS-1/2 data to produce the third principal component of eight winter time images. The blue areas show regions of higher soil moisture adjacent to the streams that one would expect in a natural drainage basin, the Zwalm catchment in Belgium. The orange areas are predominantly the hill tops and better drained soils with lower soil moisture. (Courtesy of Niko Verhoest, Laboratory for Hydrology and Water management, University of Gent, Belgium)

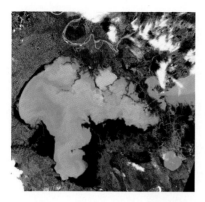 

Colour Plate 10.A. Classification results of two images in a time series of Landsat MSS and TM data for the Peace-Athabasca Delta, Northern Alberta, Canada. The top image shows the extent of drying as observed in August, 1985. The lower images highlight the areal extent of flood water in July, 1974

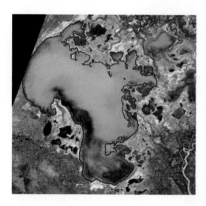 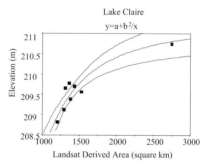

Colour Plate. 10. B. Vectors depicting the lake boundaries using a raster to vector conversion. The resulting plot shows Landsat derived areal extent versus lake elevation and the corresponding fit with the 90% confidence limits using a simple power function for Lake Claire in the Peace-Athabasca Delta, Alberta, Canada

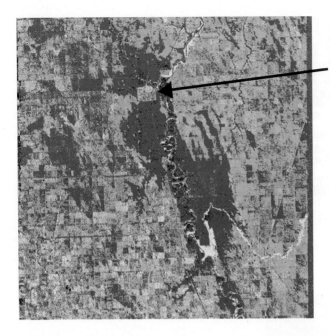

Town of Morris, Manitoba, Canada protected by a ring dike.

Colour Plate 10.C. Radarsat standard beam colour composite image, consisting of imagery from March 23, April 25 and May 9. Dark tones are flooded on both April 25 and May 9. The red areas (west side of image) represent flooding only on April 25, while the blue-green areas (north of Morris) are flooded on May 9 only (Courtesy of Terry Pultz, Canada Centre for Remote Sensing)

Colour Plate 10.D. Radarsat-1 (C-HH, May 5,11 and 18, 1996) scenes of the flood in the Peace-Athabasca Delta, Canada. Dark areas represent open water and very bright grey regions are flooded willow tree stands. Blue areas show flooded vegetation on May 18$^{th.}$ (Image provided by the Terry Pultz, Canada Centre for Remote Sensing)

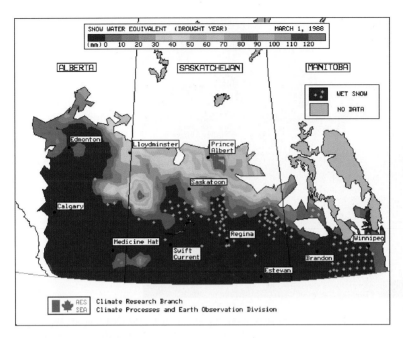

Colour Plate 11.A. Passive microwave-derived snow water equivalent for 1 March 1988 (drought year) on the Canadian prairies

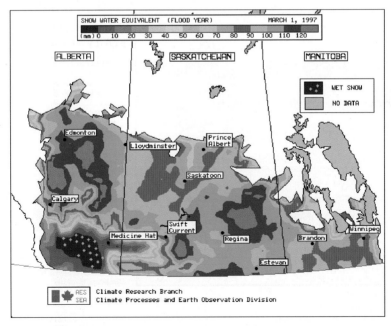

Colour Plate 11.B. Passive microwave-derived snow water equivalent for 1 March 1997 (flood year) on the Canadian prairies

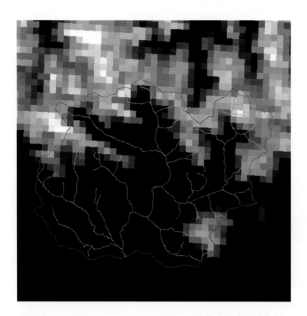

Colour Plate 11.C. NOAA-AVHRR derived snow cover for the Cinca basin (798.1 km^2) in the central Pyrenees of Spain on 10 March 1998. The different gray levels correspond to different percents of snow cover in each NOAA-AVHRR pixel

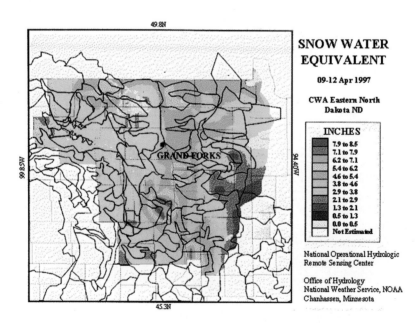

Colour Plate 11.D. Integration of satellite, airborne, and ground-based snow data (9-12 April 1997) using a geographic information system to geographically depict snow water equivalent (1inch = 2.54 cm) across Eastern North Dakota during the 1997 Red River of the North snowmelt flood

12 Soil Erosion

Jerry C. Ritchie

USDA-ARS Hydrology Laboratory, BARC-West, Bldg-007, Beltsville, MD 20705 USA

12.1 Introduction

Soil erosion is a natural process caused by water, wind, and ice that has affected the earth's surface since the beginning of time. Soil erosion and its off-site, downstream damages are major concerns around the world (Lal 1994b) causing losses in soil productivity and degradation of landscape (Walling 1983). Many of man's activities have accelerated soil erosion (Sombroek 1995; Walling 1983). Oldeman (1994) has estimated that human-induced soil degradation has affected 15% of the world's arable land surface. Estimates of global soil erosion rates range from 0.088 mm yr^{-1} (Walling 1987) to 0.30 mm yr^{-1} (Fournier 1960). These values led to estimates of 17.4 to 58.1 x 10^9 Mg of soil loss from the land surface (Walling 1990) which is carried downstream into lakes, reservoirs and estuaries where it reduces storage capacity and affects water quality, navigation, and biological productivity. On the land surface, soil erosion decreases organic matter, fine grained soil particles, water holding capacity, and rooting depth leading to loss of soil productivity and quality. The economic consequences from soil erosion on loss of productivity, land degradation, and off-site, downstream damages from eroded soil particles on water quality are a major concern. Pimentel et al. (1995) estimated the economic cost of soil erosion and subsequent sediment deposition to be $400 billion per year worldwide. While this economic estimate of the cost of erosion has been questioned (Sombroek 1995), concerns about soil loss and the estimated cost point to the need for new and improved methodologies and techniques for monitoring and quantifying soil erosion effectively and efficiently across the landscape so that effective land management practices can be applied to the land surface to control and reduce soil erosion.

Determining spatially distributed soil erosion on the landscape using classical soil erosion measurement techniques is difficult, time consuming, and expensive (Brakensiek et al. 1979; Lal 1994a). These techniques involve field studies with experimental plots to establish basic principles and measure rates of erosion. Small unit-source catchments are used for relating these plot studies to larger areas (Mutchler et al. 1994). These types of studies involve continuous measurements of soil and water movement from plots/catchments and require long time periods (years) of data collection to provide statistically significant results. Such studies are limited to establishing information about processes and rates and can seldom be used to determine spatial patterns of soil erosion. Others techniques involve measuring sediment loads at the outlet of catchments and relating these measured sediment loads to soil erosion on the catchment using sediment delivery ratios (Walling 1983). However, estimating soil erosion rates from sediment loads by sediment delivery ratios has long been recognized as a problem (Roehl 1962; Walling 1983) and provides no information on spatial patterns. Radioactive fallout (^{137}Cesium) and naturally occurring radioactive elements (i.e., ^{210}Pb) have also been used to measure soil erosion and sediment deposition (Ritchie and McHenry 1990). The use of radionuclides allows the measurement of both eroding and depositing sites within a field and thus provides better information on actual soil loss from a field and the spatial patterns of erosion and deposition within a field. Detailed information on classical field techniques for measuring soil erosion can be found in a field manual of methods for agricultural hydrology (Brakensiek et al. 1979) and a recent review of soil erosion measurement techniques is presented by Lal (1994a).

In addition to field measurements of soil erosion, empirical and process-based mathematical equations/models have been developed to estimate soil erosion (Foster 1991; Lane et al. 1992). The most widely used soil erosion model is the Universal Soil Loss Equation (USLE) which is an empirically based equation developed from data collected from standard soil erosion plots on "typical" soils of the United States east of the Rocky Mountains (Wischmeier and Smith 1978). The USLE has been extensively used and „misused" in the United States and around the World. Even with its limitations, the USLE is the most widely used, powerful and practical management tool for estimating sheet and rill erosion on agricultural landscapes. A Revised Universal Soil Loss Equation (RUSLE) using the same form and factors with revised coefficients is available with applications to a wider range of conditions and locations than the original USLE but is still based on empirical relationships (Renard et al. 1991, 1997). The form of the equation for the RUSLE (and USLE) is:

$$A = RKLSCP$$

Where A is the average annual soil loss per unit area; R, the rainfall factor; K, the soil erodability factor; L, the slope length factor; S, the slope steepness factor; C, the crop management factor; and P, the soil erosion control practice factor. Numerical coefficients for these factors were determined empirically from field and laboratory data (Wischmeier and Smith 1978; Renard et al. 1997). Many other efforts to develop empirical and physically-based models of soil erosion and its off-site effects have been made that have had varying degrees of success and applications in management and research. A general discussion of soil erosion models with a more in depth view of their applications and limitations is given by Foster (1991), Lane et al. (1992) and Lal (1994a).

The limitations of current measurement techniques and models to provide information on the spatial and temporal distribution of soil erosion across catchments and on the movement of eroded particles into streams limit our ability to develop cost-effective land management strategies for erosion control. Such limitations point to the need to investigate other techniques to supplement current techniques for monitoring and quantifying soil erosion rates and patterns and the need for techniques that can identify potential as well as actual source areas of soil erosion on the landscape.

Soil erosion causes both physical (i.e., gullies, rills) and visible (i.e., exposure of different colored soil layers) changes in the surface properties of soils on the landscape. Such changes can be measured both spatially and temporally using remote sensing techniques with a variety of sensors and sensor platforms. Remote sensing techniques can measure qualitative and quantitative information on changes in the soil surface roughness and on visible features. Remote sensing techniques to measure eroded material once it leaves the land surface and enters a stream or a water body are discussed in Chap. 13 of this book.

12.2 Basis for using Remote Sensing

Cihlar (1987) concluded that assessing and monitoring soil erosion occurs in two dimensions that he called stage and time. He uses stage to refer to the type of soil erosion (i.e., sheet, rill, or gully) and time to refer to actual or potential soil erosion. It is important to evaluate both the stage and time dimension if we are to develop management plans to reduce or control soil erosion. Both the stage and time of erosion can affect the physical and spectral properties of soil surfaces. Since remote sensing techniques measure spectral and physical properties, we can use remote sensing to provide information on these changes in surface properties of the soil caused by erosion. Remote sensing techniques can provide spatial and temporal data on these properties that will allow us to develop better management plans for large areas to reduce soil erosion.

Remote sensing techniques that measure the spectral properties of the landscape are most commonly applied to study soil erosion patterns and rates. Ground-based and aerial photographs are still the most popular remote sensing technique used to study and map soil erosion using either photointerpretation or photogrammetric techniques. These techniques provide information on spectral differences in the soil surface and are interpreted to delineate areas affected by soil erosion. Photointerpretation of images made from digital radiance data from satellite sensors (Landsat, SPOT, IRS, AVHRR, etc.) is also widely used to map areas of soil degradation. Photographs and digital imagery have been used for mapping actual/potential soil erosion areas, for determining spectral patterns and differences in the surface soils related to soil erosion, and for determining land cover and conservation practices for input to soil erosion models.

Photogrammetric techniques are used with stereo photographs to measure physical changes (i.e., roughness) in soil surface elevation. Sequential photographs can be used to determine changes in areas of erosion or deposition over time. Ground-based and airborne lasers are also used to measure landscape surface roughness and topography. These laser systems can measure physical changes in the soil surface to within a few millimeters from ground platforms and within a few centimeters from aerial platforms allowing estimates of soil loss or estimates of surface roughness that allow us to better understand soil erosion. Synthetic aperture radar (SAR) also has been used to measure the landscape topography and may have potential applications for evaluating large-scale soil erosion patterns and rates.

While remote sensing data can be used alone to access soil erosion rates and patterns, a combination of remote sensing data, soil erosion models, and Geographic Information Systems (GIS) has much to offer in understanding and measuring the effects of soil erosion on the landscape. GIS tools allow researchers or decision makers to combine the spectral, spatial, and temporal characteristics of remotely sensed data with ancillary data about the landscape and use these many sources as input to erosion models to quantify rates and patterns of soil erosion. GIS tools provide the spatial pattern of erosion needed by decision makers for carrying out management practices to reduce soil erosion on the landscape.

Many early applications of remote sensing for mapping soil erosion and degradation patterns were discussed by Dubucq (1986). He found publications from 1973 using Landsat Multispectral Scanner (MSS) data to map soil erosion prone areas. Most of the papers cited by Dubucq used photointerpretation techniques to make maps of potential and actual eroded areas.

12.3 Applications

Photointerpretation remains the most commonly applied remote sensing technique for studying and mapping soil erosion across the landscape. Photointerpretation has been defined as the science of identifying and describing objects imaged on a photograph and determining their significance (Philipson 1997). Some argue that this is as much art as science. Photointerpretation is based on the ability of the observer to identify and delineate objects based on pattern, texture, and color from a photograph. Photointerpreted objects are often noted on the photograph and later transferred to maps. Such maps are analyzed to determine the eroding area and changes in eroding area (or other interpreted objects). These changes in eroded area can be determined by photointerpreting sequential images and plotting changes on a map. More information about photointerpretation can be obtained from many sources (e.g. Philipson 1997).

Photogrammetry is also widely use for studying and mapping soil erosion. Photogrammetry has been defined as the science, art, and technology of obtaining reliable information about physical objects and the environment from photographs (Greve 1996). The principle of the photogrammetric technique is shown in Fig. 12.1. Photogrammetry has been widely used to measure changes in landscape surfaces for topographic mapping and to determine gully erosion, rill development, and erosional losses. Using photogrammetric techniques on sequential photographs allows the determination of eroding and depositing area on the landscape. More information about photogrammetric techniques can also be obtained from many sources (e.g. Greve 1996).

A wide range of digital data with differing spectral and spatial resolution are available from aircraft and satellite platforms. These data can be used to make images for photointerpretation or can be analyzed spectrally to evaluate soil erosion rates and patterns. These digital data need to be georeferenced and corrected for atmospheric interference for best results. Computer software programs are available to correct,

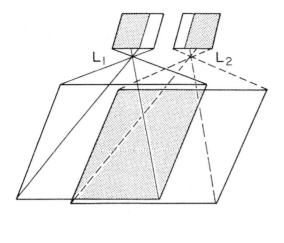

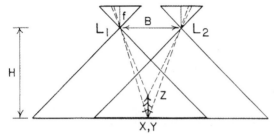

Fig. 12.1. Basic principles for using stereo photographs to develop stereomodel (adapted from Welch et al. 1983)

analyze, and classify the digital spectral data based on spectral properties into information about the characteristic of the landscape surface properties related soil erosion rates and patterns. Image classification and vegetation indices are widely used for applications in hydrology and erosion. Other sources (e.g. Jensen 1996) are available that discuss classification technique in detail.

Ground and airborne laser distancing systems can provide data on the physical properties of the landscape surface which can be analyzed to provide information on the erosional properties of the soil surface. Both profile and imaging laser systems are available that can be use in erosion research (Ritchie 1996). Synthetic Aperture Radar can also provide data related to topography which can be used as input for erosion models. As techniques improve for analyzing laser and SAR data, these data will allow the development of improved digital elevation models (DEM) for use in erosion modelling, research and management.

Once remotely sensed data is georeferenced and classified it can be used for mapping or as input for a data layer in a Geographical Information System (GIS). Figure 12.2 shows schematically how different remote sensors might provide data to be used with the USLE or RUSLE to determine erosion rates and pattern. Other models could be substituted for the USLE in this schematic to provide the output needed. Many remote sensors are potential suppliers of data for the USLE and GIS layers. Photointerpretation and image classification of photographs and spectral images are commonly used to provide input for the crop management factor (C) and soil erosion control practice factor (P) of the USLE. Photogrammetric techniques provide input for calculating the slope length factor (L) and steepness factor(S). Airborne lasers and Synthetic Aperture Radar (SAR) could also provide data for calculating the slope length factor (L) and steepness factor(S). Photographs and spectral images could provide data on soil properties and provide the basis for spatially distributing the erodability factor (K). While for the USLE the rainfall factor (R) is based on typical rainfall patterns and rates for an area, for other models radar could provide information on rainfall rates and patterns needed for input for calculating erosion on a storm event basis. Information on all the factors for the USLE is available from non remote sensing sources. However, by combining remote sensing data with conventional data a better understanding of the spatial patterns of the factors may be determined. By using the different GIS data layers determined conventionally or with remote sensing as input to soil erosion models, soil erosion rates and patterns can be spatially distributed with GIS techniques. Other techniques can also be used to combine the GIS data layer to evaluate soil erosion rates and patterns. Examples of the use of GIS and remote sensing data in natural resource management are given by Ripple (1994).

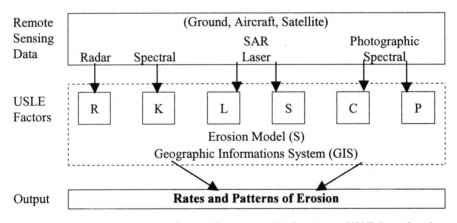

Fig. 12.2. Schematic representation of potential remote sensing data input to USLE. Input for other erosion/hydrology models would be similar

12.4 Case Studies

The application of remote sensing techniques to study and monitor soil erosion can be divided into the following general approaches: photointerpretation / photogrammetry, model/GIS inputs, spectral properties, and topographic measurements. Examples of studies will be used to describe each approach.

12.4.1 Photointerpretation/Photogrammetry

Photointerpretation techniques have been used extensively to map soils, vegetation cover, drainage patterns, soil degradation, and other erosion related factors. Many of these applications have been used for mapping erosion patterns and estimating erosion rates. The Food and Agriculture Organization (FAO 1979) routinely uses photointerpretation techniques with aerial photographs and satellite images to assess and map soil degradation in developing countries around the world. The FAO maps are used to determine areas of critical erosion and soil degradation in a series of 1:50000 and 1:100000 scale maps for developing countries.

Aerial photographs and satellite images were used to map changes in the shape and size of the Rosetta promontory in the Nile Delta (El-Raey et al. 1995). Analyses of data from 1955 to 1991 found that photographs/satellite imagery could be used to follow the erosion and deposition patterns and changes in the size of the promontory that agreed with ground measurements. They concluded that the Rosetta promontory had been eroding since the building of the Aswan High Dam. They also found that remote sensing techniques were a more accurate, efficient, and cost effective tool for monitoring such changes than were ground measurements. Samarakoon et al. (1993) used Landsat TM images in a similar manner to monitor areas susceptible to landslides and concluded that TM data was a useful tool for monitoring and mapping potential landslide areas.

Panchromatic and Color Infrared (CIR) aerial photographs (see Color Plate 12.A) have been used to identify and map rill and gully soil erosion (i.e., Morgan et al. 1978; Frazier and McCool 1981; Stephens and Cihlar 1982; Cihlar 1987). Many different photographic scales have been used. Frazier et al. (1983) used ground (see an example of their photography, Color Plate 12.B) and aerial photographs taken with a 35-mm camera to assess and map rill development and soil erosion patterns in the steep hills of Palouse area of the western United States. They found that aerial photographs were the most effective tool to study the development of rill and soil erosion patterns and to determine the effectiveness of management practice for erosion control. Aerial photographs allowed them the overview that was necessary for mapping rill development that was not possible from the ground. Using stereoscopic photointerpretation techniques, Stephens (Stephens and Cihlar 1982) established an erosional classification system and mapped erosion patterns for 100 km^2 area in New Zealand using 1:10,000 aerial photographs. He found that remote sensing techniques were more accurate and cost 30% less. Using the same technique in Canada, Stephens and Cihlar accurately mapped rill and gully erosion cost effectively. Several researchers (Stephens et al. 1982; Cihlar 1987) have used sequential photographs taken over periods of weeks to years to map and study the changing patterns of rill development and soil erosion. By comparing maps prepared from these sequential aerial photographs, changes in the pattern of erosion and estimates of rates were determined.

Cihlar (1987) describes techniques for using satellite data (MSS, TM, SPOT) for monitoring potential sheet and rill soil erosion and conservation practices on agricultural lands. However, he found that image-specific predictive relationships were required to achieve accurate quantitative results. In most studies, photointerpretation of satellite images has been most useful for repetitive monitoring and large area reconnaissance (Johannsen and Barney 1981; Johannsen and Saunders 1984) rather than gathering information about specific sites. Phillips et al. (1986) discuss the many potential applications of the Thematic Mapper (TM) to conservation assessment and concluded that TM is a valuable tool for managing the landscape. With the improving spectral and spatial resolution of current satellites, satellite data will become even more important for monitoring pattern and rates of soil erosion across the landscape.

Pelletier and Griffin (1988) used photointerpretation of aerial photographs to identifying conservation practices on agricultural lands. They concluded that CIR aerial photographs were most useful (see Color Plate 12.A for an example of one of the photographs used). They estimated that more than 90% of the soil conservation practices used in the United States could be identified at a photographic scale of 1:10000 and that 29% could still be identified at a scale of 1:80000. Stephens and Cihlar (1982) provided an interpretation key to identify agricultural practice and soil erosion features from CIR photographs taken in Canada. They also concluded that most conservation practices could be identified. Wilson et al. (1989) concluded that SPOT panchromatic data (10 meters) could be visually interpreted to develop an effective potential soil loss map while TM imagery could be used to update crop management practices during the growing season. Such monitoring of conservation practices with remotely sensed data provides a basis for evaluating the effectiveness of conservation practices, managing soil erosion control practices on the landscape, and providing data for input for soil erosion models.

Photogrammetric techniques applied to ground and aerial photographs have been used to measure changes in surface topography and quantify soil loss rates. Welch and others (Welch and Jordan 1983; Welch et al. 1983; Thomas and Welch 1988) describe the application of photogrammetric techniques to quantify soil loss using ground and aerial photographs. They estimated that photogrammetric techniques (Fig. 12.1)with aerial stereo-pairs could provide X, Y, and Z terrain coordinates with vertical accuracies of ±25 mm at contour intervals of 150 mm. In a three-year study, they made quantitative estimates of the amount and pattern of soil eroded or deposited in a 5.34 ha field. The same photogrammetric techniques have been used to measure stream channel degradation (Collins and Moon 1979) and the development of rills and ephemeral gullies in agricultural fields (Thomas and Welch 1988).

Photogrammetic techniques were used with photographs taken at 15 to 30 m above the ground (Spomer and Mahurin 1984; Spomer et al. 1986) to quantify gully development and hillslope soil erosion. They measured elevation with an accuracy of ±15 cm and estimated net soil displacement (erosion minus deposition) to be 46, 260, and 605 Mg ha^{-1} respectively for 1972-1974, 1974-1978, and 1978-1984. They estimated that sediment delivery at the end of the catchment was 53% based of the photogrammetric study as compared with 21% computed from USLE estimates. They concluded that remote sensing provided better estimates of soil loss. Using sequential photographs, Spomer and Mahurin (1984) developed topographic maps of an expanding gully area and estimated soil loss of 14 tonne yr^{-1} for a nine-year period from the area. Dymond and Hicks (1986) used the same photogrammetric principles on a series of historical aerial photographs for a period between the early 1940's and 1980 and estimated total catchment soil erosion for areas in New Zealand. Several studies (Sneddon and Lutze 1989; Kirby 1991) have used close range photogrammetry techniques to quantify soil erosion rates.

These selected examples show the application and benefits of photointerpretation and photogrammetric techniques in mapping and quantifying patterns and changes in patterns of soil erosion on the landscape. Such applications have provided information that has helped to develop a better understanding of the soil erosion process.

12.4.2 Model/GIS Inputs

Remotely sensed data have temporal, spatial, and spectral characteristics that can provide information on landscape variables needed for input to soil erosion and other natural resource models (Foster 1991; Lane et al. 1992; Agassi 1995). Pelletier (1985) reviewed the principles of the application of remotely sensed data to USLE. Traditionally, most inputs for soil erosion models (and other natural resource models) have come from historical records, soil surveys, and field surveys of land use and conservation practices. Remotely sensed data, digitized soils data, DEM, and ancillary data used for input to erosion models in a GIS framework can provide valuable insights for understanding and managing soil erosion, water quality, and natural resources. Studies in recent years have shown that many inputs for these models can be derived from aerial photographs or satellite data (Fig. 12.2). As spatially distributed models of the erosion process become available, remote sensing techniques may be the only way to collect the spatially distributed data necessary for input. Many recent studies have shown the potential for using remote sensing data for input to soil erosion models. A few examples are be given.

Stephens et al. (1985) used aerial CIR to delineate homogeneous soil erosion areas and to estimate the cropping factor (C), management practice factor (P), slope factors (L, S), and soil factors (K) for input in the USLE. Using CIR they were able to estimate soil loss faster and with the accuracy (88 ± 1.2%) needed for farm planning. Cihlar (1987) used Landsat TM and SPOT data to determine cropping practice factors (C) for input to the USLE.

Jakubauskas et al. (1992) discussed the application of remotely sensed data for soil erosion model and used remotely sensed data for input to the Agricultural Nonpoint Source (AGNPS) model. They concluded that SPOT multispectral data with supervised classification was the most cost effective source of data for catchments greater 141 km^2. Photointerpretation of aerial photographs was best for catchment less than 141 km^2. They also concluded that only on larger catchments does the use of digital satellite data become economically effective. Fraser et al (1995) concluded that classification of Landsat TM data provide a rapid, objective technique and an attractive alternative to photointerpretation for developing data bases for large catchments. Other researchers have used either photographs or satellite data to determine inputs for soil erosion models with good success (i.e., Wilson et al. 1989; Morgan and Nalepa 1982).

Pelletier (1985) provided a framework for combining soil erosion models with remotely sensed data in a Geographic Information System to expand the applications of soil erosion models and to improve our understanding of soil erosion processes across the landscape. Pelletier showed that remotely sensed data could be used more objectively and efficiently with a GIS to evaluate the landscape for problem areas of soil erosion.

Jürgens and Fander (1993) use the USLE, remote sensing data, and a GIS to develop soil erosion risk maps (see Color Plates 12.C and 12.D). Using their system they simulated soil protection measures on the landscape and determine their effectiveness based on their soil risk maps. They concluded that remote sensing data in combination with GIS was the most efficient way to simulate soil protection measures and select sites for agricultural production with the least soil erosion risk. Similarly, Hession and Shanholtz (1988) used the USLE, remote sensing data, and a GIS to estimate sediment loading from the Chesapeake Bay catchment and developed a method for targeting efforts on the catchment where soil conservation practices should be placed to reduce soil erosion.

12.4.3 Spectral Properties

Soil erosion removes the surface soil layer exposing different layers with different spectral properties. Based on this concept, Robinove et al. (1981) proposed using albedo differences between Landsat overpasses to monitor arid land soil erosion and degradation. They calculated albedo from Landsat MSS digital data and found that decreases in albedo was related to improved land use patterns (more soil moisture, organic matter, and increased vegetation productivity) and increases in albedo related to soil degradation (erosion, low soil moisture, organic matter, and productivity). Using albedo difference between different Landsat images they were able to identify areas of soil degradation and erosion in cold desert areas of the southwestern United States. Frank (1984) also used albedo differences from Landsat to assess change in the surface characteristics of a semiarid rangeland in Utah and reached the same conclusion as Robinove. Landsat radiance parameters have also been used to distinguish soil erosion, stability, and degradation in arid central Australia (Pickup and Chewing 1988). These techniques have been used extensively by international organizations (FAO) to map soil degradation in developing countries.

Seubert et al. (1979) used laboratory measurements of spectral properties of the surface soil to distinguish between slightly, moderately, and severely eroded areas in agricultural lands. Latz et al (1984) expanded on this study and showed that these spectral differences were due to the amounts of iron oxide and organic matter in the soil surface. They found that the differences in spectral intensity were related to the severity of soil erosion and suggested that Landsat data would be useful in detecting soil erosion. Pelletier and Griffin (1985) used Landsat data to map soil erosion by using spectral data to identifying areas with higher iron oxide concentration (reddish color) of the subsoil soil. They used the presence of iron oxide to map eroded areas (subsoil soil at the surface). They found this technique to be useful for bare soils but severely limited under condition with significant vegetation.

Price (1993) in a study of Pinyon-Juniper woodland found that spectral measurements from TM were better indicators of soil erosion patterns than estimates from USLE. He concluded that spectral properties of remotely sensed data gave an integrated view of soil erosion of the landscape and that remotely sensed data should be considered when developing soil erosion models for broad geographic areas of semiarid and arid land.

While such simple relationship between radiance measured in a satellite band have been related to soil erosion, most studies have used a vegetation index (i.e., NDVI, BI) or other band ratios or combinations to estimate soil erosion areas from satellite data. Mathieu et al. (1997) using SPOT data found that bands and band ratios provided a reliable means of determining a soil surface affected by erosion. However, they were not able to extrapolate these findings into predictive indicators of erosion. In a study in Norway, Leek and Solberg (1995) used CIR, SPOT, and ERS-1 SAR data to monitor autumn tillage and concluded that remotely sensed data was useful for monitoring soil exposed to rainfall soil erosion. They also found that a combination of optical and microwave data was effective in monitoring tillage areas because the microwave allowed data collection during cloud cover. By mapping these bare soil, they were able to delineate areas of potential soil erosion problems.

Since surface soil erosion is highly correlated the amount of surface residue, techniques to estimate residue cover are needed to better determine soil loss. Daughtry et al (1996) have shown that reflectance and fluorescence can be used to determine the amount of residue on the soil surface. They concluded the fluorescence techniques were better and have developed methods to estimate surface residue in the field. Whiting et al. (1987) used small format photographs to estimate residue cover and concluded that the technique was useful for monitoring conservation tillage and soil exposed to rainfall soil erosion. They found the method could reduce field time needed to monitor surface residue and soil erosion. Biard and Baret (1997) developed an algorithm (CRIM - Crop Residue Index Multiband) to estimate surface residue. They showed "reasonably good" estimates of residue using field measurements of reflectance in the TM bands. The algorithm could be used with any set of wave bands. Biard and Baret (1997) also provide a good background on studies on residue measurements using spectral data from ground and satellite sensors.

12.4.4 Topographic Measurements

Topographic maps have long been used in studies of soil erosion. However, the need for more up-to-date, accurate and rapid measurements and assessments of land surface terrain features to estimate land surface roughness, water movement, and soil erosion has led to the application of laser distancing technology for *in situ* measurements of soil surfaces from a laser mounted 10-20 cm above the surface (Bertuzzi et al. 1990; Huang and Bradford 1990). These *in situ* studies have been used to quantify soil loss from rill and sheet soil erosion with high accuracy (millimeter accuracy) and to study development of rills on the landscape. These studies are leading to a better understanding of the mechanism of soil particle detachment and movement.

The same laser techniques have been used with an airborne laser altimeter (Ritchie 1996) to study soil erosional features of the landscape. Landscape features related to soil erosion such as topography, stream channels, gullies, and canopy cover can be measured. The airborne laser measurements provide quick and accurate data on the morphology and roughness of the landscape surface (Ritchie and Jackson 1989). Laser altimeter data are valuable for measuring gully (Fig. 12.3) and channel cross-sections and roughness, for assessing soil loss from gullies and channels, for providing input to understand gully and stream channel dynamics, and for measuring soil surface roughness (Ritchie et al. 1994). Data related to changes in landscape topography, both micro and macro changes, provide basic information for determining soil erosion rates and patterns. Any remote sensor that will allow topographic observations to be updated accurately and easily will aid in the estimation of soil erosion.

12.5 Future Directions

In a few of the applications discussed, photogrammetric techniques and laser distancing techniques were used with ground or aerial photographs to quantify the actual amount of soil erosion. However in most applications, actual soil erosion was not measured directly from remotely sensed data but changes in land use or soil surface properties/conditions were used as a surrogate for delineating areas with soil erosion

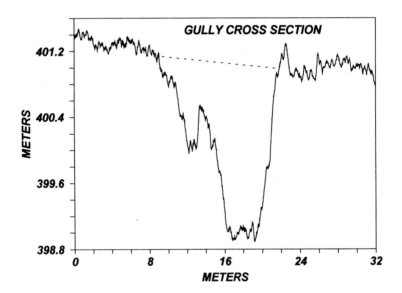

Fig. 12.3. Measurement of cross-section measured with an airborne laser. Airplane was at 300 m above the land surface. Calculated area of the gully is 16.3 m^2

or to estimate a variable for input into a soil erosion model. Future uses of remote sensing data for delineating areas with soil erosion and estimating variables should increase as new sensors (e.g. EarlyBird, Resources21, etc.; Corbley 1996) with higher spatial (1-2 meters) and spectral (100+ wavebands) resolutions are launched. The higher spatial resolution will permit measurements on smaller areas that are most important for understanding and quantifying the processes of soil erosion at the field level. Better spectral resolution may allow us to derive spectral signatures for eroded areas that are specific for degraded soils and allow better delineation of soil erosion patterns across the landscape. However, spectral patterns for eroded soils may be site specific and require a definition of soil properties to be successfully used.

The continued development of scanning laser technology with higher spatial and vertical accuracies will provide another source of data on topography that will be useful in measuring centimeter difference due to soil erosion. Krabill et al. (1995) used a scanning laser altimeter to measure the surface terrain of the Greenland ice sheet and estimated accuracies of ~20 cm for several 1000 kilometers of flight lines. These measurements were used to study the dynamics of the Greenland ice sheet. Hug (1996) used a similar scanning laser system to measure urban topography. A scanning laser altimeter is schedule to be launched on a space platform in 2002. With such measurements and updating of changes in topography, we can better quantify soil loss and have data for understanding the dynamics of water and wind flow across the landscape.

The fast developing field of Synthetic Aperture Radar (SAR) interferometry (Gens and van Genderen 1996) with applications to topographic mapping and digital elevation modeling (DEM) also holds great promise for providing a new source of data for quantifying soil erosion. SAR data are available from aircraft and satellite (ERS1/ERS2) instruments to begin research. Again the updating of changes in topography is key for quantifying soil erosion.

Combining these new sources of data with Geographic Information Systems should provide many new insights into the problems of soil erosion and provide a basis for quantifying soil loss using remotely sensed data. The future has much to offer in application of remote sensing data to understanding the role of soil erosion on the landscape. New sensors and tools will allow the researchers and decision makers to use improved spectral, spatial, and temporal characteristics of remotely sensed data and ancillary data with models and GIS for understanding and quantifying patterns of soil erosion. Management decisions can be based on actual soil erosion rates thus allowing efforts to reduce soil erosion to be targeted to critical soil erosion areas in the catchment.

References

Agassi, M. (ed.): Soil erosion, conservation and rehabilitation, New York, Marcel Dekker, Inc (1995)

Biard, F., Baret, F.: Crop residue estimate using multiband reflectance. Remote Sens, Environ. 59, 530-536 (1997)

Bertuzzi, P., Caussignac, J.M., Stengel, P., Morel, G., Lorendeau, J.Y., Pelloux, G.: An automated, noncontact laser profile meter for measuring soil roughness in situ. Soil Sci. 149, 169-178 (1990)

Brakensiek, D.L., Osborn, H.B., Rawls, W.J.: Field manual for research in agricultural hydrology. Washington, DC: USDA Agriculture Handbook No. 224 (1979)

Cihlar, J.: A methodology for mapping and monitoring soil erosion. Canadian J. Soil Sci. 67, 433-444 (1987)

Collins, S.H., Moon, G.C.: Stereometric measurement of streambank soil erosion. Photogrammetric Engr. Remote Sens. 45, 183-190 (1979)

Corbley, K. Remote sensing skies filling with satellite plans. EOM The Magazine for Geographic, Mapping and Earth Information 5, 26-28 (1996)

Daughtry, C.S.T., McMurtrey III, J.E., Chappelle, E.W., Hunter, W.J., Steiner, J.L.: Measuring crop residue cover using remote sensing techniques. Theor. Appl. Climatol. 54, 17-26 (1996)

Dubucq, M.: Télédétection spatiale et soil erosion des sols Etude Bibliographique Cah. ORSTOM, Sér. Pédol. Vol. XXII, n° 2, 1986, 247-258 (1986)

Dymond, J.R., Hicks, D.L.: Steepland soil erosion measured from historical aerial photographs. J. Soil Water Conserv. 41, 252-255 (1986)

El-Raey, M., Nasr, S.M., El-Hattab, M.M., Frihy O.E.: Change detection of Rosetta promontory over the last forty years. Internat. J. Remote Sens. 16, 825-834 (1995)

FAO: Methodology for soil degradation assessment. Rome, Italy: Food and Agriculture Organization, United Nations (1979)

Foster, G.R.: Advances in wind and water soil erosion prediction. J. Soil Water Conserv. 46, 27-29 (1991)

Fournier, F.: Climat et soil erosion. Paris: Presses Universitaires de France (1960)

Frank, T.D.: Assessing change in surficial character of a semiarid environment with Landsat residual image. Photogrammetric Engr. Remote Sens. 50, 471-480 (1984)

Fraser, R.H., Warren, M.V. Barten, P.K.: Comparative evaluation of land cover data sources for soil erosion prediction. Water Resour. Bull. 31, 991-1000 (1995)

Frazier, B.E., McCool, D.K.: Aerial photography to detect rill soil erosion. Trans. Amer. Soc. Agric. Engr. 24, 1168-1176 (1981)

Frazier, B.E., McCool, D.K., Engle, C.J.: Soil erosion in the Palouse: an aerial perspective. J. Soil Water Conserv. 38, 70-74 (1983)

Gens, R., van Genderen, J.L. SAR interferometry - issues, techniques, applications. Intl. J. Remote Sens. 17, 1803-1835 (1996)

Greve, C.W. (ed.): Digital Photogrammetry: An Addendum to the Manual of Photogrammetry. Bethesda, MD: American Society of Photogrammetry and Remote Sensing (1996)

Hession, W.C.,Shanholtz, V.O.: A geographic information system for targeting nonpoint-source agricultural pollution. J. Soil Water Conserv. 43, 264-266 (1988)

Huang, C., Bradford, J.M.: Portable laser scanner for measuring soil surface roughness. Soil Sci. Soc. Amer. J. 54,1402-1406 (1990)

Hug, C. Urban topography survey with the scanning laser altitude and reflectance sensor (SCALARS). Proc. Second Intl. Airborne Remote Sens. Conf. and Exhibition, pp. I-429 to I-438 (1996)

Jakubauskas, M.E., Whistler, J.L. Dillworth, M.E., Martinko, E.A.: Classifying remotely sensed data for use in an agricultural nonpoint-source pollution model. J. Soil Water Conserv. 47, 179-183 (1992)

Jensen, J.R.: Introductory digital image processing: A remote sensing perspective. Englewood Cliffs, NJ: Prentice-Hall, Inc (1996)

Johannsen, C.J., Barney, T.W.: Remote applications for resource management. J. Soil Water Conserv. 37, 128-131 (1981)

Johannsen, C.J., Sanders, J.L. (eds.): Remote sensing for resource management, Ankeny, Iowa: Soil Conservation Society of America (1984)

Jürgens, M., Fander, M.: Soil erosion assessment and simulation of SGEOS and ancillary digital data. Internat. J. Remote Sens. 14, 2847-2855 (1993)

Kirby, R.B.: Measurement of surface roughness in desert terrain by close range photogrammetry. Photogrammetric Record 13, 855-875 (1991)

Krabill, W.B., Thomas, R.H., Martin, C.F., Swift, R.N., Frederick, E.B.: Accuracy of airborne laser altimetry over the Greenland ice sheet. Internat. J. Remote Sens. 16, 1211-1222 (1995)

Lal, R. (ed.): Soil erosion. Ankeny, IA, Soil and Water Conservation Society (1994a)

Lal, R.: Soil erosion by wind and water: Problem and prospects, In: Lal, R. (ed.), Soil erosion, Ankeny, IA: Soil and Water Conservation Society, pp. 1-9 (1994b)

Lane, L.J., Renard, K.G., Foster, G.R., Laflen, J.M.: Development and application of modern soil erosion prediction technology - The USDA experience. Australian J. Soil Res. 30, 893-912 (1992)

Latz, K., Weismiller, R.A., Van Scoyoc, G.E., Baumgardner, M.F.: Characteristics variations in spectral reflectance of selected eroded afisols. Soil Sci. Soc. Am. J. 48, 1130-1134 (1984)

Leek, R., Solberg, R.: Using remote sensing for monitoring of autumn tillage in Norway. Internat. J. Remote Sens. 16, 447-466 (1995)

Morgan, K.M., Lee, G.B., Kiefer, R.W., Daniel, T.C., Bubenzer, G.D., Murdock, J.T.: Prediction of soil loss on cropland using remote sensing. J. Soil Water Conserv. 33, 291-293 (1978)

Morgan, K.M., Nalepa, R.: Application of aerial photographic and computer analysis to the USLE for area wide soil erosion studies. J. Soil Water Conserv. 37, 347-350 (1982)

Mutchler, C.K., Murphree, C.E., McGregor, K.C.: Laboratory and field plots for soil erosion research, In: Lal, R. (ed.), Soil erosion. Ankeny, IA: Soil and Water Conservation Society, pp.11-37 (1994)

Oldeman, L.R.: The global extent of soil degradation, In: Greenland, D.J. and I. Szaboles (eds.), Soil resilience and sustainable land use, Wallingford, UK: CAB International, pp. 99-118 (1994)

Pelletier, R.E.: Evaluating nonpoint pollution using remotely sensed data in soil erosion models. J. Soil Water Conserv. 40, 332-335 (1985)

Pelletier, R.E., Griffin II, R.H.: An evaluation of photographic scale in aerial photography for identification of conservation practices. J. Soil Water Conserv. 43, 333-337 (1988)

Phillips, K.M., Morgan, K., Newland, L., Koger, D.G.: Thematic mapper data: a new land planning tool. J. Soil Water Conserv. 41, 301-303 (1986)

Philipson, W. (ed.): The Manual of Photographic Interpretation. Bethesda, MD: American Society of Photogrammetry and Remote Sensing (1997)

Pickup, G., Chewings. V.H.: Forecasting patterns of soil erosion in arid lands from Landsat MSS data. Internat. J. Remote Sens. 9, 69-84 (1988)

Pimentel, D., Harvey, C., Resosudarmo, P., Sinclair, K., Kurz, D., McNair, M., Crist, S., Shipritz, L., Fitton, L., Saffouri, R., Blair, R.: Environmental and economic cost of soil erosion and conservation benefit. Sci. 267, 1117-1123 (1995)

Price, K.P.: Detection of soil erosion within Pinyon-Juniper woodlands using Thematic Mapper(TM) data. Remote Sens. Environ. 45, 233-248 (1993)

Ripple, W.: The GIS Applications Book: Examples in Natural Resources: A Compendium. Bethesda, MD: American Society of Photogrammetry and Remote Sensing (1994)

Renard, K.G., Foster, G.R., Weesies, G.A., Porter, J.P.: RUSLE Revised Universal Soil Loss Equation. J. Soil Water Conserv. 46, 30-33 (1991)

Renard, K.G., Foster, G.R., Weesies, G.A., McCool, D.K., Yoder, D.C. Predicting soil erosion by water: a guide to conservation planning with the Revised Universal Soil Loss Equation (RUSLE). Washington, DC, USDA Agriculture Handbook No. 537, (1997)

Ritchie, J.C.: Remote sensing applications to hydrology: airborne laser altimeters. Hydrological Sci. J. 41, 625-636 (1996)

Ritchie, J.C., Grissinger, E.H., Murphey, J.B., Garbrecht, J.D.: Measuring channel and gully cross-sections with an airborne laser altimeter. Hydrological Processes J. 7, 237-244 (1994)

Ritchie, J.C., Jackson, T.J.: Airborne laser measurement of the topography of concentrated flow gullies. Trans. Amer. Soc. Agric. Engr. 32, 645-648 (1989)

Ritchie, J.C., McHenry, J.R.: Application of radioactive fallout cesium-137 for measuring soil erosion and sediment accumulation rates and patterns: a review. J. Environ. Qual. 19, 215-233 (1990)

Robinove, C.J., Chavez, P.S., Gehring, D., Holmgren, R.: Arid land monitoring using Landsat albedo difference images. Remote Sens. Environ. 11, 133-156 (1981)

Roehl, J.E.: Sediment source areas, delivery ratios and influencing morphological factors. Wallingford, England: International Association of Hydrological Sciences, Pub. No. 59, 202-213 (1962)

Samarakoon, L., Ogawa, S., Ebisu, N., Lapitan, R., Kohki, Z.: Inferences of landslide susceptible areas by Landsat Thematic Mapper data. Wallingford, England: International Association of Hydrological Sciences, Pub. No. 217, 83-90 (1993)

Seubert, C.E., Baumgardner, M.F. Weismiller, R.A., Kirschner, F.R.: Mapping and estimating areal extent of severely eroded soil of selected sites in northern Indiana. 1979 Symposium on Machine Processing of Remotely Sensed Data, West Lafayette, Indiana: Purdue University (1979)

Sneddon, J., Lutze, T.A.: Close-range photogrammetic measurement of soil erosion in course-grained soils. Photogrammetric Engr. Remote Sens. 55, 597-600 (1989)

Sombroek, W.: Soil degradation and contamination: A global perspective. Soil and Environ. 5, 3-13 (1995)

Spomer, R.G., Mahurin, R.L.: Time-lapse remote sensing for rapid measurement of changing landforms. J. Soil Water Conserv. 39, 397-401 (1984)

Spomer, R.G., Mahurin, R.L., Piest, R.F.: Soil erosion, deposition and sediment yield from Dry Creek, Basin, Nebraska. Trans. Amer. Soc. Agric. Engr. 29, 489-493 (1986)

Stephens, P.R., Cihlar, J.: Mapping soil erosion in New Zealand and Canada. In: Johannsen, C.J., Sanders, J.L. (eds) Remote sensing for resource management, Ankeny, Iowa: Soil Conservation Society of America, pp. 232-242 (1982)

Stephens, P.R., MacMillian, J.K., Daigle, J.L., Cihlar, J.: Use of sequential aerial photographs to detect and monitor soil management changes affecting cropland soil erosion. J. Soil Water Conserv. 37, 101-105 (1982)

Stephens, P.R., MacMillian, J.K., Daigle, J.L., Cihlar, J.: Estimating Universal Soil Loss Equation factor values with aerial photography. J. Soil Water Conserv. 40, 293-296 (1985)

Thomas, A.W., Welch, R.: Measurement of ephemeral gully soil erosion. Trans. Amer. Soc. Agric. Engr. 31, 1723-1728 (1988)

Walling, D.E.: The sediment delivery problem. J. Hydrol. 65, 209-237 (1983)

Walling, D.E.: Rainfall, runoff, and soil erosion of the land: a global review, In: Gregory, K.J. (ed.), Energetics of the Physical Environment, Chichester, England: J. Wiley and Sons, Ltd., pp. 89-117 (1987)

Walling, D.E.: Sediment yield investigations: A perspective on recent developments and future needs. Proc. Fourth International Symposium on River Sedimentation, Beijing, China (1990)

Welch, R., Jordan, T.R.: Analytical non-metric close-range photogrammetry for monitoring stream channel soil erosion. Photogrammetric Engr. Remote Sens. 49, 367-374 (1983)

Welch, R., Jordan, T.R., Thomas, A.W., Ellis, J.W.: Photogrammetric techniques for monitoring soil erosion. Department of Geography, University of Georgia, Athens, GA. Research Report No. IRC 093083 (1983)

Whiting, M.L., DeGloria, S.D. Benson, A.S., Wall, S. L.: Estimating conservation tillage residue using aerial photography. J. Soil Water Conserv. 42, 130-132 (1987)

Wilson, D.A., McCourt, M.L., Humes, T.M.: Multi-temporal land use analyses for soil erosion in selected Prince Edwards Island (P.E.I.) Watersheds. International Geoscience and Remote Sensing Symposium 1989, pp. 1979-1985 (1989)

Wischmeier, W.H., D.D. Smith, D.D.: Predicting rainfall soil erosion losses. Agr. Handbook No. 537, Washington, DC: USDA (1978)

13 Water Quality

Jerry C. Ritchie[1] and Frank R. Schiebe[2]

[1] USDA-ARS Hydrology Laboratory, BARC-West, Bldg-007, Beltsville, MD 20705 USA
[2] SST Development Group Inc., 824 North Country Club Road, Stillwater, OK 74075 USA

13.1 Introduction

Water quality is a general term used to describe the physical, chemical, thermal, and/or biological properties of water. We often define it in terms of human usage for consumption, recreation, and aesthetics. In broader terms the quality of water affects all components of the aquatic ecosystem. The quality of water suitable and desirable for use by one organism may be completely unsuitable for another. Thus, water quality is a parameter that cannot be defined easily nor can standards be set that meet all uses and user needs. For example, physical, chemical, and biological parameters of water that are suitable for human consumption are quite different from those parameters suitable for a farmer irrigating a crop.

We usually define substances affecting water quality as coming from either point or nonpoint sources. Point sources are associated with substances that can be traced to a single source, such as a pipe or a ditch. Once a point source has been identified, managing the effluent from the point source requires a commitment of time and resources. Nonpoint substances are more diffuse and associated with the landscape and its response to water movement and land use. All human and natural activities contribute nonpoint substances to runoff water thus affecting its quality. For example, organic substances contained in runoff from forested lands can be considered detrimental to the hydrolimitic waters of receiving lakes if an oxygen demand is created that depletes oxygen from the water body that make it unsuitable for fish and other organisms. Agriculture, industrial, and urban areas are all major anthropogenic sources of nonpoint substances. These substances are considered pollutants when they are detrimental to the existing water quality. Determining source areas for nonpoint pollutant is a major concern and management problem. Once identified and defined, controlling nonpoint sources is a major management concern because the diffuse nature of the source makes them difficult to control. Polluting substances that may lead to deterioration of water quality (eutrophication) affects most freshwater and estuarine ecosystems in the world (Dekker et al. 1995). In the United States, off-site downstream deterioration of water quality has been estimated to cost billions of dollars per year. Timing, amount, and sources of pollutant transfer to water bodies are generally not fully understood by water resource managers responsible for developing strategies to control these pollutants. Standards have been established in many countries that recognize the water quality parameters that affect the aquatic food chain and human usage. If standards are met, a healthy aquatic habitat usually results. Reduction in the quality of water in streams, lakes, reservoirs, estuaries, and oceans is a major concern around the world (Brown 1984; Lal 1994). The knowledge base describing the effects of pollutants, the development of management techniques to control pollutants, and the education to transfer this knowledge to the end user continues to expand.

Monitoring and assessing the quality of waters in streams, reservoirs, lakes, estuaries, and oceans are critical for managing and improving the quality of the environment. Classical techniques for measuring indicators of water quality involve *in situ* measurements and/or the collection of water samples for subsequent laboratory analyses. Although these technologies give accurate measurements for a point in time and space, they are time consuming, expensive, and do not give either the spatial or temporal view of water quality needed for accurate monitoring, assessing, or managing water quality for an individual water body or for multiple water bodies across the landscape. Remote sensing of indicators of water quality offers the potential of relatively inexpensive, frequent, and synoptic measurements using sensors aboard aircraft and/or spacecraft (see Colour Plate 13.A).

13.2 Basis for using Remote Sensing

Major factors affecting water quality in fresh waters estuaries and oceans are suspended sediments, turbidity, chlorophylls (algae), chemicals, dissolved organic matter (DOM), nutrients, pesticides, thermal releases, and oils. Suspended sediments (turbidity), chlorophylls, DOM, and oils produce visible and/or thermal changes in surface waters that can change the energy spectra of reflected solar and/or emitted thermal radiation from surface waters. Such changes in the spectral signals from surface waters are measurable by remote sensing techniques from many platforms. Substances can also change the thermal properties of water thus affecting the heat content and thus water temperature that can be measured in surface water temperature remotely with thermal sensors. Most chemicals do not directly affect or change the spectral or thermal properties of surface waters. Measuring water properties affected by chemicals can only be inferred indirectly from remotely sensed measurements of other water quality parameters affected by these chemicals. Measurement of these surrogate properties may then be used in mathematical modelling and analyses to indirectly infer chemicals in water.

The strength of remote sensing techniques lies in their ability to provide both spatial and temporal views of surface water quality parameters that is typically not possible from *in situ* measurements. Remote sensing makes it possible to monitor the landscape effectively and efficiently, identifying water bodies with significant water quality problems. These water quality parameters, often, can be quantified using remote sensing techniques allowing management plans to be formulated to reduce movement of substances from catchments (see Chap. 12) to water bodies thus reducing the effects of the pollutant on water quality.

Development of remote sensing techniques for monitoring water quality began in the early 1970s. These early techniques used the visible and infrared portions of the electromagnetic spectrum to measure spectral differences and thermal infrared to measure emitted energy from water surfaces. Generally, empirical relationships between spectral properties and water quality parameters were established. Morel and Gordon (1980) discussed three general methods to determine relationships between radiance or reflectance and the concentration of constituents in water. These methods are an empirical approach, a semi-empirical approach and an analytical approach. The general form of the empirical and semi-empirical equations is:

$$Y = A + BX \quad \text{or} \quad Y = AB^X$$

Where Y is the measured radiance, reflectance, or energy and X is the water quality parameter of interest (suspended sediment, turbidity, chlorophyll, etc.). A and B are empirically derived factors or empirical factors modified based on a knowledge of the interaction of water quality parameters and optical/thermal properties of water. These relationships are often nonlinear in form.

In the empirical approach a statistical relationship is determined between measured spectral properties and measured water quality parameters. In the semi-empirical approach, information about the spectral/optical characteristic of the water quality parameter is used in statistical analyses to aid in the selection of best wavelength(s) or best model. In both approaches the empirical characteristics of the relationships limit their applications to the condition for which the data were collected. Such empirical models should only be used to estimate water quality parameters for water bodies with similar conditions.

In the analytical approach, optical properties of water and water quality parameters are used to model spectral characteristics of the water being studied. Dekker et al. (1995) proposed the following analytical model derived from the physical relationship between water quality parameters, optical properties, and remote sensing measurements.

$$R = r_i \frac{b}{a+b} = r_i \omega_b$$

Where R is reflectance, r_i is a radiance to reflectance conversion, a is absorption, and b is scattering. ω_b is defined as backscattering albedo. Dekker et al. (1995) found r_i ranged between 0.12 and 0.50 and was apparently specific for each water body.

The presence of substances in surface water can significantly change the backscattering characteristics of surface water (Jerlov 1976, Kirk 1983). Application of remote sensing for measuring water quality parameters depends on the ability to measure these changes in the energy spectral signature backscattered from water in the direction of the sensor. Visible and near-infrared light energy in specific wavelengths can indicate the presence and concentration of substances in surface waters (Schiebe et al. 1992; Gitelson et al. 1994). The optimal band (wavelength) used to measure different water quality parameters is dependent on the substance being measured and the sensor characteristics. With the coming availability of hyperspectral data, one will be able to choose an optimal band (or bands) for each water quality parameter.

Continuing development of new sensors with higher spatial and spectral (hyperspectral) resolution will improve our capability of measuring both the quality and quantity of the radiance thus improving our ability to measure and quantify water parameters. Middleton and Marcell (1983) provided a literature review of early remote sensing studies of water quality. Gordon et al. (1983) discuss applications of these techniques to ocean and estuarine waters. Kirk (1983) discusses the applications of these techniques to freshwater systems. Dekker et al. (1995) have provided a current overview of remote sensing of water quality in freshwater systems for suspended sediments and chlorophyll. Fingas et al. (1996) provided an overview of remote sensing of oils.

13.3 Application

A combination of ground (water) and remote sensing measurements are required to collect the data necessary to develop and calibrate empirical and semi-empirical models and to validate more physically based models. Water samples analyzed for the substance of interest (i.e., suspended sediment, chlorophyll) should be collected at the same time (or on the same day) that the remote sensing data is collected. Water systems are very dynamic and rapidly change so that the substances in the water are continuously changing. Certain water properties can be measured *in situ* while for other properties, water samples have to be collected and analyzed in a laboratory using standard techniques.

Location of sample sites should be determined with GPS (or other available technique) so that the correct data (pixel information) can be extracted from the remote sensing data for comparison. Often 3 or 5-pixel arrays are averaged to obtain the remote sensing data to account for the dynamic nature of the water body. Remote sensing data must be converted to radiance or reflectance data (Ritchie et al. 1988) if the algorithms developed are to be applicable to other conditions. Data from satellite sensors and high altitude aircraft sensors should be corrected for atmospheric interference. Several atmospheric correction models are available to make corrections for atmospheric interference. Many water quality studies have used the dark pixel technique to correct for atmospheric interference on the assumption that the dark pixel value in a scene is due to atmospheric interference. Care should be used in the application of the dark pixel technique in water quality studies since most often the dark pixel is from a water body thus leading to overcorrection since all waters have some reflectance. Once a data set is collected, empirical algorithms can be developed by applying standard regression techniques to the data or the data can be used to validate analytically based models.

Since the radiance/reflectance measured is predominately from the surface water, the algorithms developed are good for estimating substances in the surface water. Information about total load or depth distribution can only be derived by modelling (knowing) the relationship between surface substance distribution and total distribution of the substance in the water column.

13.4 Case Studies

Remote sensing applications to water quality are limited to measuring those substances or conditions that influence and change optical and/or thermal characteristics of the apparent surface water properties. Suspended sediments, chlorophylls (algae), DOM (humus), oil, and temperature are water quality indicators that can change the spectral and thermal properties of surface waters and are most readily measured by remote sensing techniques. Substances (i.e., chemical) that do not change the optical and/or thermal characteristics of surface waters can only be inferred by modelling using other surrogate properties (i.e., suspended sediments, chlorophylls) which may have responded to an input or reduction of chemicals. Examples of applications of remote sensing for measuring suspended sediments, chlorophylls, oils, and temperature will be given.

13.4.1 Suspended Sediments

Suspended sediments are the most common pollutant both in weight and volume in surface waters of freshwater systems (Lal, 1994). Turbidity is another term that is often used to describe water quality. Turbidity is measured by optical methods that are often difficult to quantify accurately in terms of weight or volume of substances present while suspended sediments are measured in physical terms to provide useful weight and volume measurements for management purposes. Secchi measurements also measure water clarity but as with turbidity measurements they are difficult to quantify in terms of substances present. Since neither turbidity nor secchi measurement can be accurately quantified, we have chosen to use suspended sediments in our discussion. However, the concepts for remote sensing measurements of turbidity or secchi depths would be similar. Suspended sediments increase the radiance emergent from surface waters (Fig. 13.1) in the visible and near infrared proportion of the electromagnetic spectrum (Ritchie et al. 1976). *In situ* and controlled laboratory measurements have shown that surface water radiance is affected by sediment type, texture, and color (Holyer 1978; Novo et al. 1989a; Han and Rundquist 1996), sensor view and sun angles (Ritchie et al. 1975; Novo et al. 1989b; Ferrier 1995), and water depth (Mantovani and Cabral 1992).

Airborne platforms (Hilton 1984; Dekker et al. 1992, 1995) using photography (Klooster and Schertz 1974), line scanners (Gitelson et al. 1991; Dekker et al. 1992), multispectral scanners (Dekker et al. 1992) and video (Mausel et al. 1991; Roberts et al. 1995) have all been used to study suspended sediment patterns. Since the mid 1980s remote sensing studies of suspended sediments have been made using data from satellite platforms such as Landsat (Kritikos 1974; Carpenter and Carpenter 1983; Khorram 1985; Ritchie et al. 1990; Harrington et al. 1992), SPOT (Lathrop and Lillesand 1989; Froidefond et al. 1993), IRS (Choubey and Subramanian 1992), AVHRR (Strumpf and Pennock 1991; Froidefond et al. 1993), and CZCS (Coastal Zone Color Scanner) (Amos and Toplis 1985; Mayo et al. 1993).

These studies have shown significant relationships between suspended sediments and radiance or reflectance from spectral wave bands or combinations of wave bands on satellite and aircraft sensors. Ritchie et al. (1976) using *in situ* studies concluded

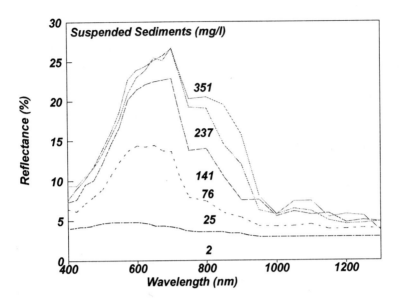

Fig. 13.1. The relationship between reflectance and wavelength as affected by the concentration of suspended sediments. Data were collected from a boat using a high resolution spectroradiometer mounted 50 cm above a lake surface. The 2 mg/l suspended sediments represent the minimum suspended sediment concentration measured during a three-year field study (Ritchie et al. 1976)

that wavelengths between 700 and 800 nm were most useful for determining suspended sediments in surface waters. However, Curran and Novo (1988) in a review of remote sensing of suspended sediments found that many wavelengths have been used in different studies but they did not identify an optimum wavelength for suspended sediments since they found that the optimum wavelength was related to suspended sediment concentration. Many studies have developed statistical relationships (algorithms) between the concentration of suspended sediments and radiance or reflectance for a specific date and site. Few studies have taken the next step and used these algorithms to estimate suspended sediments for another time or place (Whitlock et al. 1982, Curran and Novo 1988; Ritchie and Cooper 1988, 1991). Once developed, an algorithm should be applicable until some catchment event changes the quality (size, color, mineralogy, etc.) of suspended sediments delivered to the lake.

Most research that had a large range (i.e., 0-200+ mg/l) of suspended sediment concentration has found a curvilinear relationship between suspended sediments and radiance or reflectance (Ritchie et al. 1976, 1990; Khorram 1985; Curran and Novo 1988) because the amount of reflected radiance tends to saturate as suspended sediment concentrations increases. The point of saturation is wavelength dependent (see Fig. 13.2) with the shorter wavelength saturating at low concentrations. If the range of suspended sediments is between 0 and 50 mg/l, reflectance from almost any wavelength will be significantly related to suspended sediment concentrations. As the range of suspended sediments increases to 200 mg/l or higher, curvilinear relation-

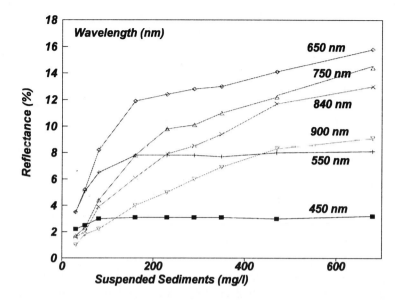

Fig. 13.2. Relationship between reflectance and suspended sediments by wavelength showing non linearity of the relationship between suspended sediments and reflectance. Adapted from Fig. 13.1 and Whitlock et al. (1981)

ships have to be developed with reflectance in the longer wavelength A physically based reflectance model with statistically determined coefficients (B_i and S_i) was successfully developed based on these data and used in the Lake Chicot project to estimate suspended sediment concentrations (Schiebe et al. 1992; Harrington et al. 1992). Their model has the form:

$$R_i = B_i \left(1 - e^{-\frac{c}{S_i}}\right)$$

Where R_i is the Landsat reflectance in wave band i, c is suspended sediment concentration, B_i represents the reflectance saturation level at high suspended sediment concentrations in wave band i, and S_i is the concentration parameter equal to the concentration when reflectance is 63 percent of saturation in wave band i. Most researchers have concluded that surface suspended sediments can be mapped and monitored in large water bodies using sensors available on current satellites. Colour Plate 13.B shows an example of the type of mapping done for turbidity.

Major concerns with this technique are usually associated with the spatial resolution of current satellite data (Roberts et al. 1995) which does not allow the detail mapping of water bodies or measurements from streams and other input regions needed for management. Detailed information on the applications, methodologies, and uses of remote sensing technology to study suspended sediments is given in reviews by Curran and Novo (1988) and Dekker et al. (1995).

13.4.2 Chlorophyll

Lakes and other water bodies depend upon their catchments for nutrients and other substances to sustain biological activities. While these nutrients and substances are required for a healthy aquatic environment, an excess of these inputs leads to nutrient enrichment and eutrophication or „aging" of the lake. The rate of eutrophication depends on topography, soils, land use, and runoff on the contributing catchment. Eutrophication of a water body can be quantified in terms of trophic level or concentration of the chlorophyll contained in the algal/plankton cells. While aging of water bodies is a normal process, better information on eutrophication (chlorophyll content) in a lake will allow better management plans to be developed to control the source in nutrient input from the catchment and thus control the rate of "aging" in lakes.

Monitoring the trophic level or concentration of the chlorophyll (algal/phytoplankton populations) is key to managing eutrophication in lakes. The algal population is a water quality parameter that if excessive can be a problem. Algal concentrations can be monitored by collecting samples, extracting chlorophyll, and measuring concentrations in the extracts by photometric techniques in the laboratory. Remote sensing has also been used to measure chlorophyll concentrations and patterns. As with suspended sediment measurements, most remote sensing studies of chlorophyll in water are based on empirical relationships between radiance/reflectance in narrow bands or band ratios and chlorophyll. Thus field data must be collected to calibrate the statistical relationship or to validate models developed. Measurements (Fig. 13.3) have been made *in situ* (Quibell 1992; Han et al. 1994; Rundquist et al. 1996) and from aircraft (Dekker et al. 1992; Gitelson et al. 1994; Harding et al. 1995), Landsat and SPOT (Carpenter and Carpenter 1983; Lathrop and Lillesand 1989; Strumpf and Tyler 1988; Dekker and Peters 1993) and CZCS (Hovis 1981; Gordon et al. 1983). These studies have used a variety of algorithms and wavelengths to successfully map chlorophyll concentrations of the oceans, estuaries and fresh waters. For example, Harding et al. (1995) used the following algorithm based on aircraft measurements to determine seasonal chlorophyll-a content in the Chesapeake Bay.

$$\mathrm{Log}_{10}[\text{Chlorophyll}] = a + b\,(-\mathrm{Log}_{10} G)$$

Where a and b are empirical constants derived from *in situ* measurements and G is $[(R_2)^2/(R_1 * R_3)]$. R_1 is radiance at 460 nm, R_2 is radiance at 490 nm and R_3 is radiance at 520 nm. Using this algorithm Harding et al. (1995) mapped total chlorophyll content in the Chesapeake Bay (see Colour Plate 13.C).

While measuring chlorophyll by remote sensing technique is possible, studies have also shown that the broad wavelength spectral data available on current satellites do not permit discrimination between chlorophyll and suspended sediments (Dekker and Peters 1993; Ritchie et al. 1994) due to the dominance of the spectral signal from suspended sediment. Some recent research has concentrated on the relationship

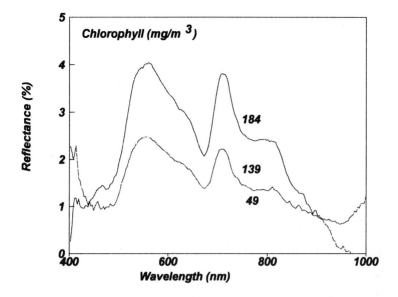

Fig. 13.3. Relationship between reflectance and wavelength for different chlorophyll concentrations. Based on measurement made *in situ* with a high spectral resolution (1 nm) spectroradiometer at 1 meter over a large tank (Schalles et al. 1997)

between chlorophyll-a and the narrow band spectral details at the „red edge" of the visible spectrum (Gitelson et al. 1994). Data has shown a linear relationship between chlorophyll-a and the difference between the emergent energy in the primarily algal scattering range (700-705 nm) and the primarily chlorophyll-a absorption range (675-680 nm). The relationship exists even in the presence of high suspended sediment concentrations that can dominate the remainder of the spectrum as seen in Fig. 13.4. These discoveries suggest new approaches for application of airborne and spaceborne sensors to exploit these phenomena to estimate chlorophyll in surface waters under all conditions as new hyperspectral sensors are launched and data become available. Data from several recently launched satellites sensors (i.e., SeaWiFS, Modular Optical Scanner (MOS), Ocean Color and Temperature Scanner (OCTS)) are now becoming available and hold great promise for measuring biological productivity (chlorophyll) in aquatic systems.

Hyperspectral and fluorescence data may also make it possible to differentiate between phytoplankton groups. Laboratory and field studies using hyperspectral data have been used to develop algorithms to estimate green and blue-green algae (Dekker et al. 1995). Hyperspectral data will probably allow better discrimination between pigments thus allowing the identification of broad algal groups. Fluorescence has also been used to identify algal and phytoplankton groups. Bazzani and Cecchi (1995) were able to identify phytoplankton species from fluorescence spectra using an excitation wavelength of 514 nm (Fig. 13.5).

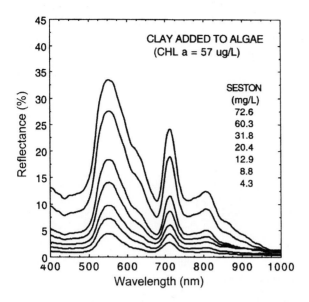

Fig. 13.4. Relative contributions of chlorophyll and suspended sediment spectra to a reflectance spectra of surface water. Based on made *in situ* laboratory measurements made 1 m above the water surface by Schalles et al. (1997)

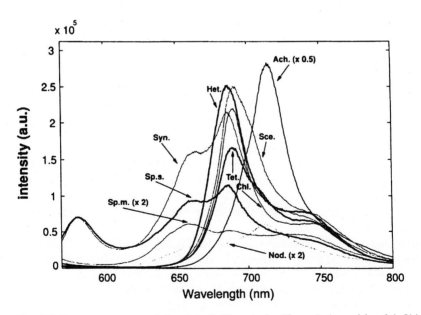

Fig. 13.5. Flourescence spectra of nine phytoplankton samples. The excitation walelength is 514.5 nm. Abbreviations on Fig., e.g., Ach., Nod etc. designate different phytoplankton species. The reader is referred to the original publication (Bazzani and Cecchi 1995; Reproduced with permission of EARSeL Publishers, Paris)

13.4.3 Temperature

From a water quality standpoint, thermal enrichment is defined as increasing the temperature of a water body by anthropogenic activity. Thermal pollution exists when biological activity in the water body is affected. The most common source of thermal pollution is cooling water releases from electrical power plants. Remotely sensed data have been used to map thermal discharge into streams, lakes and coastal waters from electrical power plants (Stefan and Schiebe 1970; Gibbons et al. 1989). Mapping thermal plumes in river and coastal waters has provided accurate estimates of absolute temperature and provided spatial and temporal patterns of thermal releases (see Fig. 13.6). Mapping thermal enrichment in streams, lakes, reservoirs, and coastal waters has been useful in managing thermal releases from electrical power plants. Management of power plant releases can be designed to reduce the impact of thermal releases based on thermal patterns determined for different flow regimes from remotely sensed data. Improved quantitative estimates of surface water temperatures also provide input for interpreting outputs from mathematical models of thermal plumes. Aircraft mounted, thermal sensors are especially useful in studies of thermal plumes because of the ability to control the timing of data collection that is critical for studying thermal releases.

Other anthropogenic activity can also lead to decreases in surface water temperature. Promotion of tree growth in riparian zones along streams can intercept solar energy preventing it from being absorbed by surface waters. One effect of suspended sediments in water is to backscatter significant quantities of solar energy that would otherwise be absorbed in a clear lake. This process, in effect, rejects solar energy and

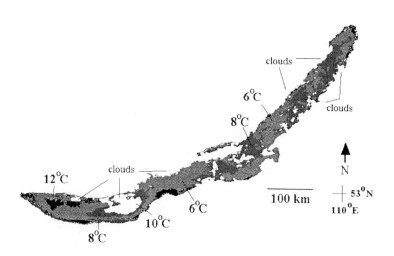

Fig. 13.6. Surface temperatures of Lake Baikai derived from AVHRR band 4 data for September 4, 1991 (Bolgrien et al. 1995; Reproduced with permission of Americal Society of Photogrammetry and Remote Sensing, Bethesda, MD)

results in turbid impoundments being cooler than its clear water neighbor (Schiebe et al. 1976).

Seasonal changes in the temperature of surface waters can be expected. Such seasonal changes of sea surface temperatures have been routinely monitored using AVHRR and other satellite platforms (Njoku and Brown 1993). Large scale remotely sensed mapping of sea surface temperatures has led to new insights into the role which oceans play in regulating weather and climate (i.e., El Nino). Such measurements also provide a basis for explaining biological activity in ocean and large freshwater systems. Bolgrien et al (1995) used AVHRR to monitor seasonal temperature in Lake Baikai (Fig. 13.6).

Miller and Millis (1989) used Heat Capacity Mapping Mission (HCMM) and Landsat TM to estimate surface water temperatures for the Great Salt Lake in Utah. They used statistical correlations to determine the relationship between surface temperature and evaporation. They concluded that satellite-derived surface water temperature along with some ancillary data could be use to estimate evaporation rates. Miller and Rango (1984) used emitted thermal energy measured by HCMM to map algal concentration in the Great Salt Lake. They found a positive correlation during the day and a negative correlation at night between emitted energy and algal concentration. Landsat TM data was used to estimate surface temperatures of an oxbow lake along the Mississippi River (Ritchie et al. 1990).

Thermal remote sensing is a useful tool for monitoring freshwater systems to detect thermal changes that can affect biological productivity. These techniques allow the development of management plans to reduce the effect of man-made thermal releases.

13.4.4 Oils

Oil spills are a common occurrence in the aquatic environment that require significant time and funds to clean up. Public concerns about these spills require that they be monitored and cleaned up as quickly and efficiently as possible. Remote sensing can play an important role in developing strategies for monitoring and cleaning up oil spills. A wide range of sensors has been used to remotely sense oils. Sensors on aircraft are commonly used to monitor oil spills to get the spatial and temporal resolution needed to monitor oil spill patterns in a timely manner. While large oil spills have been monitored with satellite data (Fingas et al. 1996), frequency of overpasses and the limited spatial and spectral resolution of sensors on current satellites limit the use of data from satellite sources. Fingas et al. (1996) published a series of papers comparing different sensors and remote sensing techniques for oils. Therefore, only an overview will be given here of commonly used sensors.

Optical sensors using cameras, scanners, and video on aircraft are the most common techniques used in monitoring oil spills. Oils increase the reflectance of surface waters in the visible and near infrared spectrum. The increase in reflectance is general across the spectrum with no single spectral feature to distinguish oil from the background (Taylor 1992). The use of visible spectrums is related to human pattern recognition rather than to automated detection by spectral algorithms. Visible techniques are used most often to document patterns because of lack of algorithms to quantify oil levels

(Fingas et al. 1996). Stringer et al. (1992) used Landsat TM data to monitor the Exxon Valdez oil spillage (see Colour Plate 13.D). Visible techniques are used because of cost and availability of aircraft to make critical measurements when and where needed.

Oils can be detected by measuring thermal energy in the 8 to 14 μm region of the spectrum. Thick oils have higher temperatures with temperatures decreasing as the thickness of the oil decreases. The minimum thickness that affects temperatures is between 10 to 70 microns (Fingas et al. 1996). Spatial resolution of the sensor becomes important for oil spills distributed in patches and wind rows (Hover 1994). Measurement of relative thickness of oil layers is important for determining the type of equipment that will be used for cleanups. Infrared (thermal) remote sensing is probably the most important tool currently used for planning oil spill cleanups.

Thin layers of oil displays high ultraviolet radiation patterns (O'Neil et al. 1983) thus making ultraviolet radiation data useful to monitor patterns of oils on water surfaces. While measurements of ultraviolet radiation are very sensitive to oil in the environment, they are not often used operationally due to the many interferences encountered from ultraviolet radiation of natural products which make interpretations difficult (Fingas et al. 1996).

Fluorosensors are useful for detecting the presence of oil in aquatic systems. While other substances (i.e., chlorophyll) fluoresce, oils fluoresce with unique spectra that allow them to be detected easily (Hengstermann and Reuter 1990). Oils from different sources have unique fluorescent signatures that allow them to be differentiated and identified (Hengstermann and Reuter 1990). Fluorosensors have a useful role in oil measurement, monitoring, and especially for identifying possible sources of the oil.

Radar can detect oils on sea surfaces because oil reduces the „sea clutter" of normal radar images of the sea surface. Many naturally occurring substances can also reduce "sea clutter" (Frysinger et al. 1992) making interpretation of radar images difficult. Even with these limitations, radar is widely used for oil spill detection and monitoring because of its ability to collect data under all weather conditions, day or night, and over large areas rapidly.

Oil is a much stronger emitter of microwave energy than water so that images of oil spill areas appear much brighter on a microwave image of water (Ulaby et al. 1986). A passive microwave sensor can detect these differences and thus patterns of oil spills can be detected although spatial resolution is usually poor. As with radar, microwave sensors have the ability to collect data under all weather conditions, day or night, and over large areas rapidly.

13.5 Future Directions

Limitations in spectral and spatial resolution of current sensors on satellites currently restrict the wide use of satellite data for monitoring water quality. New satellites (SEAWIFS, MOS, OCTS, QuickBird, Resource21, OrbView, etc.) and sensors (hyperspectral, high spatial resolution) already launched or planned to be launched over the next decade (Corbley 1996) should provide both the improved spectral and spatial resolution needed to monitor water quality parameters in surface waters in fresh water lakes and streams, estuaries, and oceans and to differentiate between different water quality parameters. Research needs to focus on understanding the relationship between water quality parameters and their effects on optical and thermal properties of surface waters so that physically based models can be developed (Jerome et al. 1996). Hyperspectral data from the new satellite platforms will allow us to discriminate between water quality parameters and to develop a better understanding of light/water/substance interactions. Such information should allow us to move away from empirical approaches now being used and develop algorithms that will allow us to use the full resolution electromagnetic spectrum to monitor water quality parameters. Research using derivative spectra and similar techniques to analyze spectral data from estimating water quality properties from high spectral resolution sensors has potential for helping us understand these interactions (Han and Rundquist 1996; Malthus and Dekker 1995).

Future uses of remote sensing to monitor water quality will improve with the launch of new satellite sensors. These sensors with higher spatial and spectral resolution should improve our ability to measure understand changes in water quality, thus allowing us to develop management plans to improve the quality of surface waters.

References

Amos, C.L., Toplis, B.J.: Discrimination of suspended particulate matter in the Bay of Fundy using Nimbus 7 Coastal Zone Scanner. Can. J. Remote Sens. 11, 85-92 (1985)

Bazzani, M., Cecchi, G.: Algae and mucillagine monitoring by fluorescence lidar experiments in field. EARSeL Advances in Remote Sensing 3, 90-101 (1995)

Bolgrien, D.W., Granin, N.G., Levin, L.: Surface temperature dynamics of Lake Baikal observed from AVHRR images. Photogrammetric Engr. Remote Sens. 61, 211-216 (1995)

Brown, L.R.: The global loss of topsoil. J. Soil Water Conserv. 39, 162-165 (1984)

Carpenter, D.S., Carpenter, S.M.: Monitoring inland water quality using Landsat data. Remote Sens. Environ. 13, 345-352 (1983)

Choubey, V.K., Subramanian, V.: Estimation of suspended solids using Indian Remote Sensing Satellite-1A data: a case study from central India. Internat. J. Remote Sens. 13, 1473-1486 (1992)

Corbley, K.: Remote sensing skies filling with satellite plans. EOM The Magazine for Geographic, Mapping and Earth Information 5, 26-28 (1996)

Curran, P.J., Dungan, J.L., Macler, B.A., Plummer, S.E.: The effect of red leaf pigment on the relationship between red edge and chlorophyll concentration. Remote Sens. Environ. 35, 69-76 (1991)

Curran, P.J., Novo, E.M.M.: The relationship between suspended sediment concentration and remotely sensed spectral radiance: A review. J. Coastal Res. 4, 351-368 (1988)

Dekker, A.G., Malthus, T.J., Hoogenboom, H.J.: The remote sensing of inland water quality. In: Danson, F.M., Plummer, S.E. (Eds.) Advances in Remote Sensing. Chichester: John Wiley and Sons (1995) pp. 123-142

Dekker, A.G., Malthus, T.J., Wijnen, M.W., Seyhan, E.: The effect of spectral bandwidth and positioning on the spectral signature analysis on inland water. Remote Sens. Environ. 41, 211-225 (1992)

Dekker, A.G., Peters, S.W.M.: The use of the Thematic Mapper for the analysis of eutrophic lakes: a case study in the Netherlands. Internat. J. Remote Sens. 14, 779-821 (1993)

Ferrier, G.: A field study of the variability in suspended sediment concentration-reflectance relationship. Internat. J. Remote Sens. 16, 2713-2720 (1995)

Fingas, M., Brown, C., Fruhwirth, M.: An assessment of sensors for oil spill applications. Proceedings Second International Airborne Remote Sensing Conference and Exposition, pp. III689-698, ERIM, Ann Arbor, MI, USA (1996)

Froidefond, J.M., Castaing, P., Jouanneau, J.M., Prud'homme, R., Dinet, A.: Method for the quantification of suspended sediments from AVHRR NOAA-11 satellite data. Internat. J. Remote Sens. 14, 885-894 (1993)

Frysinger, M., Asher, W.E., Korenowski, G.M., Barger, W.R., Klusty, M.A., Frew, N.M., Nelson, R.K.: Study of ocean slicks by non linear laser processes in second-harmonic generation. J. Geophys. Res. 97, 5253-5269 (1992)

Gibbons, D.E., Wukelic, G.E., Leighton, J.P., Doyle, M.J.: Application of Landsat Thematic Mapper data for coastal thermal plume analysis at Diablo Canyon. Photogrammetric Engr. Remote Sens. 55, 903-909 (1989)

Gitelson, A.A., Garbusov, G.P., Lopatshenko, L.L., Mittenzwey, K.H., Makhotenko, A.N., Mudrogelenko, I.V., Penig, J., Sukhorukov, B.L.: Rapid methods for the determination of the quality of surface water by means of spectrometry. Acta Hydrochimica et Hydrobiologica, 18, 397-408 (1991)

Gitelson, A., Mayo, M. Yacobi, Y.Z., Paroarov, A., Berman, T.: The use of high spectral resolution radiometer data for detection of low chlorophyll concentrations in Lake Kinneret. J. Plankton Res. 16, 993-1002 (1994)

Gordon, H.G., Clark, D.K., Brown, O.B., Evans, R.H., Broenkow, W.W.: Phytoplankton pigment concentrations, in the Middle Atlantic Bight: a comparison of ship determinations and CZCS estimates. Appl. Optics 22, 20-35 (1983)

Han, L., Rundquist, D.C.: Spectral characterization of suspended sediments generated from two texture classes of clay soil. Internat. J. Remote Sens. 17, 643-649 (1996)

Han, L., Rundquist, D.C., Liu, L.L., Fraser, R.N., Schalles, J.F.: The spectral responses of algal chlorophyll in water with varying levels of suspended sediment. Intern. J. Remote Sens. 15, 3707-3718 (1994)

Harding, L.W., Itsweire, E.C., Esaias, W.E.: Algorithm development for recovering chlorophyll concentrations int Chesapeake Bay using aircraft remote sensing, 1989-91. Photogrammetric Engr. Remote Sens. 61, 177-185 (1995)

Harrington, Jr., J.A., Schiebe, F.R., Nix. J.F.: Remote sensing of Lake Chicot, Arkansas: Monitoring suspended sediments, turbidity and secchi depth with Landsat MSS. Remote Sens. Environ. 39, 15-27 (1992)

Hengstermann, T., Reuter, R.: Lidar fluorosensing of mineral oil spill on the sea surface. Appl. Optics 29, 3218-3227 (1990)

Hilton, J.: Airborne remote sensing for freshwater and estuarine monitoring. Water Resour. 18, 1195-1123 (1984)

Holyer, R.J.: Toward universal multispectral suspended sediment algorithms. Remote Sens. Environ. 7, 323-338 (1978)

Hover, G.L.: Testing of infrared sensors for the U.S. Coast Guard oil spill response applications. Proceeding Second Thematic Conference of Remote Sensing for Marine and Coastal Environments, pp. I47-58, ERIM, Ann Arbor, MI, USA (1994)

Hovis, W.A.: The NIMBUS 7 coastal zone scanner. IN Gower, J.F.R. (Ed.) Oceanography from Space, New York: Plenum Press (1981) pp. 213-226

Jerlov, N.G.: Marine optics. Amsterdam: Elsevier (1976)

Jerome, J.H., Bukata, R.P., Miller, J.R.: Remote sensing reflectance and its relationship to optical properties of natural water. Int. J. Remote Sens. 17, 3135-3155 (1996)

Khorram, S.: Development of water quality models applicable throughout the entire San Francisco Bay and delta. Photogrammetric Engr. Remote Sens. 51, 53-62 (1985)

Kirk, J.T.O.: Light and photosynthesis in aquatic ecosystems. Cambridge: Cambridge University Press (1983)

Klooster, S.A., Scherz, J.P.: Water quality by photographic analysis. Photogrammetric Engr. Remote Sens. 40, 927-935 (1974)

Kritikos, H., Yorinks, L., Smith, H.: Suspended solids analysis using ERTS-A data. Remote Sens. Environ. 3, 69-78 (1974)

Lal, R.: Soil Erosion, Ankeny, Iowa: Soil and Water Conservation Society (1994)

Lathrop, Jr., R.G., Lillesand T.M.: Monitoring water quality and river plume transport in Green Bay, Lake Michigan with SPOT-1 imagery. Photogrammetric Engr. Remote Sens. 55, 349-354 (1989)

Malthus, T.J., Dekker, A.G.: First derivative indices for the remote sensing of inland water quality using high spectral resolution reflectance. Environ. Internat. 21, 221-232 (1995)

Mantovani, J.E., Cabral, A.P.: Tank depth determination for water radiometric measurements. Internat. J. Remote Sens. 13, 2727-2733 (1992)

Mausel, P.W., Karaska, M.A., Mao, C.Y., Escobar, D.E., Everitt, J.H.: Insights into secchi transparency through computer analysis of aerial multispectral video data. Internat. J. Remote Sens. 12, 2485-2492 (1991)

Mayo, M., Karnieli, A., Gitelson, A., Ben-Avraham, Z.: Determination of suspended sediment concentrations from CZCS data. Photogrammetric Engr. Remote Sens. 59, 1265-1269 (1993)

Middleton, E.M., Marcell, R.F.: Literature relevant to remote sensing of water quality. NASA Technical Memorandum 85077, Greenbelt, MD: NASA Goddard Space Flight Center (1983)

Miller, W., Millis, E.: Estimating evaporation from Utah's Great Salt Lake using thermal infrared satellite imagery. Water Resour. Bull. 25, 541-550 (1989)

Miller, W., Rango, A.: Using Heat Capacity Mapping Mission (HCMM) data to assess water quality. Water Resour. Bull. 20, 493-501 (1984)

Morel, A., Gordon, H.R.: Report of the working group on water color. Boundary Layer Meteorol. 18, 343-355 (1980)

Novo, E.M.M., Hansom, J.D., Curran, P.J.: The effect of sediment type on the relationship between reflectance and suspended sediment concentration. Internat. J. Remote Sens. 10, 1283-1289 (1989a)

Novo, E.M.M., Hansom, J.D., P.J. Curran, P.J.: The effect of viewing geometry and wavelength on the relationship between reflectance and suspended sediment concentration. Internat. J. Remote Sens. 10, 1357-1372 (1989b)

Njoku, E.G., Brown, O.B.: Sea surface temperature. In: Gurney, R.J., Foster, J.L., Parkinson, C.L. (Eds.) Atlas of satellite observations related to global change. Cambridge: Cambridge University Press (1993) pp. 237-249

O'Neil, R.A., Neville, R.A., Thompson, V.: The Arctic Marine Oilspill Program (AMOP) remote sensing study. Environment Canada Report No. EPS 4-EC-83-3, 257 pp, Ottawa, Canada (1983)

Quibell, G.: Estimating chlorophyll concentrations using upwelling radiance from different freshwater algal genera. Intern. J. Remote Sens. 13, 2611-2621 (1992)

Ritchie, J.C., Cooper, C.M.: An algorithm for using Landsat MSS for estimating surface suspended sediments. Water Resour. Bull. 27, 373-379 (1991)

Ritchie, J.C., Cooper, C.M.: Comparison on measured suspended sediment concentrations with suspended sediment concentrations estimated from Landsat MSS data. Internat. J. Remote Sens. 9, 379-387 (1988)

Ritchie, J.C., Cooper, C.M. Schiebe, F.R.: The relationship of MSS and TM digital data with suspended sediments, chlorophyll, and temperature in Moon lake, Mississippi. Remote Sens. Environ. 33, 137-148 (1990)

Ritchie, J.C., Schiebe, F.R., Cooper, C.M., Harrington, Jr., J.A.: Chlorophyll measurements in the presence of suspended sediment using broad band spectral sensors aboard satellites. J. Freshwater Ecol. 9, 197-206 (1994)

Ritchie, J.C., Schiebe, F.R., McHenry, J.R.: Remote sensing of suspended sediment in surface water. Photogrammetric Engr. Remote Sens. 42, 1539-1545 (1976)

Ritchie, J.C., Schiebe, F.R., McHenry, J.R., Wilson, R.B., J. May, J.: Sun angle, reflected solar radiation and suspended sediments in north Mississippi reservoirs. In: F. Shahrokhi, F. (ed.), Remote Sensing of Earth Resources Vol. IV., Tullahoma, TN: University of Tennessee Space Institute, pp. 555-564 (1975)

Roberts, A., Kirman, C., Lesack, L.: Suspended sediment concentration estimation from multi-spectral video imagery. Internat. J. Remote Sens. 16, 2439-2455 (1995)

Rundquist, D.C., Han, L., Schalles, J.F., Peake, J.S.: Remote measurement of algal chlorophyll in surface waters: The case for the first derivative of reflectance near 690 nm. Photogrammetric Engr. Remote Sens. 62, 195-200 (1996)

Schalles, J.F., Schiebe. F.R., Starks, P.J., Troeger, W.W.: Estimation of algal and suspended sediment loads (singly and combined) using hyperspectral sensors and integrated mesocosm experiments. Proceedings of the 4th International Conference on Remote Sensing for Marine and Coastal Environments Vol. 1, pp. 247-248 (1997)

Schiebe, F.R., Harrington, J.A., Ritchie, J.C.: Remote sensing of suspended sediments: The Lake Chicot, Arkansas project. Internat. J. Remote Sens. 13, 1487-1509 (1992)

Schiebe, F.R., Ritchie, J.C., McHenry, J.R.: Influence of suspended sediments on the temperature of surface waters in reservoirs. Verh. Internat. Ver. Limnol. 19, 133-136 (1976)

Stringer, W.J., Dean, K.G., Guritz, R.M., Garbeil, H.M., Groves, J.E., Ahlnaes, K.: Detection of petroleum spilled from the MV *Exxon Valdez*. In Int. J. Remote Sens. 15, 799-824 (1992)

Stefan, H., Schiebe, F.R.: Heated discharges from flumes into tanks. J. Sanitary Engr. Div. ASCE, 1415-1433 (1970)

Strumpf, R.P., Pennock, J.R.: Remote estimation of the diffuse attenuation coefficient in a moderately turbid estuary. Remote Sens. Environ. 35, 153-191 (1991)

Strumpf, R.P., Tyler, M.A.: Satellite detection of bloom and pigment distributions in estuaries. Remote Sens. Environ. 24, 385-404 (1988)

Sudhakar, S., Pal, D.K.: Water quality assessment of Lake Chilka. Int. J. Remote Sens. 14, 2575-2579 (1993)

Taylor, S.: 0.45 to 1.1 µm spectra of Prudhoe crude oil and of beach materials in Prince Williams Sound, Alaska. CRREL Special Report No. 92-5, 14pp., Cold Regions Research and Engineering Laboratory, Hanover, NH, USA (1992)

Ulaby, F.T., Moore, R.K., Fung, A.K.: Microwave Remote Sensing: Active and Passive (3 vols.) Artech House, Inc., Norwood, MA, USA (1986)

Whitlock, C.H., Kuo, C.Y., LeCroy, S.R.: Criteria for the use of regression analysis for remote sensing of sediment and pollutants. Remote Sens. Environ. 12, 151-168 (1982)

14 Groundwater

Allard M.J. Meijerink
International Institute for Aerospace Surveys and Earth Sciences (ITC)
PO.Box 6, Boulevard 1945, 7500 AA Enschede, The Netherlands

14.1 Introduction

Groundwater is essentially a subsurface phenomenon. The common current remote sensing platforms record features on the surface. Most of the information for groundwater, as yet, has to be obtained by qualitative reasoning and semi-quantitative approaches. The remotely sensed information is often of surrogate nature and has to be merged with geohydrologic data to become meaningful.

Airborne geophysics, such as aeromagnetic surveys, can provide subsurface information (Reeves 1992) but the data has to be interpreted because non-unique solutions may be possible. Non-ground based micro-wave sensors have some capability of penetration, but practical results may be obtained only if the conditions are right, i.e. coarse grained materials, simple geology and fairly shallow groundwater table. It is true that in certain areas rather spectacular microwave imagery has been produced, displaying partly buried fossil drainage networks in a desert (McCauley et al.,1982) or fracture patterns on thermal infrared imagery (Warwick et al.,1979). However, such imagery are incidental.

This text is written with practical applications in mind and the use of remote sensing is discussed and illustrated according to a *general method* of assessing regional groundwater resources, which starts with:

- Conceptualization of the hydrogeology, which consists of building-up the three dimensional hydrogeological setting based on surface and subsurface geology, followed by estimation of the regional groundwater surface. After that the conceptual groundwater model can be expanded by the identification of flow systems, in particular intake areas and zones with groundwater discharge.
- Water budget. Once the conceptual model is completed, in the sense that all the available information, geologic, water table and base flow data has been considered, the use of remote sensing for the upper boundary condition is discussed, namely recharge, evapotranspiration loss and estimation of water use for irrigation from groundwater.

Quite some space is devoted to the recharge for two reasons: (1) Remote sensing offers unique possibilities in assessing the spatial patterns, and (2) Groundwater is or is becoming a scarce resource in quantity and quality and this resource can be influenced by the conditions near the surface, in particular by the land cover and land use. Hard rock terrain, such as a basement complex, has its own characteristics and is discussed in a separate section.

14.2 Conceptualization of the hydrogeology

14.2.1 The three dimensional hydrogeologic situation

The 3-D hydrogeologic situation is compiled by merging results of image interpretations and available field data, such as drilling logs, geophysical data, etc. Within the scope of this text it is not possible to treat the interpretation of the abundance of hydrogeologic situations encountered in nature. The reader is referred to the texts on photo-geology (e.g. Ray,1960; Miller and Miller,1961; Sabins,1987; Drury,1993). The examples included here illustrate the principles.

The aerial photo of Fig.14.1 of an semi-arid area shows folded rocks consisting of slate (with cultivated fields) and of quartzitic sandstone/quartzite with thinly interbedded siltstone and slate (outcrop areas with poor shrub and grasses/herbs). The interpreted structure is illustrated by the N-S geologic section. The gentle folds and normal -tensional- faults of the upper part of the photo are flanked by imbricated structures with steep dips and high angle reverse faults in the lower part. On the right hand corner remnants of a gravelly cover (quartzites) resting on a pediment surface can be seen. The vegetation resembles that of the outcrop areas, but the pattern is different and some agricultural fields have been made by clearance of the boulders and pebbles. Minor valleys traverse the area and they have been widened and partly blocked by walls along the fields. The geologic cross-section, as derived by photo-geologic interpretation and a field traverse is shown.

Hydrogeological mapping and updating the Quaternary geology. Various studies have shown how image interpretation can contribute to hydrogeological mapping and qualitative evaluation (e.g. Meijerink,1974; Moore and Deutsch, 1975; Kruck, 1979, Sahai et al., 1985; Dutartre et al., 1990a,b; Waters et al. 1990; Anon. 1990; Li Botao et al.,1990).

The Quaternary is often poorly differentiated on existing geologic maps. This is regrettable because recharge, shallow flow systems and water quality are related to these non- or little-consolidated deposits. Quaternary deposits can often be distinguished and mapped on imagery, especially in dry regions.

In the false colour (T.M) image of Colour Plate 14.A, three Quaternary units can be differentiated; (a) the slightly dissected fanglomerate with lime crusts (generally light tones) where little recharge occurs,(b) the younger gravel beds and splays (dark grey to bronwnish colours) with high recharge and (c) the clayey, lower alluvial fan deposits (blocky red fields and the light yellow colours corresponding to a groundwater fed grassland) with little recharge. It is difficult to map such relevant differences without imagery.

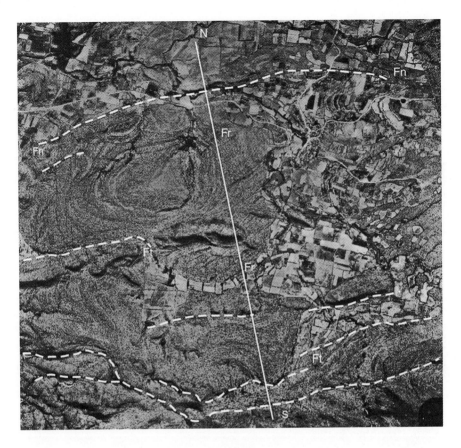

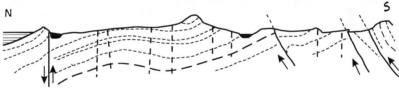

Fig. 14.1. Aerial Photo of a semi-arid area with folded quartzites and slates. Fractures (Fr) and interpreted normal faults (Fn, added in white) are promising for exploration of deeper groundwater. Shallow groundwater in the blocked valley (V) because of good recharge. Geologic cross selection (N-S) is based on photo-geologic interpretation and field check

For the Quaternary volcanic terrain in Java, for example, the landforms and the degree of dissection, which can be interpreted on imagery, relates well to the relative age. The age of the volcanics is associated with the degree of compaction, which in turn influences permeability, base flow and overall recharge. (Anon., 1990, v.d. Sommen et al., 1990).

The image of Fig.14.2 (MSS, band 7) is the near infrared band of Landsat MSS, after filtering (Laplace) for enhancement. The hydrogeologic interpretation of this complex volcanic terrain in West Java has led to an important revision of the hydrogeologic map and to the development of a conceptual hydrogeologic model which is illustrated partly by the cross-section. Fruitful image interpretation often requires local hydrogeological expertise.

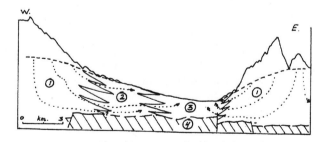

Fig. 14.2. MSS image, near-infrared band, of volcanic terrain, West Java, used for updating the hydrological map and identification of groundwater flow systems. 1. Stratovolcano, intake area; 2. Fluvio-volcanics, lahar deposits & lavas, transient & outflow; 3. Fluvio-volcanics, outflow; 4. Impermeable sedimentary rocks

14.2.2 Groundwater surface

The next important step is the reconstruction of the groundwater table at a selected time of the year, by using a topographic map, head data from wells and information on base flow, in order to differentiate influent or effluent streams with due consequences for the intersection of the groundwater table with the river beds. Head elevations have to be accurate and a common problem is to determine the altitudes if no field levelling can be done. By radar interferometry high accuracy can be obtained if corner reflectors are used. Otherwise, ground conditions (moisture content, vegetation, etc.) should not change between two recordings, a situation not often met in practice.

Image interpretations are used to locate areas where the groundwater table is *probably* shallow. The appearance of groundwater at or near the surface is caused by either the intersection of topographic depressions and a static phreatic groundwater level, or by a discharge zone of an upwelling groundwater flow system. In both cases there will be an effect on vegetation, relative soil moisture, land use (see Fig. 14.1, the fields in the valleys) or on details of the drainage system (see Fig. 14.3b), and these effects can be identified on remotely sensed imagery.

The vegetation can be used as an indicator if local knowledge is available and the types can be identified on the imagery. For example in the Mediterranean region of Europe, the presence of riverine Willows, Poplars and Oleanders, are a certain indicator of shallow groundwater. Phreatophytic vegetation along the wadi channels in the upper catchment of Wadi Nyala, western Sudan, is shown on the image (Thematic Mapper) Fig.14.3a.

Note that on the hard rocks of the interfluves no phreatophytes are present, indicating poor groundwater conditions. Other examples of vegetation indicators, related to groundwater quality are given by Kruck (1990). Terrain features can be indicators for shallow water, such as small swamps and lakes and seepage zones

Fig. 14.3. (**A**, left) Phreatophytes (dark tones) along drainage lines, Wadi Nyala, Sudan, Thematic Mapper. (**B**, right) Groundwater-fed small rivulets (enhanced) on lower part of alluvial fan with ephemeral rivers due to exfiltration (aerial photograph)

(see discussion of Fig.14.4), localized high drainage densities and transformation of river types, such as the change of ephemeral wide sandy river bed, to narrow, sinuous perennial rivulets, Fig.14.3b.

Information on piezometric heads under confined or semi-confined conditions in regional hydrogeologic studies, is often absent or available only for productive aquifers. Areas with expected differences between phreatic heads and piezometric heads may be estimated by hydrogeologic reasoning, but actual observations are required and remote sensing may contribute only in an indirect manner. It may be remarked here that reconstruction of the groundwater surface is a manual operation, unless so many data points are available that one of the interpolation methods can be meaningfully employed (Meijerink et al. 1994).

Two limitations should be mentioned. There are areas, often cultivated, where the groundwater table is at shallow depths, say, 1 to 3 meters, and where no evident recognition features can be detected on the imagery. This is true for most of the area of Fig.14.1. The other problem is to judge, in upland areas, whether the interpreted shallow groundwater table is perched and possibly seasonal or whether it is the surface of the coherent groundwater body. Merging of image interpretations with hydrogeologic field data may resolve the issue.

14.2.3 Flow systems

The theory of groundwater flow systems must be credited to Toth (1962) and it provides an "in-depth"- background for hydrogeologic image interpretations. It has been estimated that for groundwater basins 5 to 30 % of the surface area consist of groundwater discharge - or exfiltration- areas. The region with the sinuous small rivers in Fig.14.3b is such a discharge area, and the area with the grasslands in the centre of the colour image (Colour Plate 14.A).

Figure14.4 is included to illustrate how remote sensing can be used in developing a conceptual hydrogeologic model including flow systems in a data scarce area (Western Province, Zambia). The image shows the floodplains of the Zambezi River and a tributary and the gently undulating interfluve of the Kalahari Sands. The image is composed of superimposed NDVI MSS images of three different seasons. Hence, whitish tones correspond to permanent green vegetation. The area has semi-circular depressions called Dambo's. In some, intersection with the groundwater table occurs, as is indicated by perennial water. This information, together with well data, has been used for the estimation of the groundwater contours. The permanent green vegetation along the fringe of the floodplain where the topography rises, has been interpreted as an exfiltration area of a flow system. Two-dimensional modelling supports the presence of discharge zones at those places. These results are used for setting up a groundwater model, whereby the chief purpose is, at the initial stage, to guide further exploration.

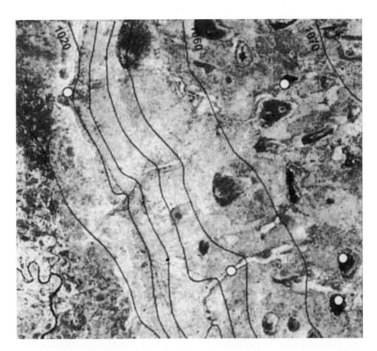

Fig. 14.4. Superimposed NDVI (MSS) images of three seasons, used for first estimation of groundwater contour lines based on visual interpretation and simple 2-D modelling. Kalahari Sands and Zambesi floodplain

Even in the absence of much hydrogeologic data, simple two-dimensional modelling along well selected cross-sections can assist the hydrogeologic image interpretations. By using various assumptions, the model results often place the image features in a different perspective than was initially assumed. Furthermore, the modelling directs the field study.

For example, the pattern of outflow of groundwater on lower volcanic slopes in Indonesia is often reflected by the presence of irrigated rice lands; the dark tones in Fig.14.2 correspond to inundated rice lands, irrigated with discharging groundwater. The upper spring-lines should coincide approximately with the upper limit of an exfiltration area. Rivers can be selected on the imagery for base flow measurements during the dry season. With this data an attempt can be made to construct the geometric input of the model, i.e. the geology of the volcano with the associated horizontal and vertical permeabilities and the groundwater table. The model is adjusted till the outflow rates match the observed ones. Applications are described by Van der Sommen et al. (1990), Anon. (1990) and Meijerink (1994).

A good agreement was found between modelled discharge areas and the vegetation responses, as identified by remote sensing for an area in Belgium (Batelaan et al.,1993). *Temperature contrasts* may exist between upwelling groundwater and surrounding dry areas and this contrast can be used for the study of exfiltration areas by using the *thermal band* of high spatial resolution satellites (e.g. band 6 of TM) or aircraft thermal imagery. In cold climates, places with groundwater emerging from some depth will be warmer than the surroundings and in summer it is the other way around. However, thermal imagery is difficult to interpret and therefore results are usually compared with those of multispectral classifications in the optical domain. A fairly good relation between the two was found by Peters and Stuurman (1989) in The Netherlands and by Bobba et al. (1992) in an area in Canada. However, vegetation may confound the situation. High transpiration results in relatively low temperatures (in fact, emissivity) but this is not necessarily restricted to discharge areas.

Groundwater leakage and upwelling in *coastal seas* can be detected on thermal imagery, if sufficient thermal contrasts exist at sea surface between upwelling groundwater with low density and surrounding sea water with higher density. Such conditions may exist when the aquifer is good and when groundwater is under pressure, rests with sufficient thickness on the salt water interface, and can escape through localized pathways, such as faults, fractures, karst openings.

14.3 Aspects of water budgets

The above stages lead to a regional hydrogeologic overview. Based on this, areas with groundwater potential can be identified for further study as well as evaluation of the groundwater vulnerability to pollution. Numerical groundwater models may be used if sufficient data is available, whereby remotely sensed data plays a role in assessing upper boundary conditions, in particular groundwater drafts for irrigation and recharge.

14.3.1 Groundwater irrigation drafts

In many parts of the world, records for irrigation drafts are lacking and for the development of a groundwater model, both steady state and transient, this information is required. The procedure involves two steps, first the determination of the acreage involved, followed by a calculation of the crop evapotranspiration.

The size of the irrigated areas can be determined accurately on multispectral satellite data if the right time is selected. There should be appreciable contrast between green irrigated crops and non-irrigated cover.

A Normalized Vegetation Difference Index (NDVI) image is generally sufficient to discriminate the two during the dry season, Fig.14.5.

All fields with crops irrigated by groundwater in this granitic area are shown by a white tone, resulting from separating (slicing) the high NDVI values from the remainder on the histogram, which showed two separated populations (irrigated and non-irrigated) in this case.

In more complex cases, supervised multispectral classification may be required for the determination of the areas irrigated by groundwater. Irrigation by diversion of water from local rivers, mixed with irrigation from groundwater confounds the situation (as is the case in the area shown by Colour Plate 14.A) and usually field work is required for the separation of the two.

Fig. 14.5. NDVI image (IRS LISSII) of a semi arid hard rock area, showing irrigated fields for water balance calculations

The calculation of the crop evapotranspiration is usually done by definition of the crop stages and assigning a crop coefficient to each stage for each crop, following the method of Doorenbos and Pruitt (1977). To this, an estimate of evaporation from conveyance has to be added. It may be assumed that return-flows reach the groundwater table and can therefore be excluded from the water budget. The Chaps. 8 and 9 discuss the more physically based methods for estimation of the losses.

14.3.2 Recharge

By adopting a soil water balance method for the recharge, remotely sensed information can be used for estimation of the spatial patterns.

The water balance can be written as:

$$R = P - I_c - Q_d - (a.E_{sw} + T_r) \qquad (1)$$

where, R is recharge, P is precipitation, I_c interception, Q_d direct runoff, E_{sw} evaporation from open water surfaces and wet soils (a=1) with coefficient (a) to account for soil evaporation when the moisture is below the field capacity, T_r is evapotranspiration. Although the equation has a simple form, the difficulties in determining the actual evapo-transpiration rates (E_{sw} and T_r), as well as the interception rates, are well known problems in hydrology. Water surfaces, type and density of vegetation or bare soil/rock surfaces can be assessed by using remote sensing, as well as relative information on the direct runoff. Apart from the general water balance, there are specific features, such as run-on conditions, river bed losses, and so on, which influence the recharge. The complexity of recharge mechanisms and difficulties in estimation are described in the good overview by Lerner et al. (1990).

First, qualitative estimation of the recharge is discussed, followed by semi-empirical approaches and quantitative methods.

Qualitative approach. It must be admitted that in hydrogeological practice often an empirical estimate of the recharge is uniformly distributed over the aquifer. Although a qualitative approach for estimating the recharge remains qualitative, at least an attempt can be made to differentiate spatial patterns of recharge. By visual interpretation of remotely sensed imagery, especially stereo-aerial photographs, the geomorphology can be used for delineation of the various soil/overburden units. Textures and depths of overburden require field observations. Segmentation of terrain in physiographic units each having a set of geomorphological and soil properties, possibly with addition of vegetation, have been described by; Goosen,1967; Verstappen,1977; Way,1978; Townshend, 1981; Meijerink,1988. Once the units and their attributes have been mapped, transfer functions are needed to convert meteorologic and terrain data into quantities of recharge, and this is where difficulties are met.

One way of conversion is using data from the literature describing similar areas or by using appropriate field methods. Specific terrain features may be added, such as micro-drainage patterns, which may influence the recharge.

The approach can be best explained by continuing the qualitative example of Figure 14.1. The relative recharge can be assessed to be negligible on the slate outcrops. The depth of the weathered zone and the clayey soil varies from 40 to 100 cm. After dry periods, initial rainfall enters into the cracks of the soil which close rapidly while the subsoil is impermeable. The gravel deposits can store temporarily some infiltrated water, but they are too thin to have an effect, because they rest on the impermeable slates. The soil depth on the sandstones is negligible, as can be deducted from the fact that bedding and fractures can be seen on the photo. Consequently, the direct runoff is high. The recharge on the quartzites is judged to be only a little better than that of the area with the slates, except where the surface drainage is impeded by obstructions. Quite effective for recharge is the capturing of surface runoff by the blocked valleys. The recharge is a function of the sizes of the micro-catchments draining into each field and occasional overflow from one walled field to another. During field work it was assessed that the yield of open wells in or near such blocked valleys was more then elsewhere, in terms of irrigated acreage and frequency and duration of water application.

In general, bias is minimized when the mapping is limited to three or four relative classes only, such as units with either "high" or "low" recharge and leave the remainder as a "medium" category until further information is available. Units such as sand and gravel deposits, highly fractured rock outcrops, limestones with surface karst and so on, can be identified on imagery. In such units in dry climates the recharge can be substantial proportions of the annual rainfall, see the overview and discussion in Lerner et al. (1990). Units with no or little recharge are the exfiltration areas (see below), or sloping units with shallow soils and outcrops.

Other elements of terrain segmentation for recharge using imagery are the mapping of river beds with transmission losses.

The Colour Plate 14.A (north Iran) shows ephemeral river beds. The snowmelt runoff and seasonal rainfall runoff, during the period March to June, from the hills is the most important source of recharge of the aquifer formed by the alluvial fan deposits (fanglomerates). It is apparent from the image that most rivers do not reach the central drainage line, because the runoff infiltrates fully. The rapid response of the hydrographs of two observation wells, A and C, see Fig.14.6, both close to ephemeral rivers and at a distance of a few kilometers from the mountain front, shows the effect of the river bed infiltration and rapid percolation through the fanglomerates over a vertical distance of some 40 m. The recovery curve of hydrograph A (September to March) can be explained by the stopping of pumping for irrigation during August. Much less response is noted in the hydrograph of well B, located on an old alluvial fan with lime crusts (white tones, far away from the recharging river beds, but at the same distance from the mountain front. The differences in reflectance of the river beds are related to differences in infiltration rates. Near well A, where they are granitic in composition, the average infiltration rates of the bed materials is 1.4 m/d. The bed materials near C consists of volcanic pebbles and boulders with an average infiltration of 3.4 m/d. Both these rates are much below the permeability values adopted for a calibrated groundwater model of the main aquifer. The lower permeability of the mountain front deposits can be attributed to a poor sorting of the gravel beds, presence of clay lenses and lime crusts. A similar observation was made by Huntley (1979).

Some permeable surfaces, such as highly fractured rocks and sandy-gravelly alluvial deposits can be indentified and mapped on aerospace imagery. If isotope

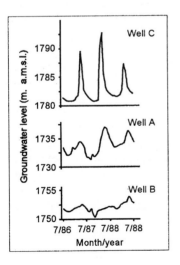

Fig. 14.6. Hydrographs of three wells, illustrating recharge on alluvial fans shown in Colour Plate 14.A

determinations are available it may be worthwhile to try to relate them to the mapping result. Mixtures of young and old water can be expected in such areas, only old water in the impermeable formations. However, the interpretation of the isotopes has to consider flow systems.

The spatial recharge, based on qualitative reasoning using an existing geological map and a spatial rainfall map, resulting from merging the sparse gauge data with vegetation patterns derived by image interpretation and by physiographic interpretation of imagery, has been described by Karanga et al.(1990).

Vegetation and recharge. Since vegetation can be recognized and mapped rapidly with reasonable accuracy by remote sensing, it may be useful to discuss briefly the effect of vegetation on the recharge evaluation. Theoretically, there are arguments pro- and contra to the hypothesis that vegetation increases the recharge. The arguments in favour are related to the higher infiltration rates under vegetation because of a favourable micro-climate near the soil surface, lower soil evaporation, increased organic matter contents and no sealing of the surface. In addition, decayed roots provide preferred pathways for the percolation flow. The arguments against it are the transpiration rates whereby infiltrated water may be fully used as well as the possible presence of deep roots taking up water from the phreatic level. Rosenzweig (1972) concluded, that for a limestone area (672 mm annual rainfall) dense thickets of natural forest evapotranspiration rates consumed all the precipitation, while areas with annual grass needed 280 mm and in this case the remainder was chiefly recharge. Finch (1990) reasoned, for an area in Botswana, that if dense vegetation is noted on NDVI's, there must be, at least temporarily soil moisture and there is a probability that a part of that reaches the groundwater surface. The effect of distribution of rainfall intensities and duration in time confound the situation.

Much depends also on the permeability of the upper zone. The Australian data for sandy environments (see Lerner et al.,1990) suggest that vegetation increases recharge, probably in the order of grass, broadleaf and pine, as long as the rainfall is not too low, say < 200 mm.

There is little comparative data for non-sandy environments. It is interesting to note that the same order is mentioned for the transpiration losses from the many paired experimental catchments, reviewed by Bosch and Hewlett (1982), although chiefly pertaining to non-sandy catchments without differentiation in direct runoff and baseflow. The higher water yields under grass and shrub as a result of lower transpiration may favour recharge, other factors being equal.

The transpiration (t_r) estimates appear in the simple equations for recharge (r) proposed by Issar et al.(1985) for areas like the coastal dunes of Israel, in the form of; $r = c (p - t_r)$, where the empirical coefficient (c) is related to direct evaporation and rainfall characteristics (0.4 in their study), and p is the mean annual precipitation and t_r the transpiration.

Semi-quantitative approaches. The "overburden" units, obtained by physiographic or geomorphological interpretation of imagery and land cover (vegetation) categories, obtained by supervised classification of multispectral imagery or synergistic microwave-optical imagery can be used in semi-quantitative approaches, whereby a water balance of the upper zone is calculated. It has been found that the method of Thornthwaite and Mather (1955) can result in recharge estimates. Usually monthly effective rainfall (Pe = rainfall minus direct runoff) and monthly potential evapotranspiration (ET0) is used but the time step can be shortened. The method consists of a simple bookkeeping procedure. The soil moisture (Sm) status during dry periods (Pe<ET0), when the water deficit is accumulated (APWL) is determined by:

$$Sm = WHC \, e^{WHC/APWL} \qquad (2)$$

The water holding capacity (WHC) is based on soil texture (assumed to be uniform in each of the terrain units) and rooting depth of vegetation classes (derived by spectral classification). The decrease of soil moisture plus the rainfall during dry periods equals the actual evapotranspiration in the method. When Pe > ET0, the actual evapotranspiration is assumed to be equal to the potential one. A delay of the calculated excess soil water can be formulated, for example by a linear reservoir, whose parameters depend on the hydrogeological situation. This approach can be implemented in a GIS, for simulation of the baseflow (Meijerink et al., 1994) and can be extended with NDVI values which are correlated to crop coefficient factors (Seevers and Ottman, 1994). It is advisable to calibrate the values of WHC and the delay with water level fluctuations in shallow wells.

Houston (1982) used - successfully - a somewhat similar method to estimate recharge for an area in Zambia. In his approach, evapotranspiration takes place at the potential rate unless the soil moisture deficit is smaller than the root constant, which must be locally estimated from the dominant vegetation type (open forests,200 mm; short vegetation,75 mm; and poor vegetation/bare soil 50 mm). This leaves room to account for water to flow along preferred paths before the root zone is saturated.

For the estimation of the direct runoff, required to determine the effective rainfall (Pe), the work of Rodier (1975) is of interest. He worked out runoff coefficients with different frequencies of rainfall in the Sahel in small catchments and presented aerial photographs showing the catchment conditions (permeable and impermeable surfaces) as a description. Similar data for other regions may be derived from local gauging data. Relative direct runoff in various units are mentioned in the descriptions of Fig.14.1 and Plate 14.B. Groundwater discharge areas can be excluded from the recharge evaluation because of the upward fluxes. A simple way of estimating the loss in vegetated discharge zones is to assume that the loss equals ET0. The rationale behind this assumption is that the capillary rise from the groundwater reaches the root zone, hence, the vegetation is "not short of water". In case of crops with shallow roots or bare conditions and prolonged dry periods, an estimate of the capillary rise may be taken, instead of the ET0 loss, based on depths of the water table and textures of the overburden. Thermal imagery could be used as a check. It is obvious that GIS procedures are suitable for the estimation, using combinations of depth from surface to groundwater, textures and land cover types. The evaporation loss can not be more than the recharge in the intake area.

Quantitative approaches for the recharge. The degree of sophistication of the quantitative methods for the spatial recharge used so far, is not necessarily synonymous with accuracy. Important sources of inaccuracy are the spatial variation, horizontally and vertically, of the hydraulic parameters of the unsaturated zone and the values of the root parameters. In addition there is the difficulty of determining the potential evapotranspiration, because the data from evaporation pans seldom agree well with those calculated, for example by the Penman-Montheith model.

By calibration of groundwater models. In hydrologic practice the spatial recharge (i.e. the flux across the groundwater table) is often estimated as a result from calibration of a groundwater model. It is useful to compare the pattern of the flux, after calibration of the simulated heads with the observed ones and with the relative recharge values based on image interpretations. Discrepancies are likely to occur and the problem then arises whether the differences are due to wrong estimates of the hydraulic model parameters and the lateral boundary conditions, in particular when specified heads or fluxes are used, or to misinterpretation of the imagery. Given that the calibration may be non-unique, there is much to say for retaining the geographic pattern of the recharge.

One-dimensional unsaturated flow models. These models simulate vertical fluxes of water, with given rainfall and potential evapotranspiration, as a function of soil textures and uptake of water by roots of vegetation. In principle, the models can be used to transform vegetation classifications in recharge quantities. However, information for root functions exist mainly for crops and not for natural vegetation. Interpolation of data from point observations for the regionalization usually does not describe well the spatial variations, as is evident from the example of Fig.14.7 (after Vekerdy, 1996).

The figure shows a small part of an alluvial region in Hungary. The drill holes are marked on the photo. The lighter tones correspond to sandy textures, the dark ones to heavier textured soil, but these can only be seen on bare fields.

It would be a step forward if by remote sensing more information on hydraulic properties of surface and subsurface conditions could be obtained. The airborne gamma ray spectrometry has shown application to surface soil mapping (Reeves, 1992), and also microwave sensors could be helpful in the regionalization which still relies on much field data.

Energy balances. Considering the difficulties, the recommendation of Schultz (1988) to base hydrologic simulations on remotely sensed input could well apply to the evaluation of recharge. The spatial recharge could be estimated from the soil water balance (eq.1), whereby evaporative losses are determined by energy

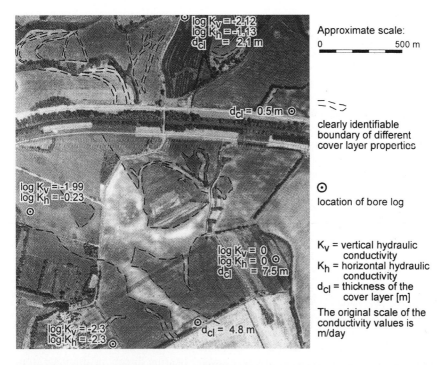

Fig. 14.7. Aerial photograph showing distribution of sandy and clayey deposits of the upper horizon, not captured by interpolation of borehole data

balance methods. Progress has been made to reduce the number of required physical field parameters, but the subject is still in the research domain.

14.4 Hard rock terrain and lineaments

Vast areas of the world consist of hard rocks (basement complexes), or limestone terrain, where porosity is restricted to secondary permeability, thus to fractures, faults (lineaments) and to weathered zones or to water filled cavities in case of karst. Most of the literature deals with the study of lineaments on imagery. However, regional *hydrogeologic zonation* is perhaps the most forceful use of imagery in hard rock terrain. This zonation makes use of natural associations of geology - petrographic, fabric and fracturing properties-, geomorphology, soils and vegetation. The mutual influences and past climatic geomorphological developments provide sufficient basis to consider each zone as a *"groundwater hydrotope"*, with specific conditions of recharge, storage and groundwater flow.

The example of Plate 14.B (T.M., Samburu district, Kenya) illustrates what is meant. There are five hydrotopes, delineated on the image; (1) the old, slightly undulating planation surface with weathered zones and soil development allowing some recharge and storage, as can also be noted from phreatophytes along the drainage. Fractures and lineaments are obliterated by the weathered zone. (2) along the strike of the rocks (gneisses) of zone 1 to the north, there is a zone which is influenced by renewed erosion. Soils are shallow, slopes are short and steep, thus runoff is much and fast, and recharge is little or absent. However, the set of processes in this hydrotope produced narrow sandy alluvium which is locally a good aquifer, recharged by bank storage and transmission losses. (3) The forested hills. This is the only zone where during 2 months there is a rainfall excess over evapotranspiration because of orographic precipitation. The few raingauges in the area are only in the lower parts where the annual rainfall varies from 400 to 700 mm. The forested hills - no gauge- must receive over 1100 mm., a good example of updating isohyetal maps by remotely sensed imagery. (see v.d. Laan, 1986; Meijerink et al.,1994). (4) The area with fairly thick colluvial and weathered zones with low savannah vegetation, local recharge and stream recharge from the adjoining hills. (5) Thin, young lavas resting on the basement, little recharge and storage of groundwater.

Photo-lineaments can be described as; Linear structural elements which are thought to have developed over fracture zones, and which are visible on remotely sensed imagery. Some of them can be seen on the Colour Plate 14.B.

Interpreted lineaments can pertain to fractures of a different tectonic nature, with or without intrusives or secondary clay fillings. Low topographic corridors of some straightness, formed by denudation initiated by open fracturing, have been identified as lineaments, as well as sharp, linear features in outcrop areas. There is general agreement that the most promising water bearing directions originate from brittle deformation caused by tensional stress related to normal faulting and strike-slip faulting (Larson, 1984; Greenbaum,1986; Castaing et al.,1989; Du Wenchai and Ye Deliao,1993). Different rocks respond to the same overall stress field with different fracture densities. Grouping such rocks as gneisses which do not fracture easily, together with granites which have the opposite behaviour, will therefore cause much unnecessary variation in the statistical analysis.

The simple global relationship of well yield - distance to lineament may be of little diagnostic significance. Indeed, the results of studies contradict each other (see Waters,1990, for a review and discussion). First of all, a segmentation according to directions should be made, and specific capacity (l/hr/m.drawdown) rather than yield (l/hr) should be used, if data permit.

The information of conventional rose diagrams can be supplemented, as Brière and Razack (1982) have shown, by directional variograms and by relating the results to the tectonic history of the area (e.g., Djeuda Tcapnga and Ekodek, 1990). Furthermore, the hydrogeologic situation around lineaments should be considered, because much water may be derived from weathered zones, with possibly appreciable specific yields, adjacent to lineaments.

Because of their varied nature the hydrogeologic significance of a lineament remains to be proven by drilling and well testing. (Carruthers et al. 1993).

In carbonate rocks (limestones, dolomites) lineaments are delineated in the hopeful expectation that they represent zones of weakness extending sufficiently in depth to have caused widening by solution by groundwater circulation. With the same expectation alignments of sinkholes attract the attention. The study of fractures on imagery of limestone terrain has attracted early attention (Bouche and Poulet, 1971; see also a discussion in Waters, 1990).

Because lineaments based on surface features, need not necessarily have a hydrogeologic significance, additional information is welcome, such as derived from airborne geophysics. Colour Plate 14.C shows an example.

A part of an aquifer in a semi-arid environment is shown. The aquifer consists of sandstones, with interbedded basalts and is overlain by fossil eolian sands. The aquifer is broken up by faults. The location and offsets have to be known for input in a groundwater model. Data of airborne magnetic surveys (along lines at regular spacings) have been processed to produce a continuous image and to show fault traces. That image has been added to a false colour image of Thematic Mapper, as shown in the Colour Plate. It can be observed that some major faults can be traced further into the region for which no geophysical data was available, showing evidence that lineaments on imagery can be associated with existing faults, covered by overburden. In the part covered by geophysics, nearly all faults are expressed as lineaments on the TM image, but a few additional lineaments have been interpreted, which could not be supported by the subsurface data.

14.5 Groundwater management and conclusions

Apart from the contribution which remote sensing can make to understanding regional hydrogeology - necessary for managing groundwater resources - perhaps the strongest application for the management is the evaluation of the recharge, the groundwater drafts for irrigation and the identification of flow systems in areas where there is a paucity of geohydrological data. Surface conditions,- soils, weathered zones, geomorphology and vegetation - determine the recharge, suitability for artificial recharge and soil and water conservation measures which can affect the recharge (see also Chap. 15).

Groundwater vulnerability to pollution is also directly related to surface conditions. Indexing methods such as the one described by (Aller et al.,1987) and similar methods, group Depth to water table, net Recharge, Aquifer media, Soil medium,Topography, Impact of the vadose zone media and hydraulic Conductivity of the aquifer (leading to the acronyme DRASTIC) into a relative ranking scheme that uses a combination of weights and ratings to produce numerical values. In most areas in the world, many of the factors are derived by analytical (physiographic or geomorphological) image interpretation.

The limitation of the indexing and ranking methods is, that lateral flows or flow systems are not considered. This can be remedied, to a certain extent, by including simple 2-D modelling in the vulnerability, provided the sections are properly selected, again using images. Furthermore, up-to-date land use classifications provided by remote sensing are essential for determining the pollution by agriculture and many point pollution sources can be interpreted on high resolution imagery, such as aerial photography.

14.6 Conclusions and future perspectives

The day-to-day, practical application of remote sensing in groundwater studies, so far, relies on qualitative approaches whereby hydrogeological experience is required. Image interpretation of stereo aerial photography, multispectral images and active microwave images have proven their worth for the compilation and updating of hydrogeologic maps and mapping of the relative recharge. With the strides in computer technology, merging of airborne and ground geophysical data, to obtain subsurface information, has become affordable. This, with satellite data for surface information, can lead to improved groundwater modelling results. The modelling approach itself can be adjusted to make better use of remote sensing. Lubczynski (1997) argues that for the variable hard rock conditions, it will be difficult, if not impossible, to avail of spatially reliable transmissivity values of the weathered and the upper fractured zone. Therefore, the spatial recharge patterns, using remote sensing data could be used as independent input and calibration of the model can be done for transmissivity. Generally it is done the other way round.

It is expected that the more physically based methods for determining and monitoring recharge based on actual evapotranspiration and soil moisture status, will become a welcome complementary addition. The methods for the calculation of the evapotranspiration using thermal bands and vegetation indices of weather satellite data, such as NOAA AVHRR have progressed much recently, e.g. Bastiaanssen (1998) and Chap. 8 of this book.

At regional scale, passive microwave satellite images have been used with reasonable success for the determination of the surface soil moisture at regional scale in southern Africa (Owe et al.1992). This data can be used in conjunction with soil moisture flux modelling for the estimation of the recharge.

Active microwave images contain, theoretically, information on the soil moisture, but effects of the status of the vegetation cover on the backscatter has to be eliminated with a-priori knowledge. This complicates the practical use as yet. More information is provided in Chap. 9.

The last decade has seen rapid advances in coupling of hydrogeological data bases to geographic information systems, digital data integration techniques and hydrogeologic models. Therefore, remotely sensed information can now be embedded in computational methods. It is to be hoped that this leads to the development and field testing of robust methods to determine recharge and groundwater outflow, two important aspects of groundwater.

References

L.Aller, T. Benett, J.H. Lehr, R.J. Petty & G.Hackett, 1987: Standardized System for Evaluating Ground Water Pollution Potential using Hydrologic Settings. U.S Environm. Protection Agency, EPA/600/2-87/035

Anon, 1990: Use of Remote Sensing for Hydrogeological Studies in Humid Tropical Areas. A Pilot Study in West Java, Indonesia. IWACO/TNO/ITC. Min. Public Works, Indonesia, 183 pp.

W.G.M. Bastiaanssen, 1998: Remote sensing in water resources management: The state of the art. IWMI, Colombo, Sri Lanka. 118 pp.

O. Batelaan, F.de Smedt & M.N. Otero Valle, 1993: Development and Application of a Groundwater Model integrated in the GIS GRASS. In: HYDROGIS'93. IAHS Publ. no. 211, 581-589

A.G. Bobba, R.P. Bukata & J.H. Jerome, 1992: Digitally Processed Satellite Data as a Tool in Detecting Potential Groundwater Flow Systems. J. Hydrol. 131, 25-62

J.M. Bosch & J.D. Hewlett, 1982: A Review of Catchment Experiments to Determine the Effects of Vegetation Changes on Water Yield and Evapotranspiration. J. of Hydrology, 55, 3-23

P. Bouche & M. Poulet,1971: Méthode et Exemple d'étude sur Photographies Aériennes de la Fracturation Naturelle des Carbonates. Rev. Inst. Fr. Pétrole, XXVI,1, 3-21

G. Brière & M. Razack, 1982: Méthode Informatique pour l'étude des Cliches Aériens de la Fracturation des Magasins Aquifères Fissures. Revue Geol. Dyn. et Geogr. Phys. 23, fasc.2, 131-142

C. Castaing, P. Dutartre, J.F. Goyet, P. Loiseau, P. Martin & T. Pointet, 1989: Etude Pluridisciplinaire d'un Réseau de Discontinuités. Image SPOT en Milieu Granitique Couvert. Implications en Hydrogéologie des Milieux Fissurés. Hydrogéologie, 1, 19-25

R.M. Carruthers, D. Greenbaum, P.D. Jackson, S. Mtetwa, R.J. Peart, & S.L. Shedlock, 1993: Geological and Geophysical Characterisation of Lineaments in south-east Zimbabwe and Implications for Groundwater Exploration. Brit. Geol.Survey, NERC, Techn. Rep.WC/93/7

H.B. Djeuda Tcapnga & G.E. Ekodek, 1990: Rélations entre la Fracturation des Roches et les Systèmes d'écoulement. In: A. Parriaux (ed.), Water Resources in Mountainous Regions. Mémoires Int. Ass. Hydrogeol. XXII, part 2, 821- 829

J. Doorenbos & W.O. Pruitt, 1977: Guidelines for Predicting Crop Water Requirements. FAO Irrigation and Drainage Paper no.24., Rome, 144 p.

S.A. Drury ,1993: Image Interpretation in Geology. Chapman & Hall, 283 pp.

P. Dutartre, E. Goachet & T. Pointet, 1990a: Implantation de Forages d'eau en Miliex Fissurés. Une Approache Integrée de la Télédétection et de la Géologie Structurale en Nouvelle Caledonie. Hydrogéologie, 2, 113-117

P. Dutartre, C. King & T. Pointet, 1990b: Utilisation de l'Image SPOT en Prospection Hydrogéologique au Burkina Faso. Hydrogéologie, 2, 145-154

Du Wencai & Ye Deliao, 1993: Methods for Recognizing and Extraction Groundwater Information from Remote Sensing Data. In: Proc. Intern. Symp. Operationalization of Remote Sensing, 9, Earth Science Applications, ITC, Enschede, The Netherlands, 105- 111

J.W. Finch ,1990: The Contribution made by Remotely Sensed Data to a Study of Groundwater Recharge in a Semi-arid Environment. In: Intern. Symp. Remote Sens. and Water Resources. IAH/Neth. Soc. R.S., Enschede, The Netherlands, 573-577

D. Goosen, 1967: Aerial Photo Interpretation in Soil Survey. FAO, Soil Bull. no.6. Rome, 55 p.

D. Greenbaum, 1986: Tectonic Investigations of the Masvingo Province, Zimbabwe. British Geol.Survey, NERC, Rep. no. MP/86/2/R

D. Greenbaum, 1992: Remote Sensing Techniques for Hydrogeological Mapping in Semi- arid Basement Terrains. Brit. Geol. Survey, NERC, Techn. Rep. WC/92/28

D. Huntley, 1979: Groundwater Recharge to the Aquifers of northern San Louis Valley, Colorado. A Remote Sensing Investigation. Colorado School of Mines, Golden, Dept. of Geol.

J.F.T. Houston, 1982: Rainfall and Recharge to a Dolomitic Aquifer at Kabwe, Zambia. Journal of Hydrology, 59, 173-187

A. Issar, J.R. Gat, A. Karniele, R. Native & E. Mazor ,1985: Groundwater Formation under Desert Condition. In: Poc. Final Meeting joint IAEA/GFS Zone of Aeration Programme. IAEA, Vienna, 35-54

F.K. Karanga, B. Hansmann, G. Krol & A.M.J. Meijerink, 1990: Use of Remote Sensing and GIS for the District Water Plan, Samburu, Samburu district, Kenya. In: Intern. Symp. Remote Sens. and Water Resources, IAH/Neth. Soc. Remote Sens., Enschede, The Netherlands, 835-848

W. Kruck, 1979: Hydrogeologic Interpretations of Landsat Imagery in Arid Zones of south and west Africa. Proc. of the Fifth Annual William T. Pecora Symp.on Remote Sensing, 408-415

W. Kruck, 1990: Application of Remote Sensing for Groundwater Prospection in the Third World. In: Intern. Symp. Remote Sens. and Water Resources, IAH/Neth. Soc. Remote Sens., Enschede, The Netherlands, 455-463

F. v.d. Laan, 1986: Landscape Guided Climatic Inventory using Remote Sensing Imagery. AGL/Misc./5/85. AGRT series 36. (FAO, Rome)

I. Larsson, 1984: Groundwater in Hard Rocks. UNESCO, Paris, 228 pp.

D.N. Lerner, A. Issar & I. Simmers, 1990: Groundwater Recharge. Int. Assoc. Hydrogeologists. Vol 8, Heinz Heise Verlag GmbH, Hannover, Germany, 345 p.

Li Botao, Dong Yuliang & Li Menghai, 1990: Application of Remote Sensing Technique to the Research of Water Resources of northern China. Intern. Symp. Remote Sens. and Water Resources, IAH/Neth.Soc. Remote Sens., Enschede, The Netherlands, 555-562

M.Lubczynski, 1997: Application of numerical flow modelling with remote sensing and GIS techniques for the quantification of regional groundwater resources in hard rock terrains. Hard Rock Hydrosystems, IAHS publ. 241, 151-157

J.F. Mc Cauley, P. Schaber, C.S. Breed, M.J. Grolier, C.V. Haynes, B. Issawa, C. Elachi & R. Blom, 1982: Subsurface Valleys and Geo-archeology of the eastern Sahara revealed by Shuttle Radar. Science, 318, 1004-1020

A.M.J. Meijerink, 1974: Photo-hydrological Reconnaissance Surveys. ITC Publ. Enschede, The Netherlands, 371 p.

A.M.J. Meijerink, 1988: Data Acquisition and Data Capture through Terrain Mapping units. ITC Journal 1988-1, 23-44

A.M.J. Meijerink, 1994: Application of Remote Sensing and Geographic Information Systems. In: IAH, Proceed. Netherlands Hydrogeol. Research in Int. Coop. Delft, The Netherlands, 27-38

A.M.J. Meijerink, H.A.M. de Brouwer, C.R. Valenzuela & C.M.M. Mannaerts, 1994: Introduction to the Use of Geographic Information Systems for Practical Hydrology. UNESCO,IHP, Paris and ITC Publ.,Enschede, The Netherlands, no. 23, 243 pp.

V.C. Miller & C.F. Miller, 1961: Photogeology. McGraw Hill, New York, 248 pp.

G.K. Moore & M. Deutsch, 1975: Erts Imagery for Groundwater Investigations. Groundwater, 13, 2, 214-226

M. Owe, A.A. van der Griend & A.T.C. Chang, 1992: Surface moisture and satellite microwave observations in semi-arid southern Africa.Water Resourc. Res.28, 829-839

S.W.M. Peters & R.J. Stuurman, 1989: Practische Toepassingen van GIS and Remote Sensing voor Grondwateronderzoek en -beheer. 47 th Meeting "Waterbeheer en R.S", C.H.O./TNO 1989. Also in: TNO/DGV report OS 90-22-A. 53 p.

R.G. Ray, 1960: Aerial Photographs in Geologic Interpretation. U.S.G.S. Prof. Pap., 373, 230 p.

C.V. Reeves, 1992: New Horizons for Airborne Geophysical Mapping. Exploration Geophysics, 23, 273- 280

J.A. Rodier, 1975: Evaluation de l'écoulement Annuel dans le Sahel Tropical Africain. Travaux et Documents de l'O.R.S.T.O.M. no.46, 121 p.

A. Rosenzweig, 1972: Study of the Differences in Effects of Forest and other Vegetative Covers on Water Yield. Final Rep., Proj. A-10-FS-13. Min. Agric., Israel

F.F. Sabins, 1987: Remote Sensing: Principles and Interpretation. 2nd. ed. W.H. Freeman. 449 p.

B. Sahai, R.K. Sood & S.C. Sharma, 1985: Groundwater Exploration in the Sauhastra Peninsula. Intern. J. of Remote Sens. 6, 3, 433-441

G.A. Schultz, 1988. Remote Sensing in Hydrology. J. of Hydrology, 100, 239-265

P.M. Seevers & R.W. Ottman, 1994: Evapotranspiration Estimation using a Normalized Difference Vegetation Index Transformation of Satellite Data. Hydr. Sciences Journal, 39, 333-345

J.J. v.d. Sommen, T. Hasudungan & A.M.J. Meijerink, 1990: Remote Sensing and Groundwater Flow System Analysis in Volcanic Terrain, West Java. Intern. Symp. Remote Sens. and Water Resources, IAH/Neth. Soc. Remote Sens., Enschede, The Netherlands, 495-513

C.W. Thornthwaite & J.R. Mather, 1955: The Water Balance. Laboratory of Climatology, Publ. no. 8. Centerton, NJ

J. Toth, 1962: A Theory of Groundwater Motion in Small Drainage Basins in Central Alberta, Canada. J.Geophys. Res., 67, 11, 4375-4387

J.R.G. Townshend, 1981: Regionalization of Terrain and Remotely Sensed Data. In: J.R.G. Townshend (ed), Terrain Analysis and Remote Sensing, Unwin, U.K. pp 109-132

Z. Vekerdy, 1996. Geographical Information System Based Hydrological Modelling of Alluvial Regions. Ph.D. thesis, ITC Publ.37., 182 pp.

H.Th. Verstappen, 1977: Remote Sensing in Geomorphology. Elsevier, Amsterdam, 214 pp.

D. Warwick, P.G. Hartopp & R.P. Viljoen, 1979: Application of the Thermal Infrared Linescanning Technique to Engineering Geological Mapping in South Africa. Quart.J.Eng. Geol.12, 159-179

P. Waters, 1988: Methodology of Lineament Analysis for Hydrogeological Investigations. Satellite Remote Sensing for Hydrology and Water management, the Mediterranean Coasts and Islands. Gordon and Breach Science Publishers, London, U.K.

P. Waters, P. Greenbaum, L. Smart & H. Osmaston, 1990: Applications of Remote Sensing to Groundwater Hydrology. Remote Sensing Reviews, 4, 2, 223-264

D.S. Way, 1978: Terrain Analysis, 2nd ed. Dowden Hutchinson & Ross,Inc. Stroudburg,U.S.A. 392 pp.

Section III

Water Management with Aid of Remote Sensing Data

15 Introduction to and General Aspects of Water Management with the aid of Remote Sensing

A.M.J.Meijerink and C.M.M.Mannaerts

International Institute for Aerospace Surveys & Earth Sciences, ITC, Enschede, the Netherlands

15.1 Introduction

Water management deals with the control, distribution and allocation of water flows and the treatment of rivers and catchments, consequently, solutions to many of the problems in water management require the use of knowledge and expertise from diverse sources. Some parts of these problems might best be solved using traditional approaches like hydrologic measurement, monitoring or simulation modelling. Other components may require information from one or more data bases from different domains such as population, law, politics, economic statistics and biophysical resources. Many problems in water resources management are however solved by qualitative reasoning and experience.

A common aspect in the many facets of water management studies is the location of the problem, the position within the catchment and the spatial interrelationships between physical catchment characteristics, land use, settlements and infrastructure. Aerial photography and satellite imagery contain spatial information of the surface and near surface features of the earth to be captured and analyzed. As explained in Chap. 2, various imaging satellite sensors are nowadays available. Remotely sensed image analysis, when applied to hydrology is best embedded in a Geographic and Hydrologic Information System (HGIS). A HGIS which can be thought of as a system coupling the following elements: data bases, i.e. thematic, spectral images and hydrologic time series data, a spatial analysis module with eventually a link to hydrologic simulation models or a rule base, which may contain heuristics or methods for multiobjective decision making (Meijerink et al., 1993).

The aim of this chapter is to introduce the practicing hydrologist or water manager to the role of remote sensing techniques for solving problems in water management. Emphasis is placed on the extraction of hydrologic knowledge from remotely sensed information layers rather than on the digital processing and handling of spectral data, for which we refer to other chapters and literature.

15.2 Potential of remote sensing in water management

Remote sensing can be used in various activity domains of the water manager, i.e. surveying and mapping, spatial analysis and prediction or forecasting and decision making in real-time. e.g. flood control (see Chap. 16), irrigation (Chap. 17).

15.2.1 Surveying and mapping

Surveying and mapping is basic to effective water management. Topographic maps have long been made by photogrammetry using stereo models of aerial photographs. Basic principles of photogrammetry can be found in Lillesand and Kieffer (1994). Topographic base maps are usually available, but map updating is often required nowadays. New technological developments make it possible to carry out map updating more efficiently than used to be the case. Geometric correction programmes for satellite imagery are a standard procedure in RS packages. Hence updating of terrain features which are liable to change, such as land cover, river courses, reservoirs, irrigation areas and so on, can be merged with the topographic base. In addition, some RS/GIS packages offer the possibility of preparing digital orthophotos of both satellite images and digitally scanned aerial photographs. This is of particular importance in land and water management, because for large parts of the world, a land use and tenure or cadastral data base does not exist, or is difficult to access. By image processing, e.g., edge enhancement filters and multi spectral classifications, at least parceled areas can be differentiated. The products derived have to be metrically as accurate as possible with corrected height displacements. The ortho-image procedure or a simpler monoplot method (ITC, 1994) allows metric registration of parcels in hilly terrain. Packages are available for the creation of a digital elevation model (DEM) using aerial photographs in digital format, based on automatic stereo matching algorithms. The accuracy can be the same or even higher than that achieved by analogue photogrammetric plotters (Ackermann & Schneider, 1992). On the same principle, a DEM can be derived from stereo satellite imagery, provided no appreciable changes in the vegetation condition occurred during the two successive recordings. The accuracy of a DEM using spectral data is at best equal to the spatial resolution of the system. The result must be checked and rectified for the so called "blunders" caused by wrong stereo matching which result in fully unrealistic parallaxes. The new generation of satellites with high resolution will make the production possible of more accurate DEM's for large catchments. A recent development is the use of airborne laser observations. Height accuracies in the order to 5 to 10 cm have been obtained for a spatial resolution of a few meters. Once a DEM at appropriate resolution is available, slopes and aspect maps can be produced, which are of course of direct interest for watershed management and small scale water development schemes. The use of digital terrain models receives more attention in Chaps. 4, 5 and 7 of this book. The mapping of areas with important seasonal changes in inundation and wetlands is essential for water management because of the effects of such areas on evaporation, flood routing and water quality. Sabins (1987) discusses examples. If a single image is used, there can, however, be confusion as to the limits of the inundation, as discussed by Imhoff et al. (1987). On panchromatic imagery, muddy water may not be separable from wet soil and swamp vegetation may not be spectrally different from dense healthy vegetation on higher terrain. The use of multi temporal and radar imagery may contribute to resolve such confusion. A simple and effective method is to prepare a false color image by NDVI transforms of three different times during a year. By using the additive color theory, together with and applied to the seasonal vegetation dynamics, the temporal changes can be deducted. The thermal channel can also be of help in differentiation wet areas from adjoining drier lands. The temperatures of moist areas at midday are often lower than those of surrounding drier regions because of evapo(transpi)ration. At night, the contrast can be reversed. Monitoring of seasonal inundation and floods have made use of these contrasts (Berg et al., 1979). Water management of alluvial plains requires information on the microtopography where height differences are too small to be determined by conventional photogrammetry. Satellite imagery during an inundation can provide valuable contour information. The results of a DEM prepared with data from field leveling traverses within a polder in Bangladesh are compared with the water limits derived from the infrared band of SPOT (Fig. 15.1), after Meijerink et al. (1994). The image was recorded at the time when the polder was inundated due to a dike failure. The water level, according to the stage records, was 3 m. above the local datum. As can be noted on Fig.15.1, the water limits compare reasonably well and this suggests that sequential imagery recorded during various known water stages could be used for constructing a digital elevation model. Most hydrologic processes are strongly influenced by vegetation. The updating and mapping of the land and vegetation cover with the aid of remote sensing and the effects of plant cover on hydrologic processes is further discussed in Chaps. 7, 8 and 19.

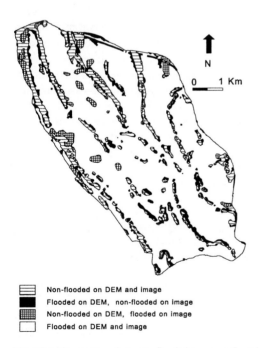

Non-flooded on DEM and image
Flooded on DEM, non-flooded on image
Non-flooded on DEM, flooded on image
Flooded on DEM and image

Fig. 15.1. Comparison between flooded areas on Spot image and digital elevation model (Megna polder, Bangladesh)

15.2.2 Spatial analysis and regionalization

Remote sensing also makes it possible to prepare a quantitative analysis of water balance components at a wide range of scales, ranging from poor water distribution problems in irrigation areas (Menenti et al., 1989) to delineations of main hydrological terrain units, which may be termed „hydrotopes" after Engelen and Venneker (1990). Terrain units group natural associations between lithology, geomorphology and soils, all of which influence the local hydrology. Conceptually these spatial units can be considered as areas with a typical set of hydrologic responses. An essential characteristic of hydrological terrain units is that their boundaries can in many cases be deduced from remotely sensed imagery, using a pragmatic, open-ended classification scheme. Knowledge of the effects of terrain factors on the hydrology should assist in formulation of criteria for their delineation. Without analytical image interpretation, the hydrological terrain units can, to a certain extent, be compiled in a GIS environment by combining a digital geologic map, topographic derivatives (e.g., slopes), hydrological features (e.g., drainage, lakes, wetlands) and a vegetation cover classification. Land use and vegetation patterns generally reflect the spatial distribution of terrain units, except where large-scale human interventions of land cover alterations have been implemented. Examples of the latter are large developments of agricultural farmland or tree plantations in many temperate or tropical countries. Descriptions of some examples (e.g., Colour Plate 14.B) will further illustrate hydrologic terrain units, how they can be differentiated on imagery and what their role in the regionalization is. For example, it is obvious that runoff, groundwater and sediment data of one of the four units of Fig. 15.4 (see Sect. 15.4.2 and illustrating the Handeni area in Tanzania), cannot be extrapolated to any one of the other units in that zone, despite the fact the total area considered is small in size. In other regions, a hydrologic terrain unit can be huge and extend over several thousands of square kilometers, for example, the vast tracts of the plateau basalts — Deccan traps — in India. This is due to uniformity in lithology, geomorphology and soil associations, apparent on the imagery. Hydrologic data and water management practices from a sub-catchment in such an environment can be meaningfully extrapolated to neighboring sub-watersheds. In complex terrain, extrapolation is much more difficult.

15.2.3 Monitoring and forecasting

The prognosis and monitoring of hydrologic phenomena by remote sensing usually rely on the use of image time series or multi temporal images from a same area. The idea is to find an empirical correlation between features measured on imagery and ground hydrometric data. If the two are correlated, the relationship can be used to reduce hydrometric ground operations, usually difficult or expensive, or to fill in gaps in the record. Obvious applications are evaporation estimations from seasonally variable swamp areas, or prediction of snowmelt runoff from snow cover.

For many parts of the world normalized differential vegetation index or NDVI maps are produced for 10-day periods in a routine fashion (Hielkema, 1990). The NDVI values are related to vegetation density and therefore to actual evapotranspiration loss but also reflect the occurrence of rainfall. Until now, the operational use of these time series, has been in the domain of crop yield monitoring and forecasting (Groten, 1993). Regionalization studies could eventually benefit from the information once correlations have been established between the NDVI values aggregated over gauged watersheds and rainfall minus streamflow figures. The hydrological significance of a change has to be assessed and the water manager is faced with the question whether a trend can be detected in the observed changes or whether the changes are within natural variations in time for the phenomenon under study. A study for the water management of Amboseli National Park (south Kenya) illustrates how remotely sensed data could contribute to the answer. Important fresh water springs in that area feed large marshes in an otherwise saline, dry plain. Large herds of herbivores depend on the marshes. In a period of a few years only, the marshes expanded and a small lake was formed, which caused problems for the tourist infrastructure. No discharge records were available. During a field study in 1992, it was found that the discharges equaled the size of the marsh times the potential evapotranspiration rate. Aerial photographs and satellite imagery of nine different dates, during the period 1950 - 1992, have been used to measure the size of the swamps and therefore indirectly the spring discharges. Because appreciable changes in swamp size were observed, it was concluded that the noted recent increase of the discharge of the springs were within the expected range of variation (Meijerink and van Wijngaarden, 1996).

Upstream-downstream problems are common in many catchments. Deforestation in the upper watersheds can cause severe problems in the downstream alluvial areas and coastal zones. The identification and quantification of the downstream damages is required to draw the attention of the decision makers involved, often this can only be done with the aid of remote sensing. An example is given of the damages to productive rice lands on an alluvial floodplain of the Komering River in South Sumatra, Indonesia. In the Komering alluvial plain, upstream of the inland delta (see Fig.15.3 in Sect. 15.3.3), loss of rice lands in the backswamps occurs due to waterlogging, and is becoming worse year by year. The waterlogging is caused by a rise of the sandy river bed, which prevents the drainage of water out of the backswamps into the river after the rainy season. The progressive aggradation of the river bed can be attributed to an increased supply of sand and gravel from the upper catchment where important deforestation took place of hilly areas consisting of sandy tuffs (Meijerink et al., 1988). Remote sensing allowed identification and mapping of the damages in the floodplain area. The relative depths of the back swamps before the changes can be accurately mapped on infrared aerial photography of 1976 (see Color Plate 15.A), because rice is planted in stages which follow the recession of the flood waters after the monsoon. At the time of photography the deeper parts of the backswamps still contained aquatic weeds in the rice fields (high reflectance). The rice, planted on the higher parts of the backswamps, had partial canopy cover, medium reflectance and the intermediate part was still under water for transplanting (no reflectance). A more recent situation was assessed using a SPOT color composite image (Colour Plate 15.A), overlaid with a relative height classification, to locate the areas (backswamps) where changes could have taken place (compare expansion of swamp areas between the two inset maps on Colour Plate 15.A). If field patterns were not present in the deeper backswamps, the area had developed in a swamp. No other interpretations or other automated spectral classifications proved to be systematically correct during the field check. Graphs and maps have been prepared to show the increase of damaged rice lands. These formed the basic input for planning of remediation measures at provincial level.

15.3 River basin planning with the aid of remote sensing

15.3.1 Introduction

Development in most countries and regions of the world stands in direct relation to their mastery and management of water resources. Rational water management should be based upon a thorough understanding of water availability and movement. The river basin, being the physical hydrologic unit, to which the basic principles of conservation of mass and energy apply, is a common and widely adopted concept in hydrology for assessing water and energy balance components. The water balance is a basic tool for analyzing the availability of water resources at national, regional or local scale. GIS has proven an excellent tool to support large scale resource planning, allowing easy aggregation, overlaying and querying between resources and demands (Keser & Bogardi, 1993). Since spatial data is needed, the mapping potential of remotely sensed imagery contributes to identification and assessment of components of the water balance over large areas. Furthermore, features can be studied of areas which will benefit from allocated water. Besides water balance studies, remote sensing has another potential for water management in large river basins. Full scenes of high resolution imagery of, e.g., Landsat TM or Spot can provide synoptic overviews of basins, and permit visual or digital interpretation and identification of landscape units, geologic features, land cover complexes, drainage patterns and geomorphology of floodplains, all essential information layers for solving water management problems. A review on the state of the art of use of remote sensing data in water resources management can be found in Bastiaanssen (1998).

15.3.2 Hydrologic monitoring & forecasting

A widely known application illustrating the use of weather satellite systems for prediction and forecasting of rainfall and flood hydrology of large international river basins is the River Nile monitoring, forecasting and simulation project. This project makes use of the low spatial resolution but high temporal resolution imagery of weather satellites (Meteosat, NOAA) which are merged with ground data in order to produce spatial rainfall estimates. These predictions are then used as inputs in a water balance and real-time flood routing and forecasting model of the Nile river (Attia et al., 1993). An alternative method for estimation of available water resources in a river basin is given in Chap. 18.

15.3.3 Upstream-downstream interrelationships in river basins

For river basin planning and management, basic knowledge on the upstream-downstream interrelationships is required. Hydrologic responses between headwaters, central basin area and the floodplain delta or estuaries are unique in every basin and their knowledge is essential for downstream long-term water use, distribution and planning. Transport and delivery of sediment from catchments are important for both downstream and within-basin considerations. Relationships between the magnitude of sediment yield of basins and climatic, physiographic and land use controls have been investigated by many researchers (Hadley et al., 1985). Knowledge of the distribution of sediment sources and sinks within a basin is essential for recommending control measures. In general, upstream areas and watersheds may be subject to important man-induced land use changes or conversions (e.g., deforestation), which might affect downstream hydrology. Occurrence of natural phenomena such as fast geologic erosion processes (i.e., mass wasting), volcanic or seismic activities can also influence the hydrologic behavior of river basins to a certain extent.

An example from Sulawesi (Verstappen, 1977), illustrates the use of remote sensing, i.e., stereo aerial photography, for detecting changes in a river regime of a tropical catchment. The deforestation in the catchment of the river, shown in Fig. 15.2, has led to an important change in the river morphology which changed from a meandering river (note the remnants on the floodplain) to a braiding river (the present one). The wavelength, the width and the gradient all have increased, the sinuosity, the radius of curvature and the meander amplitude have decreased. The cause of the changes is the sharp increase in the sediment load of the river. The width to depth ratio has increased and also in absolute terms the depth of the river may be less than during the former meandering state. Attenuation of the peak flows by overbank flow still take place and it is therefore difficult to conclude whether the peak flows have increased or not. However, the higher discharges may have become more irregular. The image provides a diagnosis for profound changes in the regime of the river during recent times.

The next example illustrates the use of remote sensing to analyze upstream-downstream processes in tropical river basins. In the lowland part of large river basins, a combination of three processes may affect long term developments, i.e., a rise of the sea level as a result of global warming, a change of river regimes caused by upstream processes, such as deforestation and urbanization, and tectonic movements. The synoptic view of small-scale satellite imagery allows for the identification of tectonism and a zonation of areas which could be affected in the near future. At least a direction is given by the image interpretation for ground-based studies. The image of Fig.15.3, a Landsat MSS infrared band shows such a region, the Musi river (M) and the inland Komering delta near Palembang (P), South Sumatra, Indonesia. The inland delta was formed after tectonic subsidence of a graben and after sea level rise during the Holocene (approx. the last 10.000 years). The faults (F) are indicated on the image. The neo-tectonism is also expressed by the east-west oriented drainage pattern (H) on the adjoining horst (uplifted and tilted block) whereas the Komering (K) river flows from west to east through the horst. The inland delta consists of large

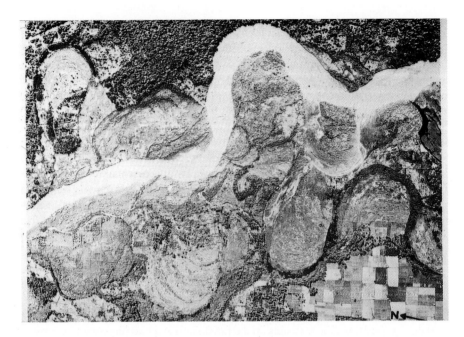

Fig. 15.2. Part of lower course of Palu river, Sulawesi, Indonesia, showing the use of airborne imagery for analysing river regimes and sedimentation hazard in river basins (after Verstappen, 1977)

backswamps (B) with clays, inundated at the time of recording, and of natural levees (L), the long sinuous ridges along the rivers, consisting of sandy materials deposited during overbank stages. The change in morphology of the levees coincides with the limit of backwater effects of the tides. Active subsidence and/or sea level rises will endanger the rice cultivation in the back swamps. A large tropical marsh and peat area is indicated by (V). As discussed earlier, the bedload of the Komering river has increased during the last few decades and this will affect a part of the drainage of the large back swamps and thus the rice cultivation. The back swamps form the vulnerable areas and the planning for safeguarding the rice cultivation should also consider a possible tectonic rise in the horst near the outlet of Palembang which will also impede overall drainage. This rise could be demonstrated by a leveling and comparing the results with a much earlier triangulation survey.

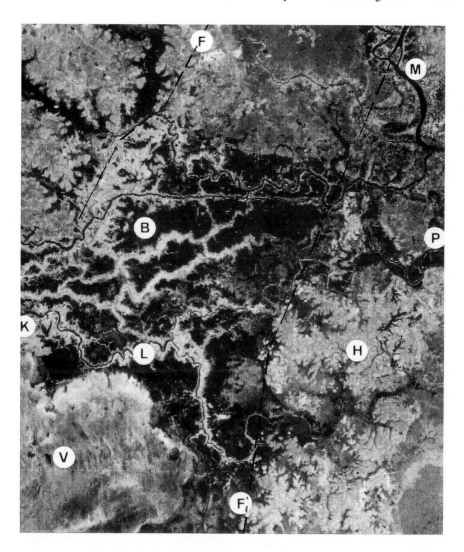

Fig. 15.3. Landsat MSS image (infrared band) of Palemberg region, south Sumatra, Indonesia. (Explanation of letter insets: M: Musi river delta, P: Palembang city, F: fault lines, H: east-west orientation of drainage pattern, K: Komering river, B: backswamps, L: natural levees, V: tropical marsch and peat areas)

15.4 Watershed management with the aid of remote sensing

15.4.1 Introduction

Watershed management contains all activities concerning sustained use, protection or rehabilitation of the water, soil and vegetation resource base in the upper parts or headwater systems of larger river basins. Various operational levels can be distinguished in watershed management, ranging from large basins, watersheds, sub-watershed to local scale (Sheng, 1990). As the detail of survey increases, one finally arrives at the farm or community level of survey and inventory. The range of levels is reflected by the range of spatial resolutions of aerospace images used. Compared to aerial photographs, satellite images such as Landsat MSS, TM and IRS LISS II, have a relative low spatial resolution. They provide overviews at regional scale, particularly of land cover. With other information they are used for zoning of areas or watersheds within larger catchments in order to list priority for treatment. In some countries, the concept of 'critical' areas is used and the priority depends on the proportion of critical areas within the catchment. Critical areas are those where the land cover which offers little protection to erosion, such as row crops or overgrazed rangelands, occurring in combination with certain lithologies — i.e. those where soils which are susceptible to erosion —, and with dissected, sloping lands. At the other end of the range, large scale aerial photographs, say, 1:10.000 or 1:20.000, are used. Apart from the details of land cover and infrastructure visible on the photographs, they offer the possibility of stereographic interpretation of the geomorphology and morphometry of the terrain. The topographic information at large scales thus derived can save much time and costs for the preparation of so called "engineering designs" for the planning of soil conservation measures which follows the priority assessment using smaller scales.

15.4.2 Hydrologic photo-interpretation for watershed management

Watershed management, and especially the planning of sustainable soil & water use, requires detailed hydrologic information pertaining to the terrain and vegetation, as well as to their interactions. The larger scales of stereo aerial photography are eminently suitable for studying the interactions by visual interpretation. These interpretations must be embedded in basic background knowledge of geology, geomorphology and soils and on the effects of such terrain factors on the hydrology of the area. This is demonstrated by the following example from Tanzania, East Africa. Figure 15.4 shows a gneiss plateau (sub-areas 1 and 2) bounded by faults in contact with a lower area of soft sedimentary shale and sandstone rocks (sub-area 3), draining into a alluvial flood plain with swamps (Handeni area, Tanzania). The lithology and geomorphological history exerts controls over the hydrology, as is shown by the three aerial photo insets, whose locations are shown on the map. Sub-area 1 is underlain by impermeable gneiss and has a thick weathered zone, associated with a planation level, which supports a forest savannah. Infiltrated water seeps down to grassy valley bottoms, which become saturated at the end of the wet season. Runoff

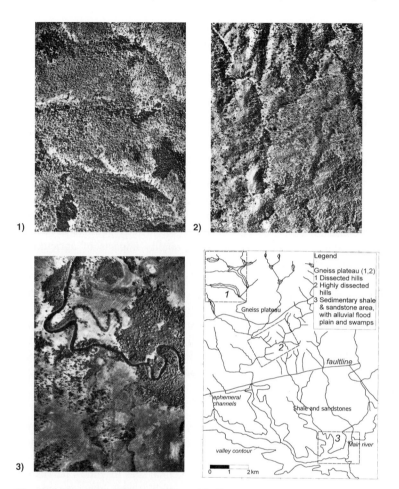

Fig. 15.4. Map of Handeni area (Tanzania), illustrating the high variation in terrain factors and hydrologic responses of neighboring landscape units. Aerial photo insets marked on the map represent: **(1)** dissected and **(2)** highly dissected hills on gneiss plateau, **(3)** soft sedimentary shale and sandstone area, draining into an alluvial flood plain with swamps

response to rainfall is much retarded, but less so during saturated conditions. In the adjoining sub-area 2, the drainage is incisive and has a higher density, slopes are shorter and steeper because of proximity to the fault controlled plateau edges. The change to grass savannah reflects also a decrease of the thickness of the weathered zone, consequently the runoff response is faster and runoff coefficients are higher than in sub-area 1. Across the fault, the hydrology changes drastically; the rivers from the plateau enter a sedimentary area of soft shales and sandstones. The channels of the smaller rivers disappear further downstream (sub-area 3) in the alluvium and swamps. There, the main river formed natural levees which causes an attenuation of the peak discharges, which do not exceed the bankfull stage. The natural levees are a sign that the river is fairly stable. It is evident that each of the areas also has different evapotranspiration losses. The highest occur in sub-area 3, the lowest in sub-area 2. This hydrologic information is of course important for development planning in the area.

15.5 Small-scale water resource development and remote sensing

15.5.1 Introduction

It can be generally expected that the attention to small-scale water resource development will continue to increase in the near future, considering the environmental problems and world-wide debate about large dams (SIWT, 1994). Also the excessive exploitation and contamination problems of large groundwater aquifers in several countries (Vbra & Zaporozec, 1994), the water quality impacts of agriculture, urbanization and industry on surface waters in general lead to an increased stress per capita on drinking water availability on a worldwide basis. Zoning and design of local water development schemes in a certain region require the study of the enormous variation of combinations of terrain and vegetation factors in nature.

Without the use of remotely sensed imagery, it is difficult to avail of sufficient information, considering the lack of soil maps and hydrometric information in large parts of the world. The following two paragraphs illustrate the use of remote sensing in some typical small-scale water resource development operations.

15.5.2 Runoff water harvesting with the aid of remote sensing

Runoff water harvesting basically is a technique by which surface runoff is collected purposely in a smaller infiltration area, where it can be used for crop production or plant growth. The technique of collecting rainwater was already practiced during early civilizations (Hillel, 1967). The determination of potential sites for runoff harvesting or irrigation require besides knowledge of the rainfall regime characteristics, a detailed evaluation of surface topography, surface soil properties and other site environmental parameters. Remote sensing techniques combined with geographic information systems have been applied to screen larger areas for potential runoff irrigation sites in the Sahelian region (Tauer & Humborg, 1992). For more detailed surveys, large scale stereo-images enable the identification and mapping of some of the relevant parameters. Relative runoff estimation and capture of runoff are discussed in some of the examples (Tanzania, Iran) and in Chap. 14. The discussions also mention terrain units with actual water losses (percolation and seepage) and which are therefore unsuitable as source areas for local runoff. The verification of the site potential and the field design of water harvesting structures remains based on field survey, experience and the use of current hydrologic estimation methods (Mannaerts, 1992). Enlargements of aerial photographs can be conveniently used for plotting the positions of the water harvesting structures.

15.5.3 Flood spreading and groundwater recharge

The following example illustrates the use of remote sensed imagery for locating a flood-spreading scheme for shallow groundwater recharge, to be used for local irrigation. Colour Plate 15.B shows a color composite image of Landsat 5 Thematic Mapper bands 4, 7 and 1 of a study area east of Fasa, Shiraz, Southern Iran. Episodic flood runoff in this region is lost to playas or to the sea. Some of that water can be intercepted using a floodwater spreading scheme for artificial recharge, if three criteria are satisfied: (a) the catchment must have an adequate size to generate sufficient runoff, but not be too large to deal with high discharges for a simple diversion, (b) the area of infiltration should be close to the ephemeral river and be underlain by permeable deposits, and (c) the infiltrated water should recharge a shallow aquifer from which the water can be pumped for irrigation. As can be seen on the image, the criteria are met in this case. Peak flows from the moderately sized catchment (shown in part) are diverted from the river where it flows on an alluvial fan (A), into parallel diversion channels which feed - sandy - infiltration basins (B), described in detail by Kowsar (1989). The evidence of an aquifer can be inferred from the presence of groundwater irrigated fields at the lower part of the alluvial fan. Geomorphological interpretation of the image leads to a differentiation of deposits of small local fans (D), not of interest, the upper sandy part of the main alluvial fan (A) suitable for a recharge scheme, and the lower part (C) with heavier soil textures. The latter part may be less suitable for artificial infiltration, but has fewer losses of irrigation water. A groundwater model was used for the assessment of the recharge.

15.6 Irrigation water management and remote sensing

It is estimated that the world's irrigated area is at present in the order of 270 million ha. This is only 17% of the world's total cropped area but accounts for about one third of the world's food harvest (Smedema, 1993). Despite this important contribution to agriculture, the performance of the irrigated agriculture sector has, in general, been disappointing. A basic reason is the low efficiency in the use of available water resources. In some projects, 60% of the diverted water does not actually contribute to crop water requirements. Technical problems arise because irrigation water supplies have not well been distributed. At the farm level, water supply may be unreliable, supply and demands seem rarely to coincide or farmers practice poor and inefficient irrigation methods.

Remote sensing techniques permit a quantitative analysis of problems associated with poor water distribution in irrigation perimeters (Menenti et al., 1989; Bastiaanssen & Molden, 1998). Inadequate water supply is clearly reflected in differences in cropping patterns, intensities and crop development; features which can be conveniently detected and mapped by satellite images. Feasibility studies for improving water distribution can make excellent use of this information.

It has been estimated (Umali, 1993) that about one-third of the irrigated land in the world is under serious salinization risk. Mapping of salt affected or water logged areas in large irrigation schemes, making conjunctive use of remote sensing data and a GIS environment, permits the combination of spectral information with environmental terrain data such as elevation, soil properties, irrigation channel locations (e.g., for seepage) and other environmental factors. Integration of a multi spectral vegetation with soil data and a predictive salinization model, enables early stages of land degradation and associated productivity losses due to salinization to be detected. Simple cropping patterns in irrigated land can be mapped using a single image recorded during the growing season. Agricultural practices and differences in crop phenology increase the spectral variability of the ground cover in irrigated areas. No straightforward and simple way of identifying crops with satellite data therefore exists, except in areas where clear phenological differences between crops strongly affect the spectral characteristics of the fields. This provides an opportunity to map cropping patterns in irrigation areas with satellite data.

Besides the use of remote sensing for monitoring in existing irrigation command areas, high-resolution imagery and aerial photography has been also used for irrigation potential assessment (Moran, 1994). We now observe a trend to view the management of irrigation schemes in a wider perspective, by considering also the conditions of the catchment i.e., water supply areas.

Remote sensing images may provide information on changes in land cover, sediment source areas and snowmelt, which may affect water and sediment yield. We refer to Chap. 17 on Irrigation and Drainage for more discussion on irrigation efficiency improvement with the aid of remote sensing.

15.7 Decision support systems for water management

15.7.1 Introduction

Many problems in water management are ill-structured because they transcend the traditional boundaries of the hydrologic sciences. Examples are current hydrologic and water management issues which need to consider the interactive coupling between computational hydrology and the environment. Solving these interdisciplinary problems in water management therefore requires the use of knowledge and expertise from diverse domains or sources. Knowledge can be of various kinds. Quantitative knowledge is usually expressed in the form of numerical models for simulation of hydrologic phenomena. Many of the problems in water management have, however, elements that are not sufficiently defined to be amenable to traditional algorithmic or numerical techniques. In these cases, a problem-solving environment that is more robust and relies on a coupling between qualitative reasoning and quantitative numerical computing has to be explored (Meijerink et al. 1993).

15.7.2 Expert and decision support systems

Expert systems. Although more complex approaches exist (Abbott, 1994), qualitative knowledge can, in the first instance, be expressed in water management using logical formalisms among hydrologic relevant objects. Objects can be physical, real-world features such as basin areas, drainage lines or river reaches, landscape units or vegetation classes. Relations can also be formulated in qualitative indicators or rating factors for evaluation of effects as is practiced in environmental impact assessment (Janssen, 1992). As such qualitative knowledge is generally contained in procedures, sets of decision rules, or logic assertions. A logical rule can be simply a conditional statement with the following format: *"if(condition) then (exist or do, action)"*. Besides this more conventional rule-based or expert system approach (Engel et al., 1988), knowledge-based applications using other formalizations of artificial intelligence are being developed (Schlumberger, 1994, Wolbring & Schultz, 1996). We refer to Abbott et al. (1994) for more information on the use of advanced information technologies and knowledge engineering in the hydrologic sciences.

Decision support systems. A decision support system (DSS) can be defined as an interactive computer-based system which permits a combination of knowledge sources from various domains in order to help decision makers to solve ill-structured or complex problems. DSS have evolved from practices in the management of information systems, particularly in the field of data processing in business sciences. When applied to water management, a DSS requires a spatial dimension and is therefore usually incorporated in a GIS, thus forming a Spatial Decision Support System (SDSS). Figure 15.5 illustrates an example of an object-oriented system architecture of DSS for use in water management. An expert shell, as, for example Nexpert Object (Nexpert, 1991), is usually used to establish relationships among the hydrology application programs, the remote sensing and geographic data analysis system, the database as well as the knowledge base.

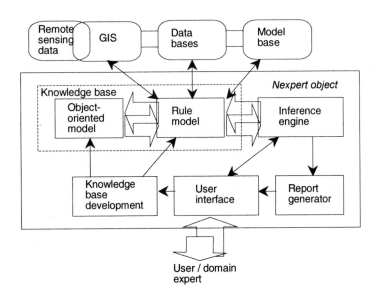

Fig. 15.5. System architecture of a decision support system for water management

Knowledge representation & processing. There are generally two types of knowledge contained in an expert system. A-priori knowledge represents the facts and the rules one knows about a specific domain prior to any knowledge processing within the system. Inferred knowledge consists of new facts or conclusions derived during information processing by the expert system. A major concern is how to represent facts and rules within the knowledge base. This involves maintaining a close correspondence between the computer and the real world facts and rules, and establishing a representation that can be easily addressed, retrieved, modified, updated and processed. Knowledge can be represented in the following ways i.e., object attribute value triplets, semantic networks, frames and rules. Knowledge processing is basically performed using a prioritized list of hypothesis, contained in an agenda. The verification of hypothesis is done using inference mechanisms, which let the expert expand the search space for relevant conclusions without exhaustively evaluating all of the rules in the knowledge-base. For example, the Nexpert Object contains a number of search mechanisms for testing an agenda with priorities. Backward chaining is provoked when a condition containing an unknown Boolean slot (in fact, a hypothesis) is encountered. All rules pointing to that hypothesis are evaluated immediately or with highest priority. Suggesting a hypothesis from, e.g., the user interface puts it on the agenda for evaluation. Suggested hypotheses have priority over others except those generated by backward chaining. When a hypothesis is used as data in a rule's left hand side condition, then the hypothesis of that rule (using the hypothesis as data) is put on the agenda immediately after evaluation of the first hypothesis. This is a forward propagation inference mechanism for evaluating hypothesis. Semantic gates are the basic mechanisms for automated goal generation or opportunistic reasoning. Furthermore, we can distinguish among right-hand side actions, volunteer and context or weak links.

We refer to Nexpert object (1991) for more details on these knowledge inference mechanisms. Figure 15.6 shows a simplified version of knowledge processing for watershed flood runoff simulation in an object-oriented data environment (Baten, 1994). The software environment is composed of Ilwis (ITC, 1994) as Remote Sensing and Geographic Information System, HEC-1 (ASCE, 1987) as flood hydrology simulation model, dBase (© Borland), as data base and the Nexpert Object as expert shell surrounding the RS/GIS, data bases and model. From Fig.15.6, the rule evaluation sequences and interaction of the rules with the object network and other system components can be seen. The remote sensing input in this application pertains to the land cover and associated curve numbers (CN), and the soil groups. Integration of knowledge processing mechanisms with hydrology and water management is usually done using an object-oriented approach, to capture the physical system e.g., a river basin or channel network system (Kim , 1990; Coad and Yourdan, 1991). This represents the more conventional application domain of knowledge engineering, i.e., object orientation, in hydrologic data base design and management. Expert systems and knowledge-based engineering should have also extensive application in conceptual model building for solving hydrologic and water contamination problems (IGWC, 1992). The solution of many hydrologic and pollution problems in the aquatic environment must be based on a broad or detailed preliminary analysis of envi-

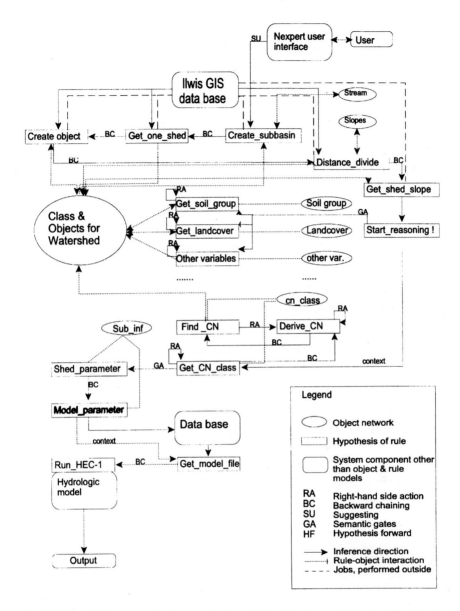

Fig. 15.6. Knowledge processing for watershed runoff simulation in an object oriented environment (after Baten, 1994)

ronmental conditions, i.e., homogeneity, scale, dimensions, processes, before a suitable method or model code can be chosen. In our opinion, model selection or assembling in hydrology could benefit from the use of new knowledge engineering approaches from information technology, in which the spatial data are provided by remotely sensed data, as advocated by Schultz (1988). Combination of remote sensing data in a GIS environment with DSS leads to Spatial Decision Support Systems (SDSS) which seem to be highly efficient in water management. SDSS combined with Expert Systems allow decisions on quantitative and qualitative information.

References

M.B.Abbott, 1994: The Question concerning Ethics or the Metamorphosis of the Object. In HydroInformatics '94, Proceedings of the First Int.Conf. On HydroInformatics, Delft, the Netherlands, 3-9., A. Verwey, A.W. Minns, V. Babovic & C. Maksimovic,(eds),1, Kema Publ., Rotterdam, Netherlands

F. Ackermann & W.Schneider,1992: Experience with Automated DEM Generation. ISPRS XVIIth Congress, Commission IV, Int. Soc.Photogrammetry. and Remote Sensing, XXIX, Part B4, 986-989

ASCE, 1987: HEC-1, Flood Hydrograph package,Users Manual. Hydrologic Engineering Centre, US.Army Corps of Engineers, Davis, California

B. Attia, M. Andjelic & W. Klohn, 1995: River Nile Monitoring, Forecasting and Simulation project. In: Use of Remote Sensing techniques in Irrigation and Drainage - FAO Water Report 4, 17-27, Food & Agricultural Organization, Rome, Italy

M.A.Baten, 1994: Object-oriented Expert System Approach for Modelling in Watershed Management. Unpublished M.Sc. thesis, International Institute of Aerospace Surveys & Earth Sciences, ITC, Enschede, the Netherlands

W.G.M. Bastiaanssen,, 1998: Remote Sensing in Water Resources Management: The State of the Art. Colombo, Sri Lanka: International Water Management Institute (IIMI)

W.G.M. Bastiaanssen, and D.J. Molden, 1998: remote Sensing for irrigated agriculture: possible applications and research needs. (Accepted by Water Resources Bulletin, Journal of the American Water Resources Association)

C.P. Berg, D.R.Wiesnet & M.Matson, 1979: Assessing the Red Rivers from the North,1987 Flooding from NOAA Satellite Data. In: M. Deutsch, B.R. Wiesnet & A.Rango (eds.) Satellite Hydrology, Proc. of American Water Resources Assessment, 309-315

P. Coad & E.Yourdan, 1991: Object-oriented Design and Analysis. 2nd edition, Prentice Hall publ.

B.A. Engel, D.B. Beasley & J.R.Barret,1988: "Estimating Soil Erosion using Multiple Knowledge Sources. Modeling Agricultural, Forest and Rangeland Hydrology, Proc. of the 1988 Int. ASEA Symp., American Society of Agric. Engineers Publ. 07-88, Michigan, US

G.B. Engelen & R.G.W.Venneker, 1990: A Distributed Hydrological System Approach to Mountain Hydrology. Memoires Int. Conf. on Water Resources in Mountainous Regions, IAH-IAHS Symposia 5-8, A. Parriaux (ed.), Laboratoire de geologie, Ecole Polyt. Fed. de Lausanne, Switserland

S.M.E. Groten, 1993: „NDVI-Crop Monitoring and Early Yield Assessment of Burkina Faso. Int.Journal of Remote Sensing,14, 8, 1495-1515

R.F. Hadley, R. Lal, C.A. Onstad, D.E.Walling & A.Yair, 1985: „Recent Developments in Erosion and Sediment Yield Studies. Technical Documents in Hydrology, International Hydrological Programme, UNESCO, Paris, 1985

J.U. Hielkema, 1990: „ Operational Satellite Environmental Monitoring for Food Security by FAO: the Artemis System. Brochure FAO Remote Sensing Centre (RSC), Rome, Italy

D. Hillel, 1967: Runoff Inducement in Arid Lands. Final Technical Report for USDA by Volcanic Research Inst. of Agric. Res. and the Hebrew Univ. of Jeruzalem, Rehovot, Israel

IGWC, 1992: Knowledge-based System for Groundwater and Soil Pollution Modelling. International Groundwater Modelling Centre, Delft, the Netherlands

M.L. Imhoff, C. Vermillion, M.H. Story, A.M. Choudburry, A.Gafoor & F.Polcyn, 1987: Monsoon Flood Boundary Delineation and Damage Assessment using Space Borne Imaging Radar and Landsat Data. Photogrammetric Engineering & Remote Sensing, 53, 405-413

ITC, 1994: ILWIS - The Integrated Land & Water Information System, Supplement to 1.4 User's Manual. Ilwis Dept. ITC, Enschede, The Netherlands
R. Janssen, 1992: Multi-Objective Decision Support for Environmental Management. Kluwer Academic Publ., Dordrecht, the Netherlands
G. Keser & J.J.Bogardi, 1993: National Water Resources Management Planning based on GIS. HYDROGIS'93, Application of Geographic Information Systems in Hydrology and Water Resources Management, K. Kovar & H.P. Nachtnebel (eds), IAHS Publication N° 211, IAHS Press, Institute of Hydrology, Wallingford, UK
W. Kim, 1990: Introduction to Object-Oriented Data Bases. Computer Systems Series, the MIT Press, 235 p.
A.Kowsar,1989: Floodwater Spreading for Desertification Control: An Integrated Approach. Publication Research Inst. of Forests and Rangelands, Ministery of Agriculture, Iran
T.M. Lillesand & R.W. Kiefer, 1994: Remote Sensing & Image Interpretation. Third ed. J.Wiley & Sons, New York
C.M.M. Mannaerts, 1992: Assessment of the Transferability of Laboratory Rainfall-Runoff and Rainfall - Soil Loss Relationships to Field and Catchment Scales: a Study in the Cape Verde Islands," Ph.D. dissertation, Agricultural Faculty, Gent State University, Gent, Belgium
A.M.J. Meijerink, W. Van Wijngaarden, Amier Asrun & B. Maathuis,1988b: Downstream Damage caused by Upstream Land Degradation in the Komering River Basin. ITC Journal 1988-1, 96-108
A.M.J. Meijerink, C.C.M. Mannaerts & C.R. Valenzuela, 1993: Application of ILWIS to decision support in water management; a case study of the Komering Basin, Indonesia. HYDROGIS'93, Application of Geographic Information Systems in Hydrology and Water Resources Management, K. Kovar & H.P. Nachtnebel (eds), IAHS Publication N° 211, IAHS Press, Institute of Hydrology, Wallingford, UK, 35-44
A.M.J. Meijerink, J.A.M. De Brouwer, C.M.M.Mannaerts & C.R.Valenzuela,1994: Introduction to the Use of Geographical Information Systems for Practical Hydrology. UNESCO - ITC Publication N° 23, International Institute of Aerospace Surveys & Earth Sciences. IHP-IV M 2.3, ITC, Enschede, the Netherlands
A.M.J. Meijerink & W.van Wijngaarden W, 1996 : Contribution to the Groundwater Hydrology of the Amboseli Ecosystem, Kenya. Groundwater / Surface Water Ecotones: Biological and Hydrological Interactions and Management Options. J.Gilbert, J.Mathieu and F.Fournier (eds.), pp.111-118, Cambridge University Press
M. Menenti, T.N.M. Visser, J.A. Morabito & A. Drovandi, 1989: Appraisal of Irrigation Performance with Satellite and Georeferenced Information. In: Irrigation Theory and Practice. J.R.Rydzewsky & K.Ward (eds)., Pentech Press, London, 785-801
M.S. Moran, 1995: Using Satellites and Aircraft for Farm Irrigation Management in Arizona. In: Use of Remote Sensing Techniques in Irrigation and Drainage - FAO Water Report 4, 107-112, Food & Agricultural Organization Publications, Rome, Italy
NEXPERT OBJECT, 1991: Nexpert Object version 2.0 - Manuals. Neuron Data Inc., California, U.S.A.
F.F. Sabins, 1987: Remote Sensing - Principles and Interpretation. Freeman W.H. & C°, New York
M. Schlumberger, 1994: Knowledge Processing: a Perspective for Consultants. In: HydroInformatics '94, Proceedings of the first Int.Conf. On HydroInformatics, Delft, the Netherlands, A. Verwey, A.W. Minns, V. Babovic & C. Maksimovic (eds)., 1, pp.13-18, Balkema Publ., Rotterdam, NL, 1994
G.A. Schultz, 1988: Remote Sensing in Hydrology. Journal of Hydrology, 100, 239-265
T.C. Sheng, 1990: Watershed Management Field Manual. FAO Conservation Guide 13/6, Food & Agricultural Organization Publications, Rome, Italy
SIWT, 1994: Second International Water Tribunal - Volume 2 : Dams. International Books, Van Arkel publ., Utrecht, the Netherlands
L.K. Smedema, 1995: Salinity Control in Irrigated Land. In: Use of Remote Sensing Techniques in Irrigation and Drainage - FAO Water Report 4, pp.141-150, Food & Agricultural Organization ,Rome, Italy

W. Tauer & G.Humborg, 1992: Runoff Irrigation in the Sahel zone: Remote Sensing and Geographic Information Systems for Determining Potential Sites. 192 p. Josef Margraf Verlag, Weikersheim, Germany

D.L. Umali, 1993: Irrigation Induced Salinity: a growing Problem for Development and the Environment. Technical Paper, World Bank, Washington D.D.

H.Th. Verstappen, 1977: Remote Sensing in Geomorphology. Elsevier Scientific Publ., Amsterdam, the Netherlands

J. Vbra & A. Zaporozec (eds).,1994: Guidebook on Mapping Groundwater Vulnerability. International Association of Hydrogeologists, IAHG Contributions to Hydrogeology, 16., 131 p., H.Heise Verlag, Hannover

F.Wolbring and G.A.Schultz, 1996: A communication support system for water authorities dealing with reservoir management, IAHS Publ. No. 231, 1996

Colour Plates of Chaps. 12–15   349

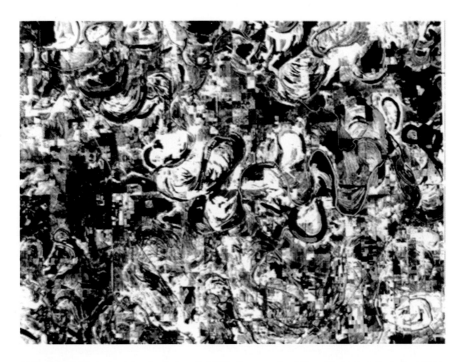

Color Plate 12.A. Aerial color infrared (CIR) photograph of Bear Creek in Mississippi USA similar to those used for determining conservation practices (Pelletier and Griffin 1988) and mapping erosion areas

Color Plate 12.B. Ground panchromatic photograph of Palouse area in Washington USA similar to those used by Frazier et al. (1983) to study and map rill development

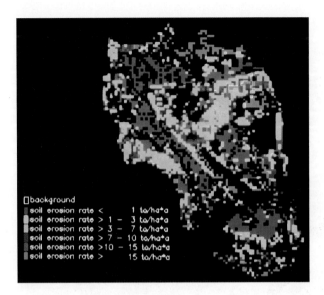

Color Plate 12.C. Long-term soil erosion rates calculated using remote sensing data, USLE, and GIS for a catchment (Jürgens and Fander 1993, Reproduced with permission of Taylor and Francis Publishers, London)

Color Plate 12.D. Soil erosion risk map developed from Color Plate 12.C and information about soil tolerance (Jürgens and Fander 1993, Reproduced with permission of Taylor and Francis Publishers, London)

23. September 1982   27. July 1987

Colour Plate 13.A. False-color images (Landsat TM bands 2, 3, and 4) of Lake Chicot in Arkansas, USA. Changes in suspended sediments in Lake Chicot in September 1982 are evident from the change of gray-blue colors (increased suspended sediment) to the dark color showing clear water. Low concentrations of suspended sediments are evident in the July 1987 image

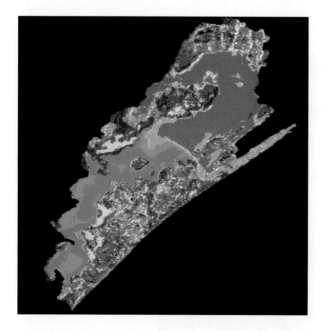

Colour Plate 13.B. Turbidity levels in Lake Chilka in India using unsupervised classification techniques with IRS-1A LISS data. Dark blues are low turbidity grading to very high turbidity in yellow (Sudhakar and Paul, 1993, Reproduced with permission of Taylor and Francis Publishers, London)

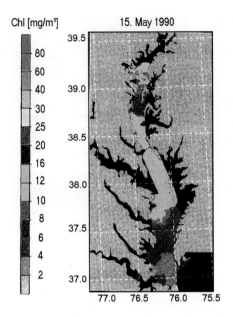

Colour Plate 13.C. Distribution of Chlorophyll-a (Chl) in the Chesapeake Bay for May 15, 1990 determined from aircraft multispectral data (Harding et al. 1995, Reproduced by permission of American Society of Photogrammetry and Remote Sensing, Bethesda, MD)

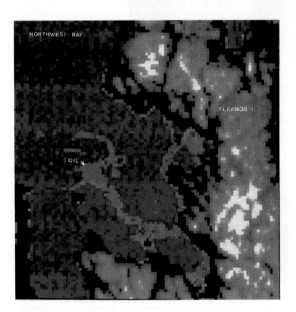

Colour Plate 13.D. Landsat TM image (April 7, 1989) showing oil spillage in Prince Williams Sound, Alaska following the *Exxon Valdez* accident. Areas with oil are Red. Landsat TM Band 5 is Red, Band 4 is green, and Band 1 is blue (Stringer et al. 1992, Reproduced with permission of Taylor and Francis Publishers, London)

Colour Plate 14.A. Thematic Mapper image of alluvial aquifer (fanglomerates), eastern Zanjan valley, N Iran, recharged by transmission losses. In central part grasslands fed by exfiltration of two, merged groundwater flow systems

Colour Plate 14.B. Thematic Mapper image of south-eastern part of Samburu District, Kenya. Various hydrogeologic terrain mapping units can be recognized on the basement complex rocks, related to geomorphic history. Forested hills (red tones) have two months of water surplus (monthly basis), remaining area has deficit

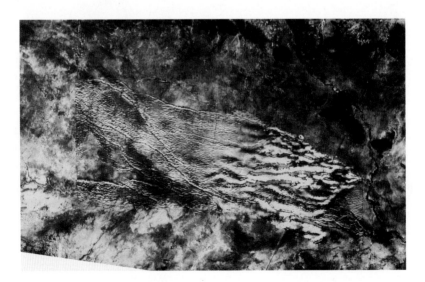

Colour Plate 14.C. Synergism of airborne geophysics and multispectral image for lineament and fault analysis. Pala Road, Botswana

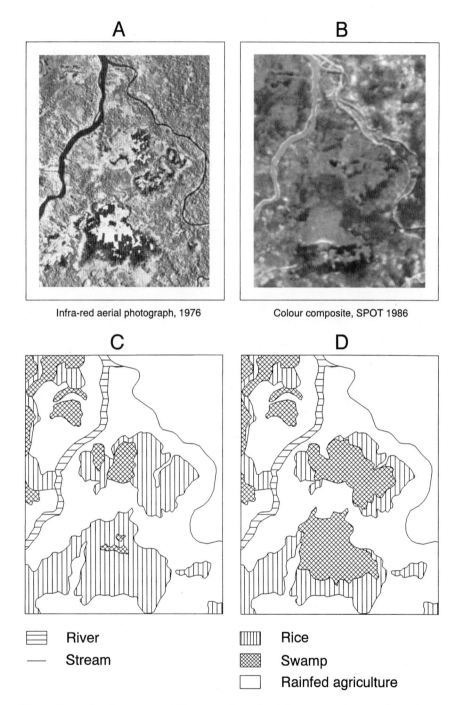

Colour Plate 15.A. Lower Part of Komering river, South Sumatra, Indonesia, illustrating extracting of swamp area extent from multi-temporal aerospace imagery (Meijerink et al., 1994)

Colour Plate 15.B. Colour composite of Landsat 5 TM (red-green-blue combination b4, b7, b1), showing a flood spreading scheme for shallow groundwater recharge near Fasa, Shiraz province, Southern Iran, September 1990

16 Flood Forecasting and Control

Gert A. Schultz

Ruhr University Bochum, 44790 Bochum, Germany

16.1 Introduction

The International Decade for Natural Disaster Reduction (IDNDR) considers floods worldwide as one of the most severe dangers to mankind. Figure 16.1 shows the dominance of floods among natural disasters (Münchner Rückversicherung 1997). This is the reason, why presently billions of dollars are spent for measures (structural and non-structural) of flood protection. This chapter deals with the potential, remote sensing has to offer in the field of flood forecasting, flood warning and flood management by reservoirs and other measures. Since runoff, particularly runoff hydrographs cannot be measured directly with remote sensing sensors it is always necessary to use mathematical models, which are capable of transforming remote sensing information into runoff. Usually the way to achieve this goes via first estimating precipitation from remote sensing data and

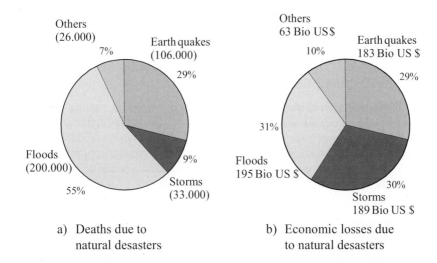

Fig 16.1. Dominance of flood dangers as compared to other natural desasters (period 1986 - 1995 worldwide) (based on Münchner Rückversicherung (1997))

then transform the rainfall hyetographs into flood hydrographs. Remote sensing data are used for both, the estimation of parameters of such models and as model input. An important issue is the fact, that flood forecasting and control can be achieved efficiently only, if the leadtime from the moment of forecast until the occurance of the event (e.g. a flood peak) is reasonably long. This means, that the relevant data (remote sensing and conventional) have to be available in real-time at the forecasting and control center and, while the flood is still in progress real-time forecasts have to be made rapidly and frequently in order to achieve a long leadtime and make the flood management measures efficient.

Since also snowmelt often plays an important role in flood generation, snowmelt runoff models based on remote sensing information are useful in flood forecasting. This topic is dealt with in Chap. 11.

In this chapter only two types of remote sensing data are used, i.e. groundbased weather radar and several types of geostationary and polar orbiting satellites.

16.2 General Approach

16.2.1 Modeling Philosophy

As mentioned above flood hydrographs have to be derived from remote sensing data with the aid of mathematical hydrological models. This has to be done in real-time in the form of flood forecasts.

Model Types. For flood forecasting purposes usually two different modeling approaches can be chosen: Stochastic models and deterministic models. The latter may be sub-divided into lumped and distributed system models. Although stochastic models can be used for forecasting purposes they will not be discussed here, since they are usually not amenable to remote sensing data. Both types of deterministic models, lumped as well as distributed may be used in connection with remote sensing data. More efficient use of the area distributed REMOTE SENSING data is made by distributed models. The application of this model type will be presented along with an example in Sect. 16.3.

Flood Forecasting Procedure. The transformation of remote sensing data into a forecast flood hydrograph may follow a procedure consisting of four consecutive steps: (1) acquistion of remote sensing data, (2) transformation of these data into hydrometeorologically relevant information (model parameter estimation as well as model input generation), (3) transformation of rainfall (or/and snowmelt) computed under (2) into a runoff hydrograph and (4) the use of such forecast runoff hydrographs for real-time control of flood protection measures, e.g. reservoirs.

The Forecasting Problem. Compared to conventional flood hindcasting procedures (computation of floods, which occurred previously with the aid of rainfall-runoff models) the forecasting problem is much more complex due to the facts, that (a) at the beginning of a flood not enough relevant information is available and that (b) the whole forecasting procedure has to be done in real-time rather quickly in order to guarantee a reasonably long forecast leadtime. Figure 16.2 shows this problem. As can be seen from the figure the input does not consist of observed rainfall/snowmelt only but also of forecast rainfall in the near future. This means that a flood forecast requires observed rainfall until the time of the forecast plus forecast rainfall, both of which are used as input into a rainfall-runoff model, which produces the required real-time flood forecast. As will be shown later both, observed rainfall and rainfall forecast can be computed with the aid of remote sensing data.

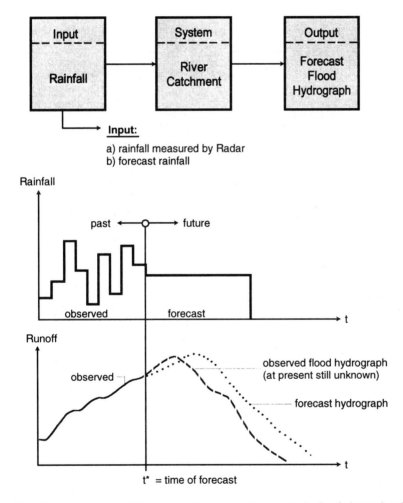

Fig. 16.2. The principle of flood forecasting in real-time on the basis of observed and forecast rainfall

16.2.2 Remote Sensing Data, Types and Acquisition

In this chapter remote sensing data will be used for (a) the estimation of rainfall runoff water parameters and (b) as basis for the computation of model input variables.

As discussed later (Sects. 16.2.3 and 16.2.4) the transformation of remote sensing data into rainfall and runoff requires knowledge of various catchment characteristics. Among these landuse and landcover play an important role since they have - among other factors - an important influence on soil water storage capacity. The relevant landuse and landcover information is obtained with the aid of remote sensing data for the whole catchment area with a high resolution in space. Landuse classification may use air photography or satellite imagery. Here particularly data from the Landsat and SPOT satellites are of importance. Relevant techniques for land cover and landuse classification are discussed in Chap. 7. In Sect. 16.2.3 relevant rainfall runoff model parameters will be estimated with the aid of Landsat satellite imagery which forms the basis for landuse classification required for the determination of soil water storage capacity.

Model input data for flood forecasting models may be derived from various remote sensing sources. Precipitation data with a high resolution in time and space may be obtained from groundbased weather radar. Also satellite imagery may be used for rainfall estimation and forecasting, e.g. data from geostationary satellites like Meteosat, GOES, INSAT or GMS. These satellites have an adequate resolution in time (30 minutes) and a reasonable resolution in space (5 km x 5 km). If the catchments considered are not too small, these data are valuable for runoff estimation (see e.g. Chap. 19). Otherwise this information is not adequate and other data (e.g. groundbased precipitation measurements, or – if available – weather radar data) have to be used. Also snow data and derived from those snowmelt runoff can be determined with the aid of satellite imagery. This topic is dealt with in Chap. 11.

For real-time flood forecasting purposes groundbased weather radar is highly adequate, since the information has not only an adequate resolution in time and space, but it is also available at one point (e.g. Met office with radar) for a large region. For detailed discussion of the use of weather radar for precipitation estimation see Chap. 6. Satellite information like Meteosat data can also be acquired in real-time and are thus available for real-time flood forecasts although their accuracy is lower than that of radar.

Other remote sensing data, e.g. Landsat or SPOT can usually not be acquired in real-time. This is, however, not necessary, since these data are usually needed only for the determination of model parameters which do not change too much with time.

16.2.3 Determination of Hydro-meteorological Information from Remote Sensing Data

Model Parameters. In Sect. 16.2.4 the structure of a deterministic hydrological model of the distributed system type will be presented, which transforms precipitation data into a forecast runoff hydrograph. Distributed models are desirable, since the spatial distribution of precipitation varies very much in space. Since this variability has a major influence on the shape of a flood hydrograph it is necessary to model the rainfall-runoff process in such a way, that it can handle the high spatial variability of precipitation, i.e. the model structure must be such, that it can represent spatial variability. Such models usually require knowledge of the area distribution of soil water storage capacity within the catchment area, within sub-catchments or within so-called Hydrologically Similar Units - depending on the specific model structure. Colour Plate 16.A shows the area distribution of soil water storage capacity in the catchment of the Prüm river in Germany. The map on the left side of Colour Plate 16.A shows soil porosity derived from a soil map. The map in the center shows root depths derived from Landuse classification, which is based on the information of Landsat-TM imagery and the map on the right side shows the soil water storage capacity of the Nims catchment distributed in space which is generated as product of root depth times soil porosity. This way a combination of remote sensing data with data from a digitized map can be merged in order to produce the new information in form of a soil water storage capacity map, which in turn forms the basis for a special set of model parameters of the model discussed in Sect. 16.2.4.

Model Input. Input into rainfall-runoff models is either precipitation or snowmelt or both. Rainfall data with high resolution in time and space can be acquired in real-time by groundbased weather radar. Colour Plate 16.B shows rainfall obtained with the aid of weather radar over a city in Germany which demonstrates the high variability of precipitation intensity in space. The transformation of the echo of the radar signal into rainfall intensity is discussed in Chap. 6 of this book. Another way of using remote sensing data for rainfall estimation uses satellite data, usually from geostationary satellites. Also this technique is briefly discussed in Chap. 6 of this book. If snowmelt is relevant, remote sensing data can be used in order to determine snowmelt runoff as discussed in Chap. 11 of this book.

Precipitation Forecasting. As discussed in Sect. 16.2.1 and shown in Fig. 16.2 it is necessary for using flood forecasts effectively to have a long lead time of the forecast, which requires not only observed precipitation, but also a precipitation forecast in real-time at the time of a flood forecast. Such quantitative precipitation forecasts (QPF) can be generated in several different ways. In Sect. 16.3.2 a method of flood forecasting is presented, which uses stochastic QPF's.

In many countries, e.g. all over Western and Central Europe the national Met offices produce routinely QPF's distributed in area over a time span up to 72 hours. These QPF's may be used together with observed precipitation for a catchment area as input into a rainfall-runoff model for the purpose of real-time flood forecasting.

A third technique of QPF is based on real-time precipitation measurements by weather radar. From a sequence of observed radar images prior to the time of the forecast it is usually possible to derive a vector indicating direction and speed of the movement of a raincell with observed precipitation intensity. The forecast of the movement of such cells towards the catchment under consideration may serve as basis for computation of rainfall input for a rainfall-runoff model.

16.2.4 Transformation of Area Precipitation into a Real-time Forecast of a Runoff Hydrograph

With the aid of a distributed rainfall-runoff model and known model input, (e.g. observed and forecast rainfall) it is possible to produce a flood hydrograph forecast in real-time.

Model Structure. There are, of course, many potential model structures which are more or less suitable for transforming precipitation measurements together with QPF into a real-time flood hydrograph forecast. If no hydrological model of the distributed system type can be applied (e.g. since no adequate data are available) it is certainly possible to use a simple lumped system model like the Unit Hydrograph. It has to be borne in mind, however, that this way the important information on spatial precipitation variability is lost and thus the forecast will be of lower accuracy and reliability. In order to take advantage of the spatial precipitation variability it is necessary to produce the flood forecast on the basis of a distributed system hydrological model. There are plenty models of this type "on the market" and here only one such model developed at the authors institute will be presented briefly as an example.

The structure of this physically based rainfall-runoff model is related to the three hydrological processes: (1) runoff generation, (2) runoff formation on hillslopes and (3) runoff concentration within the river network. For each process the catchment is sub-divided into spatial units considering the most relevant physical characteristic of each process. The heterogeneity of other relevant characteristics within these units is described by distribution functions or average parameters. In order to couple the three model components the fluxes of water from each one has to be distributed among the units of the next component. Figure 16.3 shows in form of a flow chart the coupling of the three process models which form the rainfall-runoff model. As can be seen in the top process component one of the catchment characteristics used in the model is soil storage capacity. This is determined in the way discussed in Sect. 16.2.3. More details on the model structure, estimation of initial conditions and model performance are given in the literature (e.g. Schumann and Funke, 1996). As mentioned before any other type of distributed system model may also be suitable for rainfall input obtained from weather radar or satellite imagery.

Real-time Model Parameter Updating. The quality of a flood forecast depends not only on the model quality but also on the accuracy of the parameter estimation in real-time. If, for a certain flood event, the first forecast has to be made the actual values of the model parameters are not known. It is customary therefore to start the first forecast with the aid of average parameter values known from previous forecasts. Since the flood forecast has to be updated, e.g. every hour, while new information (rainfall and runoff measurements) comes in the parameters have to be re-estimated every time a forecast is made on the basis of new information

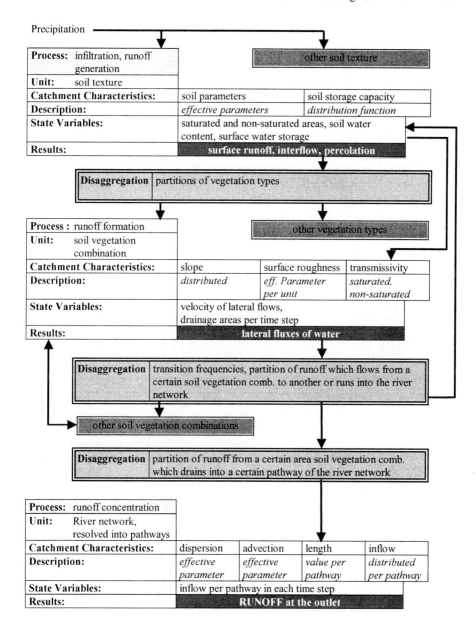

Fig. 16.3. Coupling of the three process models to form the rainfall-runoff model of the distributed type (Schumann and Funke 1996)

obtained during the last time interval. If a deviation between the computed flood hydrograph and the observed hydrograph (until the time of forecasts) is evident, it has to be assumed that this difference is due to false model parameter values. This means that the initial parameters have to be improved such that the deviation between the observed and computed hydrograph becomes minimal. This optimization problem can be solved by applying an operations research technique with an objective function which minimizes the sum of squares of deviations between observed and computed values. Figure 16.4. shows an adaptive parameter estimation, where in the diagram on the left side (first trial) the average historic parameters are used. The deviation between observed and forecast hydrographs until the time of forecast is significant. Depending on this deviation a new set of parameters of the infiltration model component has to be chosen and in an iterative procedure

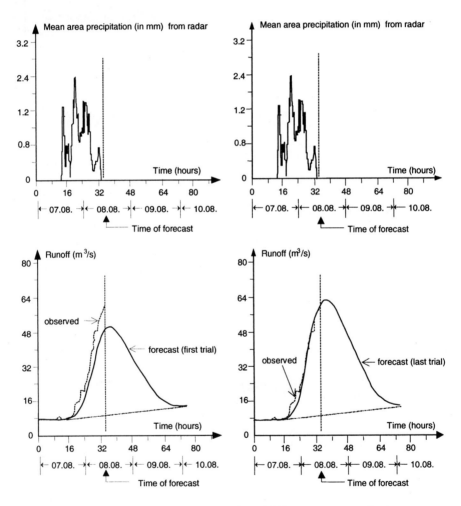

Fig. 16.4. Real-time flood forecast based on radar rainfall-measurement – adaptive parameter updating (first trial (left) and last trial (right))

the parameters have been changed in such a way that the difference between observed and computed flood hydrographs becomes minimal as can be seen in Fig. 16.4 (diagram bottom right – last trial). This optimization procedure can be done manually by trial and error or by an automatic optimization technique.

16.3 Real-time Flood Control with the Aid of Flood Forecasts Based on Remote Sensing Data - an Example

There are many ways to use remote sensing data for real-time flood forecasting and control. In order to show, how this can be done in the real world an example will be presented here. This example is not better than any other - it is shown here only because the author is more familiar with this technique than with others.

16.3.1 Basic Principle

In the field of water management real-time flood forecasts are used basically in two different ways:

- for flood warning purposes,
- for flood control purposes.

Flood warning can help to become active long before the arrival of a flood peak, thus preventing losses of lives and reducing flood damages to a certain extent. Despite the benefits of such flood warnings flood forecasts do not prevent the occurance of extreme floods and corresponding damages. These can be alleviated significantly by efficient operation of flood protection reservoirs based on real-time flood forecasts, if provided at an early stage. The following example shows the procedure to generate real-time flood forecasts followed by the presentation of optimum flood control by reservoirs on the basis of such real-time flood forecasts.

The procedure is as follows:

(1) precipitation is observed by groundbased weather radar and the observed data sets are transferred to a forecasting center,

(2) observed radar data during the last time intervalls are used for real-time quantitative precipitation forecasts (QPF) in the catchment area under consideration,

(3) a rainfall-runoff model is applied in order to transform the precipitation input (consisting of observed and forecast rainfall) in order to generate a real-time forecast of the expected flood hydrograph. The initial values at start of the model run have to be either obtained from measurements of the relevant parameters in real-time, or the model has to be started at a very early stage of the flood in order to allow the model parameters to adjust during a "warm-up period" of the model simulation run.

(4) the forecast flood hydrograph is used as input into a mathematical reservoir operation model which uses an optimization technique in order to achieve flood control in the river system under consideration in an optimal way, i.e. by minimizing flood damages. The general principle of the approach is shown in Fig. 16.5.

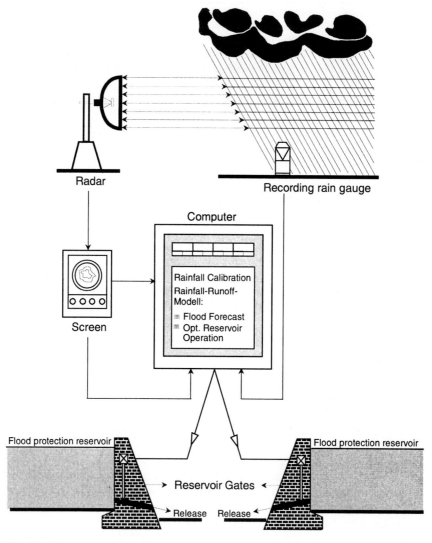

Fig. 16.5. Reservoir operation based on real-time flood forecasts with the aid of radar rainfall-measurements

The following sections show the various steps of the procedure of flood forecasting and control.

16.3.2 Radar Rainfall Measurements in the Günz River Catchment

Figure 16.6 shows the catchment of the Günz river (530 km^2) which is a tributary to the Danube river. The figure also shows the isohyetes during a storm in the year 1981, which were generated on the basis of radar measurements. The catchment area is sub-divided into area elements according to the applied polar coordinate system. The area elements have the size of 1° (Azimut angle of radar) x 1 km, thus having an area of approximately 1 km^2. The time increments are 15 minutes. The two maps of Fig. 16.6 show two consecutive radar measurements relevant for a quarter hour each, where the rainfall intensity is the average of 3 five minute measurements. The radar is a C-band radar located on the mountain Hohenpeißenberg (Bavaria). It can be seen in Fig. 16.6 that an intense storm moves over the Günz catchment area from southwest to northeast.

As mentioned above the observed rainfall during a storm alone is not sufficient information for a real-time flood forecast. Besides the already observed rainfall it is necessary - while it is still raining - to make a QPF in order to produce a model input (consisting of observed plus forecast precipitation) which is closer to reality.

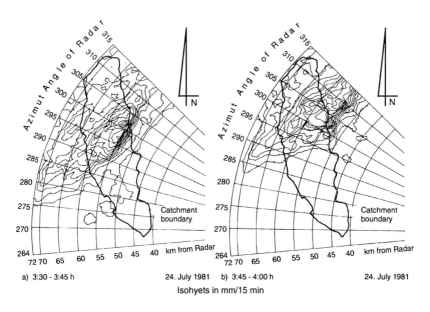

Fig. 16.6. Isohyets for the river Günz catchment, Germany, obtained from two consecutive radar measurements

16.3.3 Quantitative Precipitation Forecast (QPF)

As mentioned above there are several techniques available for producing a QPF. Since atmospheric models for medium range weather forecasting have been improved considerably during the last decade it can be expected, that such information, if available in real-time, will represent a very useful QPF to be used as input into a rainfall-runoff model in addition to the already observed rainfall at the time of the forecast.

In the example presented here, however, a stochastic QPF procedure has been applied to the Günz river catchment. This stochastic approach computes at given time intervals (e.g. every hour) the depths and duration of the future rainfall until the end of the storm event. The computed values are given along with pre-specified conditional probabilities of non-exceedence (e.g. 90 %). The condition is the quantity of rainfall which occured from the beginning of the storm event until the time of the forecast. More details of this technique are given in the literature (Klatt and Schultz, 1985). In Sect. 16.3.4 a flood forecast will be shown along with the causative rainfall consisting of observed radar rainfall data and QPF based on the stochastic approach discussed here.

Besides such stochastic QPF approaches also deterministic techniques may be applied. One such technique uses a time sequence of consecutive radar measurements from which the velocity vector of the moving rainstorm has to be identified. This process can then be extrapolated thus giving a forecast of the expected rainfall field in the river catchment of interest towards which the rainstorm is moving (Schilling, 1990). Such forecasts can be further extended if data from groundbased weather radar are combined with satellite imagery. The British FRONTIERS System is using this approach (Sargent, 1987).

16.3.4 Rainfall-Runoff-Model Application for Flood Forecasting

A real-time forecast of a flood hydrograph based on rainfall information can be made, if the following items are available:

- a rainfall-runoff model with parameters calibrated for the catchment area, for which the forecast shall be made,
- observed precipitation from the beginning of a storm until the time of forecast with an adequate resolution in time and space,
- a quantitative precipitation forecast (QPF) of future rainfall (after the time of forecast) until the end of the storm event.

The observed plus the forecast precipitation is used as input into a suitable rainfall runoff model in order to generate a flood forecast in real time with an adequate lead time. Figure 16.7 shows such a forecast for the Günz river based on radar rainfall measurements and a rainfall forecast computed with the aid of the stochastic forecasting model mentioned in Sect. 16.3.3.

In this case it is necessary to choose a probability of non-exceedence for the computed flood. Figure 16.7 shows three examples of flood forecasts computed in real-time on the basis of probabilities of non-exceedence of 60, 90 and 99 % for the forecast precipitation. Since such QPF's are usually not very accurate, it can be seen from Fig. 16.7 that also the resulting flood forecasts are not in a too good agreement with the flood hydrograph which was observed after the forecast. The forecast hydrograph with the probability of non-exceedence of 0.6 (= 60 %) underestimates the peak while the volume is in reasonable agreement with that of the observed hydrograph. The forecast hydrograph with the probability of non-exceedence of 0.99 (99 %) matches the observed peak, the volume, however, is much too high as compared with the observed hydrograph. Nevertheless such forecasts are a great help for flood warning or for the operation of flood protection reservoirs.

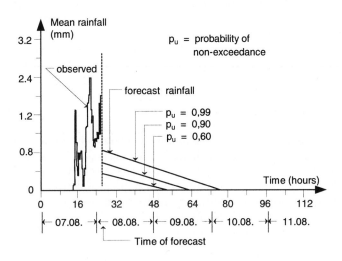

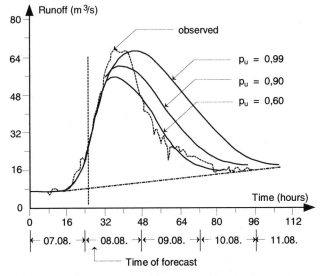

Fig. 16.7. Flood forecast based on radar rainfall measurement and rainfall forecast (QPF)

16.3.5 Optimum Reservoir Operation Based on Forecast Flood Hydrographs

One way to protect lives and property from floods in river valleys is the construction and operation of flood protection reservoirs in the upstream reaches of the river and its tributaries. The most efficient way to use such a reservoir system during flood events is the adaptive control of such reservoirs in real-time based on flood forecasts which are computed while it is still raining, i.e. with a reasonably long lead time. This topic was already briefly discussed in Sect. 16.3 along with Fig. 16.5, in which the principle of real-time flood forecasting and reservoir operation was presented. The idea of this type of reservoir operation is to provide an optimum flood control, i.e. to minimize losses of lives and damage in the river valleys below the reservoirs. It should be mentioned, that such optimum flood protection reservoir controls are very sensitive to the accuracy of the forecast flood hydro-graphs. Since flood forecasts are, however, almost always imperfect it is necessary to compute forecasts again and again during the flood event (e.g. every hour). Each such new computation of a flood forecast should be based on a new model parameter set based on the new information gained during the last time interval. Also the optimization procedure has to be computed on the basis of each new flood forecast (e.g. every hour).

This way it becomes possible to operate the reservoir system in such a way that each flood event will be controlled in an individual optimum mode. A technique which has been proven to be successful for such flood control optimization is the technique of *Dynamic Programming* (DP) from the field of Operations Research. DP is a sequential optimization technique which allows to treat non-linear objective functions and non-linear constraints (Bellman, 1957).

The procedure, how to apply DP to the optimum operation of two parallel reservoirs will be given along with an example. Figure 16.8a shows a system of two parallel flood protection reservoirs located in the upper reach of the Danube river in Germany. Figure 16.8b shows a severe flood, which occurred in the system during February 1970. In order to show the optimum operation of two parallel reservoirs the flood shown in Fig. 16.8b will be assumed to represent forecasts of flood hydrographs expected to enter the two reservoirs (Fig. 16.8a) in the Breg river (reservoir 1) and the Brigach river (reservoir 2).

The non-linear objective function which has to be minimized can be formulated in the following way:

$$OF = \sum_{i=1}^{T} (Q_{a1,i} + Q_{a2,1} - AQ)^2 = Min!$$

Where Q_{a1} = release from reservoir 1
 Q_{a2} = release from reservoir 2
 $Q_{a1}+Q_{a2}$ = total flow in the Danube river
 AQ = maximum flow in the Danube river which does not yet cause damages
 i = time increment.

The meaning of the objective function is as follows: The flow in the Danube river, $(Q_{a1} + Q_{a2})$, exceeding the harmless AQ-value has to be kept as low as pos-

16 Flood Forecasting and Control   371

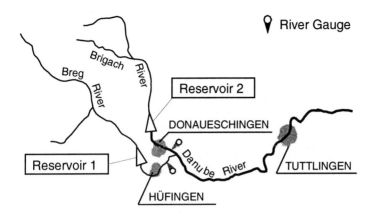

(a) System Sketch, Danube River with Tributaries, Reservoirs and Cities to be Protected

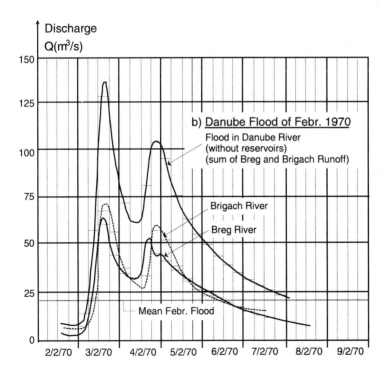

Fig. 16.8. System of two parallel reservoirs (Danube river) and historical flood (Feb. 1970)

sible for all time intervals. The quadratic objective function guarantees that during all time increments this flow exceeding AQ is kept as low as possible. Figure 16.9 shows the result of the optimization of the joint operation of the two parallel reservoirs given in Fig. 16.8a for the flood given in Fig. 16.8b. As can be seen by comparing the two diagrams above each other on the right hand side of Fig. 16.9 the floods in the Danube river are reduced from about 130 m^3/s to about 100 m^3/s by the reservoir operation. This considerable reduction could be achieved with the two rather small reservoirs only by the fact that this flood was controlled individually by the DP algorithm. During a flood event in real-time a new flood forecast and a new optimization computation has to be carried out for each new time increment due to the fact that such forecasts are never perfect. More information on this technique is given in the literature (Schultz, 1994).

16.4 Flood Forecasting and Control in an Urban Environment

The examples of flood forecasting and flood control given in this chapter have dealt with meso-scale rural catchments. In recent times flood forecasting - particularly with the aid of weather radar - is becoming rapidly more important also in the field of urban drainage systems. Since radar is capable to detect the high variability of precipitation intensity in space and time it is very suitable for computing flood forecasts in the various canals of an urban drainage system. On the basis of such forecasts it is possible to operate underground reservoirs in the canal system as well as releases into rivers in an optimum way such that the sewage treatment plant will not be overloaded during floods.

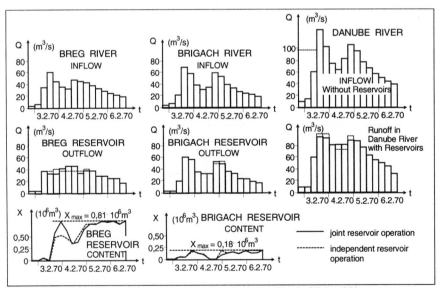

Fig. 16.9. Optimal release policy for two parallel reservoirs. Flood of Feb. 1970:
—— joint reservoir operation ----- independent reservoir operation

In present practice design and operation of an urban sewer network and the corresponding sewage treatment plant is done independently. It would be, however, much more appropriate to facilitate a joint operation of the urban sewer network and the sewage treatment plant at the end of the sewage system. This operation is based on real-time flood forecasts which are computed with the aid of radar rainfall measurements. The goal of this control is to minimize the combined negative effects of the hydraulic load and the pollution load in the receiving waters.

In order to make this control efficient, precipitation information is required with a high resolution in space and time since already small differences of precipitation input into an urban rainfall runoff model result in large differences in computed runoff. The computed real-time flood forecasts have to be rather accurate at short term (up to two hours) and maybe less accurate at medium-term predictions (based on weather radar and satellite data). For the real-time short-term forecasts the data obtained from one radar will be adequate while for the medium-term composite radar images from several radars will be required.

The optimum control of the combined sewer-sewage treatment plant system is based on multi-temporal imagery obtained from ground based weather radar systems.

Figure 16.10 shows an urban drainage system in the city of Bochum, Germany and the principle of rainfall measurement in real-time to be used for flood control.

Colour Plate 16.C shows a sequence of rainfall fields at 5-min. time increments in the same area during a storm on July 23, 1996. This information is used as input

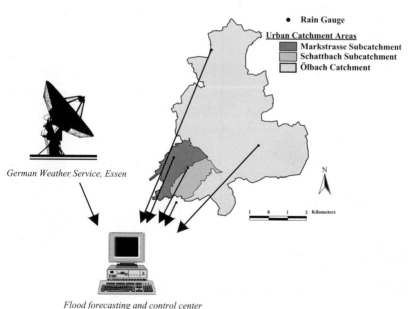

Fig. 16.10. Rainfall measurement by weather radar and raingauges. Example of an Urban Catchment in Bochum, Germany

into an urban rainfall-runoff model simulating the behaviour of the urban drainage area together with the flows in the sewer system. Such image sequences are also used for forecasting rainfall intensities during the time increments of the next couple of hours, which also serve as input to the rainfall-runoff model. The relevant pollution load forecasts are computed as a function of the forecast runoff hydrographs. The operation of the subsurface rain retention reservoirs, the flow release structures as well as the sewage treatment plant is based on these forecasts as indicated in Fig. 16.11.

The use of remote sensing data gained from weather radar increases the efficiency of a combined urban drainage system (surface water, sewer system and sewage treatment plant) significantly, both, as far as water quantity and water quality is concerned.

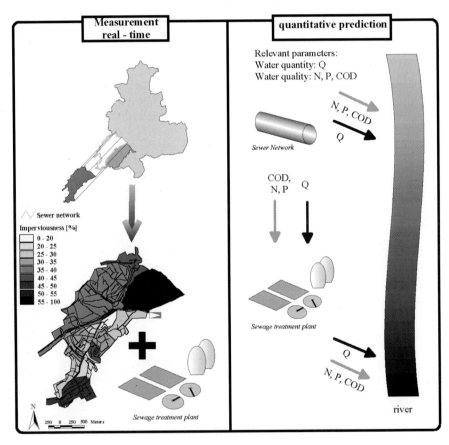

Fig. 16.11: Prediction and simulation of the hydraulic load and the pollution load in the sewer network and the sewage treatment plant

16.5 Future Perspectives

It is the idea of this chapter to show how remote sensing data may be used for flood forecasting and control. The main emphasis was on the use of groundbased weather radar for real-time flood forecasting on the basis of rainfall-runoff models. This technique can be applied, however, only if adequate radar coverage is available. At present, in Central and Western Europe an international radar system (C-band) is set up which will cover more or less the whole area of Western and Central Europe. This way it will become possible in the near future to use weather radar as basis of flood forecasts in all Western and Central European countries. Japan is already completely covered by a radar network at present. Similar developments are presently on the way in North America with the establishment of the NEXRAD system based on S-band radar.

If the existing radar system does not cover the complete area of interest (e.g. in Western Europe near the Atlantic Ocean) it is possible to add information from meteorological satellites (e.g. Meteosat, GOES, GMS) in order to extend the lead time of a flood forecast. This principle of combination of data from radar measurement and satellite imagery is incorporated in the FRONTIERS system (Sargent, 1987).

In urban drainage systems the tendency goes towards extension of flood management towards a combined management of floods (water quantity) and pollution loads (water quality).

Future flood management, in urban drainage systems as well as in larger rural catchments up to macro-scale river systems like the Rhine or Mississippi should be planned in such a way, that the construction of flood protection measures should not be done at a local or regional scale anymore, but rather in form of integrated large-scale systems. This would require consideration of potential optimum control of such systems already in the planning phase. It should, however, not be underestimated, that flood control using remote sensing and other devices requires rather sophisticated logistic structures, which usually are sensitive to power failures and other disturbances in such systems. Furthermore the accuracy of radar rainfall measurements is often not optimal (see Chap. 6!).

The direct monitoring of floods with the aid of remote sensing is discussed in Chap. 10.

References

Bellman R. (1957) Dynamic Programming, Princeton University Press, New Jersey
Klatt P. and Schultz G.A. (1985) Flood Forecasting on the Basis of Radar Rainfall Measurement and Rainfall Forecasting, in B.E. Goodison (ed.) Hydrological Applications to Remote Sensing and Remote Data Transmission, IAHS publication No. 145, pp 307-315
Münchner Rückversicherung (1997) Überschwemmung und Versicherung. Order-Nr. 2425-v-d
Sargent G.P. (1987) The FRONTIERS-Project in V. Collinge and C. Kirby (eds.) Weather Radar and Flood Forecasting, John Wiley and Sons, Chichester
Schilling W. (1990) Operationelle Siedlungsentwässerung, Oldenbourg-Verlag, München

Schultz G.A. (1994) Remote Sensing and Forecasting of Floods (Chapter 21) and Remote Sensing for Control of Floods (Chapter 36) in G. Rossi et al. (eds.), Coping with Floods, Kluwer Academic Publishers, The Netherlands

Schumann A. and Funke R. (1996) GIS-based Components for Rainfall-Runoff Models, in HydroGIS'96: Application of Geographic Information Systems in Hydrology and Water Resources Management, IAHS publication No. 235

17 Irrigation and Drainage

Massimo Menenti

The Winand Staring Centre, P.O.Box 125, 6700 AC, Wageningen, The Netherlands

17.1 Introduction

Water allocated to irrigated agriculture worldwide amounts to about 80% of total water supplies (Seckler et al., 1997). The impact of irrigation on soil water balance, vegetation and shallow groundwater bodies within the irrigated perimeter is significant (e.g. Menenti, 1990). Moreover, the generally low efficiency implies that the larger part of this water flows back to rivers and acquifers. Excess irrigation water may accumulate in areas underlain by deposits having low permeability leading to extensive water logging and salinization. Extensive irrigation has, therefore, a significant impact on regional hydrological processes (Fig. 17.1). Water diversion reduces significantly water supply to ecosystems and older irrigation schemes, while excess irrigation water may be drained elsewhere. Given the obvious link of irrigation with aridity, extensive irrigated lands become

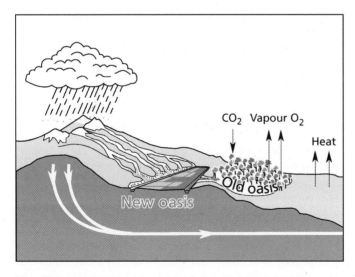

Fig. 17.1 Development of new irrigation schemes in relation with regional hydrological processes in a watershed

a major determinant of regional hydrology and of the quality of the environment. Furthermore, extensively irrigated lands in arid zones increases soil wetness and evaporation significantly over large areas leading to large impacts on the atmospheric boundary layer and increasing convective activity locally (Wang et al., 1993).

In arid environments irrigated agriculture can be either *friend or foe* of sustainable development and environmental equilibrium (see e.g. Jensen, 1996), with the outcome depending on how precisely rural planning and water management take into account hydrological constraints on development. The welfare of people is (or should be) the dominant driver towards irrigated agriculture, thereby adding complexity to sustainable land and water management.

All of the above leads to a need for improved monitoring, understanding and modelling of spatial processes related to irrigation and drainage. The potential relevance of remote sensing for this purpose is self-evident and the scope of this chapter is to provide an overview of the significant and specific body of work accumulated during the last 20 years.

Besides the watershed-oriented issues mentioned above, irrigation also includes specific management processes requiring a broad range of skills from agrometeorology to administration and management. A first and obvious need is the capability of assessing, monitoring, understanding and control of water use. For small irrigation schemes this can be easily accomplished through direct contact with farmers. For large schemes comprising towns, villages and diverse productive sectors, experience proves how difficult it is to collect accurate data on actual water use by traditional methods. Given the large difference in the water requirements of cultivated and bare land, the primary need here is the determination of irrigated area, so is not surprising (see Current Applications) that a dominant share of operational applications relate to this issue. Identification and mapping of individual crops, although a traditional area of remote sensing and irrigation studies, is more challenging and less accurate and this is reflected by literature. Useful information on water use can be obtained by observing specific spatial patterns such waterlogged or saline areas and linking such patterns with irrigation and drainage canals.

17.1.1 Current non-remote sensing approaches and limitations.

It is not feasible to compare each type of application (see Current Applications) with an equivalent non-remote sensing technique. A significant reference might be provided by considering how irrigated area is determined traditionally. In many countries farmers hold permanent water rights (assessed in area units). This provides the basis for water allocation, removing the need for accurate and up-to-date information, but often leading to significant mis-allocation of water as actual irrigated area changes. In a number of countries irrigation fees are assessed on the basis of irrigated area or of crop type: these data are typically based on statements by farmers provided to either the Irrigation Authority or the Revenue Service. The reliability of this information is questioned by many. Another approach is by using agricultural census, which although more reliable, is obviously time consuming and does not provide frequent updates. Literature provides abundant evidence on the reliability and cost-effectiveness of estimates of cropped area based on satellite data. In many instances this has reached the stage of operational use, as documented by the references mentioned in the section Current Applications.

Enforcement of legislation and of administrative procedures depends on the availability of accurate and up-to-date information on land use. Water is often allocated on the basis of water rights conditional on actual use of land for irrigated agriculture and water fees are assessed on the basis of irrigated area or crop category (e.g. water requirements being negligible, moderate or high for three broad crop categories). Moreover regular and frequent (e.g. twice yearly) determination of irrigated area and fractional cover may help fine-tuning water allocation, distribution and application, thereby improving water use efficiency.

17.1.2 Reviews of remote sensing applications in irrigation and drainage

Reviews were presented by Menenti (1990), Veerlapati-Govardhan (1993) and Vidal and Sagardoy (1995). The material given in Menenti (1990) demonstrates the feasibility of the applications listed herebelow:

- mapping of irrigated area for diverse land units (from farms to large irrigation schemes);
- integration of information on irrigated lands with GIS for assessments of water rights;
- assessment of irrigation performance;
- detection and monitoring of poorly drained areas;
- determination of height-storage relationship of natural depressions;
- appraisal of land suitability for irrigation;
- planning of irrigation development.

The more recent overview by Vidal and Sagardoy (1995) described a broader spectrum of applications:

- detection of potential groundwater-rich areas;
- evaluation and planning of surface water resources;
- assessment of irrigation potential and evaluation of the economic impact of irrigation;
- identification and mapping of irrigated areas;
- monitoring of irrigation and drainage infrastructure;
- land use and crop identification;

- management of irrigation systems;
- farm irrigation management;
- use of RS in hydrological models;
- land cover mapping;
- assessment of flood damage and flood modelling;
- detection and assessment of water logging and of salinity.

Applications may be grouped using an even simpler system of categories, based on the specialities emerged during 25 years of earth observation from space:

1. High resolution mapping of irrigated lands
2. Crop water requirements
3. Crop water stress
4. Detection of saline areas
5. Catchment hydrology
6. Irrigation management

17.2 General Approach

17.2.1 Applications versus Observables and Algorithms

Most applications listed above are typically based on a combination of observations (=what we really measure), algorithms (=how do we manipulate measurements) and a-priori knowledge (=which constraints can we place on the solutions). In some cases specific experience is necessary to retrieve the required information from the image data. To give a concise overview of the techniques used a cross-reference table has been prepared (Table 17.1).

17.2.2 Theory and conceptual approach

The remote sensing observables listed in the column headings (Table 17.1) relate to the following radiation-object interactions:

- reflection of solar radiance
- thermal emission (thermal infrared and microwave region)
- scattering of emission of microwave signals

These interaction processes have been described in Chap. 2 of this book and no additional material on theoretical aspects is presented here.

17 Irrigation and Drainage   381

Table 17.1. Remote sensing applications in irrigation and drainage vs. remote sensing observables and methods applied in data analysis

Application	Multispectral images	Multitemporal images	Reflectance spectra	Spectral indices	surface temperature	backscatter coefficient	microwave emissivity
Groundwater potential	Interpretation					interpretation	Water balance modeling
Evaluation / planning surface water	Interpretation; numeric classification	image measurements					
Irrigation potential economic impact	Interpretation			Numeric classification			
Irrigated area	Numeric classification						
Land cover / crops	Numeric classification	Interpretation; numeric classification					
Water logging	Interpretation; numeric classification						
Irrigation performance		numeric classification; GIS	Energy balance		energy balance		
Salinity	Interpretation	numeric classification	Principal components	Numeric classification; field data			
Farm irrigation management				Numeric classification	energy balance		
Management of irrigation systems	Interpretation	Interpretation; numeric classification		Numeric classification	energy balance		
Hydrological modelling		numeric classification	Semi-empirical algorithms	Numeric classification; semi-emp. algorithms	energy balance	semi-empirical algorithms; rad. transfer models	rad. transfer models
Land cover mapping	Interpretation	numeric classification					

Remote sensing methods applied in support of irrigation management are based on the type of observations and spectral regions summarized below. Taking into account the theoretical concepts presented in the preceding chapters of this book, the description of methods given in this section will be limited to an overview. The foundation for the statements given here is provided by the material presented in the section Current Applications.

A. Multispectral classification of crops. The simplest way to discriminate crops and irrigated area is by relying on reflectance spectra (0.4 through 2.5 μm); the spectral contrast between crops and soils is large, much less so between different types of crops. Spectral contrast may be improved by using transforms such the principal component analysis. In some cases textural measurements might be helpful because of the relationship of field lay-out with crop type (e.g. orientation and inter-row distance in vineyards). Agricultural fields comprise a mixture of crops and bare soil in varying fractions, so that the spectral reflectance of such mixed targets tends to depend more on the relative amount of soil and foliage than on the spectral differences between different types of crops. This is an inherent limitation in the use of automatic classification procedures (see also Chap. 7). In arid environments, discrimination of cropped irrigated patches becomes simpler because of the near-absence of green vegetation where no irrigation water is available. Such simpler classification procedures are generally adequate to obtain estimates of irrigated area.

B. Multitemporal classification of crops. Cropping patterns often comprise many crops which differ in terms of sowing date and phenology, which implies that the temporal evolution of the amount and colour of foliage is a useful feature for crop discrimination. Both the amount and colour of foliage determine the spectral reflectance, so multi-temporal sequences of satellite images can be used to measure spectral reflectance at different phenological (growth) stages. The simplest option to discriminate crops is to increase the number of attributes by using simultaneously spectral reflectances on all dates for which images have been acquired.

A better procedure is by taking phenology explicitly into account. This comprises the following steps:

- construct a crop calendar indicating the expected duration of each phenological stage for a given area;

- select the dates of acquisition of images on the basis of the crop calendar, i.e. in such a way that differences in spectral reflectance between crops are largest;

- convert the raw at-satellite observations into at-surface reflectances, i.e. taking into account sensor calibration data and correcting for atmospheric effects.

- combine the spectral measurements in spectral indices, so that just one value is associated with each image pixel on any given date. The temporal sequence of images gives then a time series I(t) comprising the values of the selected spectral index on all selected dates. The I(t)-values give a measure of crop phenology and can be used for crop identification. To this end images may be processed to enhance contrast, so that qualitative interpretation and delineation is easier, or numerical classification procedures may be applied to the I(x,y,t) images.

C. Calculation and mapping of crop water requirements. Traditionally (FAO, 1977) crop water requirements are estimated by combining climate data with crop specific coefficients, which accounts for the ratio of the maximum crop specific evapotranspiration to a reference evapotranspiration dependent on climate only. More recently (e.g. Allen et al., 1994) FAO guidelines are evolving towards the use of the Penman- Monteith (PM) equation which avoids the use of empirical crop coefficients. The rationale for this approach is that empirical values of crop coefficients, such as provided by FAO (1977), cannot account for the actual variability of canopy properties which control consumptive water use. The two approaches are less different than it may be concluded at a preliminary analysis. It can be shown, using the PM equation, that the crop coefficients are related explicitly to the parameters of the PM equation (Stanghellini et al., 1990). Accordingly, radiance measurements may contribute in different ways to the determination and mapping of crop water requirements:

- accurate determination of cropped area (see A) reduces significantly the uncertainty on crop water requirements within an irrigation scheme.

- mapping of specific crops (see B) in combination with accurate knowledge of crop phenology reduces further the uncertainty on the spatial and temporal variability of crop water requirements.

- clusters of crops having similar crop coefficients can be mapped using advanced numeric classification procedures (D'Urso and Menenti, 1996a, b);

- estimates of surface reflectance, leaf area index, surface temperature can be derived from spectral and directional radiances and applied to compute maps of crop water requirements.

- estimates of net radiation, R_n, and soil heat flux, G, can be obtained from spectral radiances, with (R_n - G) being an useful upper bound of maximum evapotranspiration and, therefore, of crop water requirements.

D. Crop stress detection using visible through near infrared radiances. Spectral reflectance of leaves is very sensitive to water absorption, with the effect being observable even in the broad bands of Thematic Mapper (Jackson et al., 1980). At higher spectral resolutions (e.g. 10 nm), water absorption is observable at multiple wavelengths (Curran, 1989). Determination of leaf water content gives a direct measure of water shortage. Relationships between properties of green foliage, e.g. leaf water content, and reflectance spectra of canopies change (e.g. Wessmann, 1992) with canopy architecture and background reflectance (soil and plant litter). Actual determination of stress level, therefore, may not be feasible for partial canopies. Reflectance spectra may be used, however, to determine the relative fractions of foliage and soil which gives a measure of irrigation uniformity.

E. Crop water stress using thermal infrared radiances. As described in Chap. 8, measurements of spectral radiances in the 0.4 through 12. μ region can be used to study the land surface heat balance and to obtain estimates of actual evaporation or of other indicators of soil water availability. Assuming weather as well as all land surface properties, except surface temperature and crop transpiration, remain constant, then measurements of radiometric temperature provide a measure of the difference between actual and maximum transpiration and, therefore, of crop water stress. This applies, for example, to a relatively homogeneous, complete canopy. Poor irrigation uniformity may well lead to a spatially variable water availability and possibly to water stress measurable through surface temperature (see e.g. Stanghellini and De Lorenzi, 1994). Spatial variability of other land surface properties such as hemispherical reflectance, vegetation height or canopy architecture will be significant for partial canopies or when considering larger, heterogeneous areas. Observed spatial variability of the radiometric temperature will be larger and related to water availability in a more complex manner. A significant amount of research has focussed on this subject (see Current Applications), but the scope for truly operational use of airborne or spaceborne sensors appears constrained by the need for timely acquisition of the observations and rapid distribution and actual use in support of irrigation management. Continuing water stress leads to observable changes in spectral reflectance (see D), which are easier to measure at the required spatial and spectral resolution.

F. Determination of soil moisture using active and passive techniques. Soil water content is related to two land surface properties:

- soil thermal capacity and conductivity;
- soil dielectric properties;

The former determine temporal changes of soil surface temperature in response to the heat balance at the soil surface. This subject has been discussed at length in the Chap. 8, devoted to evaporation. It may be recalled, however, that the soil surface can be observed at low fractional vegetation cover only. This implies that it is unlikely to obtain useful estimates of soil water content in irrigated lands on the basis of this physical principle. An alternative method, also based on the heat balance approach is by relating the ratio of actual to potential evaporation (evaporative fraction, see Chap. 8) to a water availability factor. As regards the technical feasibility, the same comments apply as given above (item E). Soil dielectric properties (see also Chaps. 2 and 9) are very sensitive to the soil water content and sensitive to soil type and salinity. Moreover both active and passive microwave instruments probe at best a limited soil depth and only the lowest frequencies currently in use (i.e. L-band) may be relevant for irrigation and drainage studies. Even in the latter case one would need to estimate soil water content in the root zone on the basis of microwave observations applying to, say, the upper 5 to 10 cm soil depth (L-band). For detection and monitoring of soil salinity this restriction in the observable soil depth may not be as severe, since salts tend to accumulate in the top soil (Taylor et al., 1996).

G. Monitoring of extensive irrigated lands. All the methods mentioned above may in principle be applied to monitor large irrigated lands. Because of the emphasis on repetitive analysis of large amounts of data, examples found in literature tend to cluster around the use of relatively simple methods. We may distinguish between applications based on data at high and low spatial resolution. In both cases most applications rely on the use of spectral reflectance measurements. Spectral indices are widely used to monitor both, conditions and extension of irrigated lands.

High spatial resolution. Determination of irrigated area using the spectral observations provided by MSS, TM, SPOT, LISS etc. has achieved a semi-operational status. Although precise mapping of individual crops remains a problem in most cases, groups of crops (e.g. summer cereals) can be mapped satisfactorily to obtain a measure of actual crop water requirements. Determination of year to year changes of irrigated area, possibly cropwise, would be a major contribution towards better irrigation management. Moreover, the same information is necessary to assess whether collected irrigation fees are paid consistently with land use. Determination of waterlogged areas and bare salt affected soils identifies the sectors within an irrigation scheme where water management has to be improved.

Low spatial resolution. Sensors on-board meteorological satellites have the capability of providing consistent observations at rather short intervals of time, although at a relatively low spatial resolution. The multispectral observations obtained worldwide with the NOAA/ AVHRR sensors at a 1 km x 1 km spatial resolution can be and have been (see examples of applications in the next section) used to assess the temporal variability of land surface characteristics relevant to irrigation management. Spatial resolution determines to a significant extent the practical relevance of remote sensing for irrigation and drainage. Literature (see Current Applications) points to an overwhelming interest for observations at high spatial resolution. In some cases, however, even data at very low spatial resolution may provide useful information on irrigated lands. Choudhury (1997), for example, determined global potential evaporation at a 270 km x 270 km resolution by combining satellite and meteorological data. Although not meant for local use, results compared well with lysimeter observations at a number of locations.

GIS applications. The information obtained with airborne and space-borne sensors provides a range of useful indicators on irrigation and drainage (see Chap. 4 and Current Applications). The scope of the remotely sensed data, however, can significantly be enhanced through integration with other types and sources of data in a GIS environment. In many cases the remotely sensed data have to be combined with measurements of flow-rates or irrigation volumes to obtain indicators of the effectiveness of irrigation (see e.g. Menenti et al., 1989, 1996). In other cases several thematic maps (e.g. soils, infrastructure) have to be combined with the remotely sensed data. This is a wide research field on itself which is covered extensively in Chap. 4.

17.2.3 Examples of applications

Some examples of applications will be described in this section to clarify the preceding description of methods, concise by necessity.

Multi-spectral classification. Irrigated area can be determined in most cases with simple methods using combinations of spectral images (Colour Plate 17.A). The spatial resolution of current sensors systems such as TM is sufficient to provide useful information at the level of individual fields. The spectral contrast between irrigated and not-irrigated plots is significant. Subtle features may also be observed, however, and may lead to appreciable errors when numeric algorithms are applied to determine the area irrigated.

Multi-temporal classification. Spectral contrast of targets comprising a mixture of crop and soil is often limited and multi-spectral classification may not perform adequately. Another solution is to take advantage of phenology to determine area under irrigation or specific crops. When dealing with rather large areas (Colour Plate 17.B) crop phenology leads to a seasonal cycle of spectral reflectances and therefore of spectral indices. A combination of few observations at well chosen dates in relation with phenology may suffice to determine land cover type. The use of multi-temporal measurements of spectral indices to classify crops and land cover has been described above (17.2.2 B). Here an example is presented based on observations of the Soil Adjusted Vegetation Index (SAVI; Huete, 1988):

$$SAVI = \frac{(1+L)(NIR - R)}{NIR + R + L} \tag{17.1}$$

where $L= 0.5$, NIR= near infrared reflectance and R= red reflectance.

The combination of low SAVI values in April (red component of the image), high SAVI values in June (green component) and again lower values in August (blue component) gives irrigated land a characteristic hue which can be used to map irrigated land and determine irrigated area.

Crop water stress. Actual evaporation can be estimated as described in Chap. 8. An example is presented in Colour Plate 17.C where the evaporative fraction (actual divided by maximum evaporation) calculated with the algorithm SEBAL (Bastiaanssen, 1995) is presented. Irrigated fields have an evaporative fraction close to unity, which indicates sufficient water supply, while bare soil has a significantly lower evaporative fraction. While this is an indication of the reliability of the method, the information is not very different from the information on land cover.

Monitoring of irrigation systems. Assessment and monitoring of the performance of irrigation systems can be accomplished by combining remotely sensed data with diverse ancillary data. The example given here is based on the approach described by Menenti et al. (1990). To measure the relation of diverted water volume with crop water requirements the following performance indicator has been calculated:

$$IP = \frac{\sum_{k=1}^{N} E_{p,k} \cdot A_{i,k}}{V_i} \qquad (17.2)$$

where $E_{p,k}$ is the potential evapotranspiration of crop k, $A_{i,k}$ is the area occupied by the crop k in the portion i of the irrigation scheme and V_i is the volume of irrigation water applied in the area A_i The summation over k gives the total crop water requirement within the area A_i. A more precise estimate of crop water requirements should be obtained as the product of $E_{p,k}$ times a crop coefficient $k_{c,k}$. These coefficients can be estimated using multi-spectral image data (D'Urso and Menenti, 1996a, b), but are omitted here for simplicity. Values of IP greater than one indicate insufficient water supply, while values less than one indicate that applied irrigation water exceeds water requirements. The data shown in Fig. 17.2 indicate that applied irrigation water has been sufficient throughout the irrigation season in the Torre de Abraham scheme, while in Peñarroya was barely sufficient in 1989 and 1991, with severe scarcity in 1996. Calculation and mapping of IP can be done with a simple GIS procedure which combines the map of fields boundaries (see Colour Plate 17.A) with a map of irrigated land obtained with e.g. a Thematic Mapper image and ancillary data on weather (to calculate $E_{p,k}$) and irrigation water volumes.

17.3 Current Applications

17.3.1 General

Most applications rely on two basic observational concepts:

A. The large difference between the red and near infrared spectral reflectance is a measure of healthy, well watered crops. This property can be used to identify irrigated crops and observed variability is a general indicator of crop conditions.

B. The radiometric temperature of canopies is an indication of crop water stress when all environmental factors but water supply are either known or constant within the area observed.

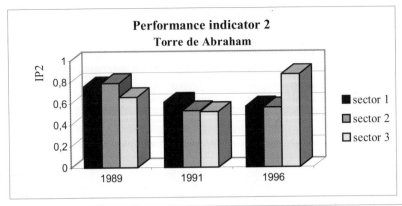

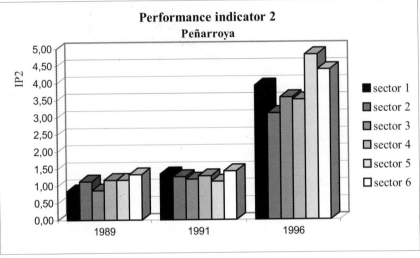

IP$_2$ Peñarroya, sector level.

Fig. 17.2. Values of the performance indicator IP (see Eq. 17.1 for definition); irrigation systems: Torre de Abraham (sprinklers) and Peñarroya (furrows) both in Spain; irrigation seasons: 1989, 1991 and 1996 (Boss and Jacobs, 1997, personal communication)

In practice, most practical approaches are case-specific and cannot be transferred to different situations without taking into account differences in characteristics of farming systems such as the cropping pattern or climatic factors. In the next pages several examples of applications are briefly described to direct readers towards literature relevant to their requirements.

17.3.2 High resolution mapping of irrigated lands

Crop identification and mapping. The feasibility of a remote sensing method to determine quantity, type and water stress of crops by means of Landsat-TM images was demonstrated by Azzali and Menenti (1989). Individual crops in two irrigation districts in the Po valley, Italy were mapped with Landsat-TM multitemporal data using a combination of spectral indices. An accurate crop calendar had to be established to perform multi-temporal analysis. Reflectance measurements in TM bands 3, 4, 5 and 7 were a more reliable indicator of water shortages than the radiometric surface temperature (obtained with TM-band 6 radiances). Three practical applications were identified: (1) estimation of crop area; (2) crop water stress; (3) estimation of water depth in rice fields.

Irrigated area. Determination of irrigated area is a mature RS application as proven by several studies in different contexts and countries. Verdin et al. (1988) used SPOT data to monitor irrigated lands in a mid latitude, high desert environment in the Newlands Project in W. Central Nevada. Merged 10- and 20-metre data sets were used to prepare 1:24 000 photographic prints, which proved suitable for delineating irrigated fields as small as 4 ha. Brightness and Greeness transformations for XS (multispectral) data were derived and used for a maximum likelihood classification of 8 crop types: Medicago sativa, pasture, fallow, maize, winter cereals, Sorghum sudanense, feed grain and vegetables. An accuracy of 91% was obtained after application of a per-field, majority-rule revision. Thelin and Heimes (1987) mapped irrigated cropland with Landsat Multi Spectral Scanner data to determine use of groundwater from the High Plains Aquifer in parts of Colorado, Kansas, Nebraska, New Mexico, Oklahoma, South Dakota, Texas, and Wyoming. Estimates of irrigated area were combined with irrigation application rates (from field data) to compute estimates of irrigation water use. Carton (1991) mapped irrigation systems in Algeria using SPOT data. The pilot study was made to develop appropriate methods for a nationwide inventory. Two SPOT images of two very different agroclimatic zones were used: (1) a mountainous area with relatively small irrigation systems using water pumped from wadis; and (2) a much more arid area where the irrigation systems are supplied from boreholes or by flooding. Digital processing of images taken in June combined with field observations gave the following results: detailed location of irrigation systems above 0.25 ha; a map of the system at a scale of 1:50 000; estimates of irrigated area and location of systems in the subwatersheds. El Kady and Mack (1992) concluded that inventories of irrigated lands in Egypt might be improved through an operational use of satellite data to supplement aerial photography. Growth in irrigated area reacts to market demand. This often occurs against regulations as for example shown by Cuevas and Gonzalez (1992) using Landsat MSS image data. Irrigated area inside the National Park of Doñana (Spain) increased by 500% from 1978 to 1988 and was lower in 1988 (a fairly wet year) than in 1989 (dry year).

Traditional irrigation. The spatial resolution of current space-borne sensors is adequate for detailed studies of relatively small irrigation schemes, even in complex environments such as the Ecuador's Andes region studied by Goulven et al. (1995) to characterise traditional irrigation systems. Practical methods were developed on the Urcuqui irrigation system (12 635 ha) using stereo-pairs of SPOT image data. Crops were identified with an accuracy of ≈70%, and 60% of infrastructures were correctly mapped. A study of old irrigation systems in Turkmenistan was presented by Volovik and Kochergin (1993).

Numeric classification algorithms are widely used in studies of land cover with numeric satellite data. Criteria have been developed to assess the performance of numeric classification techniques. The performance of several Landsat 5 Thematic Mapper (TM) image classification methods for crop estimates in an irrigation district was studied by Barbosa et al. (1996). The agricultural land cover in a 26300 ha. irrigation district was classified utilizing two Landsat 5 TM scenes. Manual and automatic selection of training areas for the classification of two single subscenes and a combined multitemporal subscene result in several differently classified images. The classifications based on the training sets selected manually provided the most precise results, with the exception of the single image spring classification of rice. D'Urso and Menenti (1996 b) presented a general approach to rank numeric classification procedures on the basis of statistical indicators of accuracy, reliability and discriminability.

17.3.3 Crop water requirements – Visible and Near Infrared

Reflectance spectra in the 400 to 2500 nm is related to green biomass, Leaf Area Index, leaf water content and chlorophyll, although this relation is not straightforward and crop-dependent. Bausch (1995) demonstrated that better estimates of crop coefficients (and therefore of crop water requirements) were obtained for a maize crop using spectral reflectance instead of traditional methods. Moran et al. (1989) evaluated various spectral indices in relation with crop water stress. The Perpendicular Vegetation Index (PVI) decreased with the stress-induced change in architecture while the near-IR:red ratio (IR/R) remained relatively constant. As a result, the near-IR:red ratio was more successful in estimating Medicago sativa biomass than was the PVI. Fernandez et al. (1994) compared the sensitivity of radiometric characteristics of a wheat crop to water and nitrogen stress. The reflectance of parts of the spectrum was modified by the applied treatments. The normalized difference vegetation index (NDVI) was the most suitable spectral combination for estimating total leaf area and leaf area index (LAI) across different treatments. Plant N content could be estimated independently of treatment through a linear combination of green and red canopy reflectance, whereas chlorophyll content could not be estimated independently of plant treatment from radiometric measurements. On the other hand Mani et al. (1991) studied the spectral reflectance of a sorghum crop as influenced by nutrient, irrigation and plant canopy. The spectral ratio, (IR/R) was highest between 55 and

65 d after sowing. The highest spectral reflectance was measured where 135 $kg_N ha^{-1}$ was applied. Spectral reflectance was related to dry matter yield. The technique could not discriminate between some of treatments in the experiment. D'Urso and Menenti (1996 a) developed an equation, based on the Penman-Monteith combination formula, to calculate crop coefficients from spectral radiances and meteorological observations.

17.3.4 Crop water stress – Thermal Infrared

Field irrigation experiments. The relation of the surface radiometric temperature of foliage T_f with crop transpiration and water stress has been demonstrated in several field studies. Kumar and Tripathi (1990) studied the impact on wheat crops of different irrigation treatments. These authors observed that environmental factors other than soil water availability influenced T_f and proposed a water stress indicator which included vapour pressure deficit. The significance of T_f as a crop water stress indicator was confirmed by other studies (e.g. Inoue, 1991) who investigated photosynthetic activity in wheat, maize, soyabeans and cotton. On the other hand the results presented by Reicosky et al. (1994) indicate a limited sensitivity of T_f to reductions in canopy photosynthesis and evapotranspiration of spring wheat. These authors observed that differences in evapotranspiration and canopy temperature were small, while canopy photosynthesis was reduced by 52% in the deficit-irrigation treatment compared with the full irrigation treatment.

Potential for irrigation scheduling. Reginato et al. (1987) reviewed the state-of-the-art on remote sensing and irrigation to conclude that radiometric measurements of T_f could be used to determine when plants are under stress and to quantify that stress for irrigation scheduling purposes. Jones (1994) evaluated different ways to enhance the sensitivity of a crop water stress indicator. A higher sensitivity was observed when T_f was compared with the temperature of either a wet or a dry reference surface. Overall sensitivity is greatest when incident radiation is high, at low ambient humidity and at low windspeeds. In another review Kadam and Magar (1994) emphasised once more the potential of improving irrigation scheduling using radiometric measurements of T_f. This conclusion relied on studies with tomatoes, wheat and chickpeas in India.

Actual transpiration and irrigation scheduling. Ben Asher et al. (1992) carried out similar experiments in California and Israel on tomato and cotton to assess the applicability of radiometric thermal infrared measurements to improve management of high frequency drip irrigation systems. They concluded that this method could not yet improve the efficiency of high frequency irrigation. Adequate measurements of T_f were only obtained when water treatments differ largely from one another. On the other hand good agreement was obtained under optimal water regime when daily lysimeter measurements were compared to TIR estimated transpiration.

Low altitude remote sensing. Fouche and Booysen (1994) described the development and field-testing of a remotely piloted aircraft. In field experiments in S. Africa, in 1989-92, soyabeans, wheat and maize received irrigation treatments which resulted in well–watered or moderately stressed conditions. The payload comprised a thermal infrared thermometer and a camera. The difference between canopy and air temperature was effective in assessing moisture stress in crops of soyabeans and maize but less so in wheat.

Satellite remote sensing. Moran (1994) presented the results of an experiment conducted in Arizona, USA, to test the feasibility of using SPOT and Landsat satellite and aircraft data for crop monitoring and irrigation management. Several shortcomings of present satellite systems with respect to providing timely information for irrigation management were identified. A simple cost/benefit analysis suggested that remotely sensed information from both, satellites and aircraft, when shared by several users in moderately-size irrigation districts could be profitable. Meteorological satellites provide observations at higher temporal resolution but at lower spatial resolution. Applications of AVHRR observations to irrigation water management were described by Vidal (1990), Vidal and Perrier (1990) for sugar cane and by A. Brasa Ramos et al. (1996) for barley.

Irrigation performance. The requirements on timeliness of observations become significantly less severe when aiming at the assessment of irrigation performance. Bastiaanssen et al. (1996) evaluated the ratio of actual to maximum evaporation of the Eastern Nile Delta with Landsat Thematic Mapper data. This ratio is a measure of the adequacy of irrigation water supply.

17.3.5 Catchment hydrology

Localized irrigation. Satellite remote sensing has been applied to assess the potential of surface runoff for irrigation in semi–arid regions (Humborg and Tauer, 1994) and to identify suitable sites (Tauer and Humborg, 1992).

Irrigation and watershed processes. The integration of hydrological simulation models with remotely sensed data is a rapidly growing area of research and applications. D' Urso et al. (1992) mapped crop water requirements in a large irrigation district by combining analysis of Landsat Thematic Mapper images with a calculation of reference evapotranspiration. The results were used in combination with a regional hydrological model to describe the interaction between surface irrigation and the groundwater system. A map showing the spatial distribution of irrigation performance indicators was also produced.

17.3.6 Detection of saline areas

Mapping of saline areas. Maps of salt affected areas have been produced using aerial photographs (Wildman, 1982) and Landsat MSS and TM imagery (Joshi and Sahai, 1993). Examples of the link with irrigation water management were presented by Younes et al. (1993) and Vincent et al. (1996). A significant amount of work has been dedicated to develop empirical correlations of spectral reflectances and indices with crop yield and vegetation cover (e.g. Wiegand et al., 1992). Besides space- and airborne multispectral scanners (e.g. Hick et al., 1984), also multispectral aerial photography was used for this purpose (Emelyanov et al., 1980). The usefulness of early MSS image data to assess changes in the salt affected area was noted by Dwivedi (1994), although several authors including Dwivedi and Rao (1992) concluded that the TM band 5 carries highly significant information for the detection and mapping of salt affected areas. To develop a reproducible procedure to measure salt affected area and its change Wiegand et al. (1994) applied numeric classification procedures. An optimal spectral index for mapping and monitoring of salt affected areas in an irrigation system was proposed by Mirabile et al. (1995).

Soil salinity. Soil dielectric properties change significantly with solute concentration. Both passive and active microwave sensors have the potential of providing measurements of soil salinity. Taylor et al. (1996) retrieved soil dielectric properties from AIRSAR data. Complex dielectric constants determined by inversion of the polarized backscatter of radar images acquired in wet conditions clearly delineated the distribution of saline soils in the Tragowel Plains Irrigation Area of Victoria, Australia. There was good agreement between the areas delineated as having anomalous dielectric constants by the radar backscatter inversion techniques with saline areas as defined by geophysics and as inferred from dielectric constants determined in the field. The magnitudes of P–band radar–determined dielectric constants were close to those expected from field determinations. The L–band–determined dielectric constants gives the best discrimination between saline and non–saline areas as seen at the surface.

Impact on crop conditions. Assessment of salinity damage to pastures using airborne remote sensing. Schaefer et al. (1990) attempted to correlate visible and near IR imagery with crop conditions and yield. Independent observations of soil salinity were necessary to assess impact on yield.

17.3.7 Irrigation management

Groundwater resources. Multi–spectral and multi–temporal satellite data from IRS LISS–I and Landsat–TM were used by Rao et al. (1993) to assess the hydrogeological characteristics as well as groundwater irrigated areas in the Anantapur district, Andhra Pradesh, India. Areas of overdeveloped groundwater were identified.

Drainage conditions. Babaev and Babaev (1994) mapped filtration lakes in the Turkmenistan deserts. Krapil'-skaya and Sadov (1987) mapped hydrogeological conditions in the region of the southern Aral Sea using satellite imagery and aerial photography. This study provided useful insights on the impact of irrigation projects in the desert area of the basin.

Inventories of irrigation systems. Landsat imagery has been used routinely since 1972 to identify and map central pivot systems in Nebraska (Rundquist et al., 1989). Data concerning the location of pivots have been collected since 1972 and summarized annually as a map publication. Inventory procedures consist of identifying and mapping occurrences of the characteristically circular pivot systems as interpreted from photographic enlargements of Landsat imagery.

Irrigation uniformity. Lahlou and Vidal (1991) demonstrated that diverse informations can be extracted from satellite data and used in irrigation water management. Mapping of irrigated crops and detection of intra–plot heterogeneity (a measure of irrigation uniformity) were of particular relevance.

Irrigation performance. Menenti et al. (1989) described the determination, mapping and interpretation of three performance indicators. All indicators are calculated with Landsat TM data and ancillary geo–referenced data. The three indicators provide respectively a measure of equity, adequacy and effectiveness (or marginal benefit) of water allocation. The implications for irrigation water management in Mendoza, Argentina, were elaborated further by Menenti et al. (1990). Broadly similar conclusions, but based on a study of four irrigation schemes in India were drawn by Thiruvengadachari et al. (1995). Satellite remote sensing is currently being employed as a operational tool to monitor and evaluate system performance and to diagnose pockets of sub–optimal performance. These techniques can provide useful data during project preparation, implementation and post–completion phases of scheme construction. The case studies demonstrate that primary data on irrigated area, cropping pattern, crop conditions and yield can be generated.

17.4 Current and future observations

The techniques and applications reviewed in the previous chapter span a period of time during which there has been a significant evolution in sensor technology. Changes in the type of measurements obtained with earth observation satellites have been limited, the only fundamentally new observations being radar backscatter with a suite of SAR instruments. A major technological evolution of airborne sensors has taken place however. The driving force of such evolution has been the development of the new International Earth Observing System (IEOS), a very diverse suite of instruments developed by space agencies: NASA, ESA, NASDA and CNES. In many cases a series of forerunners of the sensors scheduled for flight in the time frame 1999 - 2005 has been developed to test new technological solutions. Some of these developments are of direct relevance for the determination of soil moisture, e.g. the synthetic aperture microwave radiometer (ESTAR), while solid state arrays of detectors have made imaging spectrometry accessible to a larger community. The extent and direction of technological evolution can be documented by looking at three categories: current satellite sensors, current airborne sensors and future satellite sensors. For accurate and

detailed information on missions and satellites for earth observation the reader is referred to Kramer (1994). Because of the fast pace of changes in the IEOS satellite missions the information provided by this otherwise very valuable book is at times incomplete or outdated. Updates can be retrieved from the home pages set up for most missions and sensor systems. At least two consequences of technological evolution should be mentioned in the context of this chapter:
- users should be rather knowledgeable to choose the best observational approach given the complexity and diversity of available options.
- Developing countries have an increasing and active role in the development of earth observation from space.

Technological evolution has been relatively limited during the first two decennia of earth observation. A useful improvement in the most recent ones of the current sensors (e.g. MOS) is the better spectral sampling in the region where useful spectral features related to green biomass may be detected. The low spatial resolution, however, restricts the potential use of such observations to large-area surveys.

The comparison of the characteristics of the current spaceborne sensors with the characteristics of the advanced airborne sensors shows immediately that major technological developments were taking place at the same time as practical approaches were being developed on the basis of simpler measurements. The number of imaging spectrometers, for example, is increasing very rapidly and many comparable sensors are available for field studies. To a large extent this is due to the availability and rapidly decreasing cost of solid state arrays of detectors. Smaller countries and industry have developed such systems, at times with commercial applications in sight. Spectrometry from space at intermediate resolutions will be a reality soon (MODIS, 1999, while true spectrometers at high resolution will come later (PRISM, 2003). It should be noted, however, that hyperspectral observations at high spatial resolution might have been available already, had not the LEWIS mission failed in 1997.

We are past the time of generic remote sensing applications and specific approaches (i.e. based on well identified measurements) should be developed. Technological evolution may contribute to improve significantly the timeliness of observations. Light but powerful instruments may be flown on unconventional platforms such as ultra-light aircrafts at a cost attractive to true end users, such as farmers in a competitive agriculture or operators of advanced on-demand irrigation systems.

17.5 Future Directions and Potential

Irrigation and Drainage, as an area of research and application, has been a fruitful learning and development ground for a broad range of remote sensing applications. Several inherent features of irrigated lands led to this situation. Irrigated lands are generally flat or gently sloping, the contrast between irrigated and non-irrigated areas is significant especially in arid and semi-arid zones, serious water management problems tend to establish areas with open water and salt deposits, all of which clearly identifiable with simple techniques and multispectral imagery.

This potential has yet be fully brought to bearing, however. It is fair to say that land cover mapping in different forms has reached the status of an operational technique of which most irrigation professionals in most countries are fully aware. The situation is rather different as regards other techniques and applications described in this chapter. Although there are examples of applications of e.g. thermal infrared observations at a scale consistent with real applications, these more advanced techniques are still somewhat removed from the real world of irrigation and drainage. One obvious explanation is that quantitative use of thermal infrared and microwave observations require a solid scientific background and a level of specialization not readily available to irrigation professionals. Another less obvious reason is that the overriding requirement is timeliness of the observations. As regards land cover, timely observation means once or twice during an irrigation season. As regards e.g. crop water stress, timeliness means one day from data acquisition to actions undertaken. Since daily observations are only available at relatively low spatial resolution, use of this information gets out of the picture for the operation of farms and small to medium-sized irrigation schemes. Large irrigation schemes, conversely, might benefit from the use of near-real time information except that the operation of such systems is a slow and complex process.

The above remarks lead to two avenues where true applications might be developed:

A. Analyse and possibly remove the constraints which limit the use of low resolution data in extensive irrigation schemes. In the next few years several sensors will be flown to collect data at medium and low resolution with global coverage achieved in one to three days. Many sensors, e.g. MODIS and MERIS, can provide useful observations with significantly improved spectral and radiometric characteristics.

B. Develop smaller and lighter sensors usable for cheap single-user data collection and analysis. Current technology provides the opportunity of developing reasonably advanced sensors having very limited mass and size. There are already examples of multispectral cameras and radiometers flown onboard model aircrafts which have been used as a true user-oriented and affordable service to farmers.

References

Allen, R.G., M. Smith, A..Perrier and L.S. Pereira, 1994. An update for the definition of reference evapotranspiration. ICID Bull. Vol. 43(2): 1-34; 35-92

Azzali S. and Menenti M., 1989. Application of remote sensing to irrigation water management in two Italian irrigation districts. in: Calabresi G. and Guyenne D. (eds.): European coordinated effort for monitoring the earth's environment: pilot project campaign on Landsat Thematic Mapper applications 1985-87. ESA IRS Rome, Italy: 41-48

Babaev A.M. and Babaev A.A., 1994. Problems of Desert Development. 1994, No. 1, 17-23

Barbosa, P.M., M.A. Casterad and J. Herrero, 1996. Performance of several Landsat 5 Thematic Mapper (TM) image classification methods for crop extent estimates in an irrigation district. *Int. J. Rem. Sens.* Vol. **17 (18)**: 3665-3674

Bastiaanssen, W.G.M., 1995. Regionalization of surface flux densities and moisture indicators in composite terrain. Ph.D. Thesis Agricultural University of Wageningen and *Report 109*, DLO- Winand Staring Centre, Wageningen, The Netherlands.: 195 p

Bastiaanssen W.G.M., Wal T. van der and Visser T.N.M., 1996. Diagnosis of regional evaporation by remote sensing to support irrigation performance assessment. Irr. Drain. Syst. vol. 10(1): 1-23

Bausch W.C., 1995. Remote sensing of crop coefficients for improving the irrigation scheduling of corn. Agric.Water Manag. vol. 27(1): 55-68

Ben Asher J., C.J. Phene and A. Kinarti, 1992 Canopy temperature to assess daily evapotranspiration and management of high frequency drip irrigation systems. Agric. Water Manag. vol. 22 (4): 379-390

Brasa-Ramos A., Santa-Olalla F.M. de and Caselles V, 1996. Maximum and actual evapotranspiration for barley (Hordeum vulgare L.) through NOAA satellite images in Castilla-La Mancha, Spain. J. Agric. Eng. Res. vol. 63 (4): 283-293

Carton P., 1991. Inventory and mapping of irrigation systems in Algeria using SPOT data.Options Mediterraneennes. Serie A, Seminaires Mediterraneens No. 4: 139-143.

Choudhury, B.J., N.U. Ahmed, S.B. Idso, R.J.Reginato and C.S.T. Daughtry, 1994. Relations between evaporation coefficients and vegetation indices studied by model simulations. Remote Sens. Environ. vol. 50: 1-17

Choudhury, B.J., 1997. Global pattern of potential evaporation calculated from the Penman-Monteith equation using satellite and assimilated data. Remote Sens. Environ. vol. 61: 64-81

Cuevas, J.M. and F.Gonzalez,1992.Variacion temporal de las superficies cultivadas en regadio en el area del Parque Nacional de Donana mediante analisis de imagenes Landsat MSS. Investigacion Agraria, Produccion y Proteccion Vegetales. vol. 7(2): 245-252

Curran, P.J., 1989. Remote sensing of foliar chemistry. Remote Sens. Environ. vol. 30:271-278

D'Urso, G., E.P. Querner and J.A. Morabito, 1992.Integration of hydrological simulation models with remotely sensed data: An application to irrigation management. Proc. Intern. Conf. Advances in Planning, Design and Management of Irrigation Systems as related to sustainable land use. Catholic University; Leuven; Belgium: 463-472

D'Urso, G. and M.Menenti, 1996 a. Mapping crop coefficients in irrigated areas from Landsat TM images. Proc. European Symposium on Satellite Remote Remote Sensing II. SPIE, Inter. Soc. Optical Engineering, Bellingham USA, vol. 2585: 41-47

D'Urso and M.Menenti, 1996 b. Performance indicators for the statistical evaluation of digital image classifications. ISPRS J. Photogramm. Rem. Sens. vol. 51(2): 78-90

Dwivedi, R.S., 1994. Study of salinity and waterlogging in Uttar Pradesh (India) using remote sensing data. *Land Degradation and Rehabilitation* vol. 5 (3): 191-199

El Kady, M. and C.B. Mack, 1992. Use of photographic and non-photographic remote sensing techniques for agricultural inventory in Egypt. Proc. 16th ICID European Reg. Conf. Vol. 3. Methods for decision-making and applications: 39-48

Emelyanov, V.A., V.V. Gorbachev and V.A. Karitonov, 1980. Aerospace methods of evaluating the salinity of irrigated land. *Vestnik Sel'skokhozyaistvennoi Nauki.* No. 7: 120-128

Fernandez, S., D.Vidal, E.Simon and L. Sole-Sugranes, 1994. Radiometric characteristics of Triticum aestivum cv. Astral under water and nitrogen stress. Int. J. Rem. Sens. vol. 15 (9): 1867-1884

Fouche, P.S. and Booysen N., 1994. Moisture stress in crops evaluated with remotely piloted aircraft. South Africa Waterbulletin.vol. 20 (5): 6-7.

Goulven P.le, Vidal A., Ruf T. and Chaffaut I., 1995. Remote sensing and traditional irrigation in tropical mountainous regions. in:A.Vidal (ed.). Proceedings of ICID Special Technical Session on the Role of Advanced Technologies in Making Effective Use of Scarce Water Resources, Rome, Italy, 1995. vol. 2: hs16.1-hs16.14

Hick P.T., J.R. Davies and R.A. Steckis, 1984. Mapping dryland salinity in Western Australia using remotely sensed data. Proc. *Satellite Remote Sensing: Review and preview.* Remote Sensing Society: 343-350

Huete, A.R., 1988. A soil-adjusted vegetation index (SAVI). Remote Sens. Environ. vol. 25: 89-105

Humborg G. and Tauer W., 1994. Estimation of surface runoff for irrigation in semi-arid regions using information from satellite data. Proc. of International Conference on Water Resources Planning in a Changing World.: V17-V19

Inoue, Y. 1991. Remote-monitoring of physiological and ecological status of crops by means of multi-sensing. 3. Quantitative analyses of the transpiration and stomatal responses to drought stress and relationship between photosynthesis and transpiration. Bulletin of the Nat. Agric. Res. Center. No. 20: 45-71

Jackson, R.D., S.B. Idso, R.J. Reginato and P.J. Pinter, 1980. Remotely sensed crop temperatures and reflectances as inputs to irrigation scheduling. Proc. *ASCE Irrigation and Drainage Division Specialty Conference*, Boise, Idaho. ASCE, New York: 390- 397.

Jackson, T.J., D.M. LeVine, A. Griffis, D.C. Goodrich, T.J. Schmugge, C. Swift and P.E. O'Neill, 1993. Soil moisture and rainfall estimation over a semi-arid environment with the ESTAR microwave radiometer. IEEE Tran. Geosci. Remote Sensing vol. 31: 836-841

Jensen, M.E., 1996. Irrigated agriculture at the crossroads. In: Pereira L.S. et al. (eds.), *Sustainability of Irrigated Agriculture.* Kluwer Academic Publishers, Dordrecht, The Netherlands, NATO ASI series, Series E: Applied Sciences, Vol. **312**: 19-33.

Jones H.G., 1994. Use of infrared thermometry for irrigation scheduling. Aspects of-Appl.-Biol.No. 38: 247-253

Joshi, M.D. and B. Sahai, 1993. Mapping of salt affected land in Saurashtra coast using Landsat satellite data. *Int. J. Rem. Sens.* vol. 14 (10): 1919-1929

Kadam, J.R. and Magar S.S., 1994. Irrigation scheduling with thermal infrared remote sensing inputs - a review. J. Maharashtra Agric. University. vol.19 (2): 273-276

Kramer, H.J., 1994. *Observation of the Earth and its Environment. Survey of Missions and Sensors.* 2nd Edition. Springer verlag: 580 pp

Krapil'-skaya, N.M. and A.V. Sadov, 1987. Aerospace monitoring of hydrogeological and reclamational status of lands in southern Aral coastal strip. Problems of Desert Development. vol. 1: 22-26. Translated from Problemy Osvoeniya Pustyn: vol. 1: 22-27.

Kumar, A, and R.P. Tripathi, 1990. Thermal infrared irradiation for assessing crop water stress in wheat. J. Agron. Crop Sc. vol. 165 (4): 268-272

Lahlou O. and Vidal A., 1991. Remote sensing and management of large irrigation projects. Options Mediterraneennes. Serie A, Seminaires Mediterraneens. No. 4: 131-138

Mani S., V.Chellamuthu, A. Palanivel and G. Ramanathan, 1991. Spectral reflectance of sorghum crop as influenced by nutrient, irrigation and plant canopy. Agropedology. 1991, 1: 91-96

Menenti M., T.N.M. Visser, J.A. Morabito and A. Drovandi, 1989. Appraisal of irrigation performance with satellite data and georeferenced information. in:-J.R. Rydzewski and C.F. Ward (eds.). Irrigation Theory and Practice. Inst. of Irrigation Studies, Southampton Univ., UK: 785-801

Menenti, M.(ed.), 1990. Remote sensing in evaluation and management of irrigation. INCYTH-CRA, Mendoza, Argentina. 337 p

Menenti M., Visser T. and Chambouleyron J.L., 1990. The role of remote sensing in irrigation management: a case study on allocation of irrigation water. World Bank Technical Paper. No. 128: 67-81

Menenti, M., S. Azzali and G. D'Urso, 1996. Remote sensing, GIS and hydrological modelling for irrigation management. In: L.S. Pereira, R.A. Feddes et al. (eds), Sustainability of irrigated agriculture. Kluwer, Dordrecht: 453-472.

Mirabile C., R. Hudson, G. Ibanez and H. Masotta, 1995. Detection, delimitation and dynamic control of saline areas in irrigated soils through the use of Landsat TM satellite images. *FAO Water Reports* No. 4: 179-184

Moran M.S., 1994. Irrigation management in Arizona using satellites and airplanes. Irrig. Science. vol. 15(1): 35-44

Moran-MS, P.J. Jr. Pinter, B.E. Clothier and S.G. Allen,1989. Effect of water stress on the canopy architecture and spectral indices of irrigated alfalfa. Rem. Sens. Env. vol. 29(3): 251-261

Rao, R.S., M.Venkataswamy, C.M. Rao and G.V.A.R. Krishna, 1993. Identification of overdeveloped zones of ground water and the location of rainwater harvesting structures using an integrated remote sensing based approach-a case study in part of the Anantapur district, Andhra Pradesh, India. Int. J. Rem. Sens. vol.14 (17): 3231-3237

Reginato, R.J., 1987. Irrigation scheduling and plant water use. in: F. Prodi et al. (eds.). Proc. Agrometeorology. 2nd International Cesena Agricultural Conference, Cesena, 8-9 October 1987: 189-200

Reicosky D.C., Brown P.W., Moran M.S., 1994. Diurnal trends in wheat canopy temperature, photosynthesis, and evapotranspiration. Rem. Sens. of Env. vol.49 (3): 235-245

Rundquist, D.C., R.O. Hoffman, M.P. Carlson and A.E. Cook, 1989. The Nebraska center-pivot inventory: an example of operational satellite remote sensing on a long-term basis. Photogr. Eng. Rem. Sens. vol. 55(5 Pt 1): 587-590

Schaefer N.L. and Slavich P. and Barrs H.D., 1990. Assessment of salinity damage to pastures using airborne remote sensing. in: Humphreys E., Muirhead W.A. and Van der Lelij A. (eds.). Management of Soil Salinity in South East Australia. Australian Society of Soil Science Inc. Canberra; Australia: 39-47

Seckler, D., U. Amarasinghe, D. Molden, R. De Silva and R. Barker, 1998. *World water demand and supply 1990 to 2025: Scenarios and issues.* Research Report 19. International Irrigation Management Institute, Colombo, Sri Lanka

Stanghellini C., A.H.Bosma, P.C.J. Gabriels and C.Werkhoven, 1990. The water consumption of agricultural crops: how crop coefficients are affected by crop geometry and microclimate. *Acta Horticulturae*, **278**: 509-515.

Stanghellini and F. De Lorenzi, 1994. A comparison of soil- and canopy temperature-based methods for the early detection of water stress in a simulated patch of pasture. *Irrig. Sci.* **14**: 141-146.

Tauer, W. and G.Humborg, 1992. Runoff irrigation in the Sahel zone: remote sensing and geographical information systems for determining potential sites. Verlag Josef Margraf; Weikersheim; Germany: 192 pp.

Taylor G.R., Mah AH, Kruse FA, Kierein Young KS, Hewson RD and Bennett BA, 1996. The extraction of soil dielectric properties from AIRSAR data. Int. J. Rem. Sens. vol. 17(3): 501-512

Thelin, G.P. and F.J. Heimes, 1987. Mapping irrigated cropland from Landsat data for determination of water use from the High Plains Aquifer in parts of Colorado, Kansas, Nebraska, New Mexico, Oklahoma, South Dakota, Texas, and Wyoming. Bulletin USGS No. 1400-C: C1-C38

Thiruvengadachari S, Jonna S, Raju PV, Murthy CS and Harikishan J, 1995. Satellite Remote Sensing (SRS) and Geographic Information System (GIS) applications to aid irrigation system rehabilitation and management. in: A.Vidal (ed.). Proceedings of ICID Special Technical Session on the Role of Advanced Technologies in Making Effective Use of Scarce Water Resources, Rome, Italy. vol. 2: hs5.1-hs5.13

Veerlapati-Govardhan, 1993. Remote sensing and water management in command areas (Ed. 1). International Book Distributing Co; Lucknow; India: xxii + 353 pp.

Verdin, J.P., D.W. Eckhardt and G.R.Lyford, 1988. Evaluation of SPOT imagery for monitoring irrigated lands. Proc. SPOT-1, utilisation des images, bilan, resultats. Cepadues Editions, Toulouse, France: 81-91.

Vidal A., 1990. Estimation de l'evapotranspiration par teledetection. Application au controle de l'irrigation. Collection Etudes du CEMAGREF, Serie Hydraulique Agricole. No. 8: 180 pp.

Vidal, A., 1995. Proceedings of ICID Special Technical Session on the Role of Advanced Technologies in Making Effective Use of Scarce Water Resources, Rome, Italy, 1995. Volume 2. 1995, hs17.1-hs17.10;

Vidal-A and A.Perrier, 1990. Irrigation monitoring by following the water balance from NOAA-AVHRR thermal infrared data. IEEE Transactions on Geoscience and Remote Sensing. vol. 28 (5): 949-954

Vincent B., A.Vidal, D.Tabbet, A. Baqri and M. Kuper, 1996. Use of satellite remote sensing for the assessment of waterlogging or salinity as an indication of the performance of drained systems. Proc. Workshop Sustainability of irrigated agriculture: evaluation of the performance of subsurface drainage systems. 16th ICID Congress, Cairo, Egypt, 15-22 September 1996: 203-216

Volovik, V.V. and D.V. Kochergin, 1993. Studying the old irrigation lands based on materials of multizonal space photography. Problems of Desert Development. vol. (1): 53-70. translated from Problemy Osvoeniya Pustyn vol. 1: 16-21

Wang, J.M., K. Sahashi, E. Ohtaki, T. Maitani, O. Tsukamoto, Y. Mitsuta, T. Khobayashi, H. Zheng, I. Li and Z. Xie, 1993. Energy and mass transfer characteristics of soil-vegetation-atmosphere system in oasis area. Outline of the biometeorological observation period (BOP). Proc. Int. Symposium on HEIFE. Natural Disaster Research Institute, Kyoto, Japan: 507-514

Wessmann, C.A., 1992. Imaging spectrometry for remote sensing of ecosystem processes. Adv. Space Res. vol. 12 (7): 361-368

Wiegand, C.L., J.H. Everitt and A.J. Richardson, 1992. Comparison of multispectral video and SPOT-1 HRV observations for cotton affected by soil salinity. Int. J. Rem. Sens. vol. 13 (8): 1511-1525

Wiegand, C.L., J.D. Rhoades, D.E. Escobar and J.H. Everitt, 1994. Photographic and videographic observations for determining and mapping the response of cotton to soil salinity. Rem. Sens. Env. vol. 49 (3): 212-223

Wildman, W.E., 1982. Detection and management of soil, irrigation and drainage problems. in: Johannsen and Sanders (eds). Remote sensing for resource management: 387-401

Younes, H.A., A.Gad and M.A. Rahman, 1993. Utilization of different remote sensing techniques for the assessment of soil salinity and water table levels in the Serry Command Area, Egypt. Egyptian J. Soil Sci. vol. 33 (4): 343-354

18 Computation of Hydrological Data for Design of Water Projects in Ungauged River Basins

I. Papadakis[1] and G. A. Schultz[2]

[1]Consulting Engineer, Werksstr. 15, 45527 Hattingen, Germany
[2]Ruhr University Bochum, Institute of Hydrology, Water Resources and Environmental Techniques, 44780 Bochum, Germany

18.1 Introduction

Socio-economical development on regional, national and international basis requires intensive investigations of water resources. Water differs from other resources by the time variability of the amount of water available for use at any given moment. Non-standard seasonal distribution of hydrometeorological parameters, their fluctuations from region to region and from basin to basin make the planning of water projects difficult.

Purposes to be served by water management projects may include: water supply, flood control, irrigation and drainage, hydropower, industrial use, water pollution monitoring, environmental protection, ground water etc. More than 1,5 billion people are living today on the Earth without any access to drinking-water and about 10 million deaths per year are registered as caused by waterborne diseases, flood disasters and famines following long drought periods. Furthermore, since the human population grows rapidly, there is extreme increase in water demand, making efficient management of the available water resources essential for human life.

Planning of water projects requires knowledge of the quantity of water available at any given time unit. This can only be determined on the basis of adequate sets of hydrometeorological data for the area under consideration. It is well appreciated, that sets of hydrometeorological data which extend back far enough to include typical long-term fluctuations is a prerequisite for the successful planning and design of any water project. The minimum period of collecting records for reliable analysis is considered to be 20 to 30 years. If such data sets do not exist, or are not sufficiently long, or not accurate enough, the proposed water project may be either under-designed with accompanying risk of failure and consequent damage or may be over-designed which will render them uneconomic. For example, the estimated mean annual flow based on measured flow data observed in a short-term period, may be significantly different from that of flow data records over a long period of observation. The deviation may be positive or negative, depending on the cycle of the hydrometeorological phenomena to which the observed records belong (wet period, dry period, transition period). In the case of a multipurpose reservoir the negative deviation in the estimated mean annual flow would result in a smaller storage capacity than actually required. A smaller irrigated area and less hydro-electric-power production than expected would be the consequence. In the case of positive deviation in the estimated mean annual flow, all structures would be overestimated. Dam height would be higher than necessary, the reservoir would very seldom be filled up to the required water level. The consequence would be that the structures necessary for water power production such as the power station, conduits etc. would be much larger in size than required. The power production would be below the planned target, and some machines would then remain idle or work with low efficiency. For the planned irrigation schemes and water supply networks there would be insufficient water to irrigate and supply the whole area and the channels and other structures would be of disproportionate size.

Often the design of water management systems in many countries in the developing world suffers from inadequate or non-available sets of hydrometeorological data. If hydrometeorological records are available, they are either too short or, if they are sufficiently long, they include numerous gaps or discontinuities which make these useless for further analysis (Gyau-Boakye, 1993; Gyau-Boakye and Schultz 1994). The consequence is that very often water projects in these countries have to be designed under circumstances which are insufficiently representative for the project area. Instead of calculation of design data based on detailed knowledge of the hydrological behaviour of the area under consideration, data are prepared based on formulas, the validity of which has been proven for entirely different areas.

Different mathematical-statistical and deterministic methods have been developed in the past which enable the designer to estimate missing data, generate synthetic sequences and derive estimates for design purposes (Gyau-Boakye, 1993). The success of these methods, however, depends primarily on the length of the available data considered for the identification of the system in question since computation of systems reliability depends on long time series.

The remote sensing technology provides very useful methods of monitoring and identifying many hydrological parameters and variables which are necessary for the estimation of the project reliability. With the aid of remote sensing techniques, relevant hydrological data can be obtained in a short time, at periodic intervals and covering extended areas. Satellite imagery has been used in the fields of meteorology, watershed management, hydrologic modelling etc. Mathematical techniques in a multidimensional feature space allow an accurate land use classification as well as estimation of vegetation indices (Su, 1996). Changes in land use can be quantified on the basis of multi-spectral and multi-temporal satellite imagery. Also the identification of the potential location of dams in river valleys can be supported by remote sensing. A number of scientists have already used digital spectral data from meteorological geostationary satellite systems to estimate precipitation quantities with success (e.g Arkin 1979; Griffith 1987; Huygen, 1989; Stout et al. 1979; Sorooshian 1997). Furthermore, the estimation of runoff values for flood forecasting, irrigation and other purposes may be based on remote sensing (Hardy et al. 1989, Rott et al., 1986; Tiwari et al. 1991; Kite, 1991; Koren et al. 1994, 1995).

For more than 20 years now, different satellite systems have monitored the globe and will continue to do so, collecting a large amount of data in different spectral bands. These data are stored and are available for further analysis at any time. On the basis of the capabilities of remote sensing briefly mentioned above and taking into consideration the availability of long-term satellite data, a method is presented here by which historical hydrological data, particularly rainfall and runoff values can be generated on the basis of multi-spectral and multi-temporal satellite imagery for the period when satellite information is available.

18.2 General Approach

Planning and design of water management systems (e.g. multipurpose reservoirs) with small risk of failure requires long-term series of hydrometeorological data. In most countries of the developing world observed hydrological data contain either many gaps or are not available at all. Thus hydrologists are urgently seeking new ways of augmenting their conventional data supplies. Satellite remote sensing is being explored as one possible answer to the data acquisition problem.

The general approach presented here is as follows (Papadakis 1994): At the beginning of the project planning after the feasibility study, if hydrological data are not available, a hydrological network has to be installed which collects hydrometeorological data over usually two to three years. Thus on the basis of the collected data and the information obtained from satellite imagery over the same period of time, a mathematical model can be developed which connects the observed hydrometeorological data with data obtained from satellite imagery. The parameters of the mathematical model can be calibrated on the basis of simultaneous satellite data and ground truth. After the calibration it becomes possible to reconstruct historical river flows with the aid of the mathematical model on the basis of the satellite data alone for the period of time for which relevant satellite information exists. This way the very short series of hydrological data (collected during the planning period) can be extended considerably into the past, thus allowing an estimate of the future performance of the water project. It is expected that the technique presented here will become a valuable planning aid for design with short-term series of hydrological data. Figure 18.1 gives an indication of the basic principles of the method.

The approach presented here consists of three consecutive MODULS (Fig. 18.2).

18.2.1 MODUL I: Satellite system, data processing

1. Satellite system

For the choice of a satellite system suitable for the problem on hand, the following criteria should be considered:

Time resolution of the satellite system. On the one hand, for high accuracy a good resolution in time is required, while on the other hand it may not be necessary to process too many data since, for most planning purposes, monthly data are sufficient. Geostationary satellites such as Meteosat and GOES-W/E etc. have a

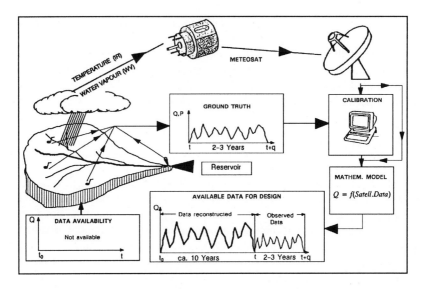

Fig. 18.1. Generation of historical river flows on the basis of long-term satellite data and short-term ground truth

repetition rate of 30 minutes, while polar orbiting satellites, such as NOAA/AVHRR produce only two images per day. Four images per day are available if two NOAA/AVHRR satellite systems are considered.

Satellite systems such as Landsat TM, Spot and ERS-1 with repetition cycles of 16, 26 and 35 days, respectively, are not suitable since the temporal resolution is not appropriate for highly dynamic hydrometeorological processes such as rainfall

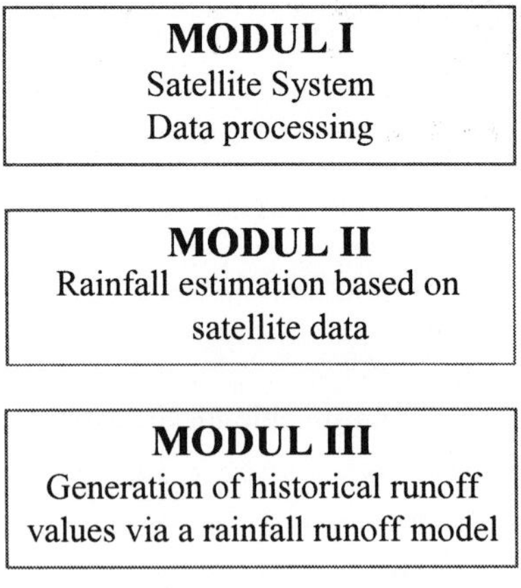

MODUL I
Satellite System
Data processing

MODUL II
Rainfall estimation based on
satellite data

MODUL III
Generation of historical runoff
values via a rainfall runoff model

Fig. 18.2. The technique for estimation of historical flow time series on the basis of satellite imagery

from convective clouds.

Spatial resolution. The resolution of the data in space is correlated with the area of the river basin under consideration. Geostationary satellite systems have a spatial resolution between 2.5 km by 2.5 km and 8.0 km by 8 km, while the spatial resolution of available polar orbiting satellite systems lies between 10 m by 10 m and 1.0 km by 1.0 km, or even larger. The unfavourable spatial resolution of Meteosat and GOES implies that these systems can be considered only for larger catchment areas (>2.500 km^2).

Spectral resolution. GOES, NOAA/AVHRR, Meteosat and other meteorological satellite systems transmit their images in the visible (0.4 μm - 1.1 μm) and thermal infrared band (10.5 μm - 12.5 μm) through the "atmospheric windows". Furthermore, Meteosat transmits images in the water vapour absorption band (5.1 μm - 7.1 μm).

On the basis of brightness characteristics of satellite images taken in the visible spectral band or temperature characteristics taken in the thermal infrared band it is possible to distinguish precipitating clouds from those without precipitation (Barrett and Martin, 1981). Non-precipitating high, thin ice cirrus clouds are very transparent (low brightness) in the visible spectral region, while in the thermal infrared region they are very bright (very cold). Furthermore, it is well known that two convective precipitating cloud cells of comparable size yield different precipitation amounts, depending on the mean relative humidity of the cells' environment. A mean relative humidity is defined as the arithmetic mean of the relative humidity values in different atmospheric layers.

Using the water vapour spectral band of the Meteosat satellite system it is possible to obtain information on the humidity conditions of atmospheric layers situated above a certain level. For a given temperature profile an increase in humidity leads to a decrease of the transmittance towards space. For example, under tropical atmospheric conditions the 600 mb-level should be considered as the lower limit of the layer while the 300 mb-level corresponds to the upper limit of the layer of the relevant water vapour contribution function (Poc et al., 1983). An existing relationship between radiance in the water vapour channel and mean relative humidity can be used for the assessment of the state of the humidity in the near environment of convective clouds.

2. Data processing

In order to adapt the spectral data obtained from the satellite imagery for hydrological purposes a number of preparatory efforts has to be made:

- conversion of the geographical coordinates of the catchment area under consideration into image pixel coordinates;

- normalization of the satellite images in the visible spectral band for sun-angle if the visible information is to be used;

- conversion of the digital counts of the thermal infrared and water vapour images into temperature and mean relative humidity in the upper troposphere respectively;
- atmospheric corrections are not necessary since it has been shown (Rott et. al. 1986) by means of a atmospheric simulation model "LOWTRAN – 5" (Kneizys et. al. 1980) that atmospheric correction is important only for low levels in the atmosphere. Levels below 700 hPa are of main influence for atmospheric transmittance. However, since for rainfall estimation purposes only higher cloud levels are of interest, the radiance of the thermal infrared channel can be converted into temperature without considering atmospheric effects. Uncertainties about cloud emissivity cause larger errors than the effects of atmospheric transmittance.

For image processing techniques applicable here see Chap. 3.

18.2.2 MODUL II: Assessment of the monthly area precipitation on the basis of multi-temporal satellite imagery

A number of scientists have already used data from geostationary satellites to compute precipitation from convective Clouds (e.g. Arkin 1979; Griffith et al., 1978; Hardy et al. 1989; Sorooshian 1997 and others). Some of their schemes are, however, based on the individual tracking of cloud entities throughout their lifetimes. These methods require substantial computing capacity and are applicable only over limited areas in space and time periods. For the reconstruction of long hydrometeorological time series the method for deriving rainfall estimates has to be relatively simple. Arkin (1979) has shown that the fractional cloud cover, which is colder than a certain cloud top threshold temperature, is proportional to the accumulated rainfall amount below (Fig. 18.3).

Hardy et al. (1989) used the duration of cold cloud (CCD) as an indication of rain and the rainfall total within a time period.

There are, however, some limitations in the Arkin and Hardy approach. The threshold cloud top temperature is the only one parameter taken into account, although two cloud cells with the same cloud top temperature and with comparable sizes produce different rainfall amounts depending on the state of their environmental humidity (Krüger and Schultz, 1982). Clouds in a moist environment produce considerably more rain than those in a relatively dry environment. The influence of the humidity on the rainfall intensity is also emphasized by Adler and Mack (1984). The mean relative humidity in the upper troposphere obtained by the water vapour channel of the satellite system Meteosat can be used as an indicator for the humidity state of the environment of clouds or cloud systems.

The principles of the techniques mentioned above for estimating rainfall from temperature of cloud tops are to associate rainfall with cold (high) clouds assuming that these are the tops of active convective cells or convective systems.

Figure 18.4 shows examples of the correlation between rainfall rate and cloud top height (i.e. cloud top temperature) for various regions in North America.

18 Computation of Hydrological Data   407

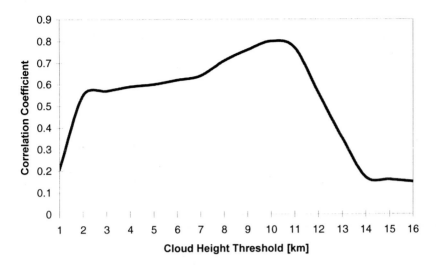

Fig. 18.3. Correlation between rainfall accumulation and fraction of area covered by clouds above various height (temperature) thresholds (acc. to Arkin, 1979)

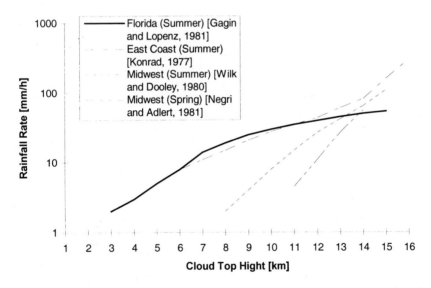

Fig. 18.4. Maximum rainfall rate as a function of cloud top height in various regions of North America (acc. to Adler et al. 1984)

The application of the technique requires that various parameter values and the mathematical approach have to be specified prior to computations:

- time resolution of the images (given by sensor system, may be chosen as multiples thereof)
- spatial and time resolution for area precipitation computations
- cloud top temperature threshold value relevant for precipitation occurrence
- mathematical relationship between satellite information and area precipitation

The choice of the above factors depends on the requirements of the user.

Based on the technique of Arkin the determination of the Fractional Cloud Cover Index (FCCI) may be carried out in the following two consecutive steps, namely:

Step 1: Identification of the cloud cell areas which are colder than a certain threshold cloud top temperature value, covering the river basin under consideration either totally or partially in the thermal infrared spectral band.

Step 2: Addition of all pixels for every cloud cell as obtained in Step 1 within the river basin under consideration.

The result of this procedure consists of a discrete Fractional Cloud Cover Index (FCCI) for each image having values between 0 and 1 representing the fraction of area (or relative area) of the total river basin under consideration, for which a certain information (Cloud top temperature) relevant for precipitation may be obtained in one or more spectral bands of a satellite system. The relative area is expressed by the number of relevant satellite image pixels divided by the total number of satellite image pixels within the river basin.

The AFCC (Accumulated Fractional Cloud Cover) which is of interest for the hydrological computations (rainfall estimation) is simply the sum of the FCCI's for a given time interval (e.g. a day or a month).

In order to estimate the rainfall in a time interval, an empirical relationship of best fit may be determined using the least-squares method. This means that different numerical values depending upon the radiance, e.g. the threshold cloud top temperature (from the thermal infrared spectral band) have to be varied such that an optimum agreement between observed rainfall and spectral information in the same time interval is achieved. The mathematical function $P = f(AFCC)$ is derived specifically for different climatic regions. For the test catchment in West Africa the equation is shown in Sect. 18.3.2.1.

18.2.3 MODUL III: Estimation of runoff values

The final goal of the approach discussed here is the estimation of historical river flow data on the basis of satellite information alone. This requires a mathematical transformation into runoff values of the rainfall information obtained from the satellite imagery by means of a simple parametric rainfall runoff model. A major constraint to the application of general rainfall runoff models is the non-availability of data required. More sophisticated physically-based distributed models require large quantities of data which are usually not available in the countries of the developing world. Therefore the most complex model is not necessarily best for all purposes, particularly where input data are sparse or nonrepresentative.

A simple model which can be applied under tropical environments is described by Higgins (1981). This model relates runoff to rainfall by employing a single equation and a single constraint and was used for estimation of monthly data. The equation assumes that runoff in period i consists of a portion r_i of the rainfall in period i, a smaller proportion r_{i-1} of the rainfall in the previous period $i-1$, and so on, to a total of n periods. Provision is made for a time lag of T periods before first runoff appears after rainfall. The portions r_i are related to each other through an exponential decay and are limited such that the simulated runoff as a portion of rainfall is equal to the observed long-term runoff/rainfall ratio A. The model is mathematically described by the following equation:

$$Y_i = \sum_{k=0}^{n-1} \left(v \cdot b^{k-T} \cdot X_{i-T-k} \right) + \alpha_{i-T} \qquad (18.1)$$

where:
- Y_i: runoff in period i
- X_i: rainfall in period i
- T: lag between rainfall and runoff
- n: number of time periods contributing to Y_i
- α_i: error component of Y_i not explained by the model
- v, b: recession parameters

The parameters v and b of the recession are related by:

$$b^{0-T} + b^{1-T} + b^{2-T} + \ldots + b^{n-1-T} = A/v \qquad (18.2)$$

where:
- A: the observed long-term runoff/rainfall ratio

Since b can be estimated from v or vice versa, using the equations above, there are only three independent parameters to be evaluated, namely, n, v (or b), and T. T will be equal to zero except in cases of unusual catchment shape or soil conditions or when very short time periods are applied.

The model applied has to be calibrated with the aid of the measured runoff and simultaneous rainfall data collected during the planning period of e.g. two or three years to produce the best overall fit. After calibration it is possible to compute historical river flows on the basis of satellite information alone. In order to find out how much satellite information is really required in order to meet the hydrological goal, a sensitivity analysis has to be done by which the amount of satellite data to be used can be optimized.

18.3 Application

In order to show, how the approach described in Sect. 18.2 functions in practice an example will be presented.

Due to the fact that the approach discussed here is mainly meant for planning of water management systems in countries of the developing world, it seems appropriate to give an example from the tropics. On the basis of an existing cooperation between the Institute of Hydrology, Water Management and Environmental Techniques of the Ruhr-University Bochum and the Water Resources Research Unit, CSIR, in Ghana, West Africa, the Tano river basin in Ghana was chosen.

18.3.1 Study area and data used

The Tano river basin comprises a catchment area of about 16.000 km^2, in which daily records from 23 raingauges for the period of 1983-1984 are available. The raingauge density is 1 gauge per 700 km^2 and their distribution over the catchment is reasonably uniform. Additionally measured monthly runoff values and monthly rainfall for the periods from 1969 - 1972, 1975 - 1976 and 07/1984 - 04/1985 are also available. The observed daily rainfall data are accumulated in monthly values, since for planning purposes of water management systems monthly data are usually adequate. The required monthly area precipitation is computed from the available point measurements with the aid of the reciprocal distance weighting method.

In West Africa, where the test catchment is located, rainfall intensity and distribution in space is dependent on the migration of the ITDZ (Inter Tropical Discontinuity Zone). The ITDZ is the front where two air masses of different origin and characteristics converge (Tropical Continental Air Mass and Tropical Maritime Air Mass). The migration of the ITDZ gives rise to two well-defined seasons (wet and dry). The months July and August are dry although the ITDZ is farthest north (wet season) which means that the atmospheric water vapour is high. This situation is defined as a climatic anomaly over West Africa. Especially for the Tano river basin two main seasons are defined, namely the wet season in which the whole catchment area under consideration is influenced only by maritime air masses excluding the months July and August (short dry season) and the dry or relatively dry season in which the catchment area is influenced mostly by continental air masses (Ojo, 1977). Figure 18.5 shows the location of the catchment in a semi-arid area in Ghana, West Africa.

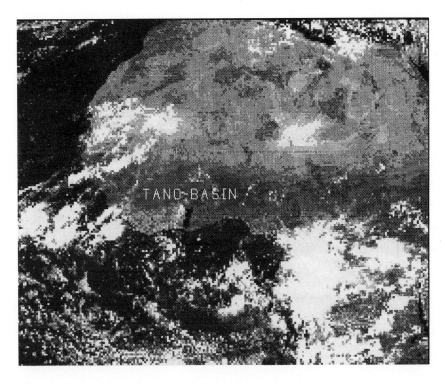

Fig. 18.5. Location of the Tano river basin in Ghana, West Africa. Meteosat VIS Image

18.3.2 Assessment of the monthly area precipitation with the aid of multi-temporal B2-Meteosat satellite imagery

Most of the existing cloud indexing techniques are based on a threshold value of a certain cloud top temperature determined from the thermal infrared spectral band. Griffith et al. (1978) states e.g. that precipitating cloud systems usually have a top black body temperature of less than -20°C and a brightness of more than 80 DC (Digital Counts) in the visible spectral band. Arkin (1979) gives a threshold value for cloud top temperature of precipitating cloud systems with 235 K (-38°C). In our case for the Tano river catchment sensitivity analyses showed that a threshold value of -20°C provides optimal results. The technique presented here uses the IR spectral band from the Meteosat in order to estimate the monthly area precipitation.

Since there is a trade-off between required model accuracy and amount and costs of the data acquisition and processing, a compromise has been made in the technique presented here. The so-called B2 Meteosat data were used which contain only every sixth pixel in a line and every sixth line in all three Meteosat spectral channels. The time resolution was chosen to be 3 hours. This way the data costs were reduced significantly without much loss of accuracy as a sensitivity analysis has shown.

The relationship between AFCC (Accumulated Fractional Cloud Cover) and monthly area precipitation. Three-hourly Fractional Cloud Cover Indices (FCCI) were determined as a function of a chosen threshold cloud top temperature (-20°C in this case). The AFCC which is of interest for the computation of the monthly area precipitation is simply the sum of the three-hourly FCCI for the whole month. Colour Plate 18.A shows the cloud development over the Tano River basin, Ghana, West Africa by means of two successive Meteosat images of the IR spectral channel for a rainfall event in August 1984.

To estimate the monthly area precipitation, the empirical relationship of best fit is determined using the least-squares method. Different relationships with numerical values depending upon the radiance and specific local climatological conditions, e.g. the threshold cloud top temperature (from the thermal infrared spectral band) and the migration of the ITDZ (Inter-Tropical Discontinuity Zone), were varied in order to find out an optimum agreement between observed monthly rainfall and spectral information from Meteosat. The AFCC and the monthly area precipitation for the test catchment in Ghana are related by:

wet season

$$P = 101.4 \cdot AFCC_{T_s}^{0.13} \qquad (18.3)$$

dry season

$$P = 2.03 \cdot AFCC_{T_s}^{1.2} \qquad (18.4)$$

where:
        P: Monthly area precipitation
        AFCC: Accumulated Fractional Cloud Cover
        T_s: Threshold value of the cloud top temperature (here -20°C)

The equations described above and developed in the calibration procedure with the aid of simultaneous satellite data (AFCC) and ground truth (observed precipitation) may be used for reconstruction of historical monthly area precipitation in the catchment area on the basis of satellite data alone. As can be seen from Fig. 18.6 the agreement between observed and computed monthly area precipitation values is reasonably good.

Moreover, according to a sensitivity analysis of the time resolution, two satellite images per day taken at 20 and 23 GMT (Greenwich Meridian Time) are enough to estimate monthly area precipitation for hydrological purposes in the Tano river basin without significant loss in accuracy as compared to the use of all 48 images per day.

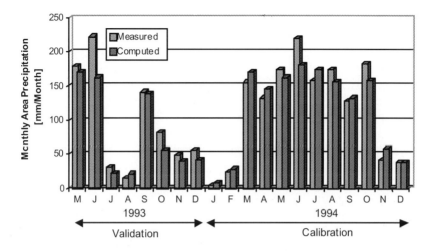

Fig. 18.6. Comparison between observed and computed monthly rainfall on the basis of one spectral channel (thermal infrared), Tano river catchment, Ghana, Westafrica

18.3.3 Rainfall - Runoff Model

The task of the rainfall-runoff model is to convert the time series of monthly precipitation values derived from remote sensing data into monthly runoff volumes.

In the case of surface runoff from the Tano river basin, an accurate application of a hydrological approach would require a detailed topographical, soil and geological survey as well as determination of many other parameters. In order to avoid these difficulties, the model described by Higgins (1981) was applied here. According to Higgins wet tropical environments allow the use of relatively simple rainfall-runoff relationships. Because rain occurs regularly throughout the year (migration of the ITCZ), soil moisture levels are seldom far from field capacity, and actual evapotranspiration approximates potential evapotranspiration. Related to this, soil moisture deficiencies are likely to be rare and of short duration when they do occur. This simplifies the soil water movement process. Time lags between the occurrence of rainfall and the appearance of surface flow are short, because the high rainfall intensities regularly exceed the maximum infiltration rates of the soil profile below the surface.

The objective is to solve the problem of calibrating the parameters of the model for the river basin under consideration. The data of rainfall and corresponding runoff observed from 1962 - 1972 are used. To estimate the monthly runoff values, the model parameters of best fit are determined using the least-squares method. The observed rainfall and runoff values are related by the equation given in Sect. 18.2.3:

wet season (months: J, J, A, O, N)

$$Q_i = vP_i + v \cdot b \cdot P_{i-1} + v \cdot b^2 \cdot P_{i-2} \tag{18.5}$$

with: v=0.112
b=0.507
T=0
P_i = precipitation in time interval i.

dry season (J, F, M, A, M, S, D):

$$Q_i = vP_i + v \cdot b \cdot P_{i-1} + vb^2 \cdot P_{i-2} + vb^3 \cdot P_{i-3} \tag{18.6}$$

with: v=0.063
b=0.582
T=0
P_i = precipitation in time interval i.

Consequently, the rainfall based on satellite spectral information is transformed into monthly runoff for the period when both satellite data and runoff values were available, namely from 07.1983 until 04.1985. The resulting fit is shown in Fig. 18.7 and is found to be satisfactory. This means that the technique is suitable for practical applications in data scarce tropical areas.

18.4 Further Applications

Not only design of water management systems but also river management (monitoring, forecasting and simulation of flows) requires adequate hydrometeorological data. Because of the frequent lack of such data, remote sensing information may be used to obtain better temporal and spatial resolution of relevant data over large river basins.

Based on remote sensing data, a technique was developed by Koren et al. (1995) to monitor, forecast, and simulate (MFS) flows along the Nile river, where no relevant data were available. The principle of this technique is shown in the flow chart of Fig. 18.8.

The MFS System consists of three subsystems: the Primary Data User System (**PDUS**) which provides the continuous input of Meteosat satellite data, the Nile Forecasting System (NFS) which resides on workstation and the High Aswan Dam (HAD) Decision Support system. **The hydroclimate data base** consists of observed hydrometeorological data and Meteosat raw imagery data. **The Preprocessor Component of the NFS** converts incoming raw data into precipitation estimates required by hydrologic models. Because very little observed precipitation data in the Nile basin are available, satellite estimates become a major source of rainfall data. A unique Hybrid Climatological Technique (Schaake and Green-Newby, 1993) was developed which uses Cold Cloud Duration (CCD) data for different temperature thresholds. CCD means the number of hours any one Pixel

18 Computation of Hydrological Data 415

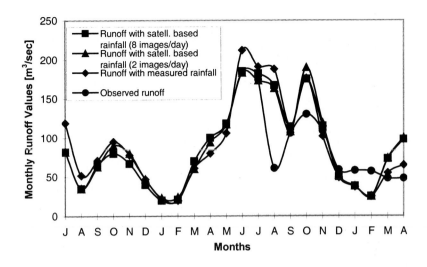

Fig. 18.7. Observed and estimated runoff values on the basis of measured rainfall and rainfall based on satellite imagery for the period July 1983 until April 1985 (acc. to Papadakis. 1994), Tano river, Ghana, Westafrica

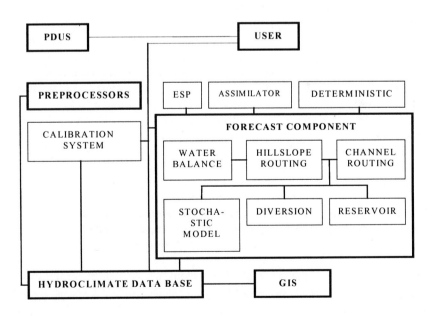

Fig.18.8. The MFS system (after Koren; et al. (1995))

(5x5 km resolution) passes a designated threshold for one day. Analogue gridded fields are created using raingage data only. Observed daily precipitation is interpolated. **The Forecast Component** consists of hydrologic models and software to produce flow and stage hydrographs based on inputs supplied by the Preprocessor. **The User Interface** displays satellite imagery data, gridded data, and time series data; it also serves as the interface between the forecaster and the NFS software, data base and models. The results gained by this technique were used for monitoring and forecasting purposes in the Nile river system.

18.5 Summary and Discussion

Since it is not possible to measure runoff directly with the aid of remote sensing information an indirect approach is suggested which uses remote sensing data for the estimation of rainfall and transform it into runoff with the aid of a rainfall-runoff-model. This way, a long time series of monthly river flows can be generated which allows the estimation of the expected future performance of a planned water project in terms of reliability indices.

The approach presented here shows that the use of multi-temporal satellite imagery allows the estimation of monthly area precipitation values essential for the transformation into runoff values by means of a rainfall runoff model.

The application of the approach presented here, however, is limited. Catchment characteristics, the rainfall type, the area extension of the catchment and the geographical transferability form the main constraints.

The relationship between rainfall and runoff depends mainly on the hydrological behaviour of the catchment area. Therefore, the reconstruction of historical river flows based on a rainfall runoff model calibrated with the aid of a short observed flow record has to take into consideration possible changes in the catchment characteristics, such as residential areas, land use, etc. These changes could be quantified using satellite systems such as NOAA/AVHRR, Landsat or SPOT.

The use of spectral information obtained from the visible, the thermal infrared and the water vapour spectral region allows estimation only of convective rainfall events. This means that only regions with mainly convective rainfall activity can be contemplated. Such regions are concentrated around the Equator.

There is a relationship between the spatial resolution of satellite data and the extension of the catchment area under consideration. By using geostationary satellite systems only areas greater than 10 000 km^2 can be modelled adequately.

The relationship between spectral cloud information and rainfall is empirical and is valid only for the area in question. Previous investigations (Wylie, 1979; Griffith et al. 1981; Adler and Mack, 1984) have shown that the direct transfer to other geographical regions is not possible without further analysis for the specific area. However, the use of the water vapour channel seems promising for overcoming this problem.

Additional spectral information such as passive microwave data could contribute to removing the limitations mentioned above, because the rainfall and passive microwave information have a direct physical relationship to each other. The disadvantage, however, is mainly the large size of the receiving antenna required for the reception of the weak passive microwave radiance emitted. Furthermore, the spatial resolution of the microwave information depends on the size of the receiving antenna, which means that in the future the equipment for the reception of passive microwave information must be considerably improved. New systems (TRMM) can fulfil the requirements for hydrological purposes. The TRMM (Tropical Rainfall Measuring Mission, a joint scientific satellite project between NASA and the NASDA (the Japanese Space Agency)) payload contains five instruments. It includes the Precipitation Radar (PR), a TRMM Microwave Imager (TMI), a Visible and Infrared Scanner (VIRS), a Cloud and Earth Radiant Energy System (CERES), and a Lightning Imaging Sensor (LIS). The main objective of TRMM is to monitor and document the rainfall and energy release at monthly time scales over 500x500 km areas. TRMM was launched on November, 27, 1997. This means that it will take many years until a long time series of data is available as required for the purpose of planning water projects as discussed in this chapter.

References

Adler, R. F. and R. A. Mack (1984): Thunderstorm Cloud Height-Rainfall Rate Relations for Use with Satellite Rainfall Estimation Techniques. Journal Climate and Applied Meteorology, Vol. 33, pp. 280-296

Arkin, P. A. (1979): The relation between fractional coverage of high cloud and rainfall accumulations during GATE over B-Scale array. Monthly Weather Review, Vol. 107, No. 10, pp. 1382-1387

Barrett, E. C. and D. W. Martin (1981): The Use of Satellite Data in Rainfall Monitoring. Academic Press

Green, JoAnna, L. and V. Koren (1995): Results using a simple weighting method to merge satellite and raingage data in the Blue Nile river basin for input into a distributed hydrological model. Conference on Hydrology, 15-20 January, 1995, Dallas Texas, Published by the American Meteorological Society, Boston, MA, pp. 173-177

Griffith, C. G.; W. L. Woodley and P. G. Grube (1978): Rain Estimation from Geosynchronous Satellite Imagery - Visible and Infrared Studies. Monthly Weather Review, Vol. 106, pp. 1153-1171

Griffith, C. G.; J. A. Augustine and W. L. Woodley (1981): Satellite rain estimation in the U.S. High Plains. Journal of Applied Meteorology, Vol. 20, pp. 53-56

Griffith, C. G. (1987): The Estimation from Satellite Imagery of Summertime Rainfall over varied Space and Time Scales. NOAA Technical Memorandum ERL ESG - 25, pp. 102

Gyau-Boakye, P. (1993): Filling Gaps in Hydrological Runoff Data Series in West-Africa. Schriftenreihe, Hydrologie/Wasserwirtschaft, Ruhr-Universität Bochum, Heft 10

Gyau-Boakye, P. (1994): Filling gaps in runoff time series in West Africa. Hydrological Sciences - Journal- des Sciences Hydrologiques, 39, 6, pp. 621-636

Hardy, S.; G. Dugdale; J. R. Milford and J. V. Sutcliffe (1989): The use of satellite derived rainfall estimates as inputs to flow prediction in the River Senegal. IAHS Publ. no. 181, pp. 23-30

Higgins, R.J. (1981): Use and Modification of a Simple Rainfall-Runoff Model for Wet Tropical Catchments. Water Resources Research, Vol. 17, No.2, pp.423-427

Huygen, J. (1989): Estimation of rainfall in Zambia using Meteosat - TIR data. Report 12, The WINAND STARING CENTRE, Wageningen, The Netherlands

Kite, G. W. (1991): Use of satellite data for water resources modelling. A canadian example. Water Resources Development, Vol. 7, No. 1, pp. 21-29

Kneizys, F. X. et al. (1980): Atmospheric transmittance/radiance: computer code LOWTRAN - 5. AFGL - TR - 80 - 0067, Environmental Res. Papers, No. 697, Air Force Geophys. Lab., Hansom AFG, Mass., U.S.A.

Koren, V. and Barrett, E.C. (1994): A satellite based river forecast system for the Nile River. In: "Water policy and management: solving the problems" by D.G. Fontana and H.N. Tuvel (Eds.), Proceedings of the 21st Annual Conference, Denver, CO, pp. 9-12

Koren, V. and E.C. Barret (1995): Satellite based distributed monitoring, forecasting, and simulation (MFS) system for the Nile river. Application of Remote Sensing in Hydrology: Proceedings of the second international workshop. NHRI Symposium No. 14, October, 1994, pp. 187-200

Krüger, L-R; G. A. Schultz (1982): Ermittlung abflußwirksamer Niederschläge aus Satellitendaten. Wasserwirtschaft, 72(1), pp. 1-5

Ojo, O (1977): The Climates of West Africa. Heinemann Educational Book Ltd.

Papadakis, I. (1994): Berechnung historischer Abflüsse mit Hilfe multispektraler und multitemporaler digitaler Satellitenbilder. Dissertation, November 1993. Schriftenreihe Hydrologie/Wasserwirtschaft, Lehrstuhl für Hydrologie, Wasserwirtschaft und Umwelttechnik, Ruhr-Unviversität Bochum

Papadakis, I.; J. Napiorkowski; G. A. Schultz (1993): Monthly runoff generation by non-linear model using multispectral and multitemporal satellite imagery. Adv. Space Res. Vol. 13, No. 5, pp. 181-186

Poc, M. M. and M. Roulleau (1983): Water Vapour Fields Deduced from METEOSAT-1 Water Vapour Channel Data. Journal of Climate and Applied Meteorology, Vol. 22, pp. 1628-1636

Rott, H.; A. Aschbacher; K. G. Lenhart (1986): Study on River-Runoff Prediction Based on Satellite Data. European Space Agency, Contract Report, Contract No. 5376/83/D/JS CSC

Schaake, J.C. and J. Green-Newby (1993): Satellite Estimation of Rainfall Using the Nile Climatoligcal Method. Nile Technical No. 06, 13 pp

Sorooshian, S. (1997): Precipitation estimation from remotely sensed information using artificial neural network models (PERSIAN). GEWEX News, Vol. 7, No. 2

Stout, J. E.; D. W. Martin and D. N. Sikdar (1979): Estimating GATE Rainfall with Geosynchronous Satellite Images. Monthly Weather Review, Vol. 107, pp. 585-598

Tiwari, K. N.; P. Kumar; M. Sebastian and D. K. Pal (1991): Hydrologic modelling for runoff determination. Remote sensing techniques. Water Resources Development, Vol. 7, No. 3, pp. 178-184

Turpeinen, O. M. and J. Schmetz (1989): Validation of the Upper Tropospheric Relative Humidity Determined from METEOSAT Data. Journal of Atmospheric and Oceanic Technology, Vol. 6, No. 2, pp. 359-364

Wylie, D. P. (1979): An Application of a Geostationary Satellite Rain Estimation Technique to an Extratropical Area. Journal of Applied Meteorology, Vol. 18, pp. 1640-1648

19 Detection of Land Cover Change Tendencies and their Effect on Water Management

A.H. Schumann and G.A. Schultz

Institute of Hydrology, Water Management and Environmental Techniques,
Ruhr University Bochum, 44780 Bochum, Germany

19.1 General Remarks

Land cover, i.e. the cover of the earth's surface with soil, vegetation, water, cities etc. depends on natural factors and human activities. The anthropogenic influence on the land cover is related to the land use for agriculture, forestry and urbanisation. As an example Fig. 19.1 shows the impact of agriculture on the global land cover (after Imhoff, 1994).

Land Cover Change Detection at the Global Scale. The importance of land cover change detection at the global scale results from two phenomena (IPCC, 1996):
- Land use change has a large impact on the emission of greenhouse gases. The carbon dioxide emissions during the period 1860 to 1994 amounted to about 360 GtC (gigatonnes of carbon) from which a third (120 GtC) was caused by

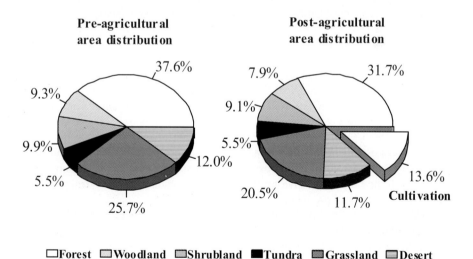

Fig. 19.1. Impact of Cultivation on Global Land Cover (related to total ice-free land area) (after Imhoff, 1994)

deforestation and land use change. The management of forests, agricultural lands and range-lands can play an important role in reducing current emissions of carbon dioxide, methane and nitrous oxide and enhance carbon sinks.
- Model projects show that the land cover of the earth could be changed significantly as a consequence of possible changes in temperature and water availability under double equivalent CO_2 equilibrium conditions:
    * a substantial fraction of the existing forested area of the world (a global average of one-third) will undergo major changes in broad vegetation types,
    * the extent of desertification - land degradation in arid, semi-arid and dry sub-humid areas- will increase
    * the altitudinal distribution of vegetation will be changed.

Land cover can change because of human impact or natural forces. As the rates of change in many places have accelerated a comprehensive view of land cover and ecosystem change is needed. In the Land-Use and Land-Cover Change (LUCC) project, a core project of the International Geosphere- Biosphere Programme (IGBP) and the international Human Dimension of Global Environmental Change Programme (HDP), regional case studies in different areas of the world have begun. The LUCC will aim for global coverage of land cover at 1 km spatial resolution. Satellite observation techniques have been developed that can determine changes in land cover type, as well as spatial and seasonal changes in vegetation (Defries and Townshend, 1994). Landsat imagery was used e.g. to measure deforestation in the Brazilian Amazon Basin. The area deforested increased from 78,000 to 230,000 km^2 from 1978 to 1988 (Skole & Tucker, 1993)

Land Cover Change Detection at Continental and National Scale. In many countries activities were launched to estimate their land cover characteristics:

The U.S. Global Change Research Programme includes an initiative to classify and inventory North American land cover by analysing Landsat data from 1970 to the present.

A U.S. Land Cover Characteristics Data Set 1990 prototype was set up at 1 km resolution by the U.S. Geological Survey, which was developed from a combination of multi-temporal NOAA-AVHRR- data with a variety of earth science data sets.

In Europe the CORINE programme (Co-ordination of Information on the Environment) was established in June 1985 by the European Community's Council of Ministers. One of its objectives consists in the creation of a number of digital databases with the purpose to give information on the status and the changes of the environment. The final product in the CORINE Land cover project is a digital land cover database. The smallest mapping unit covers 25 hectares (which corresponds to a resolution of 500m x 500m). The method adopted by most of the countries that have conducted CORINE land cover projects is visual interpretation of satellite images (Landsat TM) at a scale of 1: 100.000 (Ahlcrona, 1995). In some countries (e.g. in Sweden and the UK) the available Landsat data will be used to estimate the land coverage with a finer resolution (e.g. by generalization into 5 hectares in Sweden (Ahlcrona, 1995) or a 25m raster in Great Britain (Fuller & Brown, 1996).

Land Cover Changes and Hydrology. Changes in land cover include changes in the hydrological cycle and in most of the mass and energy fluxes that sustain the biosphere and geosphere. Due to agriculture, forest management and urbanisation the physical characteristics of the landsurface and upper soil as well as the evapotranspiration process which is strongly related to the type of vegetation are changed. As a result the amount of runoff, the soil moisture and groundwater recharge are strongly affected by land use changes. The evapotranspiration of crops in the mid latitudes can e.g. be assessed with 40-50 % of the yearly precipitation amount, but coniferous forests produce 70 % (Baumgartner, 1970). Measurements of Robinson et al. (1991) show a strong increase in evapotranspiration and (as a result) a reduction of the yearly runoff by 40 % for re-forested agricultural areas in southern Germany. Flood runoff from urbanized areas can be two or three times greater than that stemming from the surrounding vegetated areas. As a result of the large percentage of impervious areas and the artificial drainage network the flood hydrograph from urbanized areas is much steeper than from natural areas. The increase of runoff values coincides with a decrease of the groundwater recharge by urbanisation. As the land use determines not only the hydrological processes but also the material flows, land use changes have also strong impacts on water quality. In many countries of the world the intensification of agriculture is coupled with the utilization of chemical substances like pesticides, nutrients and phosphor. In Germany e.g. the amount of nitrogen fertiliser used per area was increased from 60 kg/ha in 1950 to 200 kg/ha in 1985 (Mehlhorn & Röhrle, 1990). These developments have a strong impact on water quality conditions, e.g. in fresh water reservoirs which are endangered by eutrophication.

In summary it is necessary to estimate the temporal and spatial development of land use changes to assess their impact on water quality and quantity. In most of the cases topographic maps are not sufficient to mirror the dynamics of land use changes as they were not updated in sufficiently small time steps. The land use statistic which is provided for taxation purposes by different state agencies gives us only a lumped information about the changes in general. As land use is only one of many factors which determine the local water balance its impact assessment demands a consideration of other local characteristics as geomorphological, hydro-meteorological, soil and geological parameters. The impact of land use changes on catchment hydrology depends on interactions of all physical characteristics which determine the water and energy fluxes (e.g. Dunn & Mackay, 1995). The effects of alterations in land use can be intensified or reduced in relationship to the other physical characteristics of a river basin. The impact of urbanization for example on the local water balance depends mostly on the soil type which is paved. If a soil with low permeability is paved the effects of the newly built impervious areas will be much smaller than if a permeable soil surface will be sealed. This example shows that it is not sufficient to estimate land use changes in total. The temporal and spatial highly detailed land use data which are needed for such analysis can be provided most efficiently by remote sensing. Also distributed hydrological models should be used for an impact assessment of land cover changes since they allow consideration of the spatial variability in land cover.

19.2 Hydrological Modelling and Land Cover Change

Hydrological models which are used for hydrological impact assessment of land use changes should represent the hydrological processes at the land surface physically based. Many hydrological models consider the specific land use characteristics in their parameterization. Different approaches to consider the spatial heterogeneity of landuse characteristics in hydrological modelling are possible. Examples are:
- in the widely used Soil Conservation Service Method (SCS-model) (Maidment, 1993) various types of vegetation and crops, land treatments and crop practices, paving and urbanization are considered in relationship to one of four soil types to calculate a composite runoff curve number for a drainage basin,
- a statistical approach to describe the spatial heterogeneity of the storage capacity of the upper, rooted soil zone by a distribution function which can be derived directly from an overlay of the vegetation and soil map within a GIS was proposed by Schumann (1993) (e.g. Fig. 16.3),
- in the Precipitation-Runoff Modeling System (PRMS) (Leavesley & Stannard, 1995) the vegetation type is one of many characteristics (such as slope, aspect, elevation, soil type and precipitation characteristics) which are used for partitioning of the watershed in so-called Hydrological Response Units which are modelled separately,
- in the distributed SHE- model the catchment is represented by an orthogonal grid network, where the vegetation type characterises a grid square (Abbott et al., 1986).

In many hydrological models the parameters of vegetation related model components (e.g. for interception and evapotranspiration) are chosen dependent on the land use. For urbanised areas mostly special developed model components are used to consider the specific runoff formation and concentration processes in these areas. If these models are used to describe the hydrological effects of land cover changes it seems easy to alter their parameters. But two problems have to be considered in such model analysis:

The first problem concerns the model approach:

Is the used hydrological model really able to describe the complex changes of hydrological conditions for the specific catchment of interest ?

Klemes (1986) proposed the following methodology to answer this question:

Find a gauged basin where a similar land use change has taken place during the period covered by the historic record and calibrate the model on a segment corresponding to the original land use and validate it on a segment corresponding to the changed land use. For practical purposes it is very difficult to fulfil this demand. Not only are gauged basins with the specific land use changes rare (with exception of urbanized catchments) also the climatic conditions of the validation and calibration periods should be comparable. Otherwise the ability of the model to reflect the hydrological impacts of land use change is hidden by its sensitivity to different hydro-meteorological conditions.

The second problem concerns the general possibilities to detect the hydrological effects of land use changes. Uncertainties in determination of the areal extent and location of these changes and also the inaccuracy of measured meteorological and hydrological data limit our capability to detect their sometimes relatively small effects. Nandakumar and Mein (1997) estimated in a case study the uncertainty of model predictions which results from the random characteristics of model parameters and climatic data. Under consideration of this level of background 'noise' in some cases extreme changes of land use (between 20 and 65% clear cutting of a forest) would be necessary to estimate for example detectable flow increases at the 90% prediction level.

Under the aspects of uncertainty the hydrological impacts of land use changes are mostly discussed for two types of changes which can be identified relatively easily and which are connected with specific changes in rainfall- runoff processes and water balance: urbanisation and deforestation.

Remote sensing reduces the uncertainty of estimations of land cover changes significantly. To estimate the hydrological effects of land cover changes by remote sensing a very careful classification of the satellite scenes or air photography is needed. This necessity can be explained by an example which was given by Allewijn and Baker (1993). These authors estimated the relative differences of the classification results of two Landsat TM scenes for 15 catchments between 26.0 and 184.1 km^2 in the Vechte River Basin in the Netherlands. For agricultural crops, forest types, nature areas, urbanized areas, water and bare soil the similarity between the two classifications varies between 40% to more than 80% if they were compared on catchment scale base although the time difference of the two Landsat scenes was only two years (from 1984 to 1986). The authors explain these differences with random and systematic errors caused by classification but also by "real world differences" in satellite recording dates, harvesting and crop rotation. If these data are used in hydrological modelling uncertainties in areal extent and location of land use units may result in uncertainties in areal evaporation and runoff estimation. Especially distributed models demand a high accuracy in the estimation of the location of land cover changes as their hydrological effects may be increased or decreased by the other aspects (e.g. soil, aspect etc.).

If land use changes within a catchment are identified by remote sensing two possibilities exist to characterise their hydrological effects:
− trends in runoff can be estimated by time series analysis,
− the hydrological effects of these changes can be simulated by application of a hydrological model where its parameters are changed.

If time series analysis is used to identify the effects of land use changes on runoff also possible changes in the time series of meteorological variables (esp. precipitation) should be considered. The effects of land use changes could be hidden by changes in these time series. By utilization of a hydrological model a physical explanation of its parameters in relationship to land use characteristics is necessary. The ability of the model to describe the changes in the hydrological characteristics induced by detected land cover changes can be tested by an analysis of the temporal development of the error of computed runoff values in relationship to measured discharge values at a gauge if the model parameters were calibrated for a time period before the land use change has occurred. If the errors don't show a tendency the errors of the measured discharge series are too large in relationship to the signal of the change in runoff or the model is not sensitive enough to represent these changes.

The main application of hydrological models in relationship to land use changes consist in their prognostic use. If a model is available which is able to describe the complex changes of hydrological conditions caused by land cover changes the planning of watershed management can be improved significantly. By utilization of distributed models the planned land use changes could assessed in their hydrological aspects by consideration of their specific locations.

19.3 A Case Study: Land Use Change Detection by Remote Sensing in the Sauer River Basin, Western Europe

In this study, a common research project of the Institute for Hydrology, Water Management and Environmental Techniques of the Ruhr University Bochum and the Federal Institute for Hydrology in Koblenz (Germany), two questions should be answered:

– How large are the changes in land use which can be detected by remote sensing data within a river basin during the last decades ?

– If such changes in land use became evident is it possible to estimate their hydrological effects ?

As test catchment the Sauer river basin was chosen, which has a drainage area of 4259 km^2 and is located in Western Europe. It covers almost all of Luxembourg in its eastern expansion, a part of Germany, part of Belgium in the west and north and a small part of France in the south. As the river basin is divided between four countries the land use data are not available in a consistent manner. Especially the available land use maps differ in scale and actuality. Only by application of remote sensing it was possible to get a consistent set of land use maps for this river basin as a whole. Remote sensing data were also essential to estimate the temporal development of land use changes in this river basin over last twenty years. To answer the first question mentioned above land use characteristics at different points in time were estimated.

Multi-spectral satellite data used for estimation of land use. Under consideration of the heterogeneity of the Sauer river basin in its orographic, climatic and pedological characteristics, only satellite data with high spatial resolution could be used to determine land use and its changes. Such satellite data for this river basin were available from Landsat satellites. Eight Landsat scenes from the period 1975

Table 19.1. Landsat scenes used in the Sauer River Study

Satellite/Sensor	Acquisition Date	Scene Centre	
		Latitude (North)	Longitude (East)
Landsat 2/MSS[1]	29.08.75	50°14'	6°35'
Landsat 3/MSS	13.05.80	50°16'	6°25'
Landsat 5/TM[2]	30.07.84	50°28'	7°46'
Landsat 5/TM	22.08.84	50°28'	5°84'
Landsat 5/TM	25.05.89	50°28'	7°46'
Landsat 5/TM	20.08.89	50°28'	5°92'
Landsat 5/TM	17.11.89	50°17'	7°50'
Landsat 5/TM	24.05.92	50°28'	5°92'

1: MSS = Multispectral Scanner; 2: TM = Thematic Mapper

until 1992 were selected, two of which were derived from the Multispectral Scanner (MSS), the others originate from the Thematic Mapper (TM). In order to estimate the land use changes the images should be temporally distributed over the time period of 20 years in which hydro-meteorological data were available. In Table 19.1 the selected 8 scenes which cover the whole Sauer basin are listed. The Landsat MSS scenes from 1975 and 1980 have a resolution of 80 m, all TM-scenes 30 metres. The scenes were geometrically corrected and reassembled to 80 m (MSS scenes) and 30 m (TM scenes) resolution respectively. After calculating statistics of the training classes classification was carried out for each scene using a maximum likelihood classifier.

An accuracy assessment was done for the classified TM-scene from 1989 by comparing it with topographic maps in the scale of 1:20.000 and 1:25.000 on the basis of a 2 km x 2 km raster resolution and with 31 aerial photos in the scale of 1:5000 on the basis of 500 m x 500 m raster resolution. By comparison with topographic maps the agreement with the classified scene for the sum of all raster elements was 96.1 % and for each 2 km x 2 km raster the average accuracy was 91.2%. The agreement between the aerial photographs and the classified scene from 1989 was 92.7 % for the sum of all raster elements. The comparison grid cell by grid cell gave an accuracy of 88.1 %. Under consideration of the need to classify the elements of maps and aerial photographs which contain mostly mixed land use classes as one single class the results of this comparison show a good agreement between land use data derived from these different sources. Satellite data are a cost effective way of estimating actual land use data for a large area. The classification of the four Landsat scenes mentioned above led to the results given in Table 19.2.

In Fig. 19.2 these results are visualised. The major changes in land use for the Sauer river basin which were detected in the period from 1975 until 1992 are:
– the built-up area was increased by 2,9 %,
– the cropland was decreased by 5,8 %,
– the forest was increased by 3,3 %.

Table 19.2. Land use classification results (in %) of four Landsat scenes for the Sauer River Basin (4259 km^2)

Land use class	MSS scene 212/25 on 29.08.1975	MSS scene 212/25 on 20.08.1980	TM scene 197/25 on 22.08.1984	TM scene 197/25 on 20.08.1989
Water	0.1	0.1	0.1	0.1
Built-up Area	2.9	4.3	5.5	5.8
Coniferous Forest	8.7	6.9	5.9	8.5
Deciduous Forest	15.7	13.5	19.7	17.4
Mixed Forest	8.8	14.5	9.7	10.6
Crop Land	28.6	24.1	21.7	22.6
Greenland	32.3	33.7	34.7	32.2
Park and Garden etc.	2.9	2.8	2.7	2.9

Figure 19.2 shows that these tendencies in land use changes become only evident if the seven land use classes which were considered first are summarised into three classes. Especially the uncertain differentiation between three types of forest or between crop land and agricultural greenland disturb otherwise the estimation of trends for these land use classes.

A comparison of the land use changes derived from the difference between remotely sensed data of 1975 and 1989 and the lumped land use data for this time period of the official government statistics for the Sauer basin (Table 19.3) shows a good agreement between the estimated changes for the land uses forest, cropland and built-up areas.

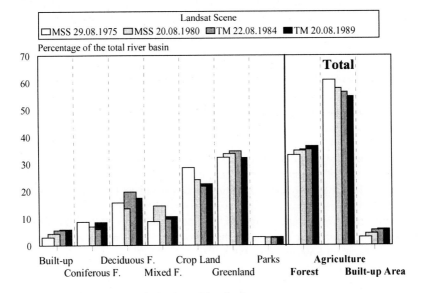

Fig. 19.2. Land use changes in the Sauer River basin

Table 19.3. Land use changes in % for the Sauer basin between 1975 ad 1989 estimated by official land use statistics and remote sensing

Land use	Land use statistics	remote sensing
forest	+ 3,49 %	+ 3,3 %
cropland	- 4,15 %	- 6,0 %
greenland	+ 0,42 %	+ 0,1 %

The land use maps for the Sauer basin which were derived from the Landsat scenes of 1975 and 1989 are displayed in Colour Plate 19.A. A detailed comparison of both maps shows that the land use changes are unevenly distributed within the Sauer river basin. Large changes of the built-up areas become evident in the southern part of the river basin where the city of Luxembourg is located. Also the reduction of cropland becomes more obvious in the South-western part of the catchment. For more detailed consideration of the heterogeneity of land use change within the river basin it was sub-divided into eight sub-catchments with drainage areas between 234 and 948 km^2 (Fig. 19.3).

For each sub-catchment the differences in land use between 1975 and 1989 were estimated by comparison of the classification results of the two Landsat scenes (Table 19.4). In the catchment of the Alzette river, a tributary of the Sauer river the land use changes by urbanisation were most evident (Colour Plate 19.B).

The estimated increase of the built-up area from 7.2 to 16.3 % is partially due to the different resolution of the different land use scenes, e.g. the line structures of the highways, which can be seen in the scene of 1989 are hidden by the coarser resolution of the MSS-scene from 1975. The effect of the spatial resolution of the land use classifications is discussed in greater detail below.

Scale problems in land use change detection. For large catchments the high resolution of the Landsat TM data provides an enormous amount of data. Therefore for hydrological modeling a spatial aggregation of the original data becomes necessary. The accuracy of land use classification depends to a certain degree on the spatial resolution (raster or pixel size) of the analysis. To demonstrate this scale effect the Landsat TM scene of 1989 with an original resolution of 30 m was reassembled to 50 m, 250 m, 500 m, 1000 m, 2000 m and 5000 m (Colour Plate 19.C). For aggregation the dominating type of land use within each new raster element was used to classify the total pixel. The effects of an aggregation of land use characteristics is mostly affected by their spatial heterogeneity (Table 19.5). Small land use units disappear, while the portion of the main land use classes is increased.

This effect may be significant for hydrological modelling as it limits the number of available land use classes. If we combine for example the two classes "greenland" and "crop land" to an new class "Agricultural land" the percentage of this main class is less affected by the scale effect than those from the two sub-classes. The portion of the land use class "Built- up area" depends strongly on the spatial

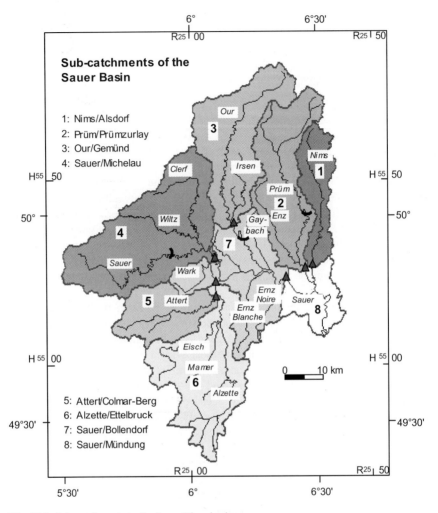

Fig. 19.3. Sub-catchments in the Sauer River basin

resolution of land use classification as a coarser resolution reduces the number of pixels which belong to this class significantly. This example shows, that a consideration of smaller land use patterns demands a higher resolution of the remote sensing data and of the results of land use characteristics. If a coarser differentiation is seen as sufficient with regard to the hydrological model to be used and/or the general aim of the model application an aggregation of the input data seems adequate.

Table 19.4. Differences in land use for each of the 8 sub-catchments of the Sauer River between 1975 and 1989 resulting from two Landsat scenes 1975 and 1989

Land use (in km²)	Alsdorf Nims		Prümzurlay Prüm		Gemünd Our		Michelau Sauer		Colmar-Berg Attert		Ettelbruck Alzette		Bollendorf Sauer		Mündung Sauer	
	L75*	L89*	L75	L89	L75	L89	L75	L89	L75	L85	L75	L89	L75	L89	L75	L89
Water	0.0	0.0	0.3	0.3	0.1	0.1	2.4	2.8	0.0	0.0	0.2	0.2	0.7	1.0	0.1	0.4
Built-up	9.5	15.1	8.8	17.1	7.6	10.5	13.0	27.4	3.9	12.9	48.8	110.2	8.7	22.4	9.7	14.4
Coniferous forest	23.0	23.9	61.5	60.9	112.9	110.2	106.0	112.0	12.3	9.9	23.6	8.5	34.6	28.4	5.6	4.0
Deciduous forest	29.4	32.8	70.7	86.8	76.9	84.7	161.9	146.9	53.8	53.1	99.4	143.1	108.6	128.0	25.6	41.9
Mixed forest	19.2	22.2	55.7	64.3	56.9	63.8	82.4	105.2	22.2	36.4	54.6	56.2	59.9	73.6	22.9	23.9
Crop land	74.1	74.2	174.0	129.5	119.4	129.8	222.0	206.9	121.1	73.0	220.5	130.5	182.9	130.7	88.5	72.3
Greenland	98.1	84.2	184.8	191.7	220.5	191.3	325.5	307.5	98.4	127.7	205.4	211.6	159.0	173.5	70.7	68.0
Park and Garden	8.0	8.8	18.9	24.0	18.0	21.8	25.6	30.1	5.9	4.5	22.7	15.0	14.6	11.4	7.3	5.4
Sum:	261.2	261.2	574.6	574.6	612.2	612.2	938.9	938.9	317.5	317.5	675.1	675.1	569.1	569.1	230.3	230.3
Land use (in %)	L75	L89	L75	L89	L75	L89	L75	L89	L75	L89	L75	L89	L75	L89	L75	L89
Water	0.0	0.0	0.1	0.1	0.0	0.0	0.3	0.3	0.0	0.0	0.0	0.0	0.1	0.2	0.0	0.2
Built-up	3.6	5.8	1.5	3.0	1.2	1.7	1.4	2.9	1.2	4.1	7.2	16.3	1.5	3.9	4.2	6.2
Forest	27.4	30.2	32.7	36.8	40.3	42.2	37.3	38.8	27.8	31.3	26.3	30.8	35.7	40.4	23.5	30.4
Agriculture	69.0	64.0	65.8	60.1	58.5	56.1	61.0	58.0	71.0	64.6	66.5	52.9	62.7	55.5	72.3	31.4

*L75: land use 1975 (Basis MSS), *L89 : land use 1989 (Basis TM)

Table 19.5. Percental area extension of land use classes for the Sauer River Basin for different aggregation levels of the original 30m x 30m Landsat TM scene of August 1989

Land use	Raster width in m				
	30 m	50 m	200 m	500 m	1000 m
Built-up Area	5.8	5.8	3.5	3.0	2.7
Forest	36.5	36.4	38.1	40.0	42.3
Greenland	32.2	35.1	37.1	38.5	40.4
Crop Land	22.6	22.6	21.2	18.4	14.5
Agricultural Land	54.8	57.7	58.3	56.9	54.9

Hydrological modelling. After estimation of the land use changes within the Sauer river basin their hydrological effects were described by a hydrological model. This model was also used for scenario analyses of possible future land use changes to describe their hydrological aspects. The model structure based on a sub- division of a watershed into area elements of "Hydrologically Similar Units (HSUs)" in analogy to the Hydrological Response Units of the PRMS-model. As criteria of similarity the following 5 characteristics were selected:
– land use
– elevation
– soil texture classes
– slope
– exposition.

These different characteristics were organised into classes, e.g. the land use was represented by the six classes: water, build-up areas, coniferous forest, deciduous forest, cropland and greenland. The classification which was used for the Sauer river basin is described in Table 19.6. The number of HSUs which was derived by combination of all these classes amounted to 542. For each of these Hydrologically Similar Units the water balance was computed in a daily time step. The model structure is shown schematically in Fig.19.4.

Table 19.6. Classes of similarity for different physical characteristics within the Sauer River Basin

Similarity Attributes	number of classes	Classes
Elevation	5	< 300 m, 300-400 m, 400-500 m, 500-600 m, > 600 m
Slope	3	0-4°, 4-12°, ≥ 12°
Exposition	3	0-60° and 300°-360°, 120°-240°, 240°-300°
Soil	3	soil texture: sandy loam, silty loam, clay loam
Land use	6	water, built-up areas, coniferous forest, deciduous forest, cropland, greenland

In order to identify the different HSU's an overlay of the five attribute types was performed within the Geographic Information System in which these data were stored. By application of spatially distributed land use data it became possible to consider the specific hydrological effects of changes dependent on the specific location where these changes occurred. In this way it was e.g. possible to consider the fact that the hydrological effects of urbanisation depend strongly on the soil type which is paved. Obviously lumped modelling and characterisation of land use change would not be appropriate to describe these spatially variable impacts. By consideration of the spatial distribution of land use changes the non-linear impact of such changes on the hydrological processes can be considered, e.g. depends on the reduction of the evapotranspiration as a result of deforestation mostly from the energy and water amounts which are available for this process. In order to specify the impact of changes in land use realistically the specific characteristics: soil, elevation, slope and exposition which are relevant for the energy and water fluxes must be known.

The model was used in three steps. First it was calibrated for the period 1978 to 1983. Then two different land use characteristics, one of 1975 and the other of 1989, were used to compute the water balance of the total period from 1.1.1970 to 31.12.1987. Finally the estimated land use changes of the period from 1975 to 1989 were extrapolated into the future. For extrapolation a time step of 20 years was chosen. As an example in Colour Plate 19.D the land use map for the catchment of the river Alzette of 1989 is shown together with a scenario of further de-

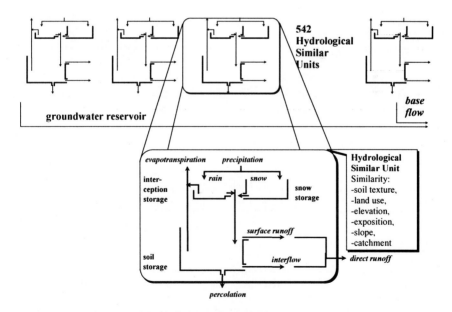

Fig. 19.4. Structure of the hydrological model for the Sauer River Basin

velopment for the year 2009. This catchment is affected strongly by urbanisation as the city of Luxembourg is located in it. In Fig. 19.5 the resulting changes in the seasonal runoff distribution are shown. The runoff in summer (April to September) is increased by a reduced infiltration capacity which leads to an increase of surface runoff in these months with high precipitation intensities. In the winter (October to February) the runoff is slightly reduced. The computed changes in the runoff distribution were not observed in the measured runoff data. The signal of changes in land use between 1975 and 1989 in its effects on the water balance seems to be hidden by errors of discharge and precipitation measurements.

19.4 Summary

Under the aspects of Global Change the concern about land cover changes and their impact of hydrolgical conditions is growing. Only by remote sensing data it is possible to analyse the changes in land use in nearly real- time and with adequate spatial resolution. Remote sensing offers unique opportunities to get consistent land use characteristics from large areas. Especially for international river basins the problems of different national mapping procedures and different scales and actuality of these maps can be overcome. Land use data derived from remote

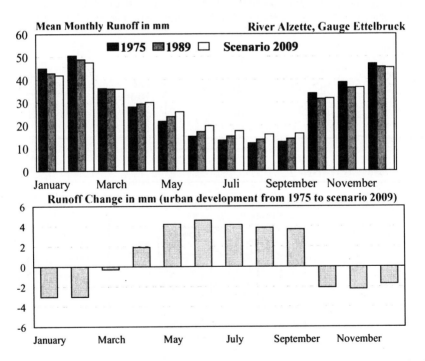

Fig. 19.5. Simulated changes of the seasonal runoff distribution as a result of urbanisation in the catchment of the river Alzette

sensing are an ideal data base for distributed hydrological modelling. Nevertheless some limitations should be considered. The spatial resolution of land use data should be suitable for the scale of the expected land use change. If a coarse spatial resolution is chosen land uses of a relatively small extent may be hidden. If the spatial resolution of classification is too fine the huge amount of data complicates the data management. Another problem of land cover change detection consists in the limited possibility to characterise their hydrological impacts. Usually the hydrometric networks are not sufficient to estimate the resulting hydrological changes. By utilization of physically based models the possible hydrological effects of land cover changes can be assessed. The precondition of such assessments is a hydrological model which represents the resulting changes in hydrological processes with adequate accuracy. Only distributed hydrological models are able to consider the complexity of hydrological effects which may result from changes of land use. These effects can be strengthened or mitigated by the specific characteristics of the locations where these changes occur. Under these aspects a high accuracy of land use classification is needed to represent not only the total amount of land cover changes but also their spatial distribution.

References

Abbott, M.B., Bathurst, J.C., Cunge, J.A., O'Connel, P.E., Rasmussen, J. (1986) An introduction to the European Hydrological System – Systeme Hydrologique "SHE". *Journal of Hydrology* 87, pp. 45-59

Ahlcrona, E. (1995) CORINE Land Cover – A pilot project in Sweden. In: Askne (ed.) *"Sensors and Environmental Applications of Remote Sensing"*, Proceedings of the 14th EARSel Symposium, Göteborg, Sweden, 6-8 June 1994, Balkema, Rotterdam

Allewijn R., Baker H.C. (1993) Comparison of Landsat TM land use classifications and spatial aggregation as input for an environmental hydrological model. In: Winkler (ed.) *"Remote Sensing for Monitoring the Changing Environment of Europe"*, Balkema, Rotterdam

Baumgartner,A. (1970) Water- and energy balance of different vegetation covers. IAHS- Proceedings of the Reading Symposium "World Water Balance"

Defries, R. S., Townshend J. R. G. (1994) NDVI-Derived Land Cover Classifications at a Global Scale. *International Journal of Remote Sensing* 15, pp. 3567-3586

Dunn, S.M., Mackay, R. (1995) Spatial variation in evapotranspiration and the influence of land use on catchment hydrology. *Journal of Hydrology* 171, pp. 49-73

Fuller,R., Brown,N. (1996) A CORINE map of Great Britain by automated means. Techniques for automatic generalization of the Land Cover Map of Great Britain. *International Journal of Geographical Information Systems* 8

Imhoff, M.L. (1994) Mapping Human Impacts on the Global Biosphere. *BioScience* 44, pp. 598

IPCC (1996) Climate Change 1995, *The Science of Climate Change*, Cambridge University Press

Klemes, V. (1986): Operational testing of hydrological simulation models. *Hydrological Sciences Journal* 31, 1

Leavesley G.H., Stannard L.G. (1995) The Precipitation- Runoff Modelling System PRMS. In: Singh, V.P. (Ed.) *Computer Models of Watershed Hydrology*, Water Resources Publications

Maidment, D.R. (1993) (ed.) *Handbook of Hydrology*. Mc Graw-Hill Inc.

Mehlhorn,H., Röhrle,B. (1990) Die Nitratbelastung der Grundwasservorkommen und Maßnahmen zur Reduzierung der Belastung. *Wasserwirtschaft* 80, H.10

Nandakumar, N., Mein R.G. (1997) Uncertainty in rainfall- runoff model simulations and the implications for predicting the hydrologic effects of land use change. *Journal of Hydrology* 192, pp. 211-232

Robinson,M., Gannon,B,, Schuch,M. (1991) A comparison of the hydrology of moorland under natural conditions, agricultural use and forestry. *Hydrological Sciences Journal* 36, 6

Schumann, A.H. (1993) Development of conceptual semi-distributed hydrological models and estimation of their parameters with the aid of GIS. *Hydr..Sciences Journal* 38, 6

Skole, D., and C. Tucker (1993) Tropical Deforestation and Habitat Fragmentation in the Amazon: Satellite Data from 1978 to 1988. *Science* 260, pp. 1905-1910

Colour Plates of Chaps. 16–19 435

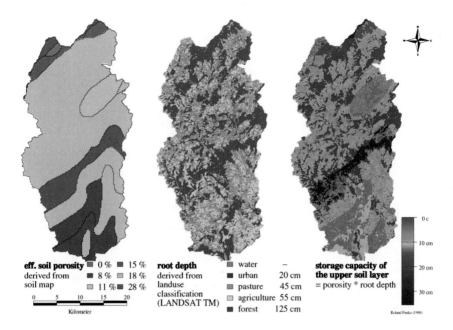

Colour Plate 16.A. Maximum soil water storage capacity (right) of the Prüm river basin (Germany) as the product of effective soil porosity (left) and root depth (center)

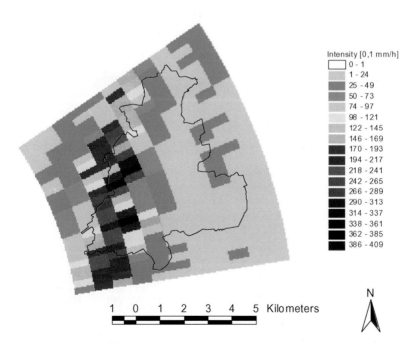

Colour Plate 16.B. Radar image transformed into rainfall intensities over a small urban catchment. Hourly value (16:30-17:30 h) on 13 September 1996 (Courtesy DWD)

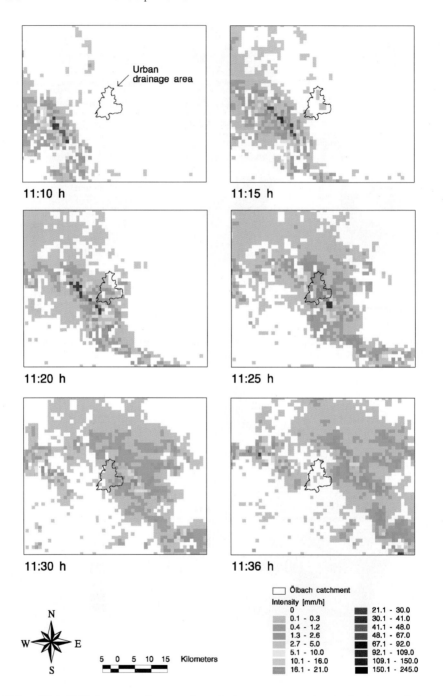

Colour Plate 16.C. Sequence of consecutive radar images over an urban catchment area during a storm on July 23, 1996. Time increments: 5 min, pixel size 1km x 1km

Colour Plates of Chaps. 16–19   437

Colour Plate 17.A. Colour composite of Thematic Mapper band 3 (blue) 4 (red) and 5 (green); Torre de Abraham irrigation scheme, Spain, August 8th, 1997. Boundaries of individual fields indicated by the overlain map

Colour Plate 17.B. Colour composite of multitemporal images of the Soil Adjusted Vegetation Index (SAVI) calculated with AVHRR Ch1 and Ch2 reflectance at 1 km spatial resolution, Aral Sea Basin 1992: SAVI (April)= red; SAVI (June)= green; SAVI (August)= blue; area is 1568 km x 1232 km; greenish areas = irrigated lands (courtesy of N.E. Di Girolamo, NASA, GSFC)

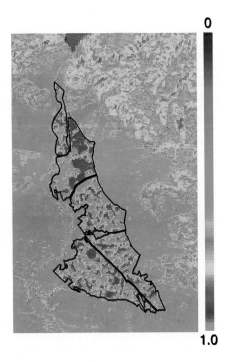

Colour Plate 17.C. Ratio of actual evaporation to net radiation minus soil heat flux (evaporative fraction) estimated using the algorithm SEBAL (Bastiaanssen, 1995); Torre de Abraham irrigation scheme, Spain, July 18th 1996

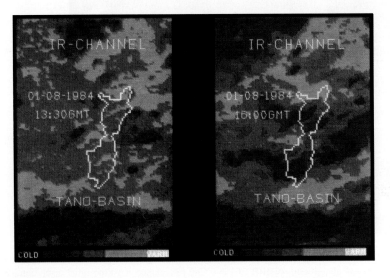

Colour Plate 18.A. Cloud development, Tano River basin, Ghana, West Africa. Two successive Meteosat images of the IR spectral channel

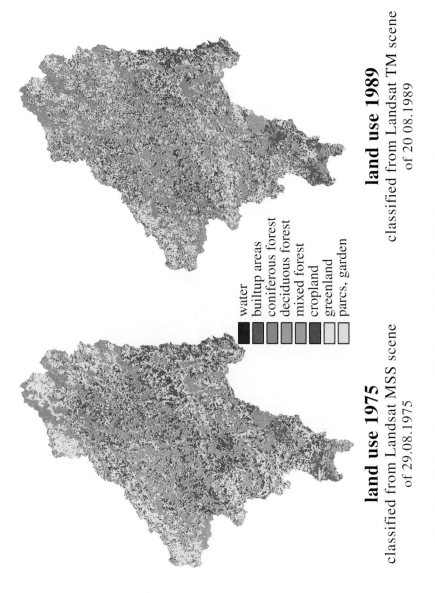

Colour Plate 19.A. Land use maps of the Sauer river basin derived by the Landsat Scenes of 1975 and 1985

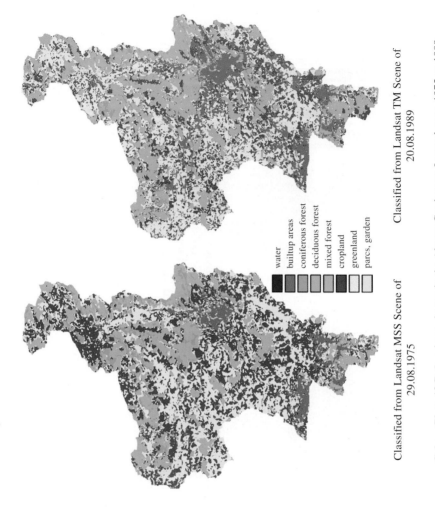

Colour Plate 19.B. Land use change in the Alzette Catchment, Luxembourg, 1975 vs 1989

Colour Plates of Chaps. 16–19 441

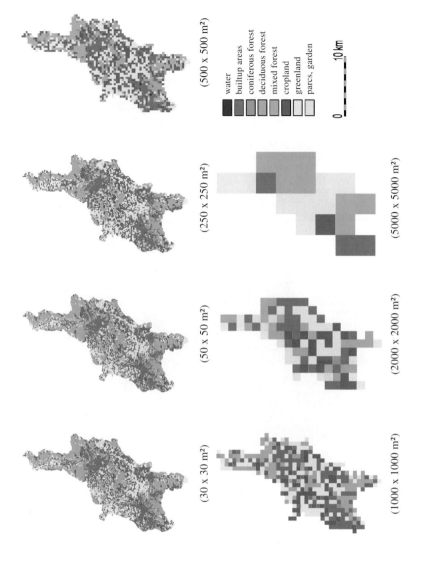

Colour Plate 19.C. Differences in landuse characteristics as a result of a reassembling of a Landsat TM scene into grid elements with different resolution (subcatchment of the Alzette river)

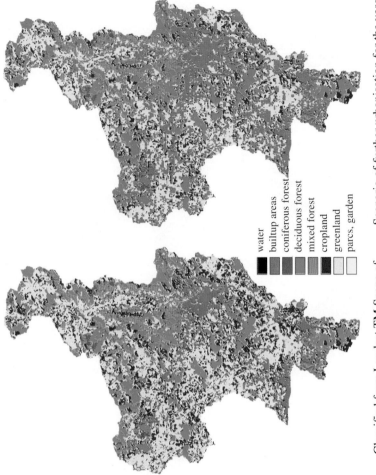

Colour Plate 19.D. Scenario for the further urbanisation of the Alzette catchment until the year 2009

Section IV

Future Perspectives

20 Future Perspectives

Edwin T. Engman[1] and Gert A. Schultz[2]

[1]Hydrological Science Branch, Code 974, Laboratory for Hydrospheric Processes-
NASA/Goddard Space Flight Center, Greenbelt, MD 29771, USA
[2]Ruhr University Bochum, 44780 Bochum, Germany

20.1 Introduction

Except for the first four chapters each chapter of this book showed, how remote sensing data may be used in the fields of hydrology and water management. Furthermore some perspectives were given on what can be expected in the foreseeable future when new sensors and platforms and new types of hydrological models will be available. It is the idea of this last chapter of the book to give a some more general perspective of future developments in the field of remote sensing and the expected use of remotely sensed data in the fields of hydrology and water management. It has to be kept in mind, however, that in various fields the level of application of RS data is rather different. For example, in snowmelt runoff estimation (Chap. 11) RS techniques have been used operationally already for several years, and in flood forecasting and control (Chap. 16) techniques have been developed, which could be used operationally today. However, in other fields research is still in full development, e.g. the determination of soil moisture profiles with the aid of RS data (Chap. 9) and in others RS data can be used only as an auxiliary measure in conjunction with the use of other types of information (e.g. in the field of groundwater exploration (Chap. 14)). Many of the advances in using remote sensing for hydrology have come from new areas of hydrologic analysis, areas where existing methods were unsatisfactory or limiting and areas where sufficient data were sparse or nonexistent. These areas include measurements of soil moisture, estimating evapotranspiration, advances in snow hydrology, and land-surface parameterizations in General Circulation Models. Equally impressive advances can be expected in the field of water management, particularly if RS data can be made available in real-time.

Remote Sensing, when it was introduced into hydrology in the seventies, held a great deal of promise for hydrology. In spite of this promise, applied or engineering hydrology has been slow to embrace remote sensing as a useful source of data, presumably because existing techniques and data have been satisfactory for limited applications. Although we have seen a somewhat cool acceptance of remote sensing, its future impact on hydrology and water resources is likely to be great for several reasons:

- The ability to provide spatial data, rather than point data,
- The potential to provide measurements of hydrological variables not available through traditional techniques such as soil moisture and snow water content,
- The ability, through satellite sensors, to provide long-term, global-wide data, even for remote and generally inaccessible regions of the earth,
- The possibility, to acquire RS data for larger areas with a high resolution in space and time at one spot (e.g. weather radar receiving station, satellite data center) and in real-time, which may serve as basis for water management decisions in real-time.

This chapter addresses some of the future issues related to hydrology and how remote sensing may help fulfill these. Central to these issues are the data needs and how new sensors and platforms may help fulfill the promise for hydrology.

20.2 Status of Hydrologic Research and Modeling

Hydrologic research has progressed rapidly in understanding the various physical processes and developing a large number of sophisticated analysis techniques and produce deceptively elaborate outputs. However, we apparently have not been able to demonstrate consistently improved accuracy or reproducibility or that the complex and sophisticated process based models work any better then the older, simple models.

For example, Naef (1981) compared the success of simple and complex models in reproducing measured discharge. His conclusions are based on two projects the World Meteorological Organization Intercomparison of Conceptual Models used in operational hydrological forecasting and on a study of rainfall-runoff models using data from small basins in Switzerland. The results show that simple models can give satisfactory results: however, neither the simple nor the more complex models tested were free from failure because none of them adequately describe the rainfall process. In addition, it could not be proved that complex models give better results than simpler ones.

Another study by Loague and Freeze (1985) presented model-performance calculations for three event-based rainfall-runoff models on three data sets involving 269 events from small upland catchments. The models include a regression model, a unit-hydrograph model, and a quasi-physically based model. The results show surprisingly poor model efficiencies for all models on all data sets on an event-by-event basis. The poor performance of the quasi-physically based model could probably be ascribed to a combination of model error and input error. They speculated that the primary barrier to the successful application of physically based models in the field may lie in the scale problems that are associated with the unmeasurable spatial variability of rainfall and soil hydraulic properties. The simpler less-data-intensive models provided as good or better predictions than the physically based model.

In a recent series of papers, Grayson et al (1992) question the value of distributed parameter models to truly represent the processes if the fundamental algorithms and assumptions cannot be validated. In their conclusions they state "the misperception that model complexity is positively correlated with confidence in the results is exacerbated by the lack of full and frank discussion of a models capability and limitations and the reticence to publish poor results". They go on to conclude that "the seductive attraction of the more complex models is their ability to provide information about points within the catchment, but it is concluded that the representations used in current process based models are often too crude to enable accurate, a priori application to predictive problems".

In an even more recent paper, Jakeman and Hornberger (1993) question "what limits the observed data place on the allowable complexity of rainfall-runoff models". They further state "conceptual and physically based models developed and used for describing rainfall-runoff processes tend to be over parameterized. They are no more useful for prediction than are simpler models whose parameters are identifiable from available data" (Jakeman and Hornberger, 1993).

From these examples, one can conclude that the development of more complex and sophisticated models and innovative analysis techniques has not resulted in an overwhelmingly improved ability to predict runoff. Often the question is asked, why is it that more complex and more physically based models do not give us better runoff predictions? To the authors this question seems to be ill posed for the following reasons:

— It is by no means logical, that a more complex and more physically based hydrological model will yield more accurate results than simpler structured models. The complex models usually contain a higher number of parameters and thus may cause more potential modeling errors and often they do not have an adequate data base which is needed for achieving accurate results,

— The reason for developing models of a higher complexity which are distributed in area, is usually not the desire for more accurate results, but the wish to apply the model also under changing conditions, e.g. future landuse changes, which is usually not possible with the more simple structured models,

— Often the more complex structured hydrological models working on the basis of high resolution in space and time are fed with input data (e.g. precipitation, snow, radiation) with a very coarse resolution in space, and often also in time. The model complexity is not capable of making up for insufficient resolution in space and time of the input data.

— The idea to develop a general hydrological model for the whole water cycle, or parts of it, which functions optimal for all conditions is still utopian. At the present stage of scientific hydrology we still have to tailor the model for the purpose. In each case we have to decide, which model would be most suitable for which process. The model complexity has to be adapted to the model purpose and available data and it should not be a goal by itself.

Remote sensing can provide new types of data that can help make the complex models easier to use as well as improve their performance. For example, remote sensing may be the only viable approach to incorporate spatial variability of watershed properties and – equally important – the spatial variability of the required model input data (precipitation, snowmelt, radiation etc.). This is based on the uniqueness of remote sensing to obtain spatially distributed information as well as some entirely new forms of measurement.

Looking back at the history of the development of hydrological models it becomes quite clear that the structure of the models was chosen such that existing hydrological data (particularly precipitation, runoff, radiation) are used for model calibration and validation instead of structuring a model such, that the hydrological process under consideration is represented in an optimum way. A typical example for this deficiency of existing models lies in the fact, that the central hydrological variable, i.e. the soil water storage or the soil moisture profile, distributed in space and time, is considered as a residual in these models. It would be logical to structure these models such that observed soil moisture (or soil water storage content) would be the basic model variable, since it decides, how much water will flow from an area element in vertical direction (evapotranspiration, infiltration, percolation) or laterally (surface runoff, interflow, groundwater movement). A primary reason for the nonexistence of such models lies in the non-availability of spatial soil water data almost everywhere in the world. If the new RS techniques of soil moisture measurements with the aid of microwave data from space become operational, then such models could be developed and there can be no doubt, that such – obviously complex – models would also provide a much better accuracy than present complex models can achieve.

20.3 Water Management

In the field of scientific hydrology, the application of remote sensing techniques has led to methods which are frequently used in practice. Unfortunately, in the field of water resources management, we observe a certain reluctance to apply these techniques operationally, although the potential benefits to be gained from the use of RS seems rather obvious. The reason for this reluctance lies certainly in the conservatism of many water agencies, which not only have to operate their water management systems technically efficiently and economically, but they also have to guarantee reliability of their systems. The introduction of completely new techniques leads to a dependence on such new methodologies including unknown uncertainties and deficiencies that are always inherent in new technologies. Although this reluctance of water managers was understandable in the early phases of RS time seems to have come now that the obvious benefits of RS should be exploited. Hydropower companies in Norway, e.g. use RS techniques for the exploration of snow covered areas (satellite data) and snow water equivalents (data from airplanes). This information allows forecasting of expected Spring and early Summer flows in rivers. These forecasts lead to better hydropower scheduling providing more efficiency, reliability and higher economic benefits.

The use of RS data for water management includes the application of hydrological models, the parameters of which and the input into which is, at least partially, based on remote sensing information. One of the main advantages of RS for water management lies, however, in the fact, that relevant data can be gathered over large areas with high resolution in time and space at one spot, e.g. at the control center of a water authority. Furthermore this information, containing large quantities of spatial and temporal data can be acquired in real-time or almost real-time. Figure 20.1 is a schematic illustrating all components of a remote sensing information system for real time water management.

This allows the use of this high quality information for management decisions in real-time. This is of great value for decisions on e.g. releases from dams for water supply purposes, for irrigation scheduling, but also for water quality control and improvement. The benefits of such data are invaluable for real-time flood management in river systems as well as in urban drainage systems. Here the processes are so highly dynamic that time intervals as short as 5 minutes are required for a process lasting only a few hours or less (see e.g. Chap. 16).

One problem, which is not so relevant for hydrological applications of RS data but is crucial in the field of real-time operation of water systems, is the question, what shall be done, if there is a partial or complete failure of the information system used. This problem also exists for the use of conventional data. If the decision support system of a water authority were based on remote sensing data it has to be clarified in advance, how the real-time operation should be carried out in case of a failure of the RS data supply. As long as we have no answer to this problem the reluctance of water managers to use RS data will continue to exist. It should be, however, not pose too much difficulty, to solve this problem in the near future, be it by combination of RS infosystems with conventional data acquisition systems or by using RS data from different sources. Also, as usual, certain emergency rules should be designed for these cases. It is hoped, that in the near future these problems will be overcome by appropriate research activities in order to enable water managers to benefit from the significant advantages remote sensing data offer as compared to conventional data.

20.4 Data Issues in Hydrology and Water Resources Management

If we accept the arguments that the current stage of model development has not given us better answers in spite of their complexity and that the new hydrological sciences are demanding even more answers from very complex systems, what is it that is needed most if we are to successfully answer the challenges of modern hydrology? We believe the single most needed item is more, better and different data. Not just numbers of measurements but the correct measurements with high

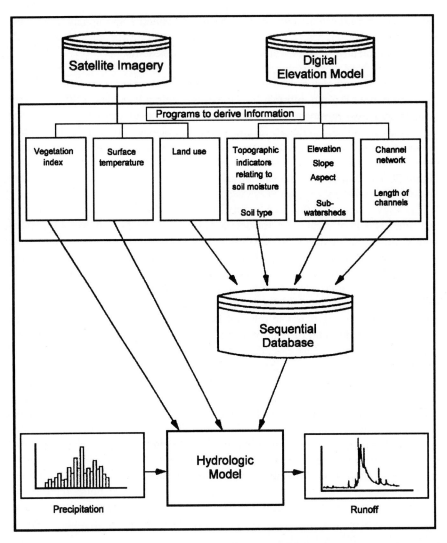

Fig. 20.1. Illustration of the various components in a sequential data base necessary to support parameter estimation for a distributed hydrological model. (After Schultz, 1993)

resolution in space and time. How can we truly believe a so-called physically based, distributed model that supposedly accounts for all the processes but is driven only by rainfall and validated only by a measured hydrograph?

The data issues have been addressed in Opportunities in the Hydrological Sciences (NRC, 1991) and base its discussions on the premise that "hydrologic science is currently data limited". The report discusses the general aspects of data needs conceding that traditionally hydrological data have been collected to address water resources problems and not the merging hydrologic science. Dozier (1992) correctly points out that our historical data have not been able to represent the wide variety of spatial and temporal scales encountered in hydrology and the resulting models reflect a simple, homogeneous view of the natural world. He concludes that "this forced oversimplification impedes scientific understanding and management of water resources".

If the emerging Hydrological Sciences are to break away from the traditional engineering hydrology, a number of general and specific data needs are going to have to be addressed and solved. Again, quoting from the "Opportunities in the Hydrological Sciences" (NRC, 1991) some of these needs are:

— "Hydrologic data are needed to measure fluxes and reservoirs in the hydrologic cycle and to monitor hydrologic change over a variety of temporal and spatial scales.

— Detection of hydrologic change requires a committed international long-term effort and requires also that the data meet rigorous standards for accuracy.

— Synergism between models and data is necessary to design effective data collection efforts to answer scientific questions.

— A fundamental block to progress in using most hydrologic data is our poor knowledge of how to interpolate between measurement points."

A very convincing case is made for the need to not make hydrologic data collection an after thought. In addition the issue of data accessibility and management is addressed as part of the general data issue. This is a particularly important issue not only because all the data in the world are useless unless one can access them but also technology for storage, transmitting, displaying and analyzing data is changing very rapidly. Other than to encourage hydrologists to keep abreast of these rapidly changing developments, further discussions of this topic is beyond the scope of this book.

The question must now be raised if one accepts the premise that "hydrologic science is data limited", what do we do about it? Two issues that traditional hydrologic instrumentation cannot address is the spatial variability of hydrologic processes and wide disparity of space and time scale that scientific hydrology must address. Remote sensing does meet these needs. Remote sensing can address the spatial heterogeneity and the scale disparity problems. For example, remote sensing may be the only viable approach to handle spatial variability of drainage basin properties and hydrologic process because the basic data are spatial in nature. Space borne instruments also have the ability to make measurements that span the many scales from near point processes to global. However, to realize the potential of remote sensing a number of carefully planned and executed studies must be carried out that demonstrate the relationship between traditional point data and remote sensing data.

20.5 Intensive Field Campaigns

In many ways the conduct of several of the existing and planned intensive field campaigns may provide the best combinations of remote sensing and traditional hydrologic data for model development and validation. The intensive field campaigns provide an opportunity to plan and carry out a data collection program that specifically addresses the needs of hydrologic modelers. The schematic shown in Fig. 20.2 illustrates the full range and scale of a well planned field campaign.

Unfortunately this is not generally the case with applications of satellite data to hydrology. In the latter case, all too frequently one has to make concessions and use data that are not designed for a specific hydrologic application in the time interval between measurements, the spatial resolution, or the sensor wavelength.

The intensive field campaigns have the advantage of providing an opportunity to collect data where and when it is needed from a purely hydrologic need. In addition the opportunity exists to use some of the newly developed sensors that have not yet progressed to a satellite platform. A good example of this is the measurement of soil moisture with the PBMR in the MONSOON 90 and the ESTAR in the MONSOON 91 and the WASHITA 92 campaigns (Jackson, et al., 1995). The data would not be possible from satellite because no long wave microwave instruments with a short revisit time are currently flying. These data sets were possible only with aircraft mounted instruments.

The value of such data sets cannot be overlooked. Although they eclipse only a few days over a limited area, they do provide a preview of the types of data that

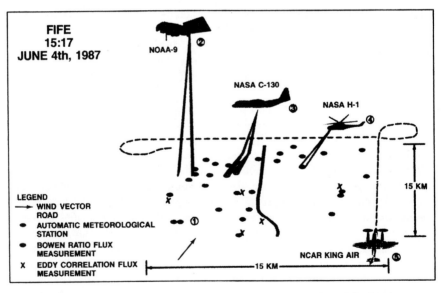

Fig. 20.2. Schematic illustrating the various types of measurements, from point instruments on the ground to satellite instruments, involved in intensive field campaigns

may be readily available in the future. These data sets, together with the concomitant detailed hydrologic data create a foundation for evaluating the hydrologic value of these new measurements and for developing hydrologic models that can maximize the information content in the remote sensing data. Successfully demonstrating the value of these new instruments in hydrology is one of the necessary steps that leads to a satellite borne sensor.

There are a number of international and regional field campaigns currently underway or being planned. Table 20.1 lists some of the better recognized large scale hydrologic field campaigns that also are designed around the use of remote sensing measurements. Even more information is available from an American Geophysical Union web page: http://www.agu.org/eos_elec/97035e.html

Table 20.1 Descriptive information on large scale hydrologic field campaigns

Project	Summary	Date
GCIP	Mississippi River Basin: Study the terrestrial-atmospheric coupling at the regional to continental scale to improve quantitative predictions of coupled atmospheric and hydrological phenomena	1995 - 2000
BALTEX	Baltic Sea Region: Study coupled hydrological processes between land, sea and ice and the atmospheric circulation to determine the energy and water balances of the Baltic Sea and related river basins	1997 - 2001
NOPEX	Scandinavian Boreal Forest: Study the hydro-meteorological exchange processes between the patchy forests, agricultural fields and lakes	1995 -
BOREAS	Boreal Forest of Canada: Improve process models of energy, water, carbon and trace constituents between the boreal forest and the atmosphere and to develop methods to apply these over large areas	1994 - 1997
MAGS	MacKenzie River Basin: Coordinated hydrological modeling and process studies of water and energy balances of the Canadian Arctic	1994 - 1997
LBA	Amazonia: To study coupling of energy, moisture and carbon budget with atmospheric circulation of region and will link with intensive eco-climate studies	1997 - 1998
GAME	Asian Monsoon Region: Study of the total atmospheric and land surface water and energy balances in four diverse climate regions in the eastern Eurasian continent	1998 - 1999

20.6 Existing Sensors and Platforms

Continuing high spatial resolution data from the Landsat and SPOT satellites, passive microwave data from the Special Sensor/Microwave Imager (SSM/I) and continuing meterological satellite coverage from the NOAA, GOES, GMS and Meteosat series all mean that the remotely sensed techniques can continue to be employed and expanded upon. However, new sensors, particularly in the microwave region, promise great potential for hydrologic applications. There are several satellites, such as ERS-1/2 launched by the European Space Agency, the JERS-1 launched by the Japanese, and RADARSAT launched by the Canadians that will provide useful data for hydrologists. All carry single polarization, single-frequency SARs. An additional satellite that was recently launched will have considerable hydrologic interest is the Tropical Rainfall Measurement Mission (TRMM) (Simpson et al., 1988).

While, space borne sensors are ideal for long term, large area and global observations, there are a number of cases in which aircraft sensors can be extremely useful for hydrologic applications. Situations in which airborne sensors may be optimal are for extremely high resolution sensing, detailed topographic mapping, and flying of new sensors that are not currently on spacecraft. A good example of the later is the AVERIS hyperspectral instrument. Additional uses for airborne instruments are intensive, but short duration field campaigns discussed above or one time detailed mapping of an area or special situation such as flooding or an environmental spill.

In the field of ground-based remote sensing the use of weather radar data for measuring precipitation with a high resolution in space and time is most important. This information is particularly useful as input to hydrological rainfall-runoff models for real-time flood forecasting. The implementation of ground-based weather radar stations is in progress on all continents. Very advanced systems covering the whole country are already operational in the UK and in Japan. In western Europe and in North America such integrated weather radar networks are under construction and partially operational. It can be expected, that in many countries and even continents, these integrated weather radar networks will be accessible to flood forecasting centers. Use of these radar data will be available in real-time to make online flood forecasts for all relevant points in the river network, for small catchments as well as large river basins.

As far as the satellite borne sensor systems are concerned, not all existing and approved sensors are ideal for hydrology. The currently available instruments do provide, however, very useful data for hydrological applications even though none of the currently available satellite sensors have been designed with hydrological applications as the primary goal. The result is that we often have to use less than ideal data for a given application, but in spite of this they have been very useful under limited circumstances. The listing of currently available satellite systems and sensors characteristics has been given in the Appendices 20.1 and 20.2 at the end of this chapter.

20.7 Planned and Proposed Sensors and Platforms

The currently available satellite sensors have not been used optimally in hydrology. However, it is possible that with the future satellite sensors we will really see an advancement in the hydrologic sciences. This will be accomplished by improved spatial resolution, narrower and more specific spectral bands, but most importantly more sensors operating in the microwave region of the spectrum. The microwave region of the spectrum provides the opportunity to make unique measurements of system states under all weather conditions.

There are a number of existing or soon-to-be launched sensors that may provide useful hydrologic measurements. These include sensors that in principle at least can measure just about all hydrologic variables. Unfortunately being able to make a measurement is only part of the story with satellite remote sensing. One also has to consider the time frequency of measurement as well as the spatial resolution. For example, the JERS-1 SAR at 1.4 GHz should be a useful instrument for measuring soil moisture, but its 46 day repeat cycle renders it essentially useless for obtaining useful information on soil moisture. Similarly, the 19 GHz channel on the SSM/I platform should provide useful information on snow depth, but its large footprint greatly reduces its capabilities for measuring snow in alpine regions.

NASA's Earth Observational System (EOS) (Butler et al., 1988), and its International partners from Europe and Japan will lead to considerable advances in the understanding of all the earth sciences, including hydrology. The EOS instruments of most interest to hydrologists would include the MODIS and AMSR, the latter is a Japanese microwave instrument with a C-band radiometer which should provide interesting measurements of the land surface soil moisture conditions and deep snow conditions. EOS also includes the organization of a data information and archiving system which is extremely important. This data system will allow many types of data to be used simultaneously to calibrate or be assimilated into numerical models. Figure 20.3 shows conceptually the complexity of the Earth Observing System.

The trend for future Earth observing missions is to smaller, cheaper and more reliable satellite systems. The large, complex and expensive multi-sensor platforms such as the EOS-AM and PM are no longer viable. This trend is supported by the NASA Earth System Science Pathfinder (ESSP) and ESA Living Planet Explorer programs. The hydrological science community is also becoming involved in the

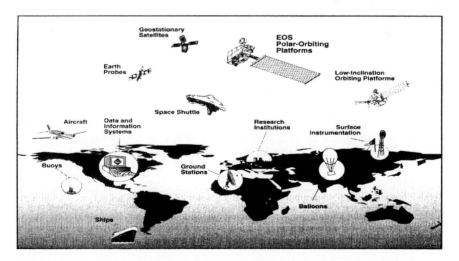

Fig. 20.3. Schematic demonstrating the complexity of the Earth Observing System

long range planning of future missions.

Appendices 20.1 and 20.2 list those future satellite/sensors that can be expected in the near future and hoped for in the more distant future. Some of these are well along in the planning and approval states whereas others are only in the proposal stage.

20.8 Remote Sensing and Future Needs in Hydrology

There are potentially many new and exciting observations of the hydrologic cycle that are going to be available from new satellite systems. For example, HYDROSTAR (a NASA ESSP) and SMOS (an ESA Living Planet Explorer proposal) are planned small satellites that would carry a L-band radiometer that uses aperture synthesis to achieve frequent global soil moisture measurements at about 30-50 km spatial resolution. If one of these missions is chosen for flight, it will be the first satellite designed purely for hydrologic applications.

New models to allow the data to be analyzed and address previously intractable problems will have to be developed specifically to use the new data types such as soil moisture or snow wetness. Remote sensing can provide many of the necessary data to supplement conventional data to expand hydrology in new directions and also provide entirely new data types and forms that will help hydrologists tackle previously unsolvable questions.

The future progress in the hydrological sciences will depend a great deal upon the availability of adequate data for model development and validation. Remote sensing can and should play a pivotal role in this progress. Without it, it is very possible that future progress in the hydrological sciences will be severely retarded if not completely stopped. With it hydrological sciences should be able to advance rapidly and to successfully address some of the previously intractable problems in hydrology and water management at all relevant scales, i.e. from microscale to global.

References

Chahine M.T. (1992) The hydrologic cycle and its influence on climate, Nature, Vol. 359, pp 373-380

Dozier J. (1992) Opportunities to improve hydrologic data, Reviews of Geophysics, 30, 4, pp 315-331

Federal Council for Science and Technology (1962) Scientific Hydrology, Washington, D.C.

Grayson R.B.; Moore I.D. and McMahon T.A. (1992) Physically Based Hydrologic Modeling. 1. A Terrain-Based Model for Investigative Purposes, Water Resources Research, Vol. 28, No. 10, pp 2639-2658

Grayson R.B.; Moore I.D. and McMahon T.A. (1992) Physically Based Hydrologic Modeling. 2. Is the Concept Realistic?, Water Resources Research, Vol. 26, No. 10, pp. 2659-2666

Hornberger G.M. (1994) Data and analysis note: A new type of article for Water Resources Research, Water Resources Research, Vol. 30, No. 12, pp 3241-3242

Hornberger G.M. (1992) Hydrologic Science: Keeping Pace with Changing Values and Perceptions, Proceedings: Sustaining Our Water Resources, Water Science and Technology Board, Tenth Anniversary Symposium

Jackson T.J.; LeVine D.M.; Swift C.T.; Schmugge T.J. and F.R. Schiebe (1995) Large Area Mapping of Soil Moisture Using the ESTAR Passive Microwave Radiometer in Washita '92 (accepted by) Remote Sensing of Environment

Jakeman A.J. and Hornberger G.M. (1993) How Much Complexity Is Warranted in a Rainfall-Runoff Model? Water Resources Research, Vol. 29, No. 8, pp 2637-2649

Loague K.M. and Freeze R.A. (1985) A comparison of rainfall-runoff-modeling techniques on small upland catchments, Water Resources Res., Vol. 21, No. 2, pp 229-248

Naef F. (1981) Can we model the rainfall-runoff process today? Hydrologic Sci. Bull., Vol. 26, No. 3, pp 281-289

NRC (National Research Council) (1991) Opportunities in the Hydrological Sciences, Washington, D.C., National Academy Press

Schultz, G.A., 1993. Hydrological modeling based on remote sensing information. Adv. Space res., Vol 13, No.5, pp 149-166

Appendix 20.1. Existing and future remote sensing satellites and sensors relevant to hydrological applications

Satellite (Agency)	Launch Date(s)	Sensors	Orbit	Orbit Details
ADEOS II (NASDA)	1999	AMSR, GII, POLDER, Sea-Winds	Polar	4 days (57 revisit) circular sun synchronous 1030 ± 15 min LST 802.9 km 101 min
DMSP (NOAA)	1965-1997	SSM/I	Polar	sun synchronous 830 km
Electro-GOMS N1 (RSA)	1994	BTVK	Geostationary	Geostationary
ENVISAT 1 (ESA)	1999	AATSR, ASAR, DORIS-NG, GOMOS, MERIS, MIPAS, MWR, RA-2, SCIAMACHY	Polar	35 day repeat 780-820 km 100.59 min
EOS AM1/PM1 (NASA)	1999 2000	ASTER, CERES, MISR, MODIS, MOPITT, AIRS, AMSR-E, AMSU, HSB	Polar	sun synchronous 1330 LST (ascending) 1030 LST (descending) 705 km 99 min
ERS 1/2 (ESA)	1991 1995	AMI-SAR image mode AMI-SAR wave mode AMI-Scatterometer mode ATSR, ATSR-2, MWR, RA, GOME	Polar	35 day repeat 777 km 98.5° inclination 100.5 min
ESSP (NASA)	2000	VCL	Polar	390-410 km 55° inclination
FY-2 (China (CMA))	1996	Multispectral Visible & IR Scan Radiometer (3 channel)	Geostationary	Geostationary
GMS (NASDA)	1995	VISSR	Geostationary	Geostationary
GOES (NOAA)	1974 – Present	VISSR IMAGER	Geostationary	Geostationary
HYDROSTAR (ESSP)	2003	HYDROSTAR	Polar	3 Day Repeat
INSAT (ISRO)	1992 1993	VHRR BSS & FSS Transponders	Geostationary	Geostationary

Satellite (Agency)	Launch Date(s)	Sensors	Orbit	Orbit Details
IRS (ISRO)	1988 - Present	LISS-I, LISS-II, LISS-III, LISS-IV, MOS, PAN, WiFS	Polar	22-24 day repeat sun synchronous 817-904 km 101.35-103 min 1030 LST 98° - 99° inclination
IKONOS 1 (Space Imaging)	1999	IKONO	Polar	1-1.5 Day Repeat sun synchronous 681 km 98 min 1030 LST 98.1° inclination
JERS-1	1992	OPS SAR	Polar	44 day repeat sun synchronous 568 km 96 min
LANDSAT (NOAA)	1972 to Present	MSS TM ETM+	Polar	16 day repeat sun synchronous 705 km 0945-1015 LST 98-99 min
METEOR-2 Operational Series (ROSHYDROMET)	Continuous	MR-2000 MR-900	Polar	80% earth coverage in 6 hours by satellite pairs
Meteosat (EUMETSAT)	1977 – Present	MVIRI	Geostationary	Geostationary
METOP 1,2,3 (EUMETSAT)	2001 2006 2010	ASCAT GOME-2 GRAS IASI MHS	Polar	sun synchronous 0930 LST (descending) 840 km 101.7 min
MOS-1 (Japan)	1987 1990	MESSR, VTIR, MSR	Polar	909 km
NEMO (US Navy)	2000	COIS PIC	Polar	Sun Synchronous 1030 LST (ascending) 605 km 7 Day Repeat
Ocean-01 N7 (ROSHYDROMET)	1994	BRK, BTVK, RLSBO, RMS, RM-0.8	Polar	Near polar 650 km 98 min 82.6° inclination
OrbView 1,2,3,4	1995 1997 1999 2000	OrbView 1,2,3,4	Polar	1-3 day repeat 470-740 km
POES (NOAA)	1978 – Present	AVHRR, ERBE, HIRS/2, MSU	Polar	833 or 870 km

Satellite (Agency)	Launch Date(s)	Sensors	Orbit	Orbit Details
PRIRODA (RSA)	1996	ISTOK-1, MSU-M, MSU-S, MSU-SK, MSU-E2, MOS, MOMS-2P, Travers SAR	Polar	MIR Space Station 380-420 km 51.6° inclination
QuickBird 1 (EarthWatch)	1999	Panchromatic Multispectral	medium-inclination	1-5 day revisit non-sun-synchronous 600 km 66° Inclination
RADARSAT (CSA)	1995	SAR	Polar	24 day repeat sun synchronous 793-821 km 1800 h LST 98.6° inclination
RESOURCE-01 N3/N4 (ROS-HYDROMET)	1994 1998	MR-900M, MSU-E1, MSU-SK, MSU-E, MSU-SK, SFM-2	Polar	Sun synchronous 670 km 98° inclination
SeaWiFS (NASA)	1997	SeaWiFS	Polar	sun synchronous 705 km 98° inclination
SMOS (ESA)	2004	L-band 2-D Aperture Synthesis	Polar	Sun Synchronous 2-4 day repeat
SPOT (CNES)	1986 1990 1992 1998 (2002)	HRV HRVIR VEGETATION	Polar	26 day repeat sun synchronous 822 km 1031 LST 98° inclination
TOPEX/ POSEIDON (NASA)	1992	ALT POSEIDON TMR	Polar	Non sun-synchronous 1336 km 66° inclination
TRMM (NASA)	1997	LIS, PR, TMI, VIRS, CERES	Polar	350 km 34-35° inclination

Appendix 20.2. Specifications for sensors listed in Appendix 20.1

Sensor	Mission(s)	Spectral Bands	Spatial Resolution	Revisit Capab.	Swath Width
AATSR Advanced Along Track Scanning Radiometer	ENVISAT-1	Visible-NIR: 0.555, 0.659, 0.865 μ m, SWIR: 1.6 μ m, TIR: 3.7, 10.85, 12 μ m	IR ocean channels: 1 km x 1 km Visible land channels: 1 km x 1 km	35 day	500 km
ALT Duel-frequency Radar Altimeter	TOPEX/ POSEIDON	Microwave: 5.3, 13.6 GHz	6-7 km along track	10 days	NA
AMI-SAR Image mode	ERS-1, ERS-2	5.3 GHz (C-band), VV polarization, bandwidth 15.5 ± 0.06 MHz	30 m	35 days	100 km
AMI-SAR Scatterometer mode	ERS-1, ERS-2	5.3 GHz (C-band), VV polarization	50km x 50km at 25km intervals	35 days	500 km
AMSR Advanced Microwave Scanning Radiometer	ADEOS II	Microwave: 6.9, 10.65, 18.7, 23.8, 36.5, 50.3, 52.8 and 89 GHz	5-50 km (dependent on frequency)	4 days (57 revisit)	1450 km
ASAR Advanced Synthetic Aperture Radar	ENVISAT-1	C-band	Image, wave and alternating polarization modes: 30m x 30m, Wide swath mode: 100m x 100m Global monitoring mode: 1 km x 1 km	35 days	Image and alternating polarization modes: up to 100 km Wave mode: 5 km Wide swath and global monitoring modes: 400 km
ASTER Advanced Space-borne Thermal Emission & Reflection Radiometer	EOS PM-1	Visible-NIR: 3 channels (0.5-0.9 μ m), SWIR: 6 channels (1.6-2.5 μ m), TIR: 5 channels (8-12 μ m)	VNIR: 15 m, stereo: 15 m horizontally and 25m vertical SWIR: 20m, TIR: 90m	16 days	60 km at nadir, swath center is pointable cross-track by ± 106 km for SWIR and TIR and ± 314 km for VNIR

Sensor	Mission(s)	Spectral Bands	Spatial Resolution	Revisit Capab.	Swath Width
AVHRR/2 Advanced Very High Resolution Radiometer	NOAA 9, NOAA 10, NOAA 11, NOAA 12, NOAA 14	Visible: 0.58-0.68μ m, NIR: 0.725-1.1 μ m, SWIR: 3.55-3.93 μ m, TIR: 10.3-11.3 μ m, 11.4-12.4 μ m	1.1 km (at nadir). Compressed global Area Coverage (GAC) data recorded at 4 km resolution	2/day	3000 km (approximate), 55.4° scan off nadir
AVHRR/3 Advanced Very High Resolution Radiometer	NOAA K, NOAA L, NOAA M, NOAA N, NOAA N'	Visible: 0.58-0.68 μ m, NIR: 0.725-1.1μ m, SWIR: 3.55-3.93 μ m, 10.3-11.3 μ m, 11.4-12.4 μ m	1.1 km at nadir	2/day	3000 km (approximate) Scan off nadir: 55.4°
BTVK	Electro-GOMS N1 Electro-GOMS N2	Visible: 0.4-0.7 μ m, TIR: 10.5-12.5 μ m	1.5 km, TIR: 8 km	NA	13,500 km
CERES Clous and the Earth's Radiant Energy System	TRMM	Visible: 0.3-5.0 μ m 8.0-12.0 μ m 0.3- >50 μ m	10 km	NA	Scan Angle ± 80°
COIS Coastal Ocean Imaging Spectrometer	NEMO	VNIR: 0.4-1.0 μm with 60 0.1 μm bands SWIR: 10 – 25 μm with 150 0.1 μm bands	30 m	2.5 days	30 km
ERBE Earth's Radiation Budget Experiment	NOAA 9 NOAA 10	Visible: 0.5-0.7 μ m, UV-SWIR: 0.2-4 μ m, UV-FIR: 0.2-5.0 μ m, TIR: 10.5-12.5 μ m	200-250 km non scanning 50 km for scanning	NA	3000 km
ETM+ Enhanced Thematic Mapper	LANDSAT 7	Visible-TIR: 8 channels: 0.45-12.5 μ m Panchromatic channel: 0.52-0.9 μ m	Panchromatic band: 15 m visible, NIR and SWIR: 30 m TIR: 60 m	16 days	185 km
GLI Global Imager	ADEOS II	Visible-NIR: 23 channels (0.375-12.5 μ m) SWIR: 7 channels, TIR: 9 channels	TIR: 7 channels: 1 km Visible/NIR/SWIR channels: 250m-1km	4 days (57 revisit)	1600 km

Sensor	Mission(s)	Spectral Bands	Spatial Resolution	Revisit Capab.	Swath Width
HRV High Resolution Visible	SPOT 1 SPOT 2	Visible: 0.50-0.59, 0.61-0.68 μm NIR: 0.79-0.89 μm Panchromatic: 0.51-0.73 μm	10 m (panchromatic) or 20 m	1 to 4 days at mid-latitude	117 km (60 km + 60 km with 3 km overlap) – steerable up to ± 27° off-track
HRVIR High Resolution Visible & IR	SPOT 4	Visible: 0.50-0.59 μm, 0.61-0.68 μm, NIR: 0.79-0.89 μm, SWIR: 1.5-1.7 μm	10 m (0.64 μm) or 20 m	1 to 4 days at mid-latitude	117 km (60 km + 60 km with 3 km overlap) – steerable up to ± 27° off-track
HYDRO-STAR	HYDROSTAR ESSP	L-band H Polarization	30 km	3 day	1700 km
IKONOS	IKONOS I	Panchromatic: 0.45-0.90 μm Multispectral: 0.45-0.52, 0.52-0.60, 0.63-0.69, 0.76-0.90	Panchromatic: 1 meter Multispectral: 4 meter	1.5 days at 1 meter resolution	13 km
ISTOK-1 Infra-red Spectrometer	PRIRODA	SWIR-FIR: 4-8 and 8-16 μm	0.6-2 km (along track)	NA	FOV: 6.5 x 26 angular minutes
LISS I	IRS 1a IRS 1b	Visible: 0.46-0.52, 0.52-0.59, 0.62-0.68 μm NIR: 0.77-0.86 μm	72.5 m	22 days for individual satellite, 11 days effectively for IRS1a/1b together	148 km
LISS II Linear Imaging Self-Scanning System	IRS 1a IRS 1b IRS P2	Visible: 0.46-0.52, 0.52-0.59, 0.62-0.68 μm NIR: 0.77-0.86 μm	32 x 37 m Output sampled to 3.6 m compatible to IRS-1a/1b	24 days	132 km
LISS III Linear Imaging Self-Scanning System	IRS 1c IRS 1d	Visible: Band 2: 0.52-0.59 μm, band 3: 0.62-0.68 μm NIR: Band 4: 0.77-0.86 μm, SWIR: Band 5: 1.55-1.75 μm	Bands 2,3 & 4: 23.5 m, Band 5: 70.5 m	24 days	Bands 2, 3 & 4: 142 km, Band 5: 148 km

Sensor	Mission(s)	Spectral Bands	Spatial Resolution	Revisit Capab.	Swath Width
LISS IV Linear Imaging Self-Scanning System	IRS P5 IRS P6	Visible (green, red), NIR and SWIR	10 m	22 days	40 km
MERIS Medium Resolution Imaging Spectrometer	ENVISAT 1	Visible-NIR: 15 bands selectable across range: 0.4-1.05 μ m (bandwidth programmable between 0.0025 and 0.03 μ m)	300 m or 1200 m at SSP	3 days	1150 km
MESSR Multispectral Electronic Self-Scanning Radiometer	MOS-1	Visible: 0.51-0.59, 0.61-0.69 NIR: 0.72-0.80, 0.80-1.10	50 m	17 days	100 km
MISR Multi-angle Imaging Spectro Radiometer	EOS AM-1	Visible: 0.44, 0.56, 0.67 μ m NIR: 0.86 μ m	240 m, 480 m, 960 m or 1.92 km	2 days	Unedited, nadir camera: 370 km Unedited, non-nadir cameras: 408 km
MODIS Moderate Resolution Imaging Spectro Radiometer	EOS AM-1 EOS PM-1	Visible-TIR: 36 bands in range 0.4-14.4 μ m	Cloud cover: 250 m (day) and 1000 m (night)	Daylight reflection and day/night emission spectral imaging at least every 2 days	2300 km at 110° (± 55°)
MOMS-2P Modular Optoelectronic Multispectral Scanner	PRIRODA	MS: Visible-NIR: 0.440-0.505 μ m, 0.53-0.575 μ m, 0.645-0.680 μ m, 0.76-0.77 μ m HR & ST: Panchromatic: 0.52-0.76 μ m	MS: 15.9-18 m HR: 5.3-6 m ST: 15.9-18 m (within 350-400 km altitude range)	14 days	MS: 92-105 km HR: 44-50 km ST: 88-105 km

Sensor	Mission(s)	Spectral Bands	Spatial Resolution	Revisit Capab.	Swath Width
MOS Modular Opto-electronic Scanning Spectrometer	IRS P3	MOS-A: NIR: 4 channels in range 0.775-0.768 μm, MOS-B: Visible-NIR: 13 channels in range 0.408-1.01 μm, MOS-C: SWIR: 1.6 μm	MOS-A: 1.6 x 1.4 km MOS-B: 0.52-0.52 km MOS-C: 0.52-0.64 km	24 days	200 km
MOS Modular Opto-electronic Scanning Spectrometer	PRIRODA	MOS-A: NIR: 4 channels in range 0.775-0.768 μm MOS-B: Visible-NIR: 13 channels in range 0.408-1.01 μm	MOS-A: 2.87 x 2.87 km MOS-B: 0.7 x 0.65 km	NA	MOS-A: 80.5 km MOS-B: 82 km
MR-2000	METEOR-2 N24	Visible-NIR: 0.5-0.8 μm	1 km (at nadir)	2/day	2600 km
MR-2000M	METEOR-3 N5 METEOR-3M N1	Visible-NIR: 0.5-0.8 μm	0.7-1.4 km (at nadir)	2/day	3100 km
MR-900	METEOR-2 N24	Visible-NIR: 0.5-0.8 μm	2 km (at nadir)	2/day	Scan angle 90° (2100 km)
MR-900M	METEOR-3M N1 Reesource-01 N4	Visible-NIR: 0.5-0.8 μm	2 x 1 km (at nadir)	2/day	2600 km
MSR Microwave Scanning Radiometer	MOS-1	Microwave: 23 GHz	32 km/23 km	17 days	320 km
MSS Multispectral Scanning System	LANDSAT 5	Visible: 0.5-0.6 μm, 0.6-0.7 μm, NIR: 0.7-0.8 μm, 0.8-1.1 μm	80 m in visible and NIR channels	16 days	185 km
MSU-E,E1,E2	Resource-01 N4 E2: PRIRODA	Visible: 0.5-0.6 μm, 0.6-0.7 μm NIR: 0.8-0.9 μm	E: 45m (at nadir) E1:25m (at nadir) E2: 10m (at nadir)	NA	45 km for one scanner, 80 km for two scanners E2: 2 x 24 km
MSU-M	Ocean-01 N7 OKEAN-O SICH-1 SICH-1M	Visible: 0.5-0.6 μm, 0.6-0.7 μm, NIR: 0.7-0.8, 0.8-1.1 μm	1 x 1.7 km	NA	1900 km

Sensor	Mission(s)	Spectral Bands	Spatial Resolution	Revisit Capab.	Swath Width
MSU-SK	PRIRODA	Visible: 0.5-0.6 μm, 0.6-0.7 μm NIR: 0.7-0.8 μm, 0.8-1.1 μm TIR: 10.3 – 11.8 μm	Visible: 80 m IR: 300 m	NA	300 km
MSU-SK	Resource-01 N3	Visible: 0.5-0.6 μm, 0.6-0.7 μm NIR: 0.7-0.8 μm, 0.8-1.1 μm TIR: 10.4-12.6 μm	Visible-NIR: 170 m TIR: 600 m	35 day	600 km
Multispectral Visible & IR Scan Radiometer (3 channel)	FY-2	Visble-NIR: 3 channels in range 0.55-1.05 μm	1.25 km (visible), 5 km (IR and water vapour)	30 min	Full Earth Disk
MVIRI METEOSAT Visible & IR Imager	METEOSAT 5-7	Visible-NIR: 0.5-0.9 μm TIR: 5.7-7.1 μm (water vapor) 10.5-12.5 μm	Visible: 2.5 km Water Vapor: 5 km TIR: 5 km	30 min	Full Earth Disk
MWR Microwave Radiometer	ENVISAT 1 ERS 1 ERS 2 ESA Future Missions	Microwave: 23.8 and 36.5 GHz	20 km	36 day	20 km
OPS Optical Sensor	JERS1	Visible-NIR: 0.52-0.60 μm, 0.63-0.69 μm, 0.76-0.86 μm	18 m x 24 m	42 day	75 km
OrbView-1	OrbView-1	.777 nm	10 km	< 2 days	1300 km
OrbView-2	OrbView-2	.402-.422 μm .433-.453 μm .480-.500 μm .500-.520 μm .545-.565 μm .660-.680 μm .745-.785 μm .845-.885 μm	1 km	1 day	2800 km
OrbView-3	OrbView-3	Panchromatic: .450-900 μm Multispectral: .450-.520 μm .520-.600 μm .625-.695 μm .760-.900 μm	Panchromatic: 1m Multispectral: 4 m	< 3 days	8 km

20 Future Perspectives 467

Sensor	Mission(s)	Spectral Bands	Spatial Resolution	Revisit Capab.	Swath Width
OrbView-4	OrbView-4	Panchromatic: .450-.900 μm Multspectral: .450-.520 μm .520-.600 μm .625-.695 μm .760-.900 μm Hyperspectral: .450-2500 μm (200 channels)	Panchromatic: 1m Multispectral: 4 m Hyperspectral: 8 m	< 3 days	8 km Panchromatic and Multispectral 5 km Hyperspectral
PAN Panchromatic Sensor	IRS 1c IRS 1d	Panchromatic: 0.5-0.75 μ m	5.8 m	5 days	70 km mosaic
PIC Panchromatic Imaging Camera	NEMO	0.5-0.7 μm	5 m	2/5 days	30 km
POLDER Polarization & Directionality of the Earth's Reflectance	ADEOS II	Visible-NIR: 0.443, 0.670 and 0.865 μ m at 3 polarizations, and 0.443, 0.49, 0.565, 0.763, 0.765 and 0.91 μ m with no polarization	6 km x 7 km	1 day	2400 km (across track) x 1800 km (along track)
POSEIDON/ SSALT Solid State Altimeter	TOPEX/ POSEIDON	Microwave: 13.65 GHz	2 km antenna footprint Basic measurement: 1/ sec Raw measurement: 20 / sec	10 days	300 km between tracks at equator
PR Precipitation Radar	TRMM	Microwave: 13.796 and 13.802 GHz	Range Resolution: 250 m Horizontal Resolution: 4.3 km at nadir	~5 days	215 km (scanned every 0.6 secs) Observable range: from surface to approx 15 km altitude
QuickBird 1	QuickBird	Panchromatic: .450–.900 μm Multispectral: .450-.520 μm .520-.600 μm .630-.690 μm .760-.890 μm	Panchromatic: 1m Multispectral: 4 m	1-5 days depending on latitude	704 km
RA Radar Altimeter	ERS-1 ERS-2	K-band: 13.8 GHz	7 km	35 days	NA

Sensor	Mission(s)	Spectral Bands	Spatial Resolution	Revisit Capab.	Swath Width
RA-2	ENVISAT 1	K band: 13.575 GHz, S band: 3.2 GHz	7 km	35 days	3-5 km
RLSBO Side-Looking Real-Aperture Radar	Ocean-01 N7 OKEAN-O SICH-1M	Microwave: 3.1 cm	1.5 x 2.0 km	NA	450 km
RM-0.8	Ocean-01 N7 SICH-1M	Microwave: 0.8 cm	15 x 20 km	NA	550 km
SAR Synthetic Aperture Radar	JERS1	Microwave: 1.275 GHz	18 m x 18 m	42 days	75 km
SAR Synthetic Aperture Radar	RADARSAT RADARSAT-2	C band: 5.3 GHz, HH polarization	Standard: 25 x 28 m (4 looks)	steerable	Standard: 100 km
SeaWiFS	SeaWiFS	8 bands 412-865 nm (20 nm widths)	1.13 km	2 days	200 km
SFM-2 Solar Radiation Spectrometer	METEOR-3 N5 METEOR-3M N1 METEOR-3M N2 Resource-01 N4	UV: 0.26-0.4 μm	40 x 10°	NA	NA
SMOS	New Explorer ESA	L-band HV polarization	50-60 km	3-4 days	~800 km
SSM/I Microwave Imager	DMSP	Passive Microwave: 19.35, 22.235, 37.0 and 85.5 GHz	13 x 15 km (85 GHz) 43 x 69 km (19 GHz)	2/days	1400 km
TM Thematic Mapper	LANDSAT 5	Visible: 0.45-0.52 μm, 0.52-0.6 μm, 0.63-0.69 μm SWIR: 1.55 – 1.75 μm, 2.08-2.35 μm TIR: 10.4-12.5 μm	Visible and SWIR: 30 m TIR: 120 m	16 days	185 km
TMI TRMM Microwave Imager	TRMM	Microwave: 10.7, 19.4, 21.3, 37, and 85.5 GHz	Vertical: 2.5 km approx. Horizontal: 18 km	12 hours	8.8 kbps

Sensor	Mission(s)	Spectral Bands	Spatial Resolution	Revisit Capab.	Swath Width
TMR Topex Microwave Radiometer	TOPEX/ POSEIDON	Microwave: 18, 21 and 37 GHz	44.6 km at 18 GHz, 37.4 at 21 GHz and 23.5 at 37 GHz	10 days	120° cone centered on nadir
Travers SAR	PRIRODA	Microwave: 9.2 and 23 cm	50-150 m	NA	50 km
VCL Vegetation Caopy Lidar	ESSP-1A	Visible-NIR	5 25m footprints	No periodic repeat. Decay orbit for ground sampling	30 km
VEGETATION	SPOT 4 SPOT 5	Visible: 0.61-0.68 μm NIR: 0.78-0.89 μm SWIR: 1.58-1.75 μm Experimental Mode: Visible 0.43-0.47 μm	1.15 km at nadir	1 day	2200 km
VHRR Very High Resolution Radiometer	INSAT IIa INSAT IIb INSAT IIe	Visible: 0.55-0.75 μm TIR: 10.5-12.5 μm	2 km in visible, 8 km in IR	30 min	Full Earth Disk
VIRS Visible Infrared Scanner	TRMM	Visible: 0.63 μm, SWIR: 1.6 and 3.75 μm, TIR: 10.8 and 12 μm	2 km at nadir	1 day	720 km (45° either side of track)
VISSR Visible & IR Spin Scan Radiometer	GMS-5	Visible: 0.55-0.9 μm, TIR: 6.5-7.0 μm, 10.5-11.5 μm and 11.5-12.5 μm	Visible: 1.25 km TIR: 5 km	30 min	Full Earth Disk
VISSSR and VAS Visible & IR Spin Scanning Radiometer	GOES 7	Imaging Mode: Visible: 0.55-0.75 μm IR: 11 μm plus 2 other channels selectable from 3.9, 6.7, 7.3 and 13.3 μm	Visible: 1 km IR: 7 and 14 km	30 min	Horizon to Horizon
VTIR Visible and Thermal Infrared Radiometer	MOS-1	Visible: 0.5-0.7 μm, 6.0-7.0 μm TIR: 10.5-11.5 μm, 11.5-12.5 μm	Visible 0.9 km TIR: 2.7 km	17 days	1500 km

Sensor	Mission(s)	Spectral Bands	Spatial Resolution	Revisit Capab.	Swath Width
WiFS Wide-Field Sensor	IRS 1c IRS 1d RS P3	Visible: 0.62-0.68 μm NIR: 0.77-0.86 μm	258 m	5 days	258 m

List of Acronyms

ADEOS II	Advanced Earth Observing Satellite II: see App. 20.1
AFCC	Accumulated Fractional Cloud Cover
AGCM	Atmospheric General Circulation Model
AGNPS	AGricultural NonPoint Source model
AIP	Algorithm Intercomparison Projects
AIRSAR	Airborne Synthetic Aperture Radar
ALU	Arithmetic Logic Unit
AMSR	Advanced Microwave Scanning Radiometer: see App. 20.2
APWL	Accumulated Potential Water Loss
ASAR	Advanced Synthetic Aperture Radar: see App. 20.2
AVHRR	Advanced Very High Resolution Radiometer: see App. 20.2
AVIRIS	Airborne Visible/Infrared Imaging Spectrometer
BALTEX	Baltic Sea Experiment
BOREAS	Boreal Ecosystem-Atmosphere Study
BSQ	Band Sequential Format
CCD	Charge Coupled Detector
CCD	Cold Cloud Duration
CD	Compact Disc
CERES	Cloud and Earth Radiant Energy System: see App. 20.2
CIR	Color Infrared
CISC	Complex-Instruction-Set-Computers
CMIS	Conical Scanning Microwave Imager/Sounder
CN	Curve Number
CNES	Centre National d'Etudes Spatiales
CORINE	Co-ordination of Information on the Environment
CPU	Central Processing Unit
CRT	Cathode Ray Tube
CSA	Canadian Space Agency
CZCS	Coastal Zone Color Scanner
DEM	Digital Elevation Model
DMSP	Defense Meteorological Satellite Program
DOM	Dissolved Organic Matter
DP	Dynamic Programming
DPCM	Differential Pulse Code Modulation (Image Format)
DRASTIC	Depth to water table, net Recharge, Acquifer media, Soil medium, Topography, Impact of vadose zone media and hydraulic Conductivity
DSS	Decision Support System
DVD	Digital Video Disc
DVI	Difference Vegetation Index
DWD	Deutscher Wetter Dienst (German Weather Service)
ENVISAT	ENVIronmental SATellite (ESA) : see App. 20.1
EOS	Earth Observing System: see App. 20.1
EOS-PM	Earth Observing System – PM (afternoon equatorial crossing) : see App. 20.1
EPSF	Encapsulated PostScript (Image Format)
ERS-1, 2	European Resources Satellite 1, 2: see App. 20.1
ERTS-1	Earth Resources Technology Satellite-1 (early Landsat)
ESA	European Space Agency
ESMR	Electrically Scanning Microwave Radiometer

List of Acronymus

ESSP	Earth Science System Pathfinder: see App. 20.1
ESTAR	Electronically Scanning Terrestrial Aperture Synthesizing Radiometer
ET	Evapotranspiration
ETO	Potential Evapotranspiration
FAO	Food and Agriculture Organization
FCCI	Fractional Cloud Cover Index
FEWS	Famine Early Warning System
FGDC	Federal Geographic Data Committee
FIFE	First ISLSCP Field Experiment
FRONTIERS	A man/machine rainfall forecasting system (UK)
GAME	GEWEX Asian Monsoon Experiment
GCIP	GEWEX Continental Scale International Project
GCM	General Circulation Model (of the Atmosphere)
GCP	Ground Control Points
GEOSAT	Geodesy Satellite (USA)
GEWEX	Global Energy and Water Cycle Experiment
GIS	Geographic Information System
GMS	Geostationary Meteorological Satellite: see App. 20.1
GMT	Greenwich Meridian Time
GOES	Geostationary Operational Environmental Satellite: see App. 20.1
GOM	Geometrical Optics Model
GPCP	**Fehler! Textmarke nicht definiert.**t
GPI	GOES Precipitation Index
GPS	Global Positioning System
GRU	Geographic/Geomorphic Response Unit
GUI	Graphical User Interface
HAD	High Aswan Dam
HART	Height-Area Rainfall Threshold
HCMM	Heat Capacity Mapping Mission
HEC-1	Hydrologic Engineering Center hydrologic model – 1
HGIS	Geographic and Hydrologic Information System
HHU	Homogeneous Hydrologic Units
HRV	High Resolution Visible
HSU	Hydrologically Similar Unit
HYDROSTAR	L-band passive microwave remote sensing satellite: see App. 20.1
HYDROTEL	A distributed Hydrologic Model (Canada)
IDNDR	International Decade for Natural Disaster Reduction
IEOS	International Earth Observation System
IFOV	Instantaneous Field of View
IGBP	International Geosphere-Biosphere Program
INSAT	INterferometry SATellite: see App. 20.1
IR	electromagnetic bands: infrared
IRS	Indian Remote-sensing Satellite
ISLSCP	**Fehler! Textmarke nicht definiert.**
ITDZ	Inter Tropical Discontinuity Zone
JERS-1	Japan Environmental Resources Satellite: see App. 20.1
JPEG	Image Format
LAC	Local Area Coverage
LAI	Leaf Area Index
LAN	Local Area Network
Landsat TM	Landsat Thematic Mapper
LBA	Large-Scale Atmosphere-Biosphere Experiment in Amazonia

List of Acronymus

LEWIS	Satellite name
LIC	Local Iso-Luminance Contour
LIGHTSAR	Light weight SAR
LIS	Lightning Imaging Sensor
LISS	Linear Imaging Self-scanning System: see App. 20.2
LUCC	Land Use and Land Cover Change project
MAC	Multi-sensor Airborne Campaign
MAGS	Mackenzie River Basin GEWEX Study
MAP	Maximum A Posteriori
MERIS	Medium Resolution Imaging Spectrometer: see App. 20.2
Meteosat	Meteorological Satellite (ESA): see App. 20.1
MFS	Monitor, Forecast and Simulate
MIPS	Millions-of-Instructions-Per-Second
MODIS	Moderate Resolution Imaging Spectrometer: see App. 20.2
MONSOON	Not an acronym; just the name for a field campaign in Arizona
MOS	Modular Optical scanner
MPEG	digital video format (Moving Pictures Experts Group)
MSS	Multi-Spectral Scanner
MV	electromagnetic bands: microwave
NAPP	National Aerial Photography Program
NASA	National Aeronautics and Space Administration
NASDA	National Space Development Agency (Japan)
NDVI	Normalized Difference Vegetation IndexI
NFS	Nile Forecasting System
NOAA	National Oceanic and Atmospheric Administration
NOPEX	Northern Hemisphere Climate Processes Land Surface Experiment
NPS	Non-Point Source pollution
OCTS	Ocean Color and Temperature Scanner
PBMR	Push Broom Microwave Radiometer
PDUS	Primary Data User System
pixel	Picture Element
PM	Penman-Monteith
PMF	Probable Maximum Flood
POM	Physical Optics Model
PRISM	Panchromatic Remote sensing Instrument for Stereo Mapping
PRMS	Precipitation Runoff Modeling System
PVI	Perpendicular Vegetation Index
QPF	quantitative precipitation forecast
radar	Radio Detection and Ranging
RADARSAT	Radar Satellite: see App. 20.1
RAID	Redundant Arrays of Inexpensive Disks
RAM	Random Access Memory
RISC	Reduced-Instruction-Set-Computers
RMS height	Root Mean Square (of individual surface elevation measurements)
RMSE	Root Mean Square Error
RS	Remote Sensing
RSDPS	Remotely Sensed Data Processing System
RUSLE	Revised Universal Soil Loss Equation
RVI	Ratio Vegetation Index
SAA	Satellite Active Archive System
SAR	Synthetic Aperture Radar: see App. 20.2
SARVI	Soil Adjusted Ratio Vegetation Index

List of Acronymus

SAVI	Soil Adjusted Vegetation Index
SCS	Soil Conservation Service
SDSS	Spatial Decision Support Systems
SeaWiFS	Sea viewing Wide Field Sensor: see App. 20.1
SEBAL	Surface Energy Balance Algorithm for Land
SIR	electromagnetic bands: sort-wve Infrared
SIR-C/X-SAR	Spaceborne Imaging Radar-C/X-Band Synthetic Aperture Radar
SLURP	Simple Lumped Reservoir Parametric hydrologic model
SMMR	Scanning Multichannel Microwave Radiometer
SMOS	Soil Moisture Ocean Salinity mission
SPM	Small Perturbation Model
SPOT	Système Probatoire pour l'Observation de la Terre
SRL	Spaceborne Radar Laboratory Missions
SSM/I	Special Sensor Microwave/Imager: see App. 20.2
SVAT	Soil-Vegetation-Atmosphere Transfer Scheme
SVF	Single Variable File
TDR	Time Domain Reflectometer
TIFF	Image Format
TIN	Triangular Irregular Network
TIR	electromagnetic bands: thermal infrared
TM	Thematic Mapper
TMI	TRMM Microwave Imager: see App. 20.2
TOGA-COARE	Tropical Ocean Global Atmosphere - Coupled Ocean Atmosphere Response Experiment
TOPEX/Poseidon	Ocean Topography Experiment (USA/France)
TOPMODEL	TOPography based hydrologic MODEL
TOVS	TIROS Operational Vertical Sounder
TRMM	Tropical Rainfall Measurement Mission
TSAVI	Transformed Soil Adjusted Vegetation Index
TVX	Temperature Vegetation Index
USAID	U.S. Agency for International Development
USLE	Universal Soil Loss Equation
UTM	Universal Transverse Mercator
VHRR-IR	Very High Resolution Radiometer - Infrared: see App. 20.2
VIC-2L, 3L	Variable Infiltration Capacity Model 2 Layers, 3 Layers
VIRS	Visible and Infrared Scanner: see App. 20.2
VIS	electromagnetic bands: visible
WASHITA 92, 94	Field Campaigns
WATERSHEDSS	WATER, Soil, and Hydro-Environmental Decision Support System
WCI	Water Content Index
WCRP	World Climate Research Program
WHC	Water Holding Capacity
WMO	World Meteorological Organization
WPMM	Window Probability Matching Method
XS	Multispectral

Index

absorption 289, 295
Accumulated Fractional Cloud Cover (AFCC) 408, 412
accuracy 13, 178, 179, 389, 390
active microwave 33, 200, 202, 208, 213, 244 pp., 256, 257
active sensing 15
ADEOS II 212
aerial photographs 277 pp., 317, 329, 330, 333, 338, 340, 342, 425
aerodynamic properties 171
aerodynamic roughness 167, 176
aggregation 127, 427, 428
Agricultural Nonpoint Source (AGNPS) model 279
airborne geophysics 305, 321
airplane 240
AIRSAR 209, 211, 393
albedo 160 pp., 171 pp., 199, 239, 289
algae 288, 291, 295, 296
AMSR 212, 255, 258, 455
angle of incidence 199, 201, 208
aquatic 287, 288, 294, 295, 298, 299, 344
aquifer 312, 314, 315, 320, 321, 340, 341
Arc/Info 68 pp.
Arctic 243
area-time integral 122
artificial intelligence 70, 343
artificial recharge 341
assimilation
  Four dimensional data assimilation (4DDA) 161, 164, 181
atmospheric
  boundary layer 163, 169, 170, 181, 378
  column 164, 181
  correction 290, 406
  window 22
attenuation 113, 115, 120, 208, 240
AVERIS 454
AVHRR 8, 49, 86, 91, 93, 94, 95, 96, 179, 219, 221, 222, 223, 226, 228, 229, 230, 242, 244, 249, 253, 255, 257, 258, 291, 298, 322, 385, 392, 404, 420
backscatter 202, 205, 206, 219, 230, 232
  coefficient 213
BALTEX 128, 453
blackbody 17
blending height 170
boolean operations 73, 74

BOREAS 453
boundary layer *see* "atmospheric boundary layer"
bright-band 116, 117, 118
brightness temperature 18, 74, 114, 115, 121, 124, 125, 206, 207, 213, 222, 233, 245, 247, 248, 256
brittle deformation 320
buried fossil drainage 305
C-Band 15, 16, 120, 202, 204, 205, 212, 246, 256, 257, 367, 375, 455
cadastral 330
capacity 313, 317, 320
carbonate rocks 320
catchment 309, 317
change detection 66
chemicals 288, 291
Chesapeake Bay 280, 294
chlorophyll 289, 290, 294, 296, 299, 390
chronology 79
classification 51, 133 pp., 312, 316, 317, 360, 361, 382, 383, 386, 389, 390, 393, 423, 425, 427, 428, 430, 433
  Gaussian 146, 148
  maximum likelihood 51, 145, 149, 150, 151, 425
  multispectral 279, 312, 382, 385
  multitemporal 382
  numeric 383, 390, 393
  supervised 143 pp., 279, 312, 316
  unsupervised 143
clouds 114, 121, 123 pp., 405, 406, 407
  cloud cover 242, 244, 406
  cloud cover index 88, 228
  cloud indexing 121, 411
  cloud top height 406, 407
  cloud top temperature 406, 408, 411, 412
CMIS 212
coastline 219, 220
Cold Cloud Duration (CCD) 406, 414
Color Infrared (CIR) 277, 281
complex models 446 pp.
conditional statement 343
conjunctive use 341
conservation practices 273, 278 pp.
conservation tillage 281
contrast 53, 331
copolarization ratios 206
CORINE programme 420

correlation length 205, 206
crop 378 pp., 422, 423, 426, 427
 coefficients 161, 383, 390
 cropping factor 279
 cropping patterns 341, 342
 mapping 219, 223, 224, 232, 378, 383, 385, 394
 stress 380, 383, 384, 386, 387, 389, 390, 391, 392, 396
 water requirements 157, 181, 341, 380, 383, 385, 387, 390, 392
cross-polarization ratios 206
cross-section 306, 308
curve number (CN) 75, 344
CZCS 291, 294
data visualization 65
database 72
decision support system (DSS) 77, 342 pp., 449
deforestation 333, 335, 420, 423, 431
degradation 271 pp., 342, 420
dielectric constant 119, 200 pp., 393
differential reflectivity 120
Digital Elevation Model (DEM) 73, 152
Digital Image processing
 Hardware 41 pp.
  arithmetic coprocessor 44
  Central processing unit 41 pp.
   Mainframe 41 pp.
   Personal computer 41 pp.
   RISC 42, 43, 59
  digitization (image scanning) 49, 50
  image processor memory 42, 49
  mass storage 47, 48
   devices 47, 48
   media 47, 48
  Operating system 46 pp.
   Macintosh 61
   NT 43, 46, 61, 62
   Unix 61
   Windows 61, 98 46
  Random Access Memory (RAM) 44 pp.
  screen display 48, 49
  Workstations 41, 42
 Mode of operation 44
  batch 44
  interactive 44
  parallel 44
  serial 44
 Software 46 pp., 61
  Arc/Info 68 pp.
  Display 48, 51, 52, 61, 62
  Enhancement 51, 52 pp.

ERDAS 44, 45, 53, 58, 61
Geographic Information Systems (GIS) 5, 50 pp., 57, 65 pp., 88, 223, 250, 274 pp., 317, 330, 332, 334, 342, 343, 344, 379, 385, 387, 422
GRASS 61, 68, 70
ILWIS 76
information extraction 51, 53 pp.
integrated systems 52
Metadata 51, 54
PCI 61
photogrammetric 51, 54, 59, 60, 273, 274, 276, 278
Preprocessing 51, 52
user interface 44, 45
Utilities 57
vendors 58, 61
digital mapping 242
digital terrain models 330
disaggregation 127
dissolved organic matter (DOM) 288
distributed parameter models 447
DMSP 125, 220, 222, 244, 248, 250, 252, 258
drainage 377 pp.
drainage densities 310
DRASTIC 76, 77, 321
earth - atmosphere system 157
 boundary layer *see* "atmospheric boundary layer"
 land- atmosphere interface 159
 soil column 159
Earth Observing System (EOS) 58, 59, 65, 78, 86, 91, 98, 180, 255, 258, 394, 395, 455
EASE-Grid 248
ecosystem 287
edge detection 222
edge enhancement filter 330
effluent stream 309
electromagnetic bands 10, 11, 15 pp.
 C-Band *see* "C-Band"
 infrared (IR) *see* "infrared (IR)"
 K-Band *see* "K-Band"
 L-Band *see* "L-Band"
 microwave *see* "microwave"
 near infrared (NIR) *see* "near infrared (NIR)"
 S-Band *see* "S-Band"
 thermal infrared (TIR) *see* "thermal infrared (TIR)"
 ultraviolet (UV) 299
 visible (VIS) *see* "visible (VIS)"

W-Band 15, 16
X-Band *see* "X-Band"
electromagnetic spectrum 9, 15 pp., 112
emissivity 18, 114, 124, 200 pp.
  microwave *see* "microwave"
energy balance 90 pp., 318, 319, 334
environmental impact assessment 343
EOS-PM 212
ephemeral gullies 278
equivalent radar reflectivity 119
ERDAS 44, 45, 53, 58, 61, 70, 77
erosion 271 pp., 338
error propagation 72
ERS (1, 2) 153, 211, 212, 453
ESA 394, 455, 456
ESSP 455, 456
ESTAR 209, 210, 211, 394, 452
eutrophication 287, 294
evaporation 91, 92, 157 pp., 330, 332, 386
  actual 159, 162 pp.
  crop coefficients 161
  evapotranspiration 90, 133, 134, 136, 139, 157, 226, 421
  latent heat flux 158, 162
  transpiration 157, 159
evapotranspiration 90, 133, 134, 136, 139, 157, 226, 421
expert system 70, 74, 343, 344
Export coefficient model 77
Famine Early Warning System (FEWS) 138
fanglomerates 306, 315
feature space 141 pp.
FIFE 86, 89, 209
flood 230 pp., 241, 250, 257, 334, 357 pp., 358, 360, 361, 362, 365, 366, 367, 368, 370, 371, 372, 373, 375, 401, 402, 445, 449, 454
  duration 221
  extent 230 pp.
  flood forecast 85, 197, 241, 357 pp., 402, 445, 454
  flood plain 230, 333, 334, 335, 338
  flood protection 85, 357, 358, 375
  flood warning 357, 365, 369
  flooding 250
  frequency 111, 114, 115, 125, 221
  timing 221
  under-canopy 230
floodwater spreading 341
fluorescence 281, 295, 296
foliage 382, 383, 384 391
forecast
  flood forecast 85, 197, 241, 357 pp., 402, 445, 454
  quantitative precipitation forecast (QPF) 361, 362, 368
  rainfall forecast 359, 361, 368
  real-time forecast 358 pp.
format 47
  data exchange 69
  lattice 70
  raster 67 pp.
  vector 67 pp.
format geometric correction 330
Four dimensional data assimilation (4DDA) 161, 164, 181
fractures 312, 314, 319, 320, 321
frequency 16
frequency of observation 12, 242 pp.
freshwater 287, 290, 291, 298
GAME 453
gamma radiation 198, 199, 240 pp.
gamma ray spectrometry 318
gauge networks 116
General Circulation Model (GCM) 86, 197, 445
Geographic Information Systems (GIS) *see* "Digital Image Processing, Software"
geometric correction 71
geometrical optics model 205, 206
GEWEX Continental Scale International Project (GCIP) 91, 453
GIS *see* "Digital Image Processing, Software"
glaciers 243, 246
GMS 8, 11, 360, 375, 453
gneiss 338
GOES 8, 11, 86, 91, 92, 93, 97, 121, 175, 244, 249, 360, 375, 405, 453
GOES Precipitation Index (GPI) 121
GPCP 128
granitic area 312
GRASS 61, 68, 70
ground-based radar 111, 124, 128 *(see also* "radar")
groundwater 305 pp., 341, 393
  flow systems 305, 306, 310, 316, 321
  hydrotope 319, 320
  model 305, 310, 311, 312, 315, 318, 321, 322
  recharge 305 pp., 312 pp., 341
  table 305, 309, 310, 311
gully 274, 277 pp., 282
hail 118, 119
hard rocks 309, 319

478 Index

heat balance 162, 163, 165 pp., 384
Heat Capacity Mapping Mission (HCMM) 163, 170, 298
Height-Area Rainfall Threshold (HART) technique 123
hydrogeologic maps 308, 322
hydrologic model 75 pp., 86 pp., 211, 213, 227, 402, 414, 416, 422 pp., 430 pp., 446, 447, 453
　deterministic model 358, 360
　distributed model 75 pp., 358, 360, 361, 362, 363, 423, 424, 433, 450
　lumped system model 358, 362
　model parameter 79, 86, 358, 360 pp.
　rainfall-runoff model 75, 76, 229, 358, 361 pp., 368, 374, 375, 413, 446, 447, 454
　Soil-Vegetation-Atmosphere Transfer Scheme (SVAT) 91, 169, 170
　stochastic model 358
hydrological requirements 417
Hydrological Similar Unit (HSU) 74, 76, 88, 90, 430, 431
hydropower production 254
HYDROSTAR 212, 456
hydrotopes 332
hyperspectral 289, 290, 295, 299
ice thickness 244, 246, 247, 248
ILWIS 76
imagery
　multi-temporal imagery 373
　radar images 127, 361, 373
indicator 309
inference mechanisms 344
infiltration influent stream 309
infrared (IR) 10, 11, 15, 16, 114, 115, 123 pp., 199, 218, 228, 243, 244, 289, 331, 333, 335, 337, 405
intensive field campaign 452 pp.
interactive coupling 342
interferometry 36, 283, 309
International Satellite Cloud Climatology Project (ISCCP) 160, 161
irradiance
　solar 23
　surface 114, 115, 116, 118, 120, 121, 124, 127
irrigated area 341, 342, 378 pp., 401
irrigation 312, 340, 341, 342, 377 pp., 401, 402, 449
　drafts 312
　performance 392, 394
　scheduling 391, 449
　uniformity 384, 394

water management 158, 162, 392, 394
IRS 273, 291, 338, 393
isotope 315
JERS-1 153, 212, 219, 223, 453, 455
K-Band 15, 16, 204
karst 312, 314, 319
L-Band 15, 16, 200, 203, 204, 205, 208, 209, 212, 456
lake 220 pp., 292, 294, 297, 298
　Aral Sea 220, 393
　depth 117, 123, 222
　Lake Baikai 298
　Lake Chicot 293
　lake ice 222, 246, 247
　　freeze-up 222, 246, 247
　Lake Nakuru 225
　Lake Nasser 220, 229
　Lake Victoria 226, 227
　levels 224 pp.
　　measuring 224
　　modelling 225
　surface area 221
　volumes 222
land
　cover 244, 245, 247, 256, 380, 386, 390, 396, 419 pp.
　mapping 134
　land surface hydrology 197
　use 74, 87, 88, 133 pp., 256, 402, 419 pp.
　　changes 74, 335, 419 pp.
　　　hydrological effects of 422, 423, 424, 430, 431, 433
　　classification 89, 360, 361 (see also "classification")
　　heterogeneity of 422, 427
land- atmosphere interface 159, 165
Landsat 8, 12, 47, 72, 73, 74, 76, 77, 87, 88, 89, 219, 220, 221, 222, 223, 224, 226, 229, 230, 233, 239, 242, 243, 246, 248, 273, 274, 277, 280, 291, 293, 294, 298, 299, 308, 334, 335, 337, 338, 341, 360, 389, 390, 392, 394, 404, 416, 420, 423, 425, 427, 453
Landsat TM 47, 87, 88, 135, 137, 140, 219, 220, 221, 222, 223, 224, 230, 233, 277, 279, 298, 299, 334, 389, 394, 404, 420, 423, 427
landscape 271, 273 pp., 287, 288
　units 334, 342
laser 273, 275, 281 pp., 330
latent heat flux 98, 158, 162
LBA 453

Leaf Area Index (LAI) 140, 141, 162, 176, 390
life history 121
line following algorithm 70
lineaments 319 pp.
local probabilities 148
logic assertions 343
logical rule 343
MAC 211
MAGS 453
management practice factor 279
map
　complexity 72
　overlay 65, 72
Marshall-Palmer relationship 113, 119
melt-freeze cycle 244
mesoscale numerical model 118
Meteosat 8, 11, 76, 88, 223, 229, 334, 360, 375, 403, 405, 406, 411, 412, 414, 453
microphysical model 118
microwave 15, 16, 27, 114, 115, 123, 123 pp., 124, 125, 128, 134, 200 pp., 244 pp., 255, 256, 281 (*see also* "active microwave" *and* "passive microwave")
microwave brightness temperature 74, 123, 200
microwave radiometers 246
Mie theory 113
Mississippi River 91, 96, 298
MODIS 59, 258, 395, 396, 455
monitoring 77, 78, 271 pp., 332 pp., 385, 387, 392, 393
MONSOON 210, 452
morphology 282, 335, 336
mountain front deposits 315
MSS 77, 219, 221, 222, 223, 228, 229, 230, 242, 274, 278, 280, 308, 310, 335, 337, 338, 385, 389, 392, 393, 425
multi-spectral video 221
multispectral 312, 322, 382
natural terrestrial gamma radiation 198
NDVI *see* "vegetation index"
near infrared (NIR) 9, 10, 31, 199, 242, 244, 291, 298, 308, 383, 387, 390
NESDIS 248
Nile 3, 229, 230, 334, 392, 414
Nimbus-7 233, 250, 252
NOAA 62, 94, 138, 219, 220, 221, 222, 226, 228, 229, 242, 243, 248, 249, 258, 385, 404, 405, 416, 453
NOAA-AVHRR *see* "AVHRR"

nonpoint sources 287
NOPEX 453
numerical models 455
nutrients 288, 294
NWS National Operational Hydrologic Remote Sensing Center (NOHRSC) 241, 249, 250
object-oriented 70, 343, 344
observation geometry 167
observations 394 pp.
oils 288, 290, 291, 298, 299
optical thickness 19
ortho-image 330
overlay operation 71 pp.
paired experimental catchments 316
panchromatic 278, 330
passive microwave 11, 33, 114, 115, 125, 200, 202, 209, 210, 211, 212, 213, 244 pp., 245, 246, 247, 250, 251, 255, 256, 299, 322, 416, 417, 453
passive sensing 15
pattern recognition 142
PBMR 209, 210, 452
PCI 61, 70
penetration depth 203, 204, 208, 209
percolation 315, 316
perspective view 53, 74
pesticides 288
photo-geology 306
photogrammetry 54, 277 pp., 330, 331
photographs 242, 248, 273 pp., 314, 317
photointerpretation 248, 273 pp.
phreatophytes 309, 319
physical optics model (POM) 205, 206
physiographic units 314
phytoplankton 294 pp.
pixel 4, 67
Planck's law 17
planetary boundary layer (PBL) 168, 170
　model 170
plankton 294
plant root depth 73
plant structure 209
platform 6, 241, 244, 248, 453, 454
plumes 297
point source 77, 287
polarization 201, 202, 208, 213, 219, 245, 256, 257
pollutant 287, 288, 291
polygon-capturing 70
positional accuracy 71
precipitable water 93 pp.

480 Index

precipitation 94, 111 pp., 239, 246, 247,
    252, 361 pp., 402, 405, 406, 408,
    410 pp. (*see also* "rainfall")
  events 111
  scales 111, 112
  types 111, 113, 117, 124
probability density function 117, 122
proximity analysis 66
quadtree 58, 70
quality control 119, 127
Quaternary deposits 306
R/Z relationship 113, 119
radar 19, 111 pp., 134, 152, 153, 205, 206,
    209, 211, 213, 247, 248, 257, 299,
    309, 393, 394
  altimeters 224 pp.
  backscatter 28, 202, 205, 206, 213, 219,
      230, 247, 297, 322
  Bragg resonance 219
  cross section 113
  equation 19
  incidence angle 199, 201, 208, 219, 220,
      230, 231
  interferometry 36, 283, 309
  rainfall measurement 367, 368, 373, 375
  reflectivity 113, 115 pp.
  reflectivity factor 21, 113, 118
  resolution 111, 113, 123, 124, 126
  speckle 224
  Synthetic Aperture Radar (SAR) *see*
      "SAR"
  targets 113
  weather radar 94, 114, 119, 128, 358,
      360 pp., 372 pp.
Radarsat 153, 212, 219, 220, 223, 232,
    234, 257, 454
radiance 17, 273, 280, 281
radiation
  longwave 15 pp., 93
  solar 15 pp., 92
  terrestrial 15 pp.
radiative transfer 18, 74, 245, 246
radioisotopes 240
radiometric 125
rainfall 117, 124 (*see also* "precipitation")
raingauge adjustment 116, 120
Rayleigh theory 113
Rayleigh-Jeans law 18
recharge 341
rectification 51, 52, 71
reflectance 218
  curve 136
  diffuse 219
  open-water 218, 219

soil 11, 114, 127, 218
specular 219, 220
vegetation 11, 174, 218, 223
reflectivity
  surface 11, 114, 115, 116, 118, 120,
      121, 124, 127, 174
regionalization 332, 333
Remote Sensing
  algorithms 91, 93
  applications 72 pp., 96, 165 pp., 209 pp.,
      249 pp., 274 pp., 291 pp., 414
  Earth Observing System (EOS) *see*
      "Earth Observing System"
reservoir 220 pp., 366, 370, 372, 401
  models 229
resistance 160, 161, 164 pp., 173, 176,
    178
  aerodynamic 176 pp.
  aerodynamic properties 171
  aerodynamic roughness 167, 176
  canopy 164, 168
  effective 160, 161
  internal 165
  stomatal 176
  surface 161
resolution 405, 427, 428
  high 321, 330, 334, 360, 446, 454
  in space 10, 79, 360, 405, 446, 447, 450
  in time 12, 79, 333, 360, 446, 447, 449,
      450
Revised Universal Soil Loss Equation
    (RUSLE) 272, 276
rill 272 pp.
Rio Grande basin 239, 253
river basin snow cover extent maps 249
river bed losses 313
river ice break-up 243
RMS height 205
Rosetta promontory 277
roughness 204 pp., 273, 281, 282
run-on 313
runoff 87, 239, 252, 255, 313, 314, 315,
    316, 317, 320, 340, 341, 421, 422,
    423
  coefficients 87, 317
  GRU 88, 89
  Hydrological Similar Unit (HSU) 74,
      76, 88, 90, 430, 431
  models 13, 76, 77, 86 pp., 118, 120,
      127, 128, 246, 250, 252, 253, 255,
      256, 446, 447, 454
Runoff Curve Numbers method 133
S-Band 15, 16
salinisation 341, 342

salinity 380, 384, 393
sampling errors 126, 127
SAR 6, 87, 140, 152, 153, 209, 211, 212, 213, 220, 223, 224, 228, 230, 232, 246, 247, 248, 256, 257, 258, 273, 275, 276, 281, 283, 394, 455
satellite 6
  data 6, 114, 279, 360 pp., 424 pp.
    active microwave 33, 256, 257 (*see also* "active microwave")
    infrared (IR) 114, 115, 123 pp., 199, 244 (*see also* "infrared (IR)")
    near infrared (NIR) 31, 199, 242, 244 (*see also* "near infrared (NIR)")
    passive microwave 33, 114, 115, 125, 246, 247, 299 (*see also* "passive microwave")
    thermal infrared (TIR) 32, 199 (*see also* "thermal infrared (TIR)")
    visible (VIS) 31, 114, 115, 121, 123, 124, 199, 243, 244, 273 (*see also* "visible (VIS)")
  polar orbiting satellite 8, 9, 358, 404, 405
satellite sensors 6, 9, 290, 300, 453, 454
  AMSR 212, 255, 258, 455
  MODIS 59, 258, 395, 396, 455
  MSS *see* "MSS"
  NOAA-AVHRR *see* "AVHRR"
  SAR *see* "SAR"
  SIR-C 6, 211, 230, 256
  SMMR *see* "SMMR"
  SSM/I *see* "SSM/I"
  TM *see* "Landsat TM"
  TOVS 98, 161
satellite system 6, 247, 334
  ADEOS II 212
  DMSP *see* "DMSP"
  EOS *see* EOS
  ERS (1, 2) 153, 211, 212, 453
  ESSP 455, 456
  GMS *see* "GMS"
  GOES *see* "GOES"
  HYDROSTAR 212, 456
  IRS 273, 291, 338, 393
  JERS-1 *see* "JERS-1"
  Landsat *see* "Landsat"
  Meteosat *see* "Meteosat"
  Nimbus-7 233, 250, 252
  NOAA-AVHRR *see* "AVHRR"
  polar orbiting satellite 8, 9, 358, 404, 405
  Radarsat *see* "Radarsat"
  SPOT *see* "SPOT"
  TRMM *see* "TRMM"
scattergram 137
scattering 113, 115, 124, 125
  Rayleigh 113
SeaWiFS 295
secchi 291
sediment 289 pp.
sediment yield 335
segmentation 149, 150, 151
semantic networks 344
sensors 9, 30 pp.
  active microwave 33, 256, 257 (*see also* "active microwave")
  AMSR 212, 255, 258, 455
  AVHRR *see* "AVHRR"
  infrared (IR) 114, 115, 123 pp., 199, 244 (*see also* "infrared (IR)")
  microwave radiometers 246
  MODIS 59, 258, 395, 396, 455
  MSS *see* "MSS"
  near infrared 31, 199, 242, 244 (*see also* "near infrared (NIR)")
  passive microwave 33, 114, 115, 125, 246, 247, 299 (*see also* "passive microwave")
  SAR *see* "SAR"
  SIR-C 6, 211, 230, 256
  SMMR *see* "SMMR"
  SSM/I *see* "SSM/I"
  thermal infrared (TIR) 32, 199 (*see also* "thermal infrared (TIR)")
  TM *see* "Landsat TM"
  TOVS *see* "TOVS"
  visible (VIS) 31, 114, 115, 121, 123, 124, 199, 243, 244, 273 (*see also* "visible (VIS)")
signature 247, 256, 289
SIMPLE model 153
single source 175, 177, 287
Single Variable File 70
SIR-C 6, 211, 230, 256
SIR-C/X-SAR 153, 211, 230, 256
slope factors 279
  modelling
SLURP 226, 227, 229
small perturbation model (SPM) 205, 206
SMMR 125, 233, 245, 247, 248, 250, 252
SMOS 212, 456
snow 119, 125, 239 pp.
  cover 239 pp.
    drainage basin 239, 242, 243, 246
    northern hemisphere 239
    world 239
  depth 245, 247

extent 242, 243, 245, 250, 252, 255
line 243
snowfall 111, 118, 119, 127
snowmelt 239, 241, 242, 243, 244, 252, 253, 255, 257
snowmelt runoff 243, 244, 253, 255, 358
  forecasts 241, 250, 255
  modelling 244
  Snowmelt Runoff Model (SRM) 252 pp.
snowpack
  liquid water 245
  metamorphism 247
survey measurements 240
water equivalent 222, 240, 241, 244, 245, 246, 249, 250, 255, 257
  gamma radiation 240 pp., 250
soil 197 pp., 271 pp., 309, 313 pp., 421, 422
  erosion 65, 78, 271 pp.
  erosion risk maps 78, 280
  factor 279
  line 137
  loss 77, 271 pp., 278 pp.
  moisture 73, 74, 79, 89, 114, 127, 197 pp., 241, 309, 316, 317, 322, 384
  moisture measurements 241
  porosity 361
  texture 202, 211, 317, 318, 341
  water content 384
  water storage 73, 89, 360, 361
Soil Adjusted Vegetation Index (SAVI) see "vegetation index"
Soil-Vegetation-Atmosphere Transfer Scheme (SVAT) 91, 169, 170
solar radiation 92
Space Shuttle 6, 8, 220, 230
space-borne radar 115
Spain 253, 389
spatial
  analysis 76, 78, 332
  averaging 177, 179
  data accuracy 71
  modeling 65, 66
  patterns 158, 305, 313, 314
  resolution 10, 247, 330, 334, 338, 385, 405, 427, 428, 453, 454
  variability 177
SPAW 158
spectral bands 9, 10
  C-Band see "C-Band"
  infrared (IR) see "infrared (IR)"
  K-Band see "K-Band"

L-Band see "L-Band"
microwave see "microwave"
near infrared (NIR) see "near infrared (NIR)"
S-Band see "S-Band"
thermal infrared (TIR) see "thermal infrared (TIR)"
ultraviolet (UV) 299
visible (VIS) see " visible (VIS)"
W-Band 15, 16
X-Band see "X-Band"
SPOT 8, 11, 73, 87, 88, 221, 223, 242, 243, 244, 273, 278, 279, 281 pp., 291, 294, 331, 333, 360, 385, 389, 390, 392, 416, 453
SSM/I 86, 91, 98, 125, 222, 247, 248, 250, 252, 258, 453, 455
Stefan-Boltzmann law 17
stereo aerial photography 322, 335, 338
stereo photographs 275
stereo-pairs 278
stream channel 228, 278, 282
streams
  braided 228
  delineating 228
  discharge 228
  levels 226 pp.
  sediment 233, 271, 272, 278, 280
  stage-discharge relationship 228
structure of landscapes 178
surface 200, 204
  residue 281
  roughness 29, 204 pp., 273, 281, 282
  temperature 94, 118, 165 pp., 199, 200, 244, 298
  water 217 pp., 288pp.
    delineating 218
    inventories 219
Surface Energy Balance Index (SEBI) 173, 175
suspended sediments 288, 290 pp.
SVAT 91, 169, 170
Synthetic Aperture Radar (SAR) see "SAR"
temperature 244, 252, 253, 288, 291, 297, 298
  air 94 pp., 169
    near surface 168
    reference 163
  canopy 163, 166
  maximum 172
  radiometric 167 pp.
  rate of increase 168
  surface see "surface temperature"

daily amplitude 170
temporal 382, 386, 406, 411, 421, 424
temporal dimension 79
terrain units 332
Thematic Mapper (TM) *see* "Landsat TM"
thermal 287 pp.
thermal inertia 11, 199, 200
thermal infrared (TIR) 10, 11, 32, 169, 199, 244, 289, 305, 384, 391, 392, 396, 405 pp., 411, 412, 416
thermal releases 288, 297, 298
Thornthwaite and Mather method 317
three-dimension 79
time-area curve 76
TIROS Operational Vertical Sounder (TOVS) 98, 161
TOGA-COARE 128
TOPEX/Poseidon 224, 226, 227, 228
TOPMODEL 73, 76
topography 281 pp.
topology 79
TOVS 98, 161
training 144, 145, 147, 149, 150, 153, 390
transmission losses 315, 320
transmissivity 19, 322
transpiration 157, 159, 391
Triangular Irregular Network (TIN) 76
TRMM 8, 94, 115, 126, 128, 417, 454
Tropical Rainfall Measurement Mission (TRMM) *see* "TRMM"
turbidity 288 pp.
two-dimensional modelling 310, 311, 321
ultraviolet (UV) 299
Universal Soil Loss Equation (USLE) 77, 272, 276, 278, 279, 280, 281
unsaturated flow models 318
upstream-downstream 335
urban drainage system 372 pp.
vector 67 pp.
vegetation 206 pp., 231, 277, 306, 309 pp., 316 pp., 402
  index 135 pp.
  DVI 136, 138, 139, 141
  Leaf Area Index (LAI) 140, 141, 162, 176, 390
  NDVI 88, 94, 138, 139, 140, 141, 168, 171, 172, 176, 281, 310, 312, 316, 317, 330, 332, 390
  RVI 136, 137, 139, 141
  SARVI 141
  SAVI 141, 386
  Soil Adjusted Vegetation Index (SAVI) 386
  TSAVI 141

Vegetation Index Temperature Trapezoid (VITT) 171, 172
vegetative cover 133
visible (VIS) 10, 11, 15, 16, 31, 114, 115, 121, 123, 124, 199, 218, 220, 221, 228, 242, 243, 244, 273, 288 pp., 295, 298, 320, 383, 390, 405, 411, 416
visible imagery 243
volcanic terrain 307, 308
volumetric soil moisture 201, 202, 203, 204, 207, 213
W-Band 15, 16
WASHITA 211, 452
water
  balance 85 pp., 163, 164, 313, 334, 453
  equivalent 119
  harvesting 340
  management 3, 329 pp., 365, 401 pp., 410, 414, 419 pp., 448 pp.
  quality 76, 77, 217, 221, 287 pp., 374, 375, 421, 449
  supply 217, 229, 341, 342, 401, 402, 449
watershed 89, 392
watershed management 338, 402, 424
wavelength 10, 16
  C-Band *see* "C-Band"
  infrared (IR) *see* "infrared (IR)"
  K-Band *see* "K-Band"
  L-Band *see* "L-Band"
  microwave *see* "microwave"
  near infrared (NIR) *see* "near infrared (NIR)"
  S-Band *see* "S-Band"
  thermal infrared (TIR) *see* "thermal infrared (TIR)"
  ultraviolet (UV) 299
  visible (VIS) *see* " visible (VIS)"
  W-Band 15, 16
  X-Band *see* "X-Band"
weathered zones 314, 319, 320, 321
wetland 223 pp.
  assessment 223
  classification 224
  Meteosat 223
  monitoring 223, 224
Window Probability Matching Method (WPMM) 117, 128
World Climate Research Programme (WCRP) 86
X-Band 15, 16, 230, 248, 256
Zambezi River 310

Printing: Mercedes-Druck, Berlin
Binding: Buchbinderei Lüderitz & Bauer, Berlin